5th Annual International Conference on Material Engineering and Application (ICMEA 2018)

IOP Conference Series: Materials Science and Engineering
Volume 484

Wuhan, China
14 - 16 December 2018

Editors:

Tarek Aly Elsayed
Arnulfo Luevanos Rojas
Ke Wang

ISBN: 978-1-5108-8481-6
ISSN: 1757-8981

Printed from e-media with permission by:

Curran Associates, Inc.
57 Morehouse Lane
Red Hook, NY 12571

CURRAN ASSOCIATES INC. proceedings.com

Some format issues inherent in the e-media version may also appear in this print version.

Printed by Curran Associates, Inc. (2019)

For permission requests, please contact the Institute of Physics at the address below.

Institute of Physics
Dirac House, Temple Back
Bristol BS1 6BE UK

Phone: 44 1 17 929 7481
Fax: 44 1 17 920 0979

techtracking@iop.org

Additional copies of this publication are available from:

Curran Associates, Inc.
57 Morehouse Lane
Red Hook, NY 12571 USA
Phone: 845-758-0400
Fax: 845-758-2633
Email: curran@proceedings.com
Web: www.proceedings.com

Table of contents

Volume 484

The 5th Annual International Conference on Material Engineering and Application (ICMEA 2018)

14–16 December 2018, Wuhan, China

Accepted papers received: 7 January 2019 Published online: 19 March 2019

Preface

The 5th Annual International Conference on Material Engineering and Application

Committees of ICMEA 2018

Peer review statement

Papers

Effects of Electrochemical Etching Conditions on the Formation and Photoluminescence Properties of P Type Porous Silicon

Le Lu, Wenbing Li, Liqiang Zhang and Daohan Ge.....1

Effect of Yogurt on Gastrointestinal Motility of Dyspepsia Mice

Xiaobing Li, Fajun Li, Weiwen Zhang, Shiping Shen, Lining Wang, Huan Zhang, Mengyuan Ding, Huanhuan Chen and Mingming Huang.....6

Viscosity and Free Surface Energy as Parameters Describing the Adhesion of the Epoxy Resin to the Substrate

AS Zewczak and MS Zeląg.....12

Influence of Neutralization on Hydration of Hemihydrate Phosphogypsum Plaster

Xuemei Chen and Jianming Gao.....18

Carbon Nano-Onions Embedded CuO Nanosheets: An Excellent Stable Anode Material for Lithium Ion Battery

Zhongjing Hao, Meng Qin, Yueming Li, Xiaoju Lv, Dafeng Zhang and Qiushi Wang.....23

Size-TunableMesoporous Silica for Macromolecular Drug Delivery

Yiyu Wang, Chunqing Niu, Jiamin Gu, Chao Hu, Youning Wang and Junhua Zhang.....29

The Effect of Cone Additional Permanent Magnet Thickness on Magnetic Field Distribution and the Levitation Force of Single Domain GdBCO Bulk Superconductor

Jun Ma, Yongjie Zhou, Jinglin Li, Bohang Chen and Weiqiang Fang.....36

Rice Straw/Sewage Sludge-Based Composite Activated Carbon: Its Some Basic Properties and Adsorption Effect for Cr (VI)

Liangqian Fan, Xianda Wang, Wenxin Wan, Qin Liu, Jie Cai and Yan Wan.....42

The Low Molecular Weight Organic Acids in Root Exudates of *Leersia hexandra* Swartz and Its Role in Mobilization of Insoluble Chromium

Chancui Wu, Jie Liu, Ying Liang, Yongrong Jiang and Xuehong Zhang.....47

Preparation of Catalyst by Phosphating Slag and Its Photocatalytic Performance on Rhodamine B and Methyl Orange

Weilin Guo, Lijun Wang, Lingxue Kong, Yongmin Qiao and Suna Zhang.....55

Effect of Dietary Carbohydrate/Protein Ratios and Feeding Frequency on Carbohydrate Metabolism of Common Carp

Zhenyan Cheng, Anqi Wang, Ze Fan, Jinhui Sun, Pei Cui and Xiuting Qiao.....62

Experiment and Mechanism Analysis of Selective Non-Catalytic Reduction with the Compound Reducing Agents in a Lab-scale CFB Reactor

Jiancong Dong, Degui Bi, Tuo Zhou, Juan Yu, Xiaojiang Wu and Zhongxiao Zhang.....72

Study on Preparation of Activated Carbon from Sludge

Mingming Du, Tao Yu, Feifei Wang and Chengtun Qu.....78

Investigation of Hybrid Natural Fibre Reinforced Composite for Impact Energy Absorption

Yiou Shen, Yichun Guo and Yan Li.....83

Dependencies between Cracking Patterns and the Physico-Mechanical Properties of Microsilica Modified Cement Matrix

M Szeląg and A Szewczak.....90

Research on New Materials in Civil and Construction Engineering

Peng Jiang and Junhai Hu.....96

Chip Deformation Mechanism during Drilling of Nickel-based Superalloy FGH97

Yang Qiao, Peiquan Guo, Hongtang Chen, Shouren Wang and Hui Wang.....99

Temperature Effect on Dislocation Slip Mechanism of Nanotwinned Mg with Void Defect at the Twin Boundary

Kaixin Huang, Junping Yao, Qiyao Hu, Letian Shao and Zhong Sun.....103

Experimental Study on Preparation and Mechanical Properties of CF Knitted Fabric Composites for Nut Lining

Bin Zhao, Zhipei Tong, Qiang Zhang, Peng Zhang, Yongqi Cheng and Yousong Sun.....107

Mode I Elastic Solution of Cracked Composite Plate

Yongyong Suo, Purong Jia, Qiang Wang and Leilei Zeng.....113

The Price Linkage between Oil, Gold, Stock and Exchange Rate Based on Vine Copula

Huan Sun.....119

Experimental Investigation of NOx Reduction by Urea Solution Injection in Fuel-rich Zone of Coal Combustion

Degui Bi, Zhongxiao Zhang, Zhixiang Zhu, Jiancong Dong and Tuo Zhou.....126

Experimental Research on NOx Emission Characteristics During Coal Combustion in the Stoker-fired Boiler

Xinwei Guo, Zhongxiao Zhang, Juan Yu, Jian Zhang, Degui Bi, Jiancong Dong, Hao Bai and Junjie Fan.....132

Analysis on Characteristics of Structural Impact Response of Airborne Armored Vehicle in Landing Process

Xinhua Fu.....140

Investigation on Fusion Bridge of Au Electrical Contacts Material under Different Voltage

Jiheng Fang, Di Wu, Jinghong Chen, Xi Liu, Ming Xie, Song Chen, Jieqiong Hu, Yongtai Chen, Youcai Yang, Hongwei Ma, Jiming Zhang and Manmen Liu.....147

Numerical Analysis of Influencing Factors and Capability for Line Laser Scanning Thermography Nondestructive Testing Technology in Chemicals Corrosion Defect Detection

Zhengwei Yang, Xingyu Xie, Yin Li and Gan Tian.....155

The Research and Application of Industrial Robot in Steel Deep Processing Field

Xiaodong Hao, Yan Bian, Lu Bai, Lihua Li and Changjie Sun.....160

The Relationship between Morphology of Milling Surface and Its Corrosion Resistance for Nickel base Superalloy FGH97

Jinping Shi, Peiquan Guo, Yang Qiao, Shouren Wang and Tonghui Liu.....164

Stress and Strain of Cracked Orthotropic Sheet

Purong Jia, Gang Wang, Long Zhang and Yongyong Suo.....170

A Simple and High-performance Platform for Refractive Index Sensing based on Plasmonic Metal Disks on a Metal Mirror

Xianchao Liu, Rui Ma, Duncan Sutherland and Jun Wang.....176

Application of Lanthanum Oxide and Cerium Oxide in E4303 Electrode

Enhuai Li, Nengwei Wang and Shuquan Wan.....181

Delaunay-Based Aircraft Engine Casing Strain Reconstruction and Data Processing

Kun Wang and Kejia Liu.....188

Mechanical Property of Roof Plate Based on Engineered Cementitious Composites

Yuxin Gao, Rui Shen, Shuo Wang, Xinyi Yan, Baojun Cheng, Yaning Kong, Yao Bi and Jun Wang.....193

Effect of Pinellia Ageratum Decoction on Cisplatin-induced Vomiting and Its Mechanism

Xiaobing Li, Shu Wang, Wenjing Chen, Mengting Zhao, Xueqing Kong, Xiaoyu He, Yan Chen, Wenyu Chen, Tong Wu, Chunlei Xu, Jun Huang and Shuanzhu Song.....199

Distributed Optical Fiber Micro-vibration Sensing for Broken Fabric Bag Inspection in a Baghouse Filtration System

Bo Shi, Jun Li, Xu'an Liu, Yuquan Tang, Tao Pang and Fengzhong Dong.....205

The Development and Improvement of the Coal Chemical Circular Economy System in China: a Case Study

Shuo Yang, Yun Fu, Wanying Zhao, Lili Wu and Jianfei Wang.....210

Operation Performance Evaluation of Municipal Wastewater Treatment Plant by Analytic Hierarchy Process

Xiaoxin Zhang, Jin Huang, Ling Lin, Yueping Wang and Xinheng Zhang.....220

Technical Assessment of Industrial Solid Waste Recycling Process Based on the Willingness-to-Pay LCIA Model

Huiting Guo, Shuo Yang, Xiuteng Wang and Jing Liu.....230

Response of the Plant and Soil Features to Degradation Grades in Semi-arid Grassland of the Inner Mongolia, China

Xia Xu, Honglei Jiang, Xiaoyu Tian, Mengxi Guan and Lingfei Wang.....244

Study on Extreme Microbial Degradation of Petroleum Hydrocarbons

Hong Li, Xiaokang Li, Tao Yu, Feifei Wang and Chengtun Qu.....251

Synthesis and Degradation Properties of Heat-resistant Polymer based on Cyclotriphosphazene

Yani Guo, Liyun Liang and Chengmei Liu.....256

Development of Small Intestinal Submucosa as Biomaterial in Tissue Engineering

Yu Wang, Peng Chen and Jianbo Qin.....261

Polyurea Materials and Their Environmental Applications

Xiaojing Dong, Xubao Jiang, Shusheng Li and Xiang Zheng Kong.....266

$Co_{1-x}Zn_xIn_2S_4$ Photocatalysts for Photocatalytic H_2 Production Activity

Renzhang Wang, Zhiyun Liang, Fuying Li, Qiyou Wu, Peijing Xu and Yu Niu.....272

Electrodeposition of PPy/HNO_3 Film and its Capacitive Behaviour

Yong Chen, Xia Zhang, Juan Juan Li and Hui Xu.....276

Market-Oriented Operation Pattern of Regional Power Network Integrated With High Penetration Level of Distributed Energy Resources

Qing Duan, Jian Le, Xi Ye, Zhipeng Lv and Qian Zhou.....281

Investigation Aging Behavior of High Energy Propellant Materials

Gu Yan, Wenliang Liu, Qiong Wang, Jiaojiao Du, Fangfang Wang and Silong Yu.....292

Development of Environment-Friendly Plastering Gypsum

Xingyu Wang, Shuling Sun, Xiaohua Tian, Dongji Liu and Fengqing Zhao.....298

Impact of Riverine Pollutants on the Water Quality of Lake Chaohu, China

Wang Zhang, Kun Lei, Libiao Yang and Xubo Lv.....303

Research Progress in Water Quality Improvement

Changxin Cai, Canjun Liu and Baohui Men.....312

A Study on the Application of Green Design Theory in Environmental Art Design

Kun Bai.....317

An AHP-Bayesian Model for Operational Risk Evaluation of Integrated Energy Systems

Yongqi Yang, Wanlei Xue, Chaoyuan Li, Xin Zhao and Nan Xu.....322

Short-time Transport Stress Affected Biochemical Parameters in *Takifugu Rubripes*

Qingkui Wang, Lixiao Yang, Kezhi Xing, Xueliang Sun and Jinsheng Sun.....329

Microbiostatic and Synergistic Actions of Extract of Green Walnut Peel

Li Yang, Gu Yao, Rusi Wen and Lizhu Zhou.....338

The Development and Application of the "DALY"-Based Environmental Risk Assessment Methods with a Case Study on the Impact of PM2.5 in Beijing

Shuo Yang, Xiuteng Wang, Huiting Guo, Jing Liu and Jianfei Wang.....344

Study on Environmental Risk Assessment and Management Methods for the Recycle of Industrial Solid Waste

Shuo Yang, Huiting Guo and Xiuteng Wang.....353

Antialgal Activity of Glycoglycerolipids Derived from a Green Macroalgae *Ulva prolifera* on Six Species of Red Tide Microalgae

Yingying Sun, Shasha Dong, Ganlin Guo, Lei Guo and Yinfang Pu.....364

Analysis of Iron and Manganese Release Effect in Drinking Water Systems from a New Perspective

Huiyan Tong and Zhongyue Li.....371

Study on the Existence Form and Removal of Boron Acid

Rongrong He, Tao Yu, Jiajia Du and Chengtun Qu.....377

A Review of the Mechanism of Microbial Degradation of Petroleum Pollution

Xiaokang Li, Hong Li and Chengtun Qu.....382

Brief Analysis of Seasonal Correlations in the Power Generation Data of Photovoltaic Power Stations

Weihao Yu.....386

IOP Conf. Series: Materials Science and Engineering **484** (2019) 011001 doi:10.1088/1757-899X/484/1/011001

Preface

The 2018 5th International Conference on Material Engineering and Application (ICMEA 2018) was held from December 14-16, 2018 in Wuhan, China. Aiming to provide an excellent international academic forum for all the researchers and practitioners, the conference enjoyed widespread participation among universities and research institutes.

It provided opportunities for the delegates to exchange new ideas and application experiences, to establish business or research relations and to find global partners for future collaboration. Papers and topics concentrating on related fields of the environment, materials, material engineering and application were presented.

The ICMEA 2018 proceedings collects the most up-to-date, comprehensive, and worldwide state-of-the-art knowledge on environment, material engineering and application and other related topics. All the accepted papers have been submitted to a strict peer-review by 2-3 international expert referees, and selected based on originality, significance and clarity for the purpose of the conference. The conference program is extremely rich, profound and featured high-impact presentations of selected papers and additional late-breaking contributions. I sincerely hope that the conference would not only provide the participants with a broad overview of the latest research results in related fields, but also provide them with a significant platform for academic connection and exchange. The Technical Program Committee members have been working very hard to meet the deadline of review.

I would like to express our sincere gratitude to all the keynote speakers, editors, TPC members and organizers for their hard work, precious time and endeavor while preparing the conference. Our deepest thanks also go to the volunteers and staff for their long working hours and the generosity they have given to the conference. Last but not least, we would like to thank each and every one of the authors, speakers and participants for their great contributions to the success of ICMEA 2018.

Editor. Prof. FELLAH Mamoun
Dr. Mamoun Fellah
Prof. Tarek Aly Elsayed
Prof. Arnulfo Luévanos Rojas

IOP Conf. Series: Materials Science and Engineering **484** (2019) 011002 doi:10.1088/1757-899X/484/1/011002

Committees of ICMEA2018

Chair
Prof. Ke Wang, East China University of Technology (ECUT), China

Co-Chair
Prof. YoungPak Lee, Hanyang University, Korea
Prof. Tarek Aly Elsayed, Helwan University, Egypt

Editors
Prof. Tarek Aly Elsayed, Helwan University, Egypt
Prof. Arnulfo Luévanos Rojas, Autonomous University of Coahuila, Mexico
Prof. Ke Wang, East China University of Technology (ECUT), China

Co-Editors
Prof. Amin Riad Maouche, Mhamed Bougara University of Boumerdes, Algeria
Dr. Mamoun Fellah, Khenchela University, Algeria

Technical Program Committee
Prof. Madhav Prasad Neupane, Tribhuvan University, Nepal
Prof. Farid Abed-meraim, Paul Verlaine University, France
Dr. Mohammad Shojafar, Sapienza University of Rome, Italy
Dr. Bappa Acherjee, Department of Production Engineering, India
Dr. Adnene Arbi, University of Carthage, Tunisia
Dr. Dariusz Jacek Jakobczak, Koszalin University of Technology, Poland
Prof. Helena Cristina Vasconcelos, Azores University, Portugal
Prof. N.Ethiraj, M.G.R Educational and Research Institute, India
Dr. Ioannis Dassios, University of Limerick, Ireland
Prof. Márcio José Cuccolo Rosales, Federal University of Sao Carlos, Brazil
Prof. Doaa Abdel Nabi Abdel Aziz, Zagazig University, Egypt
Dr. Elżbieta Macioszek, Silesian University of Technology, Poland
Dr. P. Suresh, Anna University, India
Dr. Niyazi Ugur Kockal, Akdeniz University, Turkey
Dr. Zain Anwar Ali, Nanjing University of Aeronautics & Astronautics, Pakistan
Prof. Ingri Jazmin Guerrero Moreno, National Polytechnic Institute, Mexico
Dr. José Barroso Aguiar, University of Minho, Portugal
Dr. Zviad Gviniashvili, Georgian Technical University, Georgia
Dr. Corneliu Doroftei, Alexandru Ioan Cuza University of Iasi, Romania
Dr. Gbasouzor Austin Ikechukwu, Chukwuemeka Odumegwu Ojukwu University, Nigeria
Prof. N.Kanthavelkumaran, Arunachala College of Engineering for Women, India
Dr. Omar Ahmed Mohamed Amhimmid, Swinburne University of Technology, Libya
Dr. Lidjici Hamza, University of Laghouat, Algeria

IOP Conf. Series: Materials Science and Engineering **484** (2019) 011002 doi:10.1088/1757-899X/484/1/011002

IOP Conf. Series: Materials Science and Engineering **484** (2019) 011003 doi:10.1088/1757-899X/484/1/011003

Peer review statement

All papers published in this volume of *IOP Conference Series: Materials Science and Engineering* have been peer reviewed through processes administered by the proceedings Editors. Reviews were conducted by expert referees to the professional and scientific standards expected of a proceedings journal published by IOP Publishing.

IOP Conf. Series: Materials Science and Engineering **484** (2019) 012001 doi:10.1088/1757-899X/484/1/012001

Effects of Electrochemical Etching Conditions on the Formation and Photoluminescence Properties of P-Type Porous Silicon

Le Lu[1], Wenbing Li [1], Liqiang Zhang[1, 3] and Daohan Ge [1, 2, 3, *]

[1]Nano/micro science and technology centre, School of Mechanical Engineering, Jiangsu University, Zhenjiang 212013, PR. China
[2]National Laboratory of Solid State Microstructures, School of Physics, Nanjing University, Nanjing 210093, PR China
[3]School of Chemistry, The University of New South Wales, Sydney, New South Wales 2052, Australia
[*]Email: gedaohan@mail.ujs.edu.cn

Abstract. Porous silicon has been widely used in sensors, microelectronics and other fields. This material retains the characteristics of the original silicon-based material, while having good optical, electrical and mechanical properties. How to make porous silicon efficient and controllable by means of anodization has become the focus of research. This article mainly studied the influence of different current conditions and eletrolytes on the pore formation and performance of porous silicon in the electrochemical etching process of p-type silicon. It was found that porous silicon structures with controllable morphologies can be prepared by changing the etching current densities. Moreover, adding oxidants (H_2O_2) and dimethylformamide (DMF) into the electrolytes will significantly enlarge the etching parameter window of porous silicon and improve its photoluminescence properties. This will help to expand the applications of porous silicon in the field of microelectronics such as biosensors.

1. Introduction

With the rapid development of microelectronics technology, many excellent properties of porous silicon have been widely applied in various biosensors [1], micro batteries [2-4], photoelectric integrated devices [5] and other fields. At the same time, the high surface area of porous silicon has opened the possibility of a wide range of applications of porous silicon in biosensors such as drug delivery carriers [6-7].

In order to guide the controllable preparation of porous silicon, hoping that the structure and properties can be optimized during the preparation process. In the past several decades, a series of quite different models has been put forward to explain the formation of n-type porous silicon, including the famous space charge region (SCR) model, and the current burst model [8]. However, for p-type silicon, it's hard to produce and explain the pore-forming, especially the macropore formation [9]. Therefore, at present, passivation of p-type porous silicon surface or composite of other materials to improve its performance is the focus of researchers [10-12]. The disadvantage of these approaches is that operations are generally complex and require high cost and technology. For these reasons, a method for preparing P-type porous silicon with superior performance by controlling electrolyte and corrosion conditions, can not only promote the research of the formation process of the P-type porous silicon, but also greatly reduce the cost and conducive to the realization of its industrial application.

In this paper, the effects of different current densities and electrolytes on the pore formation of p-type porous silicon were systematically investigated by a series of anodizaiton experiments. Moreover, the HF/ethanol-containing electrolyte was modified with strong oxidizer (H_2O_2) and organic solvent (DMF), and special attention was paid on exploring the roles of electrolyte components and proportions in the pore morphologies and properties. We have found that the parameter widow of pore formation was enlarged under the condition of oxidants and organic solvents, along with greatly enhanced photoluminescence (PL).

2. Experiment

Porous silicon samples were fabricated by electrochemical etching of p-type silicon (1-10 Ω·cm resistivity, <100>oriented, 450±25 μm thick). The electrolyte solutions for anodization consisted of 40 wt% aqueous hydrofluoric acid (HF) and 99.7 wt% ethanol solutions with the volume ratio of 1:1 as a basic solvent. This HF-ethanol-based electrolyte was modified with two kinds of solutions, H_2O_2 and DMF. The concentration of electrolyte component is expressed in volume unless otherwise stated. Galvanostatic conditions of 30–100 mA cm^{-2} were applied from a computer controlled current source (tianjin lanli LK2500) with a tolerance down to 0.001 mA. All samples were prepared in the dark on an electrochemical etching platform in a standard decontamination chamber at room temperature (18±1°C), with the etching area of 1 cm^2. After anodization, the samples were rinsed repeatedly with deionized water and dried with nitrogen gas.

Field Emission Scanning Electron Microscopy (Shimadzu SPM9700) was used to characterize the microstructure of etched pores. Plan-view micrographs show the (1 0 0) planes, and the cross-section view micrographs, after cleaving the samples, show the pores in the (1 1 0) cleavage planes. The PL spectrums of the PS samples were measured at room temperature using an UV light from a Cary Eclipse Fluorescence Spectrophotometer.

3. Results and Discussion

First the dependence of pore size and etching rate on the applied current has been studied. For the given HF-ethanol basic electrolyte, a set of electrochemical etching experiments were performed for 5 min at different current densities ranging from 30 to 100 mA/cm^2. The results are shown in figure 1. It can be seen that uniform pores with straight pore walls can be obtained and the pore-wall roughness decreases with the increase of current density.

At 30 mA cm^{-2}, branched pores were obtained and the average diameters were about 78 nm (Fig. 1a). The depth of pores was about 3.5 μm, corresponding to the etching rate of 0.7 μm min^{-1}. While current density increased to 50 mA cm^{-2}, straight pores with average diameters of 88 nm were obtained (Fig. 1b). The pore walls were relatively straight and smooth. The pore depth was about 5.3 μm after 5 min anodization, with the etching rate increased up to 1 μm min^{-1}. Increasing current densities to 80 mA cm^{-2}, the pore diameters increased to 80 nm while the etching rate increased to 1.37 μm min^{-1}, corresponding to the etching depth of 6.8 μm (Fig. 1c). At 100 mA cm^{-2}, well-defined pores with smooth pore-walls were obtained, and the pore diameters were about 85 nm (Fig. 1d). The pore depth was about 11.9 μm, corresponding to the aspect ratio and the etching rate 170 and 2.4 μm min^{-1}, respectively. One trend is that the pore diameters increase with the increase of applied current density, and the etching rate increases from 0.7 to 2.4 μm min^{-1} when the current density increases from 30 to 100 mA cm^{-2}. Besides, some pore walls were broken or even collapsed during the etching process especially when the current density increased up to 50 mA cm^{-2}.This phenomenon became increasingly serious with the increase of the current density, and a large amount of collapse occurred when the current density reached 100 mA cm^{-2}. Therefore, it can be seen that the current parameter window for straight pore formation with the conventional HF/ethanol electrolyte is less than 100 mA/cm^2.

IOP Conf. Series: Materials Science and Engineering **484** (2019) 012001 doi:10.1088/1757-899X/484/1/012001

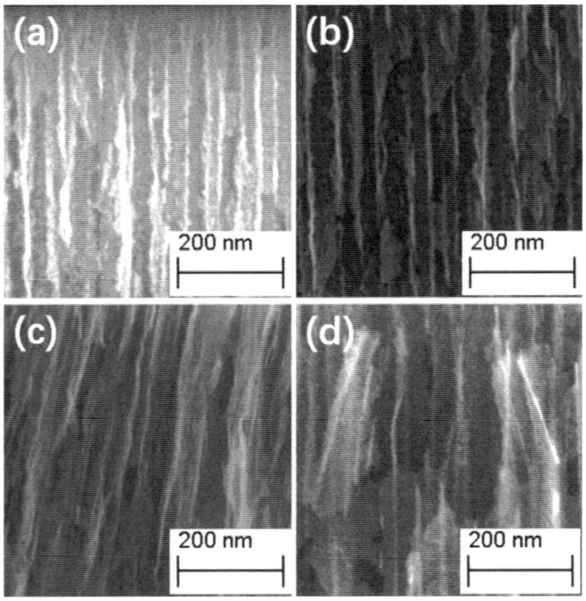

Figure 1. Cross-section SEM micrographs of pores formed on different current densities: (a) 30, (b) 50, (c) 80, (d) 100 mA cm^{-2}. All samples were anodized in HF-ethanol basic solution for 5 min.

Then, the dependence of pore formation and properties on electrolyte was investigated in detail. Two kinds of solvents, strong oxidizing agent (H_2O_2) and organic solvent (DMF), were added into the HF-ethanol based electrolyte respectively, with a volume ratio of 1:1:1. Figure 2 shows the top-view SEM images of the porous silicon formed with these two modified electrolytes at an etching current density of 100 mA cm^{-2}. It can be seen that the surface of the porous silicon was completely intact without any collapse after 5 min anodiaztion both in H_2O_2 and DMF based electrolytes. It indicates that the addition of H_2O_2 or DMF can effectively improve the parameter window for the p-type porous silicon formation. Moreover, it is also apparent from Fig. 2a that the surface of the porous silicon formed by the H_2O_2-contained electrolyte is covered with a translucent film. This interesting phenomenon is probably due to the oxidation effect. A translucent oxide layer was formed on the porous silicon surface due to the existence of the strong oxidant, and the porous structure can be clearly observed through the oxide film. As a comparison, no similar oxide layer was formed under the condition of organic solvent modified electrolyte, shown in Fig. 2b.

Finally, the PL spectrums of porous silicon formed by different electrolytes were carried out at an excitation wavelength of 320 nm, and the results are shown in Figure 3. It can be seen from Fig. 3 that the porous silicon prepared in the HF/ethanol based electrolyte shows relatively weak luminescence intensity, with the luminescence peak located at 750 nm. However, the PL intensity of porous silicon obtained in H_2O_2 or DMF modified electrolytes was greatly improved, with the same current density. The PL intensity of the porous silicon obtained in H_2O_2 modified solution is about 15 times higher than that in basic electrolyte. Besides, the position of the luminescence peak was blue-shifted to 600 nm.

We believe that the change in porous silicon structure caused by changes in the etching solution system is the main cause of changes in photoluminescence properties. According to the "inrush current model", the formation process of porous silicon is essentially the dissolution process of silicon. The rate of electrochemical etching in the HF/C_2H_5OH electrolyte is slow due to the indirect dissolution, which is the process of first forming silicon oxide and then dissolving. However, the addition of a strong oxidant accelerates the oxidation process and shortens the overall reaction time to speed up the etching process. The addition of an organic solvent is just the opposite. It can inhibit the indirect dissolution process during the reaction and effectively reduce the pore-forming time of the porous silicon.

IOP Conf. Series: Materials Science and Engineering **484** (2019) 012001 doi:10.1088/1757-899X/484/1/012001

The modification of oxidant and organic solvent accelerates the pore-forming process of porous silicon, forming porous silicon with larger porosity under the same current conditions, and the larger void ratio brings more refined silicon nanowire and silicon nanoparticles. According to the quantum size effect, the corresponding photoluminescence efficiency increases, and the refinement of the silicon nanowires and nanoparticles causes the band gap to increase the PL spectrum peak to move toward high energy, which is represented by the blue shift of the luminescence peak position. With the addition of strong oxidizing agent, the Si-H bond of the porous silicon is replaced by a more stable Si-O bond, and a layer of oxide film is formed on the surface of the porous silicon as showing in Figure 2(a), which effectively passivates the porous silicon and makes the surface of the porous silicon form more luminescent center so that the formed porous silicon has higher photoluminescence efficiency.

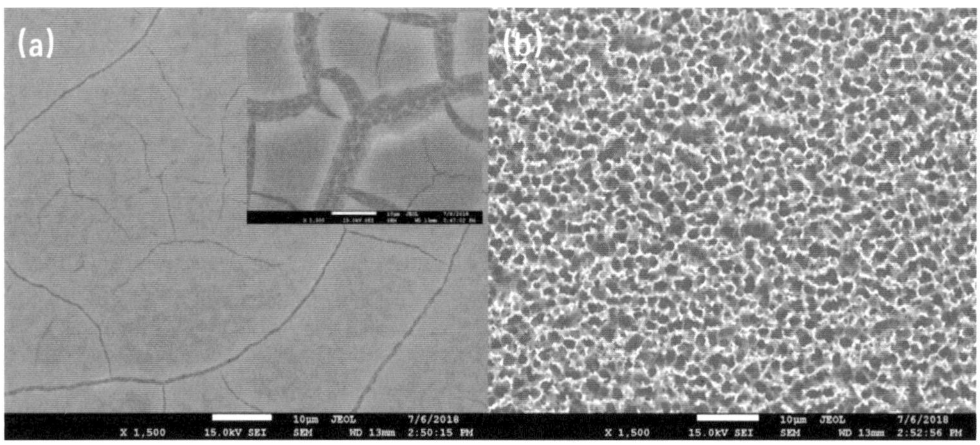

Figure 2. Porous silicon surface after addition of oxidant and organic solvent: (a) HF: ethanol: H2O2= 1: 1: 1; (b) HF: ethanol: DMF= 1: 1: 1. All samples were anodized at 100 mA cm^{-2} for 5 min.

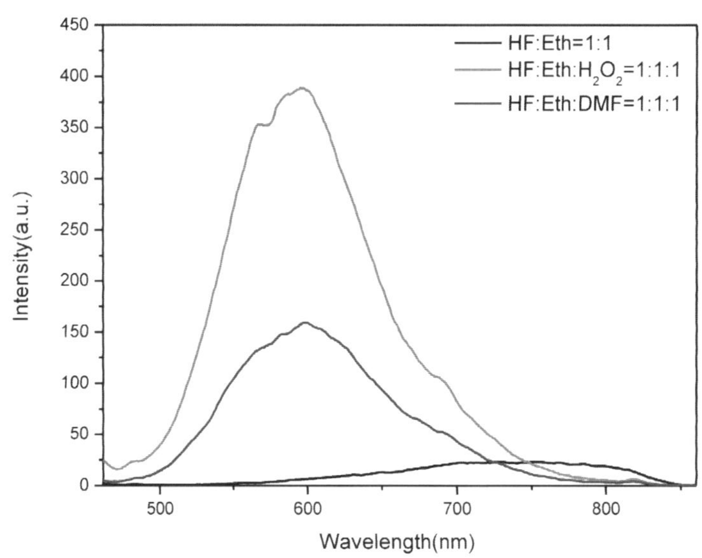

Figure 3. PL spectrum of porous silicon anodized in different electrolytes

4. Conclusion

In conclusion, straight pores were prepared on p-type silicon substrate with different current densities and electrolytes. The parameter window for pore formation is relatively small in the traditional HF/ethanol electrolyte, as well as poor PL efficiency. It is found that the addition of H_2O_2 and DMF into the electrolyte can obviously enlarge the parameter window and improve the PL. The PL intensity of the porous silicon obtained in H_2O_2 modified solution was observed 15 times higher than that in basic electrolyte. This result demonstrates an important method to enhance the PL of porous silicon and is expected to promote the application of p-type porous silicon in high-efficiency sensors and optoelectronic devices.

5. Acknowledgments

This work was supported by grants from the National Natural Science Foundation of China (11404146), the Natural Science Foundation of Jiangsu Province (BK20180098), China Postdoctoral Science Foundation (2018M630543), and the Starting Foundation of Jiangsu University Advanced Talent (13JDG021 and 13JDG020).

6. References

[1] Jiecui L, Zhengcao L, Guojing W, Chienhua C, Shasha L and Mingyang L 2016 *Phys. Chem. Chem. Phys* 18 4835
[2] Ping W, Hui W, Yawen T, Yiming Z and Tianhong L 2014 *ACS Appl. Mater. Interfaces* 6 3546
[3] Oakes L, Westover A, Mares J W, Chatterjee S, Erwin W R and Bardhan R 2013 *Sci. Rep* 3 3020
[4] Wei L, Hou Z, Wei H 2017 *Electrochimica Acta* 229 445-51
[5] Kim J, Joo S S, Lee K W, Kim J H, Shin D H , Kim S and Choi S H 2014 *ACS Appl. Mater. Interfaces* 6 20880-6
[6] Urmann K, Arshavsky-Graham S, Walter J G, Scheper T, Segal E 2016 *Analyst* 141 5432-40
[7] Kun G, Huihui R, Wentong S, Qi Z. Guang J, Aimin Z, Cuimiao Z, Jinchao Z 2016 *J. Nanopart. Res* 18 81
[8] H. Föll, M. Christophersen, J. Carstensen and G.Hasse 2002 *Mat Sci Eng R* 39 93-141
[9] Slimani A, Iratni A, Henry H, Plapp M, Chazalviel, Ozanam F and Gabouze N 2014 *Nanoscale Res. Lett* 9 585
[10] Gan L, He H, Sun L and Ye Z 2013 *Phys Chem Chem Phys* 16 890-4
[11] Ran J, Xianghao D, Weideng S, Zuyin H and Zhengran W 2015 *Scientific Reports* 5 15574
[12] Azaiez K, Benabderrahmane Zaghouani R, Khamlich S, Meddeb H and Dimassi W 2018 *Appl. Surf. Sci* 4332 30349-0

Effect of Yogurt on Gastrointestinal Motility of Dyspepsia Mice

Xiaobing Li, Fajun Li, Weiwen Zhang, Shiping Shen, Lining Wang, Huan Zhang, Mengyuan Ding, Huanhuan Chen and Mingming Huang

College of Basic Medicine, Henan University of Chinese Medicine, room A473, 156 Jinshui East St., Zhengzhou, China.
Email: baishaoyao@163.com

Abstract. With the development of the food industry and the improvement of the living standard, people's diet structure has changed gradually. Excessive diet with high calorie and high protein could cause dyspepsia. Aim of our study is to discuss the effect of yogurt on gastrointestinal motility of dyspepsia mice. Mice were randomly divided into four groups: normal group, model group, Chinese medical group (Drug group), and yogurt group. The mice of normal group were given ordinary feed, the rest of groups were given high protein and high calorie feed and administered through gastric intubation with 52% milk solution to establish dyspeptic model from day 1 to 4. the gastric emptying rate, intestinal propulsion rate, intestinal muscle tension, small intestinal contraction frequency and motilin were measured.Compared with the normal group, the gastric emptying rate, small intestinal propulsion rate, small intestine muscle tension, the small intestine contraction frequency and the motilin concentration were decreased in the model group, Compared with the model group, the intestine muscle tension in the yogurt group was increased, the difference has statistical significance($p<0.05$), in addition, the gastric emptying rate in yogurt group was increased($p<0.05$),the small intestine contraction frequency and the motilin of yogurt group were increased.Yogurt could improve gastrointestinal motility of dyspepsia mice.

1. Introduction

With the development of the food industry and the improvement of the living standard, people's diet structure has changed gradually. Excessive diet with high calorie and high protein could cause dyspepsia. Yoghurt is fermented from fresh, non-resistant whole milk by adding lactic acid bacteria. Research showed that Lactobacillus could regulate the balance of human intestinal microflora (1) and is helpful to control adult and child diarrhea (2) and it is important to maintain physiological function of stomach and intestines. Compared with drug treatment, yoghurt as the daily drinking has advantages of yummy, abundant nutrition and convenience. The aim of this study was to investigate the effects of yogurt on gastrointestinal motility in dyspepsia mice.

2. Materials and Methods

2.1. Animal

Healthy Kunming mice, weighing 20~25g, male, were provided by the experimental animal center of Henan University of Chinese Medicine.

IOP Conf. Series: Materials Science and Engineering **484** (2019) 012002 doi:10.1088/1757-899X/484/1/012002

2.2. *Experimental Materials*

2.2.1. Experimental Reagents. Mouse Motilin ELISA Kit was purchased from Shanghai Yuduo Biotechnology Company Limited(N: YD3146); Yogurt was purchased from Inner Mongolia Yili Industrial Group Co., Ltd.; High fat and high calorie feed(Mixed fish floss, flour, milk powder and bean powder together according to the quality of 1: 1: 1: 2 , added water and stirred evenly, made into the same shape as the normal feed); 52% milk(Milk and water mixed in a mass ratio of 52:48); Nutritional semisolid paste(Mixed 5 g sodium carboxymethylcellulose into 125 mL of water, then added 8 g of whole milk powder, 4 g oral glucose, 4 g of soluble starch, 2 g of activated carbon, stirred evenly).

2.2.2. Experimental Instruments. Refrigerator, BL420 Biological Function Experiment System, constant temperature smooth muscle groove, enzyme-labelling measuring instrument, electrothermal constant temperature blast oven.

2.3. *Animal Treatment*

Mice were used to set up dyspepsia model by fed high fat and high calorie diet. Mice were grouped as follows: normal group, model group, drug group, yogurt group. The mice in normal group were fed with standard diet, and the mice in the other groups were fed with high fat and high calorie diet and given intragastric administration of 52% milk from the 1st day of experiment to the 4th day(3). Mice in yogurt group were given intragastric administration of 0.3ml yogurt/mice; and mice in drug group were given intragastric administration of 0.2 ml Jianweixiaoshi tablets solution/mice. Mice in normal group and model group were given intragastric administration of 0.2ml normal saline/mice. Five days later, the mice were sacrificed.

2.4. *Observations*

*2.4.1 Gastric Emptying Rate and Small Intestine Propulsion rate (4).*Mice were given intragastrically nutritional semisolid paste (self-made, sodium carboxymethyl cellulose 5 g, dissolved in 125 mL of water. Respectively, added whole milk powder 8 g, oral glucose 4 g, soluble starch 4 g, activated carbon 2 g. Stirred well, made into 150 mL of about 150 g of black semi-solid paste).20 minutes later, mice were sacrificed by cervical dislocation, the stomach was removed and weighed. The difference between the weight of the stomach and the net weight of the stomach were taken as the weight of the residual substance in the stomach. The small intestine was removed and the distance from the pylorus to the ileocecal and the pylorus to the front of the black semi-solid paste was measured.

Intestinal propulsive rate=Charcoal powder advancing distance/ Small intestine full length*100%

Gastric emptying rate=1- (Stomach full weight- Stomach weight)/ Nutritional semi-solid paste weight*100%

2. 4.2. Small Intestine Muscle Tension and Contraction Frequenc (5). 2~3 centimeter intestine was removed, then tied lines at both end of the intestine, one end fixed on the specimen hook, the other end connected tension sensor, Immersed the intestine in 37 °C desktop liquid, constantly supply of 95% O2 and 5% CO2 mixed gas into the desktop liquid. Applied 1.0 g of the preload to the intestine, balance 20 min, one hour later intestine appeared regular and stable spontaneous contraction, recorded the contraction activity curve with the BL420 biofunctional experimental system.

2.4.3. Determination of Motilin *Levels in Mice.* Motilin levels were measured according to the kit instructions.

2.5. *Statistical Analysis*

Data are presented as the mean ± SD. Differences were evaluated using Statistical Package for Social Science 21.0. Statistical analysis was performed using One-way ANOVA followed by least-significant difference (LSD). $p < 0.05$ was considered to be statistically significant.

3. Results

3.1. Effect of Yoghurt on Gastric Emptying Rate of Dyspepsia Mice

From Table 1, the gastric emptying rate in the model group was lower than that in the normal group (*P*<0.05). Compared with the model group, the gastric emptying rate in drug group and yogurt group was increased (P <0.05).

Table 1. Effect of yoghurt on gastric emptying rate in dyspepsia mice($\bar{x}$±s)

group	gastric emptying rate(%)
normal group	47.10±1.9
model group	34.60±1.4*
drug group	59.20±8.3$^{*▼}$
yogurt group	54.60±3.1$^{▼}$

Note: Compared with the normal group **P*<0.05, compared with the model group$^{▼}$*P*<0.05.

3.2. Effect of Yoghurt on Small Intestine Propulsion of Dyspepsia Mice

From Table 2, Compared with the normal group, the model group of small intestine propulsion rate was significantly reduced (*P*<0.05). Compared with the model group, the small intestine propulsion rate in the yogurt group was higher than that in the model group.

Table 2. Effect of yoghurt on intestinal propulsion rate of dyspepsia mice($\bar{x}$±s)

group	intestinal propulsion rate(%)
normal group	49.30±5.2
model group	33.00±8.3*
drug group	41.00±1.4
yogurt group	38.80±5.6*

Note: Compared with the normal group **P*<0.05, compared with the model group$^{▼}$*P*<0.05.

3.3. Effects of Yoghurt on Intestinal Muscle Tension in Dyspepsia Mice

From table3. Compared with the normal group, the intestinal muscle tension in the model group was decreased (*P*<0.05). Compared with the model group, intestinal muscle tension in the yogurt group and the drug group was increased (*P* <0.05).

IOP Conf. Series: Materials Science and Engineering **484** (2019) 012002 doi:10.1088/1757-899X/484/1/012002

Table 3. Effects of yoghurt on intestinal muscle tension in dyspepsia mice($\bar{x}$±s)

group	Intestinal muscle tension(g)
normal group	0.15±0.03
model group	0.08±0.01*
drug group	0.16±0.04▼
yogurt group	0.13±0.01▼

Note: Compared with the normal group *P<0.05, compared with the model group▼P<0.05.

3.4. Effects of Yoghurt on Small Intestinal Contraction Frequency in Dyspepsia Mice

From Table 4, Compared with the normal group, the small intestinal contraction frequency in model group was decreased, Compared with the model group, The small intestinal contraction frequency in yogurt group and drug group was increased, and the drug group had significant statistical difference (P<0.05).

Table 4. Effects of yoghurt on the small intestinal contraction frequency in dyspepsia mice($\bar{x}$±s)

group	small intestinal contraction frequency (times / min)
normal group	6.31±0.64
model group	6.00±0.71
drug group	7.56±1.58▼
yogurt group	7.46±1.59

Note: compared with the model group ▼P<0.05.

3.5. Effect of Yogurt on the level of Motilin in Dyspepsia mice

From Table 5, Compared with the normal group, the level of motilin in the model group was decreased (P<0.05). Compared with the model group, the level of motilin in yogurt group and drug group was increased.

Table 5. Effect of Yogurt on the level of Motilin in dyspepsia mice($\bar{x}$±s)

group	the level of Motilin(ng/L)
normal group	2002.68±374.14
model group	1101.77±275.75*
drug group	1221.19±392.92
yogurt group	1105.56±271.11

Note: Compared with the normal group*P<0.05

4. Discussion

Gastric emptying refers to the process of chyme from the stomach into the duodenum, is a comprehensive manifestation of gastric motility, and delayed gastric emptying is the main manifestation of gastrointestinal motility disorders (6). During the experiments, mice were fed high-calorie high-protein foods to establish the model of dyspepsia. Excessive consumption of high-calorie high-protein foods is prone to reduce the transportation of stomach and inhibit the movement of gastrointestinal motility. The gastric emptying rate of mice in yogurt group and drug group was higher than that in the model group, which indicated that yogurt and Jianwei Xiaoshi Tablets have a positive effect and can effectively promote gastric emptying function.

Small intestinal propulsive rate, is one of the classical indexes of gastrointestinal mechanical movement, it can effectively reflect the gastrointestinal peristalsis. Clinically, the main pathophysiological basis of gastrointestinal motility disorders is gastric emptying delay and peristalsis slowing (7).According to the experimental results, the small intestine propulsion of mice in yogurt group and drug group was higher than that in the model group, which indicated yogurt and Jianwei Xiaoshi Tablets improved gastrointestinal peristalsis function.

The main function of gastrointestinal smooth muscle was regulating gastrointestinal movement. Lots of gastrointestinal diseases were also accompanied by muscle contraction function abnormal, which lead to gastrointestinal motility disorder (8).The experimental results showed that the small intestine muscle tension in mice of the yogurt group and the drug group was higher than that of the model group, which indicated that the yogurt and the drug could enhance the small intestine muscle tension of dyspepsia mice.

Motilin could stimulate gastrointestinal mechanical movement and physiological electrical activity, especially, it has an important role in the regulation of the migrating motor complex (MMC)-phase III of the gastrointestinal tract in interdigestive state (9).The experimental results showed that the yoghurt groups and drug groups was higher than that in the model group, indicated that both yogurt and JianweiXiaoshi tablet could promote the peristalsis of the gastrointestinal tract.

5. Acknowledgments

Supported by a project grant from The National Natural Science Foundation of China (Grand No.81603527), Science and technology project of Henan Province (Grand No.162102310466), Key scientific research projects of Henan Province College sand universities (Grand No.16A360010), Henan University of Traditional Chinese Medicine Scientific and technological innovation talent support program (Grand No.2015XCXRC05). Science and technology project of Zhengzhou City (Grand No.20150310), The Young Core Teacher of Henan Province (Grant No. 2016GGJS-080).

6. References

[1] Jack R W, Tagg J R, Ray B. Bacteriocins of gram-positive bacterial. Microbial. Rev, 1995, 59:171-200.

[2] BezkorovainyA. Probiotics: determinantsofsurvivalandgrowthin th gut. Am J Clin Nutr, 2001, 73:399-405.

[3] LIU T.G, YU H, ZHANG W, LV G.K, ZHEN J.H, LI X.F, GU X.H. Influences of Yinlai Tang on sIgA, TNF-α and IL-10 in intestinal mucosal tissues of mice with dyspepsia combined with influenza virus infection. Journal of Beijing University of Traditional Chinese Medicine, 2014, 37:87.

[4] XING J.F, FENG W.Y, HOU J.Y. Investigation of the Experimental Methods for Observing Mouse Gastric Emptying and Small Intestine Propulsion. Journal of Beijing University of Traditional Chinese Medicine, 2003, 26:50-52.

[5] LIANG J.L, YANG J.B, YANG D.Y, ZHOU Z, MA M.X. Effects of 17β-estradiol on small intestinal motility in ovariectomized SD rats. Chin J Gastroenterol Hepatol, 2013, 22:63.

[6] NIU X, ZHANG Z.X. Physiology. Ninth Edition. Beijing: China Press of Traditional Chinese Medicine.2015.

[7] HU X.J, HUANG S.P, DENG S.G. Effect of Decoction of Invigorating Spleen and Regulating Qi on Gastrointestinal Motility, Motilin and Gastrin in Rats with Functional Dyspepsia. Chinese Journal of Experimental Traditional Medical Formulae, 2011, 17:217.
[8] HU J. Studies on Therapeutic Substances of Semen Crotonis Pulveratum on Bowel Movement and Anti-endotoxin Activities of Wei-Chang-An-Wan. Master's thesis, Tianjin University, Tianjin, 2009:16.
[9] LIU C.Q, DENG Q, TAN Z.C, YANG H, JIANG D.J, ZENG G.R. Effect of Simo decoction on functional dyspepsia with diarrhea in juvenile rats. Central South Pharmacy, 2017, 01:8-12

Viscosity and Free Surface Energy as Parameters Describing the Adhesion of the Epoxy Resin to the Substrate

ASzewczak[1, 2] and MSzeląg[1]

[1] Lublin University of Technology, Department of Civil Engineering and Architecture, Faculty of Civil Engineering Nadbystrzycka 40, 20-618 Lublin, Poland
[2] Corresponding Author

Abstract. The polymers engineering is a dynamically developing branch of technology. The main research directions are the search for new polymers and methods for modifying existing ones. Among the extensive group of polymers, construction, coating, and adhesive polymers are the most commonly used in the construction industry. The last family includes the epoxy resins that are used to combine two or more materials with different properties. Under the influence of the hardening agent (catalyst or temperature), the polymers are cross-linking to obtain a solid form. During the gluing process, the most important aspect is the resin preparation procedure and the state of the surface to which it will be applied. The adhesion of the resin to the substrate, the effectiveness and durability of the resulting mixture depend on three factors: resin's composition, viscosity, and adhesive parameters of hardened resin.

The paper contains the research results aimed at modification of the epoxy resin using inorganic fillers: microsilica and carbon nanotubes. An epoxy resin is commonly used for e.g., reinforcing structural elements. Ultrasounds were used as a disintegrating agent of the liquid resin structure, allowing mixing it with the filler. The effect of sonication and fillers on the viscosity of the resin at 22 °C was determined. Next, the free surface energy was evaluated, which is the factor determining the final adhesion. Based on the results obtained, the phenomena occurring at the interface between the liquid phase of the resin and the solid phase of the fillers were explained.

1. Introduction

Polymers are chemical substances composed of cyclic repeating chemical sequences (the so-called mers), which form a polymer chain [1]. The mers combine with each other using various types of chemical bonds: ionic, coordinating, polarized, or hydrogen [2]. The chemical composition of the polymer and the chemical bonds occurring between the sequences determine the intended use of a polymer. Viscosity and the final adhesion to the substrate (e.g., concrete, steel [3]) determine the suitability of a polymer. The viscosity affects the possibility of spreading the adhesive on the target surface, filling its irregular ities, and the resin adhesion after hardening. The cross-linking reactions occur at different times, from several hours to several days. The course of the reactions depends largely on the presence of specific polymer groups (phenolic, epoxide, and ester) and the type of factor initiating such a reaction. In the case of chemo-hardening polymers, the amine-based hardeners are most often used [1, 2]. Construction adhesives are hardened after just a few hours, and after a period of 2 days they usually get full adhesion to the substrate.

Among the very large group of polymers encountered in civil engineering the following ones are the most commonly used in construction [1]:

- fiber-forming, e.g., strengthening and reinforcing meshes,
- structural (resin concrete, support profiles, masking elements),

- coatings (hydrophobic coatings, varnishes, impregnates), insulating construction foils,
- adhesive (glues) – epoxy, polyester, phenol, formaldehyde, urea resins and mixtures,
- special (e.g., with increased resistance to fire, UV radiation, low temperatures).

The main advantages of the construction polymers are [4]:

- very good strength parameters in relation to the weight of the element, e.g., beams,
- very good adhesion to wood, steel, concrete, and ceramics,
- constantly decreasing costs of obtaining polymers and their final price.

The epoxy resins are oligo-polymers that are the product of reaction of the epichlorohydrin and bis-phenol. In addition to the aforementioned properties, its additional features are chemical resistance, impermeability, and dielectricity. The main disadvantages include low resistance to high temperatures and UV radiation.

Fillers are additives, thanks to which it is possible to obtain specific properties and improve the parameters of resins [5, 6, 7]. Their quantity and type are selected due to the type of modification planned, the chemical composition of the polymer, the properties of the filler particles, and the required final effect. The amount of the filler is usually dependent on the resin's weight (volumetric dosing is less precise due to the sensitivity of the polymer to change in temperature and thus their density). It is also important to effectively distribute the filler in the volume of the resin. For this purpose, ultrasounds are often used in the polymer technology [8, 9, 10]. As a result of dynamic phenomena caused by the action of their energy (vibrations, change of temperature, pressure, cavitation), the filler molecules can fill spaces between polymer particles or connect with them, most often by means of temporary atomic bonds. Proper selection of the sonication energy and the amount of filler gives the possibility of appropriate change in the viscosity of the resin – the parameter that determines the initial, early adherence of the resin to the substrate. Then, the hardened polymer obtains proper adhesion to the substrate.

The paper contains the research results on the properties of epoxy resins modified with two fillers – microsilica and carbon nanotubes. The exact mixing of the fillers with the resin was obtained by the action of the sonication energy. The dependence of the early viscosity (at the time of the resin application to the target surface) on the final adhesion of the resin to the substrate was evaluated. For this purpose, the free surface energy was measured [8, 9, 11]. It is a parameter describing dependencies on the liquid-solid. The Owens-Wendt method was applied, which requires the use of two liquids with known dispersive and polar properties. The results obtained were the basis for determining the effectiveness of the modifications applied with regard to the possibilities of practical application, i.e., during strengthening and gluing of building elements. The research conducted was the first stage of research on the effectiveness of the modifications applied in the concrete reinforcing issues, i.e. strengthening by the FRP tape.

2. Materials, Devices and Methodology

The Epidian 52 resin was used, by means of which it is possible to connect concrete, ceramics, wood or metal elements. Its basic parameters are:

- state/color – thick, yellow-brown liquid with a characteristic smell,
- ignition temperature – 64 °C,
- gelation time (the Z-1 hardener) – 40 min,
- epoxide number [mol/100g]: 0.51 – 0.55,
- density (at T = 22 °C): 1.12 – 1.13 g/cm^3,
- viscosity (at T = 22 °C): 400 – 800 m·Pa·s, (1m·Pa·s = 10^{-3} Pa·s) ,
- solubility: insoluble in water, soluble in ketones, esters, alcohols, aromatic hydrocarbons,
- hardener: the Z-1 (triethylene-tetrachloro amine).

The fillers used are microsilica and carbon nanotubes. The amount of filler was related to the weight of the resin and it was determined experimentally. The recipes of the composites evaluated are shown in Table 1:

Table 1. List of recipes of composites tested

Recipe designation	Type of resin	Type of addition/modification	The amount of filler [%]	The amount of hardener [%]
EP52	epoxy	----------------	----------------	7.6
EP52+UD	epoxy	sonication	----------------	7.6
EP52+Mk	epoxy	sonication and microsilica	0.5	10
EP52+NR	epoxy	sonication and carbon nanotubes	0.1	10

In the first stage of the research, the initial parameters (viscosity and density) of the unmodified resin at 22 °C were determined. Next, the polymer was subjected to the three types of modifications shown in Table 1. For each recipe, the same time of sonication (3 minutes) was assumed (time and power of sonication was determined experimentally). After this time, the initial viscosity of the resin and its temperature was measured. Then the temperature drop and the viscosity increase was measured at intervals of 5 min until the resin reached the initial temperature. The ultrasounds source was the UP 400S stationary sonicator, which emits 24kHz waves and have a power range regulation from 0 to 400 W (Figure 1). The viscosity measurement was made using the Fungilab type H rotational viscometer with the use of the R2 spindle (measurement accuracy up to 0.1 m·Pa·s). The temperature of the resin was measured with a laboratory thermometer with an accuracy of 0.1 °C and with a range of measured temperatures from -25 to 150 °C.

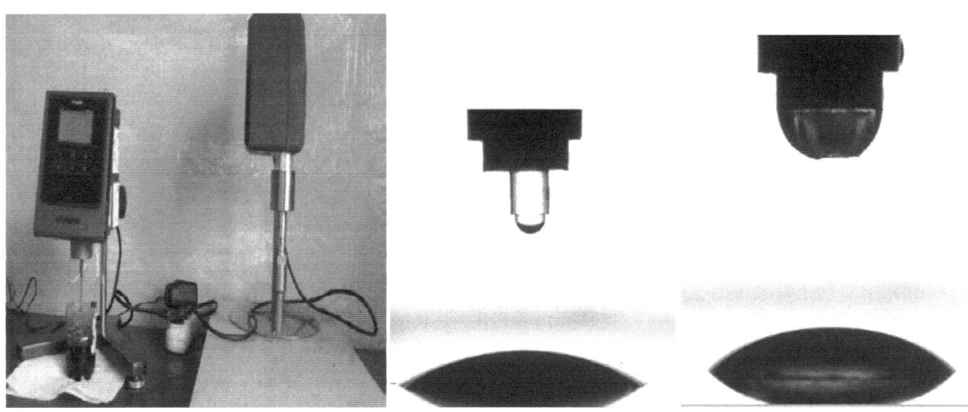

a) b) c)

Figure 1. The test stand: a) the sonicator and the rotational viscometer, b) the drop of distilled water, c) the drop of diiodomethane

The surface free energy (SFE) was determined according to the Owens-Wendt method, which consists in measuring the wetting angle of the hardened resin surface. This angle is measured between the horizontal straight line and the tangent to the drop in two variants (for two different liquids Figures 1b and 1c). These liquids were diiodomethane, which had a dispersive component value of 50.8 mJ/m^2 and zero polar component (highly dispersive), and distilled water (highly polar liquid) with a dispersive component value of 51 mJ/m^2 and polar 21.8 mJ/m^2. The SFE is determined as the sum of the dispersive and polar component for each sample. For each series, 30 SFE measurements were made. In order to perform the accurate measurements of rheological properties (viscosity and SFE), it was important to maintain a constant final temperature (22°C); The use of this temperature was necessary to be able to compare the parameters measured, and it was set according to the guidelines and recommendations for this type of research.

3. Results and Discussions

Table 2 shows the results obtained: the initial viscosity μ_0 (at the time of switching off the sonicator), the final μ (after the resin reached 22ºC), and the dispersive components γ_d, polar γ_p, and the total value $\gamma_0$of the SFE. The results obtained were compared with the results for the unmodified resin.

Table 2. Results of research

Series	EP52	EP52+UD	EP52+Mk	EP52+NR
Viscosityμ_0 [mPas]	665.6	54.4	51.2	192
Viscosityμ [mPas]	--------	758.4	585.6	1977.6
Dispersive component [mJ/m^2]	42.7	40.1	42.03	40.87
Polar componentSFE [mJ/m^2]	22.67	32.33	30.6	23.57
TotalSFE [mJ/m^2]	65.43	72.43	72.6	64.43

The results indicated that the final viscosity as well as the individual components of the SFE strictly depends on the type of modification applied. Sonication leads to the dynamic changes in the volume of the resin. An important result of this change is the rise of temperature and pressure drop of the polymer and a violation of its original structure. This changes the structure of the composite and allow the resin to connect with the filler particles and the hardener, mainly in course of the crosslinking reaction. These changes affect the way the filler molecules are arranged, which enter the polymer chain. Sonication cause temporary breaking of the polymer chains, which makes it possible to reorganize them and attach filler molecules. However, the type of such a connection closely depends on the shape of the filler molecules. The sonication energy causes the destruction of long polymer chains and the increase in the viscosity due to the presence of free ends of the individual mers; this affects the parameters of the hardened resin, i.e., a higher SFE value (due to the presence of the free chain ends, while the internal structure is ordered, especially the polar component, directly related to the presence of the free chemical bonds at the ends of polymer chains(Figure 2).The presence of microsilica changes this process; due to intermolecular interactions, microsilica entered into the polymer chains and binds to free polymer endings. This reduces the viscosity and SFE, however, it is not decreased compared to the SFE of the resin modified only by sonication. On the surface of the resin, a thin layer composed of microsilica particles and components of the polymer chain is present (Figure 2), which increases the final adhesion, e.g., to the concrete substrate. Carbon nanotubes, due to the action of ultrasounds, break down and combine with the polymer chains because they are chemically compatible with the main element that forms a polymer – carbon; this causes the initial increase in the initial viscosity. As a result of further interactions, the free polymeric ends are connected to each other by means of nanotubes. The lack of free ends of the polymer chains was noticed based on the results of the SFE value.

The method of "entering" the filler particles into the polymer chains affected the final viscosity. The microsilica is not directly connected to the resin molecules, but only enters into the open spaces between the polymer chains, which reduces the viscosity of the resin. Carbon nanotubes densify the polymer network, consequently increasing the viscosity of the adhesive.

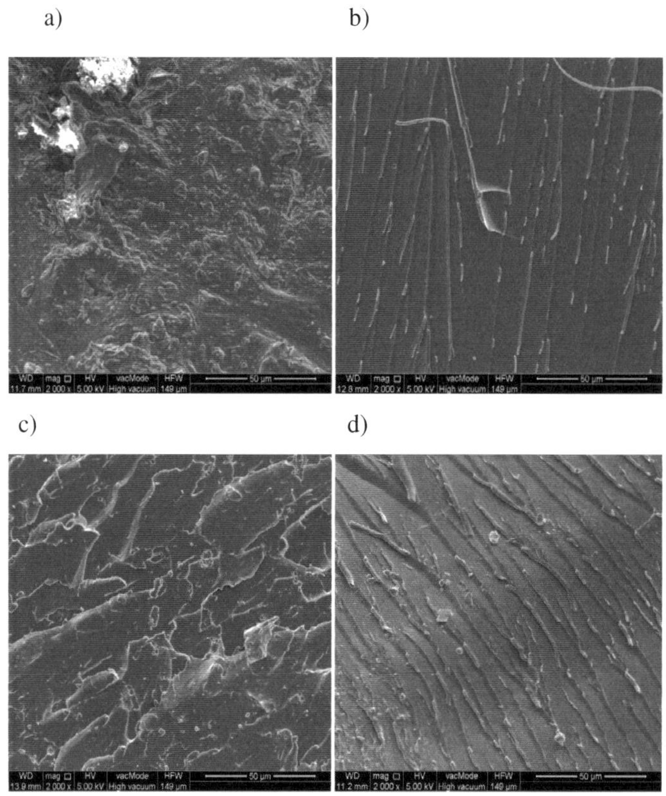

Figure 2. SEM Picture of the samples a) EP52, b) EP52+UD, c) EP52+Mk, d) EP52+NR

4. Conclusions

The results obtained gave a clear picture of how the resin subjected to the modifications behave on the target surface. The type of modification and filler used may depend on the target conditions under which the particular polymer composite will be operated.

The conclusions obtained are very important from a practical point of view. Modifications applied in the case of the EP52+UD and EP52+NR, due to the marked increase in the final viscosity, may be useful in the case of bonding materials at elevated temperatures, or in case when the surface of the substrate is smooth and even, less rough, and less adhesive. The use of the microsilica, on the other hand, may be useful in the case of bonding at lower temperatures or when the application surface is uneven and porous.

The results will be used during the next tests regarding the effectiveness of the modifications during the process of gluing the FRP tapes and reinforcing the concrete elements.

5. Acknowledgments

This work was financially supported by Ministry of Science and Higher Education in Poland within the statutory research number S/14/2016.

6. References

[1] Rabek J F 2017 *Modern Knowledge about Polymers* vol. 1 (Wyd.Naukowe PWN Warsaw) p. 3 – 20

[2] Ehrenstein G 2016 *Polymeric Materials: Structure, Properties, Application* (Wyd.Naukowe PWN Warsaw) p. 3 – 100.

[3] Sadowski Ł Czarnecki S Hoła J 2016 Evaluation of the height 3D roughness parameters of concrete substrate and the adhesion to epoxy resin *International Journal of Adhesion and Adhesives* vol. 67 p. 3 – 13.

[4] Lokensgard E 2016 *Industrial Plastics* (Cengage Learning Eastern Michigan University, Ypsilanti, Michigan).
[5] Wypych G 2016 *Handbook of Fillers* (ChemTec Publishing, Toronto).
[6] Fang-Long J Xiang L Soo-Jin P 2015 Synthesis and application of epoxy resins: A review, *Journal of Industrial and Engineering Chemistry*, vol. 29, p. 1-11.
[7] Z Jinbo W Lili Z Chuanxing S Qian 2017 Overview of polymer nanocomposites: Computer simulation understanding of physical properties *Polymer* vol. 133 pp. 272 - 287
[8] Fic S Szewczak A Barnat – Hunek D Łagód G 2017 Processes of fatigue destruction in nanopolymer-hydrophobised ceramic bricks *MATERIALS* vol. 1/10 p.44
[9] Fic S Szewczak A Kłonica M 2015 Adhesive properties of a low molecular weight polymer modified with nanosilica and disintegrated with ultrasounds for hydrophobization of building ceramics *Polimers,* vol. 11-12 pp. 730-734.
[10] Fic S Szewczak A Kłonica M 2016 Effect of hydrophobization on the durability of building ceramics *Polimers* vol. 61 pp. 46 - 48.
[11] Barnat-Hunek D 2017 *Free Surface Energy as a Factor Shaping the Effectiveness of Hydrophobization in the Protection of Building Structures* Wyd Politechniki Lubelskiej Lublin Poland p.20 – 105.

IOP Conf. Series: Materials Science and Engineering **484** (2019) 012004 doi:10.1088/1757-899X/484/1/012004

Influence of Neutralization on Hydration of Hemihydrate Phosphogypsum Plaster

Xuemei Chen and Jianming Gao*

Jiangsu Key Laboratory of Construction Materials, Southeast University, Nanjing 211189, China.

Email: chenxuemei424@foxmail.com (X. Chen), jmgao@seu.edu.cn (J. Gao)

Abstract. Effect of lime on hydration of hemihydrate phosphogypsum (HPG) plaster were investigated aiming at neutralizing HPG since it was obtained by calcining phosphgypsum without any purification. Results showed that neutralization would lead to retardation and strength degradation, particularly with small amounts of lime (0.1%-0.4%). As a result, the role of lime in HPG plaster can be described as 1) reacting with impurities in HPG and the reactant would affect the dissolution of HPG, gypsum nucleation and growth; 2) increasing ion concentration and supersaturation by excessive lime which can promote the hydration again. The gypsum crystal changed from needle-like to tiny crystal with little interlocking and then long prismatic crystal with comparable strength due to the mechanism of lime. Therefore, sufficient lime is needed for neutralization which is economic and beneficial for the application of HPG.

1. Introduction

Over 70 million tons of industrial wastes are produced annually in China from the phosphoric acid process for manufacturing fertilizers. Out of this, over 40 million tons of phosphogypsum (PG) is stacking directly without any treatment. At present, the surface stockpile of PG in China is estimated to be above 300 million tons [1-5]. PG contains impurities such as water-soluble phosphate (P_2O_5), water-soluble fluoride (F), insoluble P_2O_5 substituted in the gypsum crystal lattice, residual acid and organic matters which cause environmental concern [6-8]. The commercial importance of gypsum is due to its ability to form hemihydrate plaster, thus hemihydrate PG (HPG) is obtained by thermal dehydration from PG directly which complies with the requirements in Chinese standard calcined gypsum (GB/T 9776-2008) including radioactivity. HPG is generally used in preparation of gypsum block and plaster board [9]. Due to the fact that the cleanup PG often lead to increasing costs of production, the commercial HPG in China are still acidic, set fast and contain impurities which deteriorate the performance of HPG and thus restrict its further use in building materials.

In order to broader the application scope of HPG and improve its performance, neutralization is necessary. As reported, lime water washing is a good way to reduce, eliminate and remove impurities from PG [10]. Meanwhile, lime is a common traditional material which is also one of the hydration products of cement. How about adding lime to HPG plaster directly? Would it affect the properties of HPG plaster? The influence of lime is helpful to guide the development of HPG based composite cement. Investigations were, therefore, undertaken regarding the hydration process of HPG plaster after neutralization. It was interesting to find that a small amount of lime has adverse effect on the hydration of HPG plaster while beyond a certain lime content, HPG plaster can set and harden well.

2. Experimental Methods

2.1. Materials and Mix Proportions
HPG was from longmang Co., Ltd., in Deyang Sichuan province China and it was obtained by thermal dehydration from PG after three time's water washing. Retarder (Retardan-200P) was added to acquire a desired operation time at the rate of 0.06% by mass of HPG powder and the water to powder was 0.64 which was the water requirement of HPG for normal consistency. Lime powder ($Ca(OH)_2$) was also added by mass percent and the ratio of materials HPG:retarder:water:lime was 100:0.06:64:x (varies from 0.1 to 4.4).

2.2. Setting Time and Compressive Strength
The setting time of HPG plaster were examined by Vicat apparatus specified in GB/T 17669.4-1999. Prismatic specimens of size 40×40×160 mm^3 were prepared according to above mix proportion and cured at 20.0±1°C, 60±5% relative humidity. After curing for seven days, specimens were dried at 40°C until the weight was constant and then compressive strength was tested by reference to GB/T 17669.3-1999 in China.

2.3. pH and Hydration Degree Testing
The powder with above mix proportion (before hydration) or by grinding hydrated samples at 7 days after drying at 40°C (after hydration) were mixed with water (the ratio is 1:10). After stirring for 5 min and waiting for 25 min, upper supernatant were used for pH testing by a Digital pH Meter Model PHS-25 (Leici, Shanghai). The hydration degree was calculated through chemically combined water which was tested according to Chinese standard GB/T 5484-2012.

2.4. Hyration Heat and SEM Analysis
An eight channel TAM Air Isothermal Calorimeter was used to analyze the hydrate heat of HPG plaster by the direct method in accordance with GB/T 12959-2008. The fragments from samples after compressive strength were analyzed by SEM.

3. Results and Discussion

3.1. Setting Time and Compressive Strength
HPG is acidic arousing concern in application which would corrode the equipment and get mildewed. Addition of lime to neutralization comes to our mind firstly with respect to solve this problem. Since natural gypsum is neutral and being alkaline is beneficial to make the bestuse of admixture, a small amount of lime (0.1 to 0.4 by wt.% of gypsum plaster) were added to adjust the pH of HPG plaster in alkaline range (pH 7.19-11.08). As shown in Figure 1, it was clearly that once the lime was added, the setting time increased from 72 min to above 1000 min while the strength dropped from 10 MPa to 3 MPa. However, with the further increase of lime regardless of pH, the setting time of HPG plaster decreased greatly and presented little variation when the contents of lime were 2.4%-4.4%. In addition, the strength of HPG plaster increased with enhancement in lime content indicating that the role of lime in HPG plaster was not just neutralization. Through the pH before and after hydration in Figure 2(b), it was clear that lime had reacted with impurities and thus affected the hydration of hemihydrate to gypsum resulting in abnormal setting and hardening of HPG plaster while the excessive lime had eliminated the influence of reactants contributing to the decrease in setting time and increase in strength.

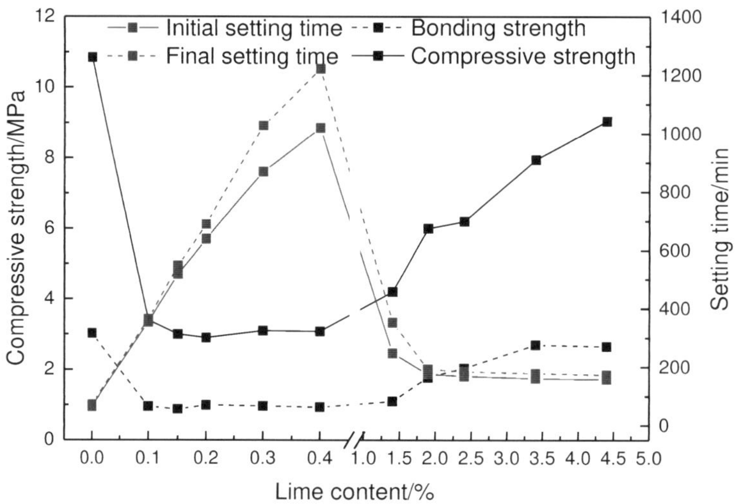

Figure 1. The setting time and strength of HPG plaster with different lime content.

3.2. Hydration Kinetics Analysis

The influences of the lime content on the hydration heat evolution rate, hydration degree and pH before and after hydration of HPG plaster were presented in Figure 2a and b. Figure 2a shown that a sharp and transient exothermic peak was formed probably owe to the fast dissolution of HPG after contacting with water. Then the heat emission rate decreased dramatically. The plaster was in a dihydrate gypsum nucleation and the supersaturation process of Ca^{2+} and SO_4^{2-}. The induction period was prolonged greatly with the increased dosage of lime within 0.4% and subsequently, shortened with further increase in lime content. Accordingly, the maximum heat evolution rate and the hydration degree were decreased at first and then increased with enhancement in lime content which was good in line with the results in Figure 1. It can be deduced that the reactants from lime and impurities have effect on the the dissolution of HPG and gypsum nucleation as well as the growth of gypsum, therefore the induction period got longer and the hydration heat evolution rate peaks came smaller while the excessive lime increased the ion concentration and supersaturation which promoted the hydration again. In addition, the pH after hydration of HPG plaster within 0.4% lime only increased a little confirming the reaction of lime and impurities. The hydration degree of HPG plaster whose lime content was beyond 2.4% was equal to that of reference, but its setting time was a little longer and strength was a bit lower. That was why the peaks of heat evolution rate were on the right side of the reference indicating the retardation of reactant. Besides, their peaks width were much longer proving that the hydration was gently promoted by crystal nucleus and higher ion concentration from excessive lime. Furthermore, the intensive heat evolution was avoided which would reduce the risk of temperature deformation and was better for the application of HPG plaster.

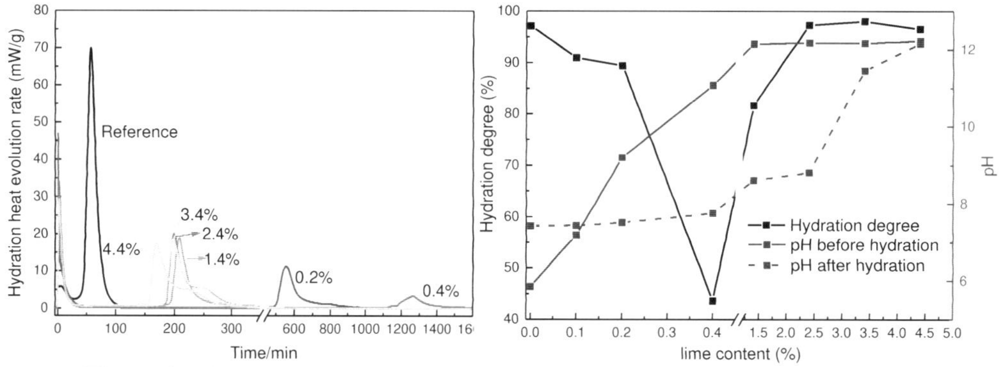

Figure 2. Effect of lime content on (a) heat flow; (b) hydration degree and pH.

3.3. SEM Analysis

The strength of gypsum is closely related with the crystal interlocking except of the influence of hydration degree. As shown in Figure 3(a), the gypsum crystal in reference was mainly needle-like which interlaced with each other tightly contributing to a dense microstructure and higher strength in return. According to the function of lime, the amount of crystal nucleus were reduced after adding lime and gypsum crystal grew slowly, thus bulky and granular gypsum crystal were observed in Figure 3(b) which were well crystallized but there is no interlocking between each other. This is why its strength was not higher than samples in Figure 3(c) in spite of its hydration degree were much higher. Fluffy and floccus substances were filled in Figure 3(c) which were composed of incompletely dissolved hemihydrate and crystallized dihydrate gypsum, since the formation of crystal nucleus and growth of gypsum crystal were strongly inhibited. Compared with Figure 3(c), gypsum crystal in Figure 3(d) grew continuously and interlaced with each other to some extent, but a few undissolved hemihydrate were still visible. Meanwhile, there was no big difference in gypsum appearance in Figure 3(e) and (f) but undissolved hemihydrate disappeared highlighting that the further increase in lime content can really promote gypsum dissolution. However, it would not have influence on hydration degree and gypsum morphology when beyond a certain content. In fact, the higher lime content would lead to higher water requirement. Therefore, the optimal addition of lime for neutralization was in the ranges of 3%-4%. Comparing with the reference sample, the gypsum crystal after adding enough lime in Figure 3(e) and (f) were thick and prismatic resulting in decrease of interlocking and increase of porosity. This was the main reason responsible for lower strength of samples after adding enough lime.

Figure 3. SEM images of HPG plaster with different lime content.

Based on above results, it can be concluded that the impurities in PG would consume a small amount of lime and the reactants would inhibit the dissolution of hemihydrate and growth of gypsum crystal inducing retardation and strength degradation. The hydration can be promoted due to the increase in calcium concentration and supersaturation after further increasing lime content. However, retardation cannot be avoided after neutralization even with high content of lime, thus the amount of crystal nucleus reduced leading to the thick crystal and a little lower strength when compared with the reference samples.

4. Conclusions

Minor amounts of lime (0.1%-0.4%) can change the pH of HPG plaster from 5.85 to 11.08, but it has deleterious effect on hydration leading to abnormal setting and strength degradation. Interestingly, further increase in lime content would promote the rehydration and the development of strength. Therefore, addition of enough lime (3%-5%) is needed to neutralize HPG plaster instead of controlling the pH.

5. Acknowledgments

The financial support of Sino Japanese Project from MOST of China (2016YFE0118200), National Natural Science Foundation of China (grant no.51578141) and the Major State Basic Research Development Program of China (2015CB6551002) were gratefully acknowledged.

6. References

[1] Chen X.M, Gao J.M, Liu C.B and Zhao Y.S 2018 *Constr. Build. Mater* 190 53-64
[2] Ma L.P, Ning P, Zheng S.C, Niu X.K and Du Y.L 2010 *Ind. Eng. Chem. Res* 49 3597-3602
[3] Du L.S, Ming D.Z, Li Z.X and Li Y.H 2010 *Tech. Dev. Chem. Ind* 39 25-28
[4] Mun K.J, Hyoung W.K, Lee C.W and So S.Y 2017 *Constr. Build. Mater* 21 1342-1350
[5] Tan T, Yan Y and Hu Z.H 2016 *Constr. Build. Mater* 115 143-152
[6] Yang L, Zhang Y.S and Yan Y 2016 *Constr. Build. Mater* 127 204-213
[7] Singh M 2005 *Cement. Concrete. Res* 33 1363–1369
[8] Sakthieswaran N and Sophia M 2018 *Constr. Build. Mater* 179 675–691
[9] Rashad A.M 2017 *J. Clean. Prod* 166 732-743
[10] Singh M 2005 *Constr. Build. Mater* 19 480–486

IOP Conf. Series: Materials Science and Engineering **484** (2019) 012005 doi:10.1088/1757-899X/484/1/012005

Carbon Nano-Onions Embedded CuO Nanosheets: An Excellent Stable Anode Material for Lithium Ion Battery

Zhongjing Hao[#, 1, 2], Meng Qin[#, 1], Yueming Li[*, 1], Xiaojun Lv[*, 2], Dafeng Zhang[2] and Qiushi Wang[2]

[1]State Key Laboratory of Metastable Materials Science and Technology, College of Materials Science and Engineering, Yanshan University, Qinhuangdao 066004, China
[2]Key Laboratory of Photochemical Conversion and Optoelectronic Materials and HKU-CAS Joint Laboratory on New Materials, Technical Institute of Physics and Chemistry, Chinese Academy of Sciences, Beijing 100190, P.R. China.
E-mail: liyueming@ysu.edu.cn

Abstract. Carbon nano-onions embedded CuO nanosheets was synthesized by hydrothermal method and this report reports the application of carbon nano-onions embedded CuO nanosheets in lithium ion battery anode materials. The introduction of carbon nano-onions leads to improved electrochemical performance. When applied to potential anode materials for lithium-ion batteries, the novel carbon nano-onions embedded CuO nanosheets deliver reversible discharge capacity of 709.7 mAh g^{-1} at 0.1 A g^{-1} and specific capacity of 167 mAh g^{-1} at 5 A g^{-1}, much higher than those of the pure CuO nanosheets with same preparation conditions. In this paper, we demonstrate that the compositing of carbon nano-onions embedded CuO nanosheets is able to improve the performance significantly.

1. Introduction

Many electronic products, such as cell phones, notebook computers, and electric vehicles, use lithium-ion batteries because of their long cycle life, high energy density, environmental friendliness, and excellent rate performance. [1]. Graphite as an anode material has a limited specific capacity (372 mAh g^{-1}). Therefore, it is very important to find a high capacity anode material. In recent years, transition metal oxides (such as MnO2 [2], Fe3O4 [3], Fe2O3 [4], CuxO [5], NiO [6], SnO2 [7], Co3O4 [8], TiO2 [9]) Because of its high energy density and higher theoretical lithium ion storage capacity, it will be regarded as the most promising anode material, which has been studied by many scholars. Among various transition metal oxides, CuO is favored by some scholars because of its abundant yield, simple preparation process, stable chemical properties, low synthesis cost, high theoretical capacity (674mAh g-1), and environmental harmlessness [10]. However, some of the inherent properties of CuO limit its practical application, for example, poor chemical stability, low electrical conductivity, and large and uneven volume changes during charging and discharging of lithium ions, resulting in rapid capacity degradation [11, 12]. The addition of some well-conducting carbon materials (such as graphene [13, 14] or carbon nanotubes [15]) to CuO is a good method to achieve a method of increasing stability. These conductive carbon materials can improve the inherent disadvantages of CuO and enhance the electronic conductivity of CuO by creating electron paths. Moreover, the conductive carbon can also act to buffer the volume change of the active particles, thereby alleviating the pulverization of the particles during charging and discharging, and ensuring good contact with the copper oxide active particles [16-18]. Compared to other forms of carbon, carbon nano-onions (CNO) can more effectively improve the electrical conductivity [19, 20]. Since carbon nano onion itself has high conductivity and small particle size of 5-10 nm and 0-D nanostructure, these characteristics make it possible for them to form a conductive network evenly distributed over the entire electrode when making electrode materials, which will be very It is beneficial to improve the electrochemical stability of the battery [21].

In this report, a novel carbon nano-onions embedded CuO nanosheets has been prepared through a simple hydrothermal method. Compared with the copper oxide nanosheet electrode without the carbon nano-onion, the carbon nano-onion and the copper oxide nanosheet combine to function as a negative electrode material for a lithium ion battery, exhibiting a high capacity.

2. Experimental

2.1. Synthesis of CNO.

CNO was synthesized by thermal annealing using a nanodiamond raw material (average particle diameter of about 5 nm), then annealed in Ar atmosphere at a heating rate of 10 °C min $^{-1}$ and insulation 1300 °C for 1 hour.

2.2. Synthesis of CNO Embedded CuO Nanosheets.

First, in order to form a uniform liquid, CNO was added to a deionized aqueous solution (50 mL) for 30 minutes to maintain ultrasonic conditions. Second, 0.8 g of NaOH and 1.2 g of Cu $(CH_3COO)_2 \cdot H_2O$ were added and stirring was continued for ten minutes. Third, the above solution was hydrothermal at 180°C for 12 hours, and naturally cooled to finally obtain a black precipitate which was washed with suction and deionized water for three times, and then dried at 80 °C for 12 hours. The CNO embedded CuO nanosheets composites with different amount of CNO synthesized by changing the amount of CNO (6, 12, 36mg), were named as CuO/CNO-1, CuO/CNO-2, CuO/CNO-3, respectively. The pure CuO was prepared used same procedure except no CNO was added.

2.3. Characterization.

Powder X-ray diffraction (XRD) was measured with a scattering angles from 10 to 80 °. Raman spectroscopy was carried out within 200 cm^{-1} to 1800 cm^{-1} using Ar^+ laser excitation with a wavelength of 532 nm. Morphologies characterization was studied with scanning electron microscopy S-4800. The TEM images were obtained with JEM-2100F equipped with anoperating at 200 kV.

2.4. Electrochemical Measurements.

The active material, conductive Super P, and polyvinylidene fluoride (PVDF) binder with mass ratio of 8:1:1 and N-methylpyrrolidone (NMP) solvent were uniformly coated on copper. The foil is prepared as an anode material. The electrode was vacuum dried at 120 ° C for 12 hours and then pressed into a battery for electrochemical testing. The active material mass load is 1.2 mg cm $^{-2}$ .The lithium sheet is used as a cathode material, and the battery is assembled in a glove box that is insulated from air. Constant current charge/discharge measurements were made for different current densities using a Arbin battery test device over a voltage range of 0.01 to 3.0 V (vs. Li / Li +). Cyclic voltammetry (CV) measurements and electrochemical impedance spectroscopy were performed using a P4000 electrochemical workstation (Princeton Applied Research, USA) in the range of 0.01 to 3 V (vs. Li / Li^+).

3. Results and Discussion

The crystallographic information for the CuO, CuO/CNO-2 and precursor-CuO/CNO-2 was characterized by XRD, and the result was shown in Figure 1. The eleven diffraction peaks at 2θ values of 32.4, 35.4, 38.6, 48.6, 53.4, 58.2, 614, 66.1 and 68.0° can be indexed to monoclinic CuO with crystal planes of (110), $(11\bar{1})$, (111), $(20\bar{2})$, (020), (202), $(11\bar{3})$, $(31\bar{1})$ and (220), which consistent with the JCPDS card NO. 48-1548 [22] .As can be seen from Figure 1, compared with the three materials, the crystallinity of CuO/CNO-2 before the hydrothermal reaction is not good, and the crystallinity of

CuO is the best. Additionally, there is no peak of CNO in the XRD patterns. This result may be attributed to the effect of small amount and small particle size (5-10 nm) of CNO.

To further prove the existence of carbon, the structure of the CuO/CNO-2 and CNO was studied by Raman Spectra (Figure 2). The composition of the product was further investigated by Raman spectroscopy. Three peaks at 298, 346, 632 and 1141 cm^{-1} (marked by "♦") are detected, consistenting with characteristic peaks of CuO [23-25]. It is found in the Raman spectrum that carbon has two characteristic peaks at about 1350 cm^{-1} (D band) and 1580 cm^{-1} (G band), demonstrating the presence of disordered carbon in the prepared material. Which is consistent with the peak of CNO [26].

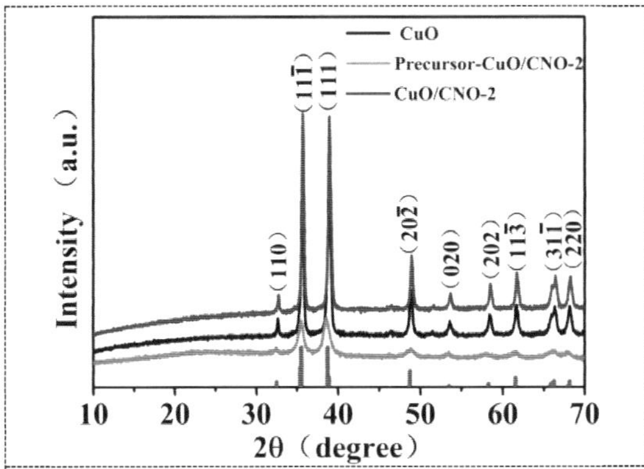

Figure 1. Typical XRD patterns of the CuO, CuO/CNO-2 and precursor-CuO/CNO-2.

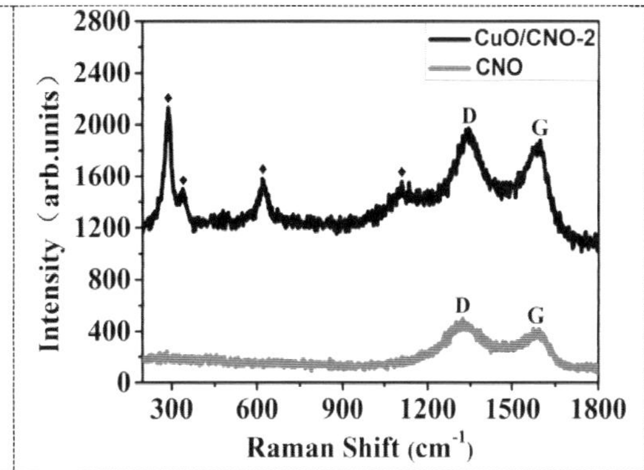

Figure 2. Raman spectra of CNO and CuO/CNO-2.

Figure 3 shows SEM images of precursor (a), the CuO (b) and the CuO/CNO-2 (c) and TEM images of the CNO (d), the CuO (e) and the CNO/CuO-2 (f). Figure 3 (a) shows the CuO/CNO-2 morphology before hydrothermal treatment, which shows an irregular particles morphology. However, after 12 hours of hydrothermal treatment, CuO sheets appeared (Figure 3 (c)). The particle size and morphology of CuO/CNO were characterized by using transmission electron microscopy (TEM). It can be seen from Figure 3 (d) that the carbon nano-onion before the composite with CuO nanosheets has a particle size of 5-10 nm, the inner layer is a diamond core, the outer layer is wrapped with several layers of graphite, and the inner layer of diamond supports the outer layer. The graphite layer is an effective electron path for lithium ion transport. Figure 3 (e) provides TEM micrographs of CuO comprising particles below 10 nm. The 0.253 nm spacing correspond to the $(11\bar{1})$ planes of monoclinic CuO. As seen at Figure 3 (f), CuO nanosheets and carbon onions still have partial been hybridized.

IOP Conf. Series: Materials Science and Engineering **484** (2019) 012005 doi:10.1088/1757-899X/484/1/012005

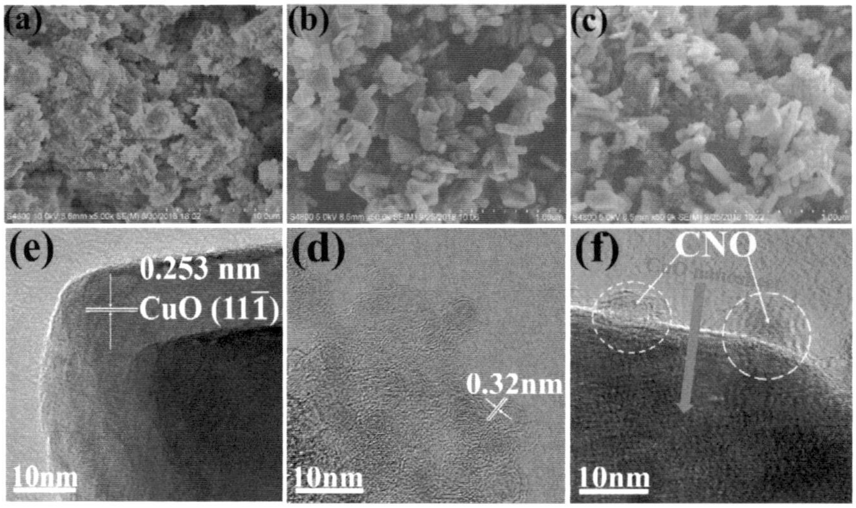

Figure 3. SEM images of (a) the CuO/CNO-2 before hydrothermal, (b) the CuO and (c) the CuO/CNO-2; TEM images of (d) the CNO, (e) the CuO and (f) the CNO/CuO-2.precursor-CuO/CNO-2.

CV curve of CuO / CNO-2 at 0.1 mV s-1 was shown in figure 4(a). It can observe from the curve that during the first scan, the prominent peak at the 1.02V position is due to the formation of the Cu_2O mesophase and the associated structural damage. The other two small cathode peaks at 0.87 V and 0.70 V correspond to further reduction of Cu_2O to metal Cu^0 in the Li_2O matrix, possibly with decomposition of the electrolyte solution. Decomposition of the electrolyte solution results in the formation of inorganic by-products in the form of polymer layers and solid electrolyte interfaces (SEI) [27-29]. The above conversion reaction can be seen as $2CuO+4Li^++4e^- \rightarrow Cu_2O \rightarrow 2Cu^0+2Li_2O$. In the anode scan, the oxidation peak at 2.46V appears to be the oxidation reaction of Cu^0 to Cu^{2+} ($Cu+Li_2O \rightarrow CuO+2Li^++2e^-$) [30], which is reserved in subsequent cycles. Since the second cycle, the phenomenon that the cathode peak potential is shifted to a higher voltage of 1.21 V proves that an irreversible phase transition occurs in the first cycle. In the anode scan, the oxidation peak appeared at 2.58V. A large number of previous articles have also proved that the electrochemical reaction has the appearance of a cuprous oxide intermediate phase and a similar mechanism [31-33]. As is well known, in the second and third cycles, the CV curves of the materials become overlapping and the area increases slightly, demonstrating that this material provides excellent reversibility in electrochemical reactions. Figure 4 (a) shows that the area of the third circle of CuO/CNO-2 is slightly larger than that of the second circle, which proves that the capacity didn't decrease. However, the integrate area of the third circle of CuO is smaller than the second circle, which proves that the pure copper oxide capacity is decreasing.

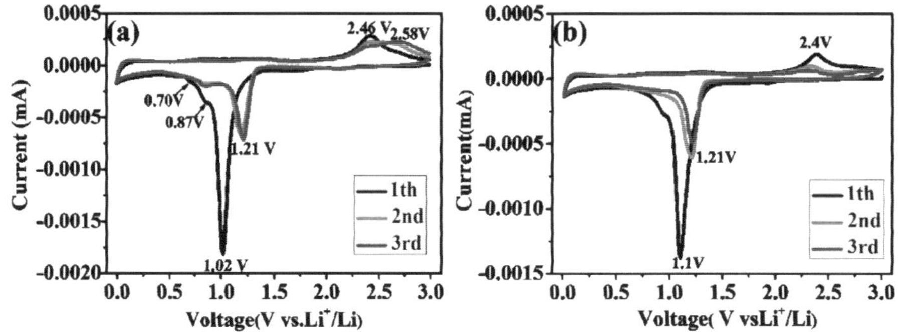

Figure 4. CV curve of (a) CuO/CNO-2, (b) CuO at 0.1 mVs^{-1}.

As shown in Figure 5 (b), the rate capabilities of bare CuO, CuO/CNO-1, CuO/CNO-2 and CuO/CNO-3 were test from 0.1 to 5 A g^{-1}. The capacity retentions of CuO/CNO-2 were 709.7, 643.4, 549, 458.9, 327 and 167.2 mAh g^{-1} from 0.1 to 5 A g^{-1} (Figure 5 (a)), respectively, exhibiting a better performance than those of bare CuO (196.3, 124.5, 89.9, 72.3, 55.6, and 26.4 mAh g^{-1}), CuO/CNO-1(456.2, 331.1, 230.5, 164.9, 115.2 and 58.6 mAh g^{-1}) and CuO/CNO-3 (515.2, 410.1, 282.2, 186.0, 103.5 and 40.4 mAh g^{-1}) upon increasing the current densities. This performance is excellent compared to the best values reported previously among Copper oxide and carbon materials for anode used in lithium ion battery [1, 13, 34-41] (Figure 5 (c)).

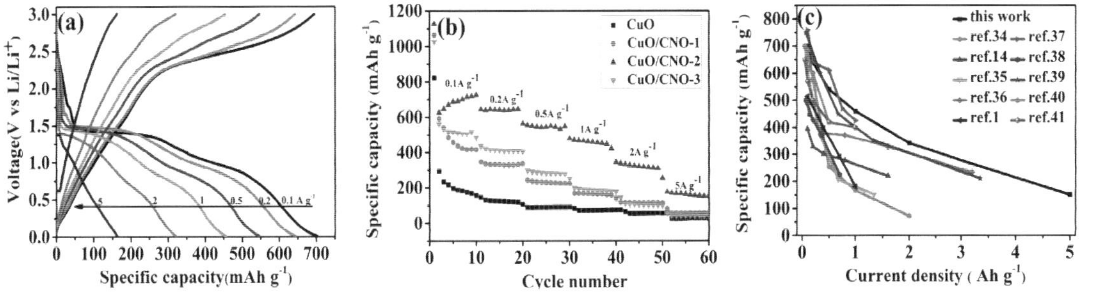

Figure 5. (a) Charge–discharge curves for CuO/CNO-2 at different current densities. (b) Cycling performance of CuO, CuO/CNO-1, CuO/CNO-2, CuO/CNO-3 at various rates from 0.1 to 5 A g^{-1}. (c) Comparison of the rate capability of the CuO/CNO-2 with that of other recently reported CuO/C composites.

The above datas demonstrate that the combination of CNO and CuO nanosheets greatly improves the cyclic performance and capacity. Main roles of the carbon shells of CNOs is that the multi-layered graphite carbon shell enables Li ion transport to be faster and more efficient, resulting in a significant increase in transmission capacity [42]. Therefore, too little carbon is not enough to play the role discussed above. But too much carbon is not good either. Since the theoretical specific capacity of carbon is lower than that of copper oxide, the amount of mixing will reduce the total capacity.

4. Conclusions

In all, carbon nano-onions (CNO) embedded CuO nanosheets has been prepared by hydrothermal method. As a potential anode material, compared with bare CuO nanosheets, the CNO embedded CuO nanosheets exhibits high reversible capacity and better rate performance. Accordingly, this research suggests that the introduction of CNO was an effective method to improve the performance of electrode material used as lithium-ion batteries.

5. Acknowledgments

This paper was supported by the foundation of State Key Laboratory of Metastable Materials Science and Technology in Yanshan University.

6. References

[1] Sun Y, Zhang P, Wang B, et al 2018 *J. Alloys Compd* 750 77-84
[2] Zang J, Ye J, Qian H, et al 2018 *Electrochim. Acta* 260 783-8
[3] Hao S, Zhang B, Wang Y, et al 2018 *Electrochim. Acta* 260 965-73
[4] Ding Y, Liu B, Cai R, et al 2018 *Nano* 13 1850018
[5] Kim A-Y, Kim M K, Cho K, et al 2016 *ACS Appl. Mater. Interfaces* 8 19514
[6] Ruan X, Yang Y, Pu K, Gao M, Liu Y and Pan H 2018 *J. Power Sources* 397 134-42
[7] Abouali S, Akbari Garakani M and Kim J-K 2018 *Electrochim. Acta* 284 436-43
[8] Li J, Li Z, Ning F, Zhou L, Zhang R, Shao M and Wei M 2018 *ACS Omega* 3 1675-83
[9] Qin M, Hao Z, Li Y, Liu D, Liu S and Li P 2018 *J. Alloys Compd* 769 463-70
[10] Zhou J, Ma L, Song H, Wu B and Chen X 2011 *Electrochem. Commun* 13 1357-60

[11] Chen K and Xue D 2013 Phys. *Chem. Chem. Phys* 15 19708-14
[12]Wang F, Tao W, Zhao M, et al 2011 *J. Alloys Compd* 509 9798-803
[13] Chen H, Feng F, Hu Z-l, et al 2012 T *Nonferr Metal Soc* 22 2523-8
[14] Ji D, Zhou H, Tong Y, et al 2017 *Chem. Eng. J* 313 1623-32
[15] Cui X, Song B, Cheng S, et al 2018 *Nanotechnology* 29 035603
[16] Huang X H, Wang C B, Zhang S Y et al 2011 *Electrochim. Acta* 56 6752-6
[17] Kim H, Seo D-H, Kim S-W, Kim J and Kang K et al 2011 *Carbon* 49 326-32
[18] Hailiang Wang, L-F C, Yuan Yang, et al 2010 *ACS* 132 13978-80
[19] Liu X G, Li B, Geng D Y, et al 2009 *Carbon* 47 470-4
[20] Liu X G, Ou Z Q, Geng D Y, et al 2010 *Carbon* 48 891-7
[21] Fleischmann S, Jäckel N, Zeiger M, et al 2016 *Chem. Mater* 28 2802-13
[22] Zhao B, Liu P, Zhuang H, et al 2013 *J. Mater. Chem. A* 1 367-73
[23] Yanying Lu, N Z, Qing Zhao, et al 2015 *Nanoscale* 2 770-6
[24] Kong C, Tang L, Zhang X, et al 2014 *J. Mater. Chem. A* 2 7306-12
[25] Wang W, Zhou Q, Fei X, et al 2010 *CrystEngComm* 12 2232
[26] Wang S, Chen M, Xie Y, et al 2016 *Small* 12 2365-75
[27] Pereira N, Dupont L, Tarascon J M, et al 2003 J. *Electrochem. Soc* 150 A1273
[28] Banerjee A, Singh U, Aravindan V, et al 2013 *Nano Energy* 2 1158-63
[29] Huogen Yu J Y S L, and Stephen Mann 2007 *Chem. Mater.* 19 (17) 4327–34
[30] Ramaraju B, Li C-H, Prakash S and Chen C-C 2016 *Chem. Commun* 52 946-9
[31] Sahay R, Suresh Kumar P, Aravindan V, et al 2012 *J. Phys. Chem. C* 116 18087-92
[32] Grugeon S, Laruelle S, Herrera-Urbina R, et al 2001 *J. Electrochem. Soc* 148 A285
[33] Zhang D W, Yi T H and Chen C H 2005 *Nanotechnology* 16 2338-41
[34] Yuan W, Qiu Z, Chen Y, et al 2018 *Electrochim. Acta* 267 150-60
[35] Yuan W, Luo J, Pan B, et al 2017 *Electrochim. Acta* 241 261-71
[36] Yin H, Yu X-X, Li Q-W, et al 2017 *J. Alloys Compd* 706 97-102
[37] Li G, He B, Zhou M, et al 2016 *Ionics* 23 607-16
[38] Zhang X, Hu Y, Zhu D, et al 2016 *Ceram Int* 42 1833-9
[39] Liu Y, Cai X and Shi W 2016 *Mater Lett* 172 72-5
[40] Lyu F, Yu S, Li M, et al 2017 *Chem Commun (Camb)* 53 2138-41
[41] Chen J-X, Zhao D-L, Yao R-R, et al 2017 *J. Alloys Compd* 714 419-24
[42] Zheng Y and Zhu P 2016 *RSC Advances* 6 92285-98

IOP Conf. Series: Materials Science and Engineering **484** (2019) 012006 doi:10.1088/1757-899X/484/1/012006

Size-TunableMesoporous Silica for Macromolecular Drug Delivery

YiyuWang[1,2], Chunqing Niu[1,2], Jiamin Gu[1,2], Chao Hu[1,2],Youning Wang[1,2] and JunhuaZhang[1,2,3*]

[1] Hubei Key Laboratory of Quality Control of Characteristic Fruits and Vegetables, Hubei Engineering University, Xiaogan 432000, No. 272 Traffic Avenue, People's Republic of China.
[2] School of Life Science Technology, Hubei Engineering University, No. 272 Traffic Avenue, Xiaogan 432000, Hubei Province, People's Republic of China.
[3] State Key Laboratory of Silicate Materials for Architectures, Wuhan University of Technology, Wuhan 430070, People's Republic of China.
[*]Corresponding author Email: andyhua-001@163.com

Abstract. A series of mesoporous silica nanospheres(MSNs)and different particle sizes rangingfrom about48 to 215nm has been successfully synthesized through modified base-catalyzed sol–gel method.The particle size distribution, structural and morphological of the samples were well characterized by dynamic light scattering technique (DLS), transmission electron microscopy (TEM), Fourier transform infrared spectroscopy (FT-IR),X-ray diffraction (XRD) and thermalgravimetry (TG).The results reveal that this MSNs exhibits good physicalproperties, including narrow size distribution,uniform pore texture in the nanospheres and good thermal stability.What's more, the size canbetuned by altering the concentration of aqueous ammonia in the silica precursor systems.The obtained MSNs are further used as a drug delivery carrier toinvestigate the in vitro drug release properties usingbovine serum albumin (BSA) as a macromolecularrepresentative model drug.It is found that this MSNsexhibit good obvious sustained release properties for macromolecular drug, suggesting its potential application in biological fields.

1. Introduction

Protein therapeutics are promising candidates for disease treatment due to their high specificity and minimal adverse side effects, such as those used in the treatment of diabetes type I and II provided by insulin, are extremelybeneficialtreatments[1]. However, effective delivery ofproteins to target sites has proven monumental challenge [2].Proteins are rapidly cleared from the body, unable to survive high and low pH environments or denaturing enzymes (proteases) [3].One of the main functions of deliveryvehicles in these systems is to protect proteins from premature degradation in a biological environment.

Mesoporous silica nanoparticles (MSNs) have been proven to be ideal candidatesfor this application due to unique characteristic such as ordered mesoporousstructure [4, 5, 6],high surface area (>800 m^2 g^{-1}), large porevolume (≈1 cm3 g-1), tunable pore diameters (2–10 nm), good biocompatibility [7, 8] and well-defined surface properties for modification [9, 10],which endow them with great potential inthe fields of protein delivery [11]. To date, there are two important kinds of MSNs inprotein delivery, SBA-15 and MCM-41. SBA-15 particles have larger pore sizes (5–30 nm), allowing them to accommodate and deliver large, therapeuticmolecules [12]. However, the particle sizes of SBA-15 aregenerally in the range of 800 nm to 2 μm, which leading to limitations in vivo due to small particle size (<200 nm) is required for efficient cellular uptake[4]. On the other hand, MCM-41-type MSN materials possess tunable sizes, with an average distribution of pores in nanoparticles. For the effective delivery of proteins in particular, special attention must be paid to the appropriate size. These developments are encouraging forfuture applications; whereas new breakthroughs are still neededto further extend the applications of mesoporous silica materials.

Furthermore, it is well known that the size and structure of MSNs have effect on the interface between the drug-carrying particle and body fluids, and could thereby affect the drug releasekinetics. Therefore, it is significant to studythe effect of different particle sizes on the loading and release of protein drugs. In particular, some recent reports indicated there are alternate strategies for controlling the particle size. The particle size,pore size, and pore structure is tunedby the concentrationof aqueous ammonia (NH4OH) in ethanol [13]. Slowing et aldemonstratedMCM-41 with alarge average pore diameter were able to host Cyt*c*, thenrelease the protein under physiological conditions[5].Thus, the design and fabrication ofMSNs with different particle sizesshould beimportant for understanding the protein drug release behaviours fromsilica carriers.

Herein, a series of mesoporous silica nanospheres(MSNs) with tunable particle size, high thermal stability, high specific surface area and open porechannels has been obtained by modified base-catalyzed sol–gel method. For the purpose of investigating the morphology and nanostructure of scaffolds MSNs, the MSNs fabricated in different conditions were observed by TEM, and DLS confirmed the particle size and distribution properties. Futhermore, the aggregation structure and thermal stability were studied by FTIR, XRD and TG. Morever, the macromolecular drug load and release feature into/from a series of MSNs were investigated in detail.

2. Experimental Section

2.1. Material

Tetraethyl orthosilicate (TEOS)andHexadecyltrimethyl ammonium bromide (CTAB)were purchased from Aladdin Reagent Co. Ltd. (Shanghai, China). Ammonium hydroxide, ethanol, methanol and hydrochloric acid were purchased fromSinopharmChemical Reagent Co. Ltd (Shanghai, China).Bovine serum albumin (BSA), and BCA Protein Assay Kit was purchased fromBeyotime Biotechnology Co. Ltd (Shanghai, China).

2.2. Synthesis of MSNs

All chemicals were of analytical grade and used without furtherpurification.MSNs used in this experiment wereprepared by the CATB-templated, base-catalyzed sol–gelmethod[14]. Briefly, different volumesof29 wt.% NH3·H2Owas added to 200 ml deionized water. And then 0.279 g of CTAB was added at 50 °C, under the same temperature, 1.394 ml TEOS was added for the synthesis reaction. After 6 h reaction and 2 h aging process, the sample was centrifuged and washed twice with deionized water and ethanol respectively. Thirdly, thesurfactant templates were removed by extraction in acidicmethanol (2 mL of HCl/100 mL of methanol, 36 h) at80 °C. The final samples were washed thoroughly by deionized water and ethanol again, and then dried in the vacuum drying chamber overnight. These samples were named as MSNs-2, MSNs-4, MSNs-6, MSNs-8 according to amount of $NH_3{\cdot}H_2O$ (Table 1). The final dried sample was weighed and the yield was calculated according to the mass percentage.

Table 1. The codes of MSNs according to amount of $NH_3{\cdot}H_2O$

Sample	MSNs-2	MSNs-4	MSNs-6	MSNs-8
$NH_3{\cdot}H_2O$ (ml)	2	4	6	8

2.3. Characterization

15 mg of samples was added into 6 ml anhydrous ethanol, and then dispersed by ultrasonic oscillation for 30 minutes. Then, the transmission electron microscopy (TEM) wasused to observe sizes and

morphologies. TEM was performed on a transmission electronmicroscope (JEM-2100, Japan) with a field emission gunoperating at 200 kV.FT-IRspectra were performed using a FTIR spectrophotometer (Nicolet 6700, USA) using KBr technique, for each measurement, each spectrum was obtained in the wave number ranging from 400 to 4000 cm^{-1}. X-ray diffraction (XRD) patterns were obtained in a X-Ray Diffractometer (Bruker D8 Advance, Germany) using Cu Kα irradiation. Scattering patterns were collected from 0.8 to 8.0° with a scan time of 0.1 s per 0.002° step. Thermogravimetry (TG) wascarried out on a thermo-analyzer(Netzsch STA 409, Germany) with a heatingrate of 10 °C min^{-1}.

2.4. Drug loading and releasing in vitro

100mg of MSNs (MSNs-2, MSNs-4, MSNs-6 and MSNs-8) were soaked in 2mlof 10mg/ml BSA solution at room temperature for 4 hand the solutions wereoscillatedat 200rpm. After the loading, suspensions were centrifuged for 10min at 5000r/min, and the MSNs were freeze-dried for 24 h.The BSA concentration of supernatant was measured by BCA Protein Assay Kit and the loading rate quantification of drug loading rate was performed and was expressed as w/w% unless otherwise specified.

The MSNs loaded with BSA were added to 8 ml 0.1 mM PBS solution. The suspension was put into the shaking table with rotating speed of 110 rpm 37 ℃.At specific time points (0.5 h,1 h,2 h,4 h,8 h,12 h,24 h,48 h,72 h and 96 h), 1ml solution was taken out for teat the concentration, and 1.0 mL fresh PBS was added into released fluid immediately. The concentrations of BSA in the released sample fluid were measured via BCA protein assay kit.

3. Results and Discussion

3.1. Preparation and Characterizations of MSNs

In order to prepare MSNs with different sizes, wemodified the synthesis method of MCM-48 silica particles reported by Kim et al [14].In this experiment, the yield of MSNs increased gradually with the increase of the amount of $NH_3{\cdot}H_2O$. The yield of MSNs-6 and MSNs-8 was very similar and the yields of MSNs-6 and MSNs-8 were about 56% and 58% respectively, which were slightly higher than the other two groups (Figure.1). It can be seen from these results that withaugment of the amount of $NH_3{\cdot}H_2O$ in the reaction systems, the yieldofMSNsalso increased gradually.

The morphologies and microstructures of the MSNs are determined by TEM, asshown in Figure 2. TEM and the corresponding inserted HRTEMimage confirmed the hexagonal array of mesoporous channels in the MSNs.It can be seen that MSNs-2 and MSNs-4 consist of monodisperse spherical particles and these particles are non-aggregated with narrow size, but with the increase of $NH_3{\cdot}H_2O$ amount, the size of particles increased and tend to aggregated. The average diameters of these MSNs are calculated to be about 48, 110, 185 and 215 nm for MSMs-2, MSMs-4, MSMs-6, and MSMs-8, respectively.

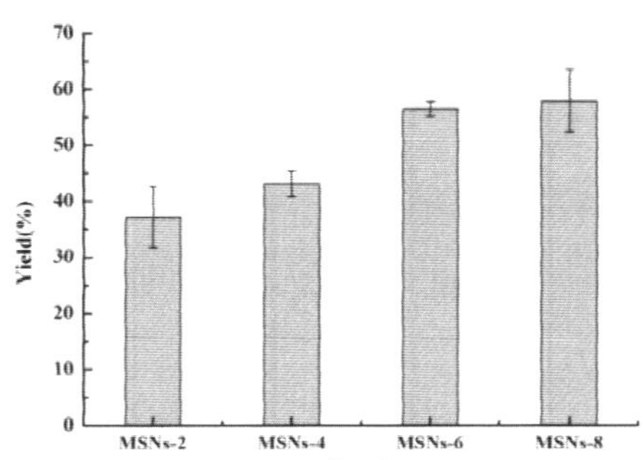

Figure1. Yield of different sample

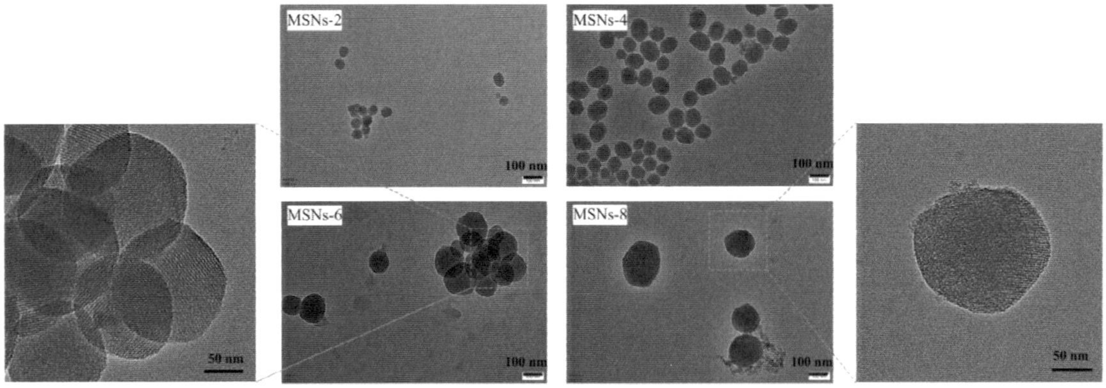

Figure 2. TEM of mesoporous silica nanoparticles of samples MSNs-2, MSNs-4, MSNs-6 and MSNs-8. The inset in micrograph is HRTEM image of the sample.

The distributional stability of these samples in water were investigated and monitored by dynamic light scattering (DLS). Hydrodynamic diameter distributions are shown in Figure 3. Volume-weighted data of MSNs-2 and MSNs-4 showed a narrow hydrodynamic particle size distributions. These data indicated that these two samples could bethoroughly suspended and well-dispersed in water. But there appeared two peaks in the MSNs-8, which indicating obvious interparticle aggregation occurred in their suspension. The particle size of MSNs in water was further measured, It was found the particle size of sample were was 123, 142, 337 and 579 nm for MSNs-2, MSNs-4, MSNs-6 and MSNs-8respectively, which were slightly larger than the core particle sizes observed by TEM (Figure 2), which is usually the case taking into account the hydration corona around the particles and aggregation or flocculation.

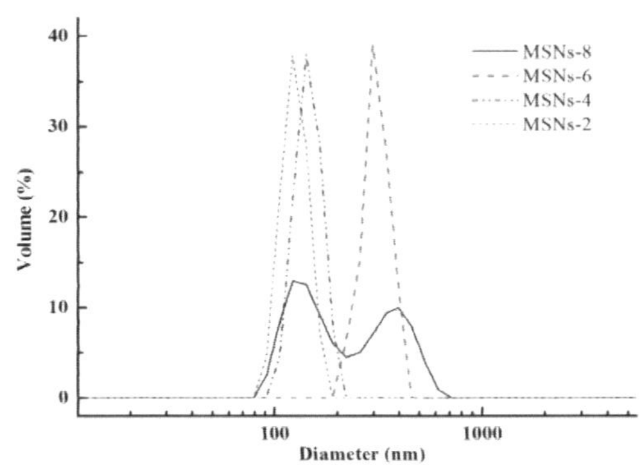

Figure 3. DLS results of samples

IOP Conf. Series: Materials Science and Engineering **484** (2019) 012006 doi:10.1088/1757-899X/484/1/012006

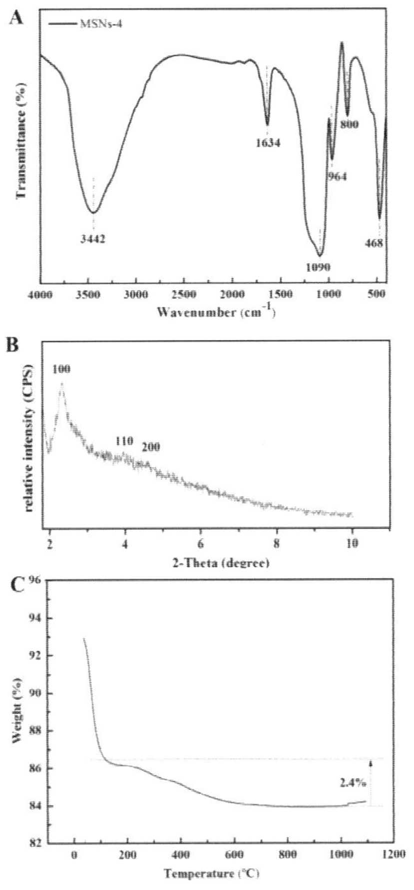

Figure 4. FTIR spectra (A), XRDpattern (B), TG profile of MSNs-4

In experiment, the structure properties and thermal stability of nanoparticles were investigated with MSNs-4 as the representative. It can be seen from the FTIR spectra (Figure 4, A), several broad absorption bands assigned to OH (3442 cm^{-1}), Si–O–Si (1090 cm^{-1}), Si–OH (800 cm^{-1}), and Si–O (468 cm^{-1}) bands can be seen clearly [15]. The strong band of OH indicates a large number of OH groups present on the surface, which are important for bonding drug molecules. The MSNs showed an ordered mesostructure like MCM-48type nanoparticles as revealed by low-angle XRD patterns (Figure 4, B), hexagonal mesoporous structure with the characteristic (100) peak between 2.14 and 2.30° and the d_{100} spacings were between 3.8 nm and 4.2 nm. The thermal stability of MSNs-4 was examined by TG analysis (Figure 4, C). It can be seen that at the beginning of the heating, there was a rapid decline with apparent weight loss, which may be due to the adsorption of a certain amount of free water on the surface of the material. After 110°C, only slight weight loss up to2.4% is observed in the temperature range from room temperature to 1100 °C, which can mainly be due to the physically adsorbed water and hydroxyl groups in the materials, indicating the high stability of this material.

3.2. Macromolecular Drug Loading and Release Properties

To study the effect of MSNs size on the drug loading capacity and drug release, BSA was used as a model of macromolecular agent. Figure 5 showed the adsorption of BSA into the mesoporous matrices with different particle sizes. The experimental results showed that the drug loading rate of mesoporousnanospheres increased with the increase of argument of MSNs particle size. It can be found in Figure5 (A) that the drug loading rate of MSNs-2 was about 1.7%, the drug loading rate of MSNs-4 significantly increased to 6.1%, and the drug loading rate of MSNs-6 and MSNs-8 increased gradually to 8.3% and 10.3%, respectively.

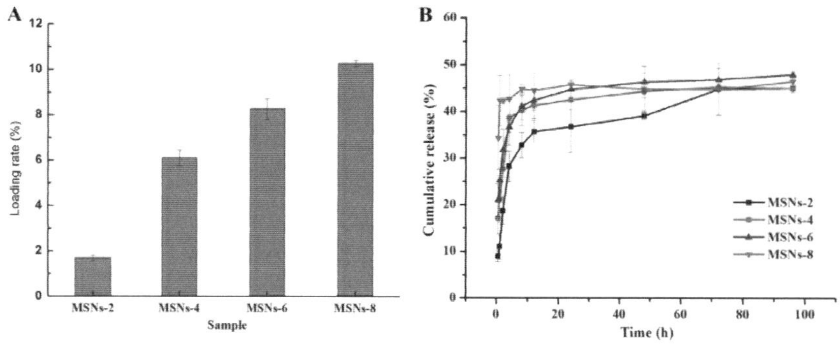

Figure 5. The loading rate of samples (A) and cumulative release curves of BSA from different MSNs

Figure 5 (B) showed the cumulative release profiles of the BSA from MSNs as a function of time in 0.1 mM PBS solution (pH=7.4).A relatively fast release within the first 12 h followed by a sustained release over 96h was observed, indicating the good sustained release properties of all the samples. In the initial 24h, the smaller particle size MSNs showed slower drug release rates, which may be due to the bigger particle size MSNs loaded more drugs. The slowest drug release rates assigned to MSNs-2, around 37 wt.% of total drug loading, while the drug release rates from MSNs-4, MSNs-6, and MSNs-8 were 43 wt.%, 44 wt.%, 45 wt.%. After 96 h, the amount of released BSA reached to 45 wt.%, 45 wt.%, 48 wt.% and 47 wt.% for MSNs-2, MSNs-4, MSNs-6, and MSNs-8, respectively.

4. Conclusion

In summary, a family of mesoporous silica nanospheres with different size has been synthesized via a facile and mild sol-gel method. The obtained MSNs possesses high thermal stability, tuneable particle size and special open pore channels, which allow easy access to the high specific area inside the pores. BSA release tests indicate this kind of MSNsexhibits good sustained release properties and the drug release rates are related to the particle size of the samples. Based on these results, it can be concluded that the MSNs-4 and MSNs-6 are better choicesfor macromolecular drug delivery system, this nanospheres presented here may have a good potential due to their perfect morphology, narrow and uniform pore size, good dispersancy and higher drug loading rate. This mesoporous silica should be promising in macromolecular drug release fields based on their good properties.

5. Acknowledgements

The authors wish to express their gratitude to the financial supports from Research Project of Hubei Provincial Department of Education (Q20182701).

6. References

[1]Leader B, Baca Q J and Golan DE 2008 *Nature Rev. Drug Dis* 7 21
[2] Vivero-Escoto J L 2013 *J. Biotechnol. Biomater*3 117
[3] Wang Y, Zhao Q, Han N, Bai L, Li J, Liu J, Che E, Hu L, Zhang Q, Jiang T and Wang S 2015 *Nanomed.Nanotechnol*11 313
[4] Vivero-Escoto J L, Slowing I I, TrewynB G and Lin V S Y 2010 *Small* 6 1952
[5] Slowing I I, Trewyn B G and Lin V S Y 2007 *J. Am. Chem. Soc* 129 8845
[6] Trewyn B G, Whitman C M and Lin V S Y 2004 *Nano Lett.* 4 2139
[7] Radu D R, Lai C Y, Jeftinija K, Rowe E W, Jeftinija S and Lin V S Y 2004 *J. Am. Chem. Soc*126 13216
[8] Lai C Y, Trewyn B G, Jeftinija D M, Jeftinija K, Xu S, Jeftinija S and Lin V S Y 2003 *J.Am.Ch em. Soc 125* 4451
[9]Song SW, Hidajat K and Kawi S 2005*Langmuir*219568
[10] Muñoz B, Rámila A, Pérez-Pariente J, Díaz I and Vallet-RegíM 2003*Chem. Mater*15 500
[11] Song Y H, Li Y H, Xu Q and Liu Z 2017 Int J Nanomedicine 1287
[12] Zhao D Y, Huo Q S, Feng J L, Chmelka B F and Stucky G D 1998 *J. Am. Chem. Soc* 120 6024

IOP Conf. Series: Materials Science and Engineering **484** (2019) 012006 doi:10.1088/1757-899X/484/1/012006

[13] Kao KC, Mou CY 2013 *Micropor. Mesopor. Mat* 169 7
[14] Kim T W, Chung P W and Lin V S Y 2010 *Chem. Mat* 22 5093
[15] Yu M, Lin J and Fang J 2005 *Chem. Mater* 17 1783

IOP Conf. Series: Materials Science and Engineering **484** (2019) 012007 doi:10.1088/1757-899X/484/1/012007

The Effect of Cone Additional Permanent Magnet Thickness on Magnetic Field Distribution and the Levitation Force of Single Domain GdBCO Bulk Superconductor

Jun Ma, Yongjie Zhou, Jinglin Li, Bohang Chen and Weiqiang Fang

College of physics and electronic information engineering, Qing Hai Normal University, Xi ning 810008, China
E-mail: mjun7302@163.com

Abstract. It has been investigated that the interaction force between truncated cone permanent magnet (PM) and a GdBCO bulk (HTSC) superconducting permanent magnet (SCPM) magnetized by a truncated cone permanent magnet in their configurations at 77K. The influence of two configurations in different thickness of truncated cone permanent magnets of single domain GdBCO bulk capture field and superconducting magnetic levitation force results. The maximum superconducting magnetic levitation force first increases as H increase, and then decreases with H when it reaches a certain value, from 5.7N to 16.6N at first, and then it decreases to 3.8N. The vertical component of the magnetic field intensity on the top surface (h=2mm) of truncated cone permanent magnet first increases, and then decreases also, from 0.20T to 0.54T, and then decrease to 0.25T. These results indicate that the levitation force of high temperature superconductor bulk can be effectively improved by scientific choosing of shape and size of permanent magnet, and reasonable designing of the system configurations, which is very important during the practical design and applications of superconducting magnetic levitation system.

1. Introduction

High temperature superconducting materials can provide large magnetic levitation force and good self-stabilizing suspension in the application of external magnetic field due to their anti-magnetic properties. It has been widely used in magnetic suspension bearing systems,[1] superconducting energy storage flywheel,[2] high temperature superconducting transmission line, electromagnetic emission,[3] superconducting generator, seismic isolator,[4] high temperature superconducting magnetic suspension system and superconducting maglev train systems.[5] In practical , how to effectively improve the magnetic levitation force between permanent magnet and superconductor is particularly important. The magnetic levitation force are mainly related to the magnetic field strength that the permanent magnet used, [6] the magnetic field distribution, [7, 8] the thickness of the superconductor, [9] the radius, [10] the grain orientation, [11]and the degree of magnetization [12] in magnetization method. [13]

In the previous work,[14] the influence of the combination form of strip-shaped permanent magnets on the magnetic levitation force about superconducting blocks was studied. The result shows that the superconducting magnetic levitation force is closely relate to the configuration form and combination mode of the magnet and the spacing between the magnets. The magnetic levitation force between superconductor and permanent magnet was also studied by the introduction of the auxiliary permanent magnet. [15,16] These results show that the superconducting magnetic levitation force were closely related with the combination of permanent magnets, the introduction of permanent magnets, the magnetic field distribution of the magnet, the magnetization of the superconducting block and the trapping magnetic field during field cooling. Cheng et al. [17] Put forward the effect of the magnetization of square-assisted permanent magnets on superconductors on the superconducting magnetic levitation force and the magnetic field of capture magnetic field. The result show that the superconducting magnetic levitation force and the trapping magnetic field related to the superconductor magnetization. However, these studies mainly focus on the effects of square permanent magnets, strip-shaped permanent magnets and combined magnets on the magnetic levitation force and the magnetic field distribution of superconductors. The influence of the thickness variation of the truncated cone permanent magnets on the characteristics of superconducting magnetic levitation forces has not been systematically studied.

In order to clarify the influence of the thickness variation about the truncated cone permanent magnet on the magnetic levitation force of the single-domain GdBCO superconductor, this paper studies the magnetic levitation force of the truncated cone permanent magnet and the single-domain GdBCO superconductor in the case of zero fields cold and axisymmetric. The influence of the thickness variation of the magnet on the magnetic levitation force of the superconductor. Some experimental conclusions with reference value are obtained.

2. Experimental

The single domain GdBCO superconductor used in the experiment was prepared by the top seed crystal melt texturing method (MTG). [18]The diameter of the sample is 20mm and the thickness is 12mm. The truncated cone permanent magnet used is magnetized along the axial direction. Figure 1 shows the macroscopic morphology of the experimental single-domain GdBCO superconductor. In order to systematically study the effect of the thickness variation of the truncated cone permanent magnet on the magnetic levitation force characteristics of the single-domain GdBCO superconductor, the experimental scheme shown in Figure 2 is designed. The three-dimensional magnetic field and magnetic force testing device is used to measure the influence of the thickness variation of the truncated cone permanent magnet on the magnetic levitation force of the single-domain GdBCO superconductor. The thicknesses are 5mm, 10mm, 15mm, 20mm, 25mm, 30mm, 35mm, 40mm, 45mm and 50mm respectively, and the corresponding upper surface diameter of the truncated-shaped permanent magnets are 32mm, 30mm, 26mm, 22mm, 18mm, 14mm, 12mm, 10mm, 8mm and 6mm, respectively. Before each experimental measurement, the truncated cone permanent magnet and the superconductor are in an axisymmetric state, and the vertical distance between the two is Z=35 mm. The liquid crystal nitrogen is used to start cooling the superconductor. After the superconductor is cooled to the liquid nitrogen temperature (77K), the truncated cone permanent magnet begins to vertically approach the superconductor along the axis. When the distance between the two is Z=2 mm, the truncated cone permanent magnet is returned along the original path and complete an experimental measurement.

Figure 1. The morphology of the sample.

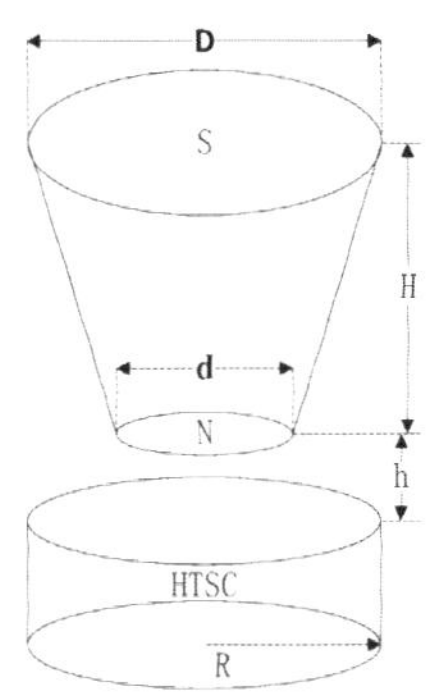

Figure 2. The force measurement diagram between the permanent magnet and the superconductor.

3. Results and Discussion

Under the zero field cold experimental condition, when the vertical distance between the truncated cone permanent magnet and the superconductor is 2 mm, the relationship between the magnetic levitation force of the superconductor and the thickness of the truncated cone permanent magnet are shown in Figure 3. It shows that when the thickness H of the truncated cone permanent magnet increases from 5mm to 50mm, the magnetic levitation force of the superconductor is closely related to the thickness of the truncated cone permanent magnet. All superconductor magnetic levitation force curves are tilted as a whole to the lower left, and there is hysteresis. As H increases, the hysteresis of the superconducting magnetic levitation force first increases and then decreases. When H are 20mm and 25mm, corresponding to the diameter of the upper surface of the truncated cone permanent magnet are 22mm and 18mm, the superconducting magnetic levitation force has a large hysteresis. When H are 5mm and 50mm, corresponding to the diameter of the upper surface of the truncated cone permanent magnet are 32mm and 6mm, the superconducting magnetic levitation force is very small.

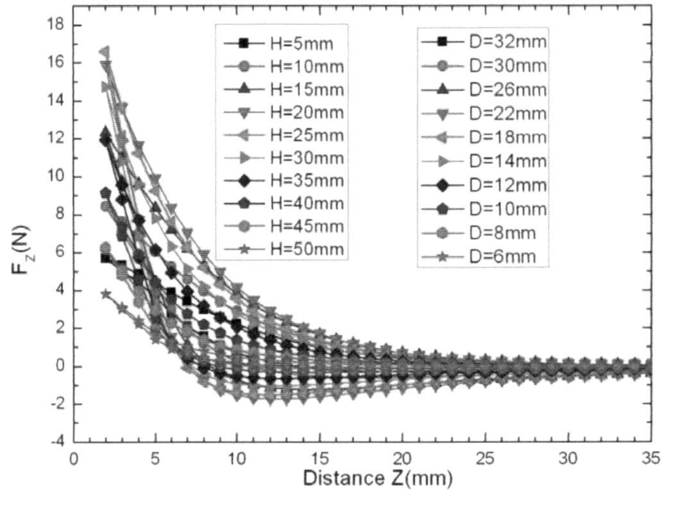

Figure 3. The magnetic levitation force (Z=2mm) with the thickness H of the truncated cone permanent magnet.

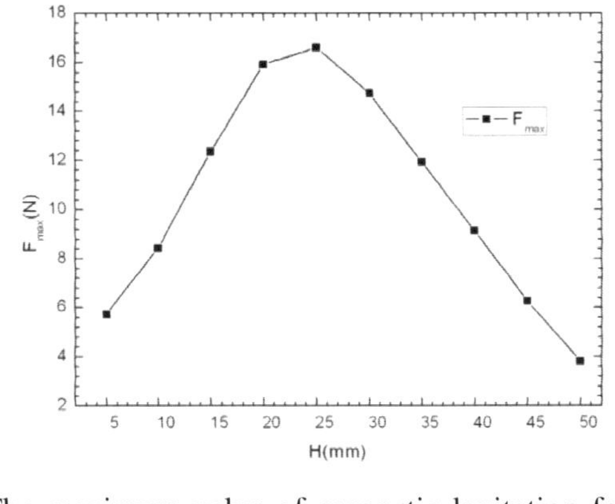

Figure 4. The maximum value of magnetic levitation force (Z=2mm) with the thickness H of the truncated cone permanent magnet.

Figure 4 is the relationship between the maximum value of superconducting magnetic levitation force (Minimum measuring gap distance Z=2mm) and the thickness H of truncated cone permanent magnet. It shows that with the increase of H, the maximum value of superconducting magnetic levitation force first increases and then decreases, from 5.7N to 16.6N, and then decreases to 3.8N. When H are 20mm and 25mm, corresponding to the diameter of the upper surface of the truncated cone permanent magnet are 22mm and 18mm, the maximum value of superconducting magnetic levitation force is larger. When H are 5mm and 50mm, corresponding to the diameter of the upper surface of the truncated cone permanent magnet 32mm and 6mm, the maximum value of the superconducting magnetic levitation force is small.

In order to further clarify the influence of the variation of the thickness H of the truncated cone permanent magnet on the magnetic levitation force of the superconductor, the low-temperature Hall probe was used to measure the vertical component Bz of the magnetic field at Z=2mm on the upper surface of the truncated cone permanent magnet under the change of thickness H. Figure 5 is the relationship between the vertical component of the magnetic field strength along the diameter direction (BZ-X) from the upper surface of the magnet (h=2mm) and the thickness of the truncated cone permanent magnets of 5mm, 10mm, 15mm, 20mm, 25mm, 30mm, 35mm, 40mm, 45mm and 50mm, the corresponding upper surface diameter of the truncated permanent magnets are 32mm, 30mm, 26mm, 22mm, 18mm, 14mm, 12mm, 10mm, 8mm and 6mm respectively. In general, the vertical component of magnetic field strength is closely related to the thickness of the truncated cone permanent magnet. There are differences in the magnitude of the magnetic field and the distribution of the magnetic field at different thicknesses.

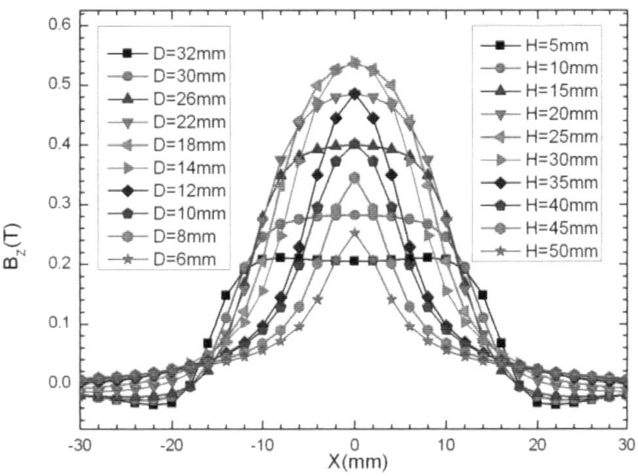

Figure 5. The vertical component of the magnetic field of the permanent magnet upper surface (h = 2 mm) with transverse displacement (B_z-X) curve.

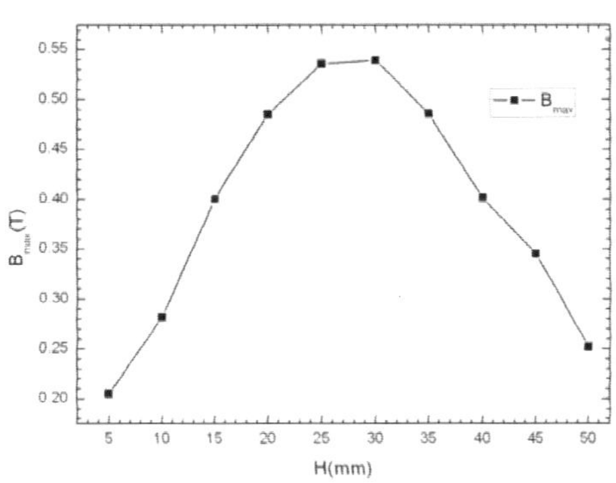

Figure 6. The maximum value of the vertical component of the magnetic field intensity on the upper surface of the permanent magnet (h=2mm) with the thickness variation (B_{max}-H)

Figure 6 shows the relation curve of the maximum value of the vertical component of the magnetic field intensity on the upper surface of the permanent magnet (h=2mm) with the thickness variation (Bmax-H). As the thickness of the magnet increases from 5mm to 50mm, the magnetic field strength first increases and then decreases. The maximum magnetic field strength increases from 0.20T to 0.54T and then decreases to 0.25T. These maximum values is obtained at x=0mm. When H=25 mm and 30 mm, the peak values of the maximum magnetic field strength are substantially equal, corresponding to the upper surface diameter d=18 mm and 14 mm. This is inconsistent with the position where the maximum value of the superconducting magnetic levitation forces in Figure 4.

In the magnetic levitation system consisting of a truncated cone permanent magnet and a superconductor, the superconducting magnetic levitation force is not only related to the magnetic field strength of the applied external magnetic field, but also interact with the gradient of the magnetic field, the effective area of the interaction between the permanent magnet and the superconductor. When the thickness of the truncated cone permanent magnets are H=20mm and 25mm, corresponding to the diameter of the upper surface of the truncated permanent magnets are 22mm and 18mm, the diameter ratio of the superconductor is 20mm, and the magnetic field strength and effective area are optimized. Matching results in a maximum magnetic levitation force at H = 20 mm and 25 mm and similar.

4. Conclusions

The effect of the thickness of truncated cone permanent magnet on the magnetic levitation of GdBCO superconductor under axisymmetric condition is studied experimentally. The results show that the maximum magnetic levitation force changes with the thickness of the truncated cone permanent magnet, which are closely related to the magnetic field strength of the applied external magnetic field, magnetic gradient and the effective area of the interaction between permanent magnet and superconductor. When the diameter of the upper surface of the truncated permanent magnet and the diameter of the superconductor reach the optimum diameter ratio, and the magnetic field strength and effective area are optimally matched, the superconducting magnetic levitation force reaches a maximum value. These results show that in the practical application of magnetic levitation system, the scientific design and reasonable magnetized superconducting bulk material can adjust adjust the magnitude superconducting magnetic levitation force, thereby improving the magnetic levitation force characteristics.

5. Acknowledgments

We gratefully acknowledge financial support from the Natural Science Foundation of Qinghai (No. 2016-ZJ-915), and the National Natural Science Foundation of China (No. 51167016).

6. References

[1] John R H, Shaul H and Tommotake M 2005 *Supercond Sci Technol* 18 S1
[2] Miyagawa Y, Kameno H, Takahata R and Ueyama H 1999 *IEEE Trans Appl Supercond* 9 996
[3] Yang W M, Li Z G, Ma J, Chao X X and Li J W 2010 *IEEE Trans Appl Supercond* 20 2317
[4] Tsuda M, Kawasaki T, Yagai T and Hamajima T 2008 *Journal of Physics:Conference Series* 97 012104.
[5] Liu C T and Lee S C 2006 *Journal of Magnetism and Magnetic Materials* 304 e454
[6] Valle N D, Sanchez A, Pardo E, Chen D X and Navau C 2007 *Appl.Phys.Lett.* 90 042503
[7] Zhang X J, Yan L, Dong Z, Jin W T, Hao S J, Zhao Y and Zhang H J 2010 *Supercond Nov Magn* 23 1007
[8] Zhu M, Ren Z Y, Wang S and Jiang He 2002 *Chinese Journal of Low Temperature Physics* 24 213
[9] Yang W M, Zhou L, Feng Y, Zhang P X, Zhang C P, Yu Z M, Tang X D, Nicolsky R and Andrade J R 2002 *Cryogenics* 42 589
[10] Koblischka A V, Koblischka M R, Babu N H, Cardwell D A, Mücklich F and Murakami M 2005 *Supercond Nov Magn* 18 469
[11] Zhou J, Zhang X Y and Zhou Y H 2009 *Physica C: Superconductivity* 469 207
[12] Ma J, Chen Z L, Xian T, Wei X G, Yang W M, Chen S L and Li J W 2018 *Acta Phys. Sin.* 67 077401
[13] Ma J andYang W M 2011 *Acta Phys. Sin.* 60 077401.
[14] Ma J, YANG W M, LI G Z, Cheng X F and Guo X D 2011 *Acta Phys. Sin.* 60 027401
[15] Ma J, Yang W M, Li J W, Wang M and Chen S L 2012 *Acta Phys. Sin.* 61 137401

[16] Ma J, Yang W M, Wang M, Wang M, Chen S L, Feng Z L 2013 *Acta Phys. Sin.* 62 227401
[17] Cheng X F, Yang W M, Li G Z 2010 *Chinese Journal of Low Temperature Physics* 32 150
[18] Chen S L, Yang W M, Li J W, Yuan X C, Ma J and Wang M 2014 *Physica C:Superconductivity* 469 39

Rice Straw/Sewage Sludge-Based Composite Activated Carbon: Its Some Basic Properties and Adsorption Effect for Cr (VI)

Liangqian Fan [1,*], Xianda Wang [1], Wenxin Wan [2], Qin Liu [1], Jie Cai [3] and Yan Wan [1]

[1] College of Civil Engineering, Sichuan Agricultural University, Dujiangyan 611830, China
[2] College of Environment, Sichuan Agricultural Universities, Chengdu 611130, China
[3] Department of Physical and Chemical Analysis, Dujiangyan Center for Disease Control and Prevention, Dujiangyan, 611830, China
Email: flqjacky@163.com

Abstract. In this study, the rice straw/sewage sludge-based composite activated carbon (RS-SCAC) was prepared from the mixture of sewage sludge and rice straw. Some basic properties of RS-SCAC, i.e., yield, iodine number, contents of heavy metals, surface functional groups, and leaching toxicity, were characterized. Furthermore, the adsorption performance of RS-SCAC for Cr (VI) was investigated. For RS-SCAC, the yield was moderate, and the iodine number was higher than that of the sewage sludge-based activated carbon (SBAC). RS-SCAC contained a variety of heavy metals, and Zn, Fe, and Cu had the relatively high content. The surface of RS-SCAC was rich in functional groups, especially for –OH, –COOH, and –COH, which was helpful for the adsorption of polar solute. For all heavy metals except Zn in RS-SCAC, the leaching concentrations were below the limits for identifying hazardous wastes, which indicated that RS-SCAC was relatively safe. Compared with SBAC, RS-SCAC had the better adsorption effect for Cr (VI). The results indicated that RS-SCAC can be used as a safety and efficient adsorbent for Cr (VI).

1. Introduction

Sewage sludge, as a by-product of wastewater biological treatment, contains heavy metals, organic toxins, and pathogens, which has the risks of environmental pollution [1, 2]. In China, the annual sewage sludge quantity (measured by wet sludge containing 80% water) is expected to reach 50 million tons in 2020 according to the 13th Five-Year Plan. This is an urgent task for the Chinese government to disposal such a huge amount of sewage sludge.

In the past decades, a large number of scholars used various methods to treat sewage sludge, such as sanitary landfill, compost, incineration, ocean dumpling, etc. [3-6]. However, these methods have some defects, such as secondary pollution and high cost. At present, the safe and low-cost treatment of sludge is attracting more and more attention. In view of this point, the preparation of activated carbon from sewage sludge has become a research hotspot. However, due to the low carbon content and high ash content of the sewage sludge [7], the performance of the sludge-based activated carbon is generally poor, not as good as commercial activated carbon, which limits its application. In recent years, sugarcane bagasse, corn straw, and coconut husk were added in sewage sludge to prepare composite activated carbons [8, 9]. These composite activated carbons displayed good performances in practical application.

In this study, rice straw was mixed in sewage sludge to prepare the rice straw/sewage sludge-based composite activated carbon (RS-SCAC). Firstly, the yield, iodine number, and contents of heavy metals in RS-SCAC were detected, and the surface functional groups of RS-SCAC were characterized by FTIR. Secondly, the heavy metal leaching toxicity of RS-SCAC was determined. At last, the difference of Cr (VI) adsorption effect between the sewage sludge-based activated carbon (SBAC) and RS-SCAC was analyzed. The objective of this study was to explore some basic properties of RS-SCAC and its adsorption effect for Cr (VI).

2. Materials and Methods

2.1. Preparation of RS-SCAC and SBAC

The sewage sludge and rice straw were grinded and sieved to 1~3 mm. Then, they were dehydrated in a drying oven (101-2AB, Zhongxinweiye Co. LTD., Beijing, China) to constant weight. After that, the sewage sludge and rice straw were mixed with a mass ratio of 8: 2 to acquire the raw material for the preparation of RS-SCAC. For the acquired raw material, 20 g sample was immersed in the 200 g/L $ZnCl_2$ solution for 12 h at room temperature. When the soaking time was out, the sample was filtered and dried to constant weight. And then the dried sample was poured into a steel reactor and carbonized in a muffle furnace (SX2-4-10, Shenyang Energy Saving Electric Furnace Factory, Shenyang, China) at 600 ^oC for 2 h. The acquired carbon material was firstly washed with 3 M HCl and then with ultrapure water for several times. The washed carbon was dried to constant weight and sieved through a 60 mesh screen. So far the preparation of RS-SCAC was finished. The prepared RS-SCAC was stored in a desiccator before use. For SBAC, the preparation method referred to that of RS-SCAC.

2.2. Yield, Iodine Number, Contents of Heavy Metals, and Surface Functional Groups

The yield was calculated with the weight of raw material and end-product. The iodine numbers was measured according to China national standards (GB/T 12496.8-2015). The contents of heavy metals (As, Cd, Cr, Cu, Mn, Ni, Pb, Fe, and Zn) were detected according to the norms of Ministry of Environmental Protection of the People's Republic of China (HJ 781-2016) by an inductively coupled plasma spectrum analyzer (iCAP-6000, Thermo Fisher Scientific Inc., Waltham, MA, USA). The surface functional groups were characterized by a Spectrum GX spectrometer (Spectrum two, Perkin-Elmer Crop., Norwalk, Ohio, USA).

2.3. Leaching Toxicity

The leachate of RS-SCAC was prepared according to the norms of Ministry of Environmental Protection of the People's Republic of China (HJ 557-2009). The concentrations of heavy metals (As, Cd, Cr, Cu, Mn, Ni, Pb, Fe, and Zn) in the leachate were detected by the inductively coupled plasma spectrum analyzer (iCAP-6000, Thermo Fisher Scientific Inc., Waltham, MA, USA). The detected concentrations of heavy metals were compared with the concentration limits in China national standards (GB 5085.3-2007) used for identifying hazardous wastes.

2.4. Adsorption Effect for Cr(VI)

In the experiment, 0.2829 g of $K_2Cr_2O_7$ was dissolved in 1 L ultrapure water to prepare the wastewater containing Cr (VI). 0.1 g of RS-SCAC and SBAC was added into 150 mL conical flask with 50 mL wastewater, respectively. The initial pH of the mixture was adjusted to 3.0. Then the mixture was reacted in a thermostatic shaker (Ts-2012c, Shanghai Baidian Instrument Equipment Co. LTD., Shanghai, China) for 24 h. In the thermostatic shaker, the temperature and shaking rate was controlled at 25 ^oC and 150 rpm, respectively. After that, the mixture was filtered. The residual concentration of Cr (VI) in the filtrate was detected according to China national standards (GB 7467-87) with a UV spectrophotometer (UV-1800, Mapada (Shanghai) Co. LTD., Shanghai, China) at 540 nm. The detected residual concentration was used to calculate the Cr (VI) removal percentage.

3. Results and Discussion

3.1. Yield, Iodine Number and Contents of Heavy Metals

The yield of RS-SCAC was 39.18%. The yield was roughly the same as that of other activated carbons reported in the previous studies [10, 11]. The result indicated that the preparation of RS-SCAC was feasible. The iodine numbers of RS-SCAC and SBAC are shown in figure 1. The iodine number of RS-SCAC was 438.97 mg/g, which was 1.46 times than that of SBAC (300.51 mg/g). For activated carbon, iodine number is used as a common indicator to assess its adsorption capacity [12]. The iodine number of RS-SCAC was significantly higher than that of SBAC, indicating that RS-SCAC has a better adsorption capacity than SBAC.

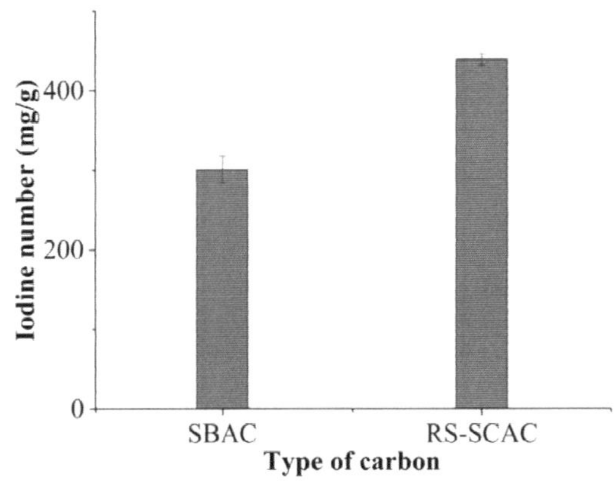

Figure 1. Iodine numbers of SBAC and RS-SCAC.

Sewage sludge often contains heavy metals, which has potential risks to environment. Due to sewage sludge as one main raw material for the preparation of RS-SCAC, the contents of heavy metals in RS-SCAC needed to be determined. Table 1 shows the contents of heavy metals in RS-SCAC. It can be found that there were many kinds of heavy metals in RS-SCAC. The order of the contents of heavy metals in RS-SCAC was Zn> Fe> Cu> Cr> Ni> Mn> As> Cd> Pb. The high Zn content was due to the used activator ($ZnCl_2$), while the high Fe content may be connected with the raw material, i.e., sewage sludge. Considering that RS-SCAC contained a certain amount of heavy metals, there was a need to assess the safety of RS-SCAC in the following section.

Table 1. Contents of heavy metals in RS-SCAC.

Metals (mg/g)	As	Cd	Cr	Cu	Mn	Ni	Pb	Fe	Zn
RS-SCAC	0.0017	0.0002	0.0653	0.2389	0.0158	0.0313	/	1.9773	12.5517

"/" represents "not detected"

3.2. Surface Functional Groups

The surface functional groups affect the chemical properties and adsorption effect of activated carbon. The surface functional groups on RS-SCAC were characterized with FTIR. The FTIR spectra of RS-SCAC is shown in figure 2. The intense absorption bands at 3423cm^{-1} was attributed to the stretching vibration of –OH [13]. The adsorption peak at 2925 cm^{-1} represented the stretching vibrations of –CH and –CH_2 [14]. The peak observed at 1614 cm^{-1} was due to the –COOH [15], and the peak observed near 1384 cm^{-1} was connected with phenol [16]. The characteristic peak at 1089cm^{-1} could be ascribed to the C–O bending vibration in –COH [17]. The band at around 780 cm^{-1} was related with the out-of-plane bending mode of O–H [18]. Overall, the surface of RS-SCAC was rich in functional groups, especially for –OH, –COOH, and –COH. The main functional groups, i.e. –OH, –COOH, and –COH, were polar functional groups, which can promote the adsorption of polar solute.

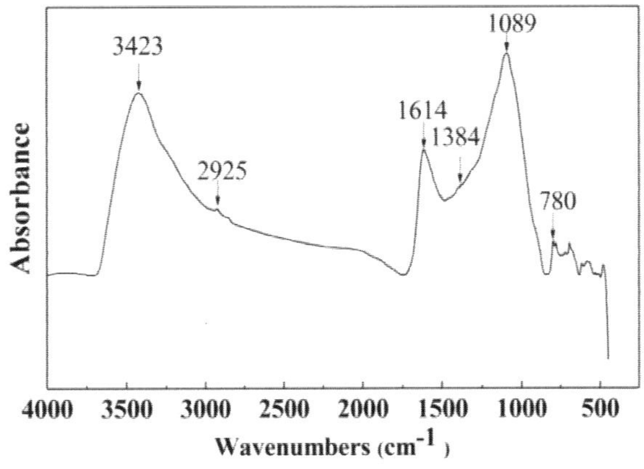

Figure 2. FTIR spectra of RS-SCAC.

3.3. Leaching Toxicity

Table 2 shows the concentrations of heavy metals (As, Cd, Cr, Cu, Mn, Ni, Pb, Fe, and Zn) in the leachate of RS-SCAC. As shown in table 2, the concentrations of As, Cd, and Ni were not detected. The concentration of Zn was much higher than that of other heavy metals. The order of the concentrations of heavy metals was Zn> Mn> Fe> Cu> Cr > Pb. Compared with the concentration limits in China national standards (GB 5085.3-2007), which were used for identifying hazardous wastes, the concentrations of all heavy metals except Zn were significantly lower than the limits. However, the concentration of Zn was only 1.46 times than its limit. The results demonstrated that RS-SCAC was relatively safe. Thus, RS-SCAC can be used as a safety adsorbent if there were no restrict limit for Zn. Moreover, the main reason for excessive Zn leaching was that $ZnCl_2$ was used as the activating agent. Therefore, repeated washing can be used in the process of RS-SCAC preparation to reduce Zn leaching, which is beneficial for RS-SCAC application if there were a restrict limit for Zn.

Table 2. Leaching concentrations of heavy metals from RS-SCAC.

Metal	As	Cd	Cr	Cu	Fe	Mn	Ni	Pb	Zn
Concentration (mg/L)	/	/	0.039	0.061	0.077	1.015	/	0.002	145.940
Concentration limit (mg/L)	5	1	15	100	-	-	5	5	100

"/" represents "not detected"
"-" represents "no limit"

3.4. Cr(VI) Adsorption Effect

The Cr (VI) removal percentages of RS-SCAC and SBAC are shown in figure 2. For RS-SCAC and SBAC, the Cr (VI) removal percentage was 41.56% and 71.66%, respectively. Overall, the Cr (VI) removal efficiency of RS-SCAC was 1.7 times than that of SBAC. For RS-SCAC, adding rice straw might lead to the increase of porosity, which further caused the rise of specific surface area. Thus the Cr (VI) removal percentage of RS-SCAC was promoted. The reason can be verified with the results of iodine number in section 3.1. Furthermore, adding rice straw might enhance the polarity of RS-SCAC, which also could improve the adsorption capacity of RS-SCAC for Cr (VI). The results in section 3.2 supported the deduction in a certain extent. Overall, RS-SCAC can be used as an efficient adsorbent for Cr (VI).

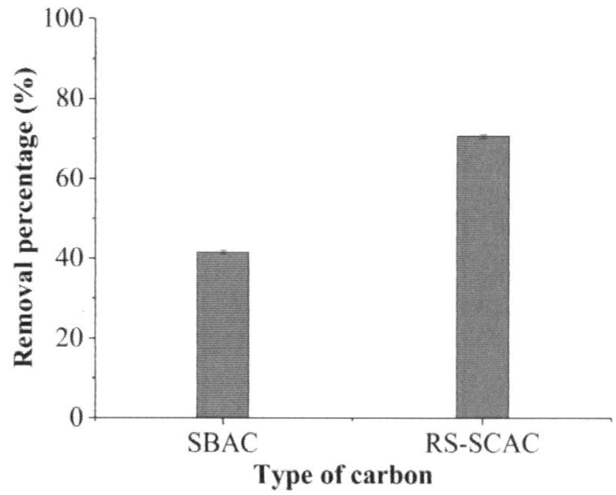

Figure 3. Cr (VI) removal percentages of SBAC and RS-SCAC.

4. Conclusions

In this study, RS-SCAC was successfully prepared by adding rice straw as the carbon additive in sewage sludge. For RS-SCAC, the yield was moderate. The iodine number significantly increased compared with that of SBAC. The main functional groups on the surface of RS-SCAC were –OH, –COOH, and –COH. RS-SCAC contained a certain amount of heavy metals. Compared with the concentration limits in China national standards (GB 5085.3-2007), which were used for identifying hazardous wastes, the leaching concentrations of all heavy metals except Zn were significantly lower than the limits. The Cr (VI) removal efficiency of RS-SCAC was 1.7 times than that of SBAC. Overall, RS-SCAC can be used as a safety and efficient adsorbent for Cr (VI).

5. Acknowledgment

This work was supported by the National Undergraduate Innovation Training Program Funded Project of Sichuan Agricultural University (No. 201610626042).

6. References

[1] Feng L, Luo J and Chen Y 2015 *Environ. Sci. Techno.* **49** 4781-2
[2] Tang J, He J, Liu T and Xin X 2017 *Rsc Adv.* **7** 33385-401
[3] Mukesh KA et al. 2016 *J. Clean. Prod.* **135** 829-35
[4] Zha J, Huang Y, Xia W, Xia Z, Liu C, Dong L and Liu L 2018 *Fuel* **229** 241-47
[5] Eisenbud M 1978 *Environ. Health Persp.* **27** 291-4
[6] Yang K, Zhu Y, Shan R, Shao Yand Tian C 2016 *J. Environ. Manage.* **189** 58-66
[7] Tay JH, Chen XG, Jeyaseelan S and Graham N 2001 *Chemosphere* **44** 45-51
[8] Tao HC, Zhang HR, Li JB and Ding WY 2015 *Bioresource Technol.* **192** 611-7
[9] Zeng F, Liao X, Hu H and Liao L *J. Air Waste Manage.* **68** 255-64
[10] Kan Y, Yue Q, Kong J, Gao B and Li Q 2015 *Chem. Eng. J.* **260** 541-9
[11] Gao Y, Xua S , Ortaboy S, Kan Y, Yue Q and GaoB 2016 *J Taiwan Inst. Chem. E.* 66 394-9
[12] Aggarwal D, Goyal M and Bansal RC 1999 *Carbon* **37** 1989-97
[13] Romanos J et al. 2013 *Carbon* **54** 208-14
[14] Zhang M, Liu Y, Li T, Xu W, Zheng B, Tan X, Wang H, Guo Y, Guo F and Wang S 2015 *Rsc Adv.* **5** 46955-64
[15] Giri AK, Patel R and Mandal S 2012 *Chem. Eng. J.* **185-186** 71–81
[16] You JH, Chiang HL and Chiang PC *Environ. Prog. Sustain.* **13** 31-36
[17] Gao Y, Xua S, Ortaboy S, Kan Y, Yue Q and Gao B *J. Taiwan Inst. Chem. E.* 66 394-9
[18] Liu QS, Zheng T, Li N, Wang P and Abulikemu G 2010 *Appl. Surf. Sci.* **256** 3309-15

The Low Molecular Weight Organic Acids in Root Exudates of *Leersia hexandra* Swartz and Its Role in Mobilization of Insoluble Chromium

Chancui Wu[1, 2], Jie Liu[3], Ying Liang[2], Yongrong Jiang[2] and Xuehong Zhang[2, *]

[1]Light industry and Food Engineering College, Guangxi University, Nanning, 530004, China;
[2]School of Life and Environmental Sciences, Guilin University of Electronic Technology, Guilin 541004, China;
[3]Guangxi Key Laboratory of Environmental Pollution Control Theory and Technology, Guilin University of Technology, Guilin, 541004, China
Email: wuchancui@163.com

Abstract. The root exudates of *Leersia hexandra* Swartz were collected to investigate their mobilization of chromic oxide (Cr_2O_3), lead chromate ($PbCrO_4$) and Cr-contaminated soil. It was observed that the root exudates can mobilize Cr_2O_3, $PbCrO_4$ as well as Cr-contaminated soil. Low molecular weight organic acids in the root exudates *L. hexandra* were analyzed by high performance liquid chromatography (HPLC). Six organic acids, including oxalic acid, tartaric acid, malic acid, lactic acid, maleic acid and citric acid, were detected in the root exudates, and their concentrations were 18.91±0.23, 130.90±1.44, 1031.34±4.38, 65.54±1.01, $0.96\pm3.67\times10^{-3}$ and 201.50±1.13 μg (g root DW)$^{-1}$, respectively. Chemical reagents of these 6 acids were used to mobilize Cr_2O_3, $PbCrO_4$ and Cr-contaminated soil. The mobilization of different organic acids to Cr-contamination soil declined in the order of oxalic acid > malic acid > lactic acid > tartaric acid > maleic acid > citric acid. These results suggested that organic acids in the root exudates play an important role in uptaking a large amount of Cr from soil by *L. hexandra*. Information obtained from this study should develop the understanding of the accumulation mechanism of *L. hexandra* to Cr in the soil, and provide insights for improving the efficiency of phytoremediation.

1. Introduction

Root exudates are plant metabolites that are released to the root surface or into the rhizosphere to enhance plant nutrient uptake [1], [2]. They are generally classified into two types, high molecular weight (polysaccharides and proteins) and low molecular weight (i.e. amino acids, organic acids, sugars, phenolics) compounds [3]. Both of them play an important role in these rhizospheric processes [4], while low molecular weight organic acids (LMWOAs) in root exudates would affect metal availability in the soil. Moreover, heavy metal bioavailability is the most important factor to be monitored in the restoration process of a contaminated soil [3]. However, the composition and quality of root exudates vary from plant to plant. Therefore, study on LMWOAs in root exudates and its role in effect on heavy metal bioavailability could provide the basis to the phytoremediation of contaminated soil.

Hyperaccumulator is an efficient technology based on the use of plants to remove metals from contaminated sites, which are potential tools for phytoremediation [5]. There are specific physiological characteristics of metal uptake in the roots of hyperaccumulators. Most plants are capable of exuding some LMWOAs, such as citric, malic and oxalic acids [6]. These organic acids can react with metal ions in both the soil solution and solid phases, and increase metal mobility in the rhizospheric environment, thereby improving the phytoavailability of metals to plants [7]. Research on *Thlaspi caerulescens*, a Zn/Cd hyperaccumulator, indicates that the plant mobilizes Zn and Cd efficiently from soils [8]. Similar results were obtained for a Ni hyperaccumulator, *Thlaspi goesingense* [9]. However, there is still a vast gap in our understanding of the involvement of the LMWOAs in the root exudates of Cr hyperaccumulation.

Leersia hexandra Swartz has been found to be a Cr hyperaccumulator with an extraordinary accumulation capacity for both trivalent Cr and hexavalent Cr [10], [11]. A great bioaccumulation capacity for Cr was observed in the leaves, stems and roots of *L. hexandra*. The maximum Cr concentrations in the leaves, stems and roots were 5430 mg kg^{-1}, 1956 mg kg^{-1} and 40599 mg kg^{-1}, respectively [12]. *L. hexandra* is suitable for phytoremediation to Cr-contaminated soil [13]. It is hypothesized that a considerable amount of Cr uptaked by *L. hexandra* from the contamination soil owe to the LMWOAs in the root exudates. In order to investigate the role of root exudates in mobilizing Cr, hydroponics experiments have been conduct to study the root exudates of *L. hexandra*. The objectives of this research were: 1) to investigate the mobilization of insoluble chromium by the root exudates of *L. hexandra*; 2) to analyze the composition and quantity of the root exudates of *L. hexandra* by HPLC, and 3) to test the mobilization of insoluble chromium by organic acids. This study will provide scientific basis for improving phytoremediation efficiency to Cr-contaminated soil by *L. hexandra*.

2. Materials and Methods

2.1. Plant Culture and Root Exudates Collection

Seedlings of *L. hexandra* were collected from the riverside of Hua River in Guilin. The seedlings were washed thoroughly (at least three times) with redistilled water and placed in the round plastic pots filled with 1.5 L 25% Hoagland's nutrient solution which renewed every 3 days. Plants were grown in the pots for 20 days in a controlled environment growth cabinet (14 h photoperiod, 25 °C day/18 °C night, relative humidity 70-75%, and a light intensity of 300 μmol $(m^2\ s)^{-1}$).

In this work, the periods of 4 hours under sterile conditions was chosen to collect root exudates. The root exudates collection media used in this work was $CaCl_2$ (0.5 mmol L^{-1}). After being filtered to 0.45 μm with PTFE membrane filters, the collection solution of root exudates were concentrated on the Rotary Evaporator (45 °C, 80-90 r min^{-1}) to 25 ml, and concentrated solution of root exudates were reserved in the refrigerator (-20 °C) for mobilization experiments. In the meanwhile, 2-3 drops of 0.05% thymol were used in the bath, collecting and reserving to control the activities of microorganisms. Additionally, the concentrated solution of root exudates was vacuum-freeze dried then dissolved with 3 ml ultrapure water and filtered (0.45 μm) into the automatic sample bottles for HPLC analysis.

2.2. HPLC Analysis of Organic Acids In the Root Exudates

Common LMWOAs secreted by plant roots [14], [15] such as oxalic, tartaric, formic, malic, lactic, acetic, maleic and citric acids were prepared individually. Single stock solutions of 8 organic acids were mixed according to an optimum proportion and diluted by mobile phase to obtain the mixture standard solution.

The HPLC system, assembled from Waters' modules, was equipped with a 717 plus Autosampler; a 515 HPLC Pump operated under isocratic conditions; a 2996 Photodiode Array (PDA) Detector. The determination condition: XSelect HSS HPLC Columns (5 μm, 250 mm×4.6 mm, Waters, USA); mobile phase was: 40 mmol/L KH_2PO_4 (pH=2.40); wavelength was 205 nm; flow velocity was 1.0 ml/min; column temperature was 25 °C and the sample size was 10 μl. Qualitative and quantitative analysis of organic acids in the root exudates were performed by external standard method and measurement of peak area respectively.

IOP Conf. Series: Materials Science and Engineering **484** (2019) 012009 doi:10.1088/1757-899X/484/1/012009

2.3. Mobilization Experiments

The soils used in mobilization experiments were collected from the constructed wetland for treating chromium containing waste water. After drying (80 °C), grinding and sieving (100 mesh), the chromium content of the tested soil was determined, which has a value of 1.80 mg g^{-1}.

Concentrated root exudates of 25 ml were mixed with 0.200 g synthetic Cr_2O_3, $PbCrO_4$ and 2.000g Cr-contaminated soil respectively in a plastic centrifuge tube. After shaking samples for 1, 2, 3, 4 and 5 h (25 °C, 150 r min^{-1}), and being filtered through the Syringe Filters (0.45 μm), the amount of mobilized Cr in root exudates were determined using atomic absorption spectrophotometer (AA-6300, Shimadzu, Japan). Working standard solutions for analysis of the elements were prepared with serial dilutions of stock standard solutions Cr (GBW (E) 080257). Ultrapure water was used to mobilize Cr_2O_3, $PbCrO_4$ and soil sample instead of the root exudates as the control. Chemical reagents of every organic acid are used to simulate root exudates and repeat the experimental steps described above. The mobilization of Cr was calculated as follows:

$$\text{mobilization}\left(\text{mg (kgCr)}^{-1}\right) = \frac{C_{\text{chromium}} \times V}{m_{\text{chromium}}} \quad (1)$$

Where $C_{chromium}$ is the concentration of the Cr in the supernatant, V is the volume of the activator (root exudates, ultrapure water or organic acid solution) and $m_{chromium}$ is the mass of chromium in the Cr_2O_3, $PbCrO_4$ and soil Cr added.

2.4. Statistical Analysis

The data presented in this paper are the average of three independent replicates ± standard error of means (SD). Each treatment was replicated at least three times. Analysis of variance (ANOVA) was performed on all data sets, and least significant difference (LSD) was used to compare treatments. Graphical work was performed using Origin Pro 8 and Excel 2003.

3. Results

3.1. Mobilization of Root Exudates on Cr-contaminated Soil And Insoluble Chromium

The effects of shaking time on Cr mobilization were investigated in Cr_2O_3, $PbCrO_4$ and soil Cr by the root exudates of *L. hexandra*. Results (Figure 1) showed that the mobilization of Cr_2O_3 and $PbCrO_4$ increased with the shaking time from 1 to 3 h and reached the greatest amount after 3 hours shaking. Moreover, the mobilization of soil Cr increased non-linearly with the shaking time and tended to steady. Shaking time ranges from 1 to 3 h enhanced the mobilization of soil Cr significantly. When the shaking time ranges from 3 to 5 h, there was no significant difference in the mobilization of soil Cr. Therefore, a shaking time of 3 h was selected for all other experiments.

The root exudates of *L. hexandra* were used to mobilize Cr_2O_3, $PbCrO_4$ and soil Cr with a shaking time of 3 h. Meanwhile, ultrapure water was used to replace the root exudates as the control. The result (Table 1) showed that the root exudates can mobilize Cr_2O_3 and $PbCrO_4$ as well as soil Cr. These tests revealed that the mobilization of Cr_2O_3, $PbCrO_4$ and soil Cr by the root exudates were all remarkably higher than that by water. The mobilization of $PbCrO_4$ by water was no detected. While the mobilization of Cr_2O_3 and soil Cr by the root exudates were 2.2 and 3.4 times higher than that by water respectively. The soil Cr is easier to be mobilized than Cr_2O_3 and $PbCrO_4$ by the root exudates.

Table 1. The mobilization capacity of chromium by root exudates of *L. hexandra* (*n*=3)

Insoluble Chromium	Mobilization (mg $kgCr^{-1}$)	
	ultrapure water	The root exudates of *L. hexandra*
Cr_2O_3	193.45±19.51	425.19±13.10
$PbCrO_4$	*n.d.*	292.12±16.05
soil Cr	201.39±78.26	694.44±133.94

3.2. Organic Acids In Root Exudates

Six organic acids in the root exudates of *L. hexandra* were measured by HPLC. The results showed (Table 2) that the contents of these 6 organic acids in root exudates were very different. The contents of malic acid was the greatest among the six organic acids assayed, which reached 1031.34 ± 4.38 μg (g root DW) $^{-1}$. The contents of the organic acids declined in the order of malic > citric > tartaric > lactic > oxalic > maleic. The contents of malic, citric, tartaric, lactic and oxalic acid are 1074.31, 209.89, 136.35, 68.27 and 19.69 times higher than that of maleic, respectively.

Table 2. The contents of organic acids in *L. hexandra* root exudates as determined by HPLC analyses (n = 5)

Organic Acids	Contents of organic acids ± SD (μg (g root DW) $^{-1}$)	Relative proportion
Oxalic	18.91 ± 0.23	19.69
Tartaric	130.90 ± 1.44	136.35
Malic	1031.34 ± 4.38	1074.31
Lactic	65.54 ± 1.01	68.27
Maleic	0.96 ± 3.67×10^{-3}	1.00
Citric	201.50 ± 1.13	209.89

3.3. Mobilization of Organic Acids on Cr-contaminated Soil And Insoluble Chromium

Results above indicated that there are organic acids in the root exudates of *L. hexandra* which can mobilize Cr_2O_3, $PbCrO_4$ and soil Cr. In order to verify the role of the organic acids in mobilizing insoluble chromium, chemical reagents of malic, citric, tartaric, lactic, oxalic and maleic acid were selected for simulation analysis of Cr mobilization of Cr_2O_3, $PbCrO_4$ and soil Cr. The results (Figure 2 to 4) showed that the Cr mobilization of three kinds of insoluble Cr are different, and vary from organic acid to organic acid. The mobilizations of soil Cr by all the 6 acids were higher than that of Cr_2O_3 and $PbCrO_4$. Considering the mobilization capacity of every organic acid, oxalic acid obtained the greatest, while that of citric acid was the lowest. The mobilization capacity of different organic acids to Cr-contamination soil declined in the order of oxalic acid > malic acid > lactic acid > tartaric acid > maleic acid > citric acid. On the other hand, the organic acids with concentration of 5mmol/L displayed the higher mobilization capacity compared to that of 1 mmol L^{-1}. The mobilization of Cr_2O_3, $PbCrO_4$ and soil Cr by 5mmol L^{-1} oxalic acid were 397.48±15.22 mg (kgCr) $^{-1}$, 431.18±18.53 mg (kgCr) $^{-1}$ and 1067.59±24.70 mg (kgCr) $^{-1}$, respectively.

According to the proportion of six organic acids in the root exudates of *L. hexandra*, a mixed solution of 6 organic acids are prepared for simulation analysis of Cr mobilization of Cr_2O_3, $PbCrO_4$ and soil Cr. The results of mobilization of insoluble chromium were compared with those of root exudates. The results (Figure 5) showed that the mobilization of Cr_2O_3 and soil Cr by the root exudates were 1.2, 2.0 and 1.4 times higher than that by the mixed organic acids respectively.

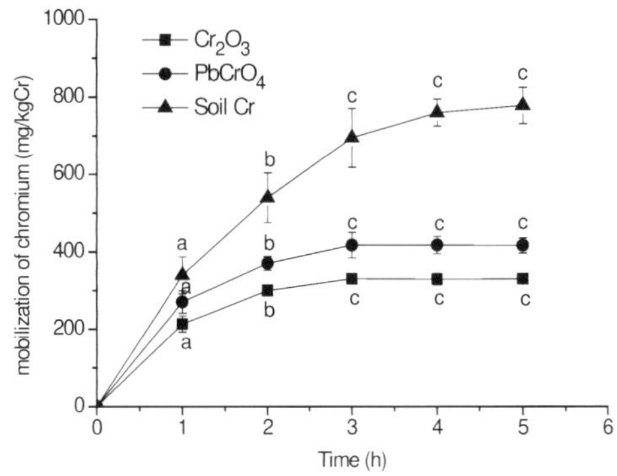

Figure 1. The effect of different shaking time on the mobilization of chromium by the root exudates of *L. hexandra* ($n = 3$)

Different letters above or below standard error bars indicated significant difference at $P<0.05$ (the same below)

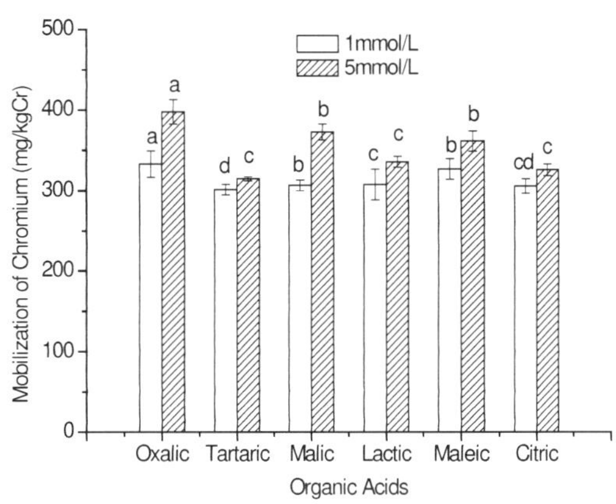

Figure 2. Mobilization of Cr_2O_3 by organic acids (n=3)

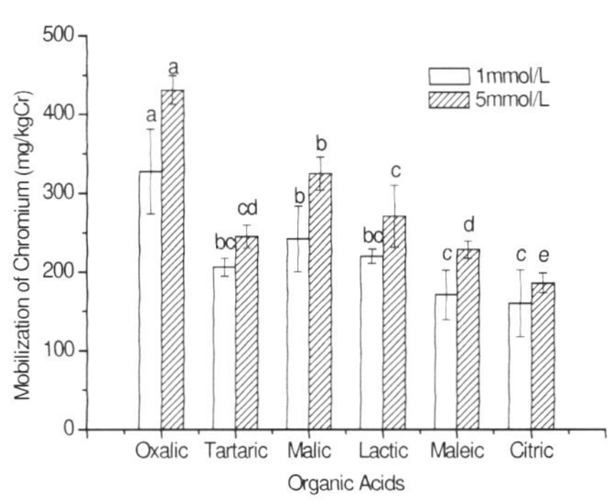

Figure 3. Mobilization of $PbCrO_4$ by organic acids (n=3)

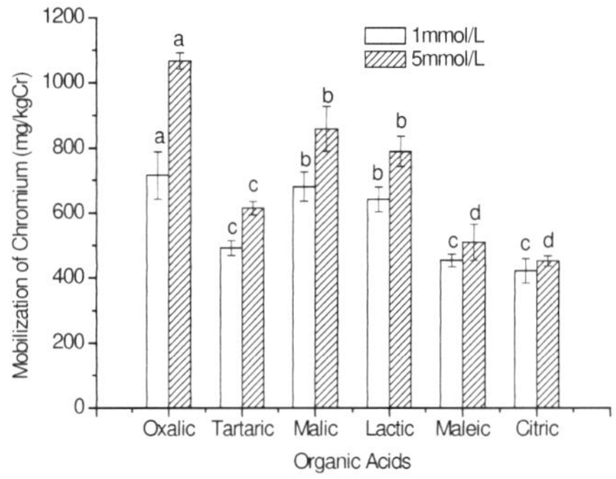

Figure 4. Mobilization of Cr-contamination soil Cr by organic acids (n=3)

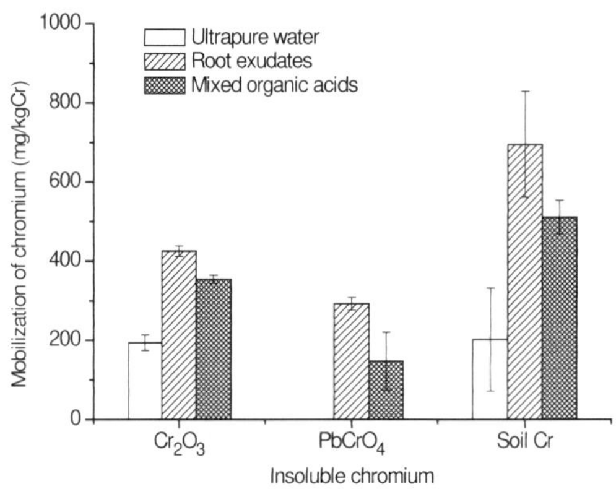

Figure 5. Comparison of mobilization of chromium by root exudates of *L. hexandra* and mixed organic acid solutions (n=3)

4. Discussion

In Cr-contaminated soil, most of the chromium exists in the forms of sparingly soluble mineral such as Cr_2O_3 and $PbCrO_4$. The available Cr concentration in soil is too low to be absorbed by plants. As a hyperaccumulator, *L.hexandra* shows a higher accumulation capacity for Cr than the non-hyperaccumulator. How is chromium in soil converted into the available form for use by plants? Our preliminary results indicated that the root exudates of *L. hexandra* can mobilize Cr_2O_3, $PbCrO_4$ and soil Cr effectively. The root exudates of *L. hexandra* solubilizing the insoluble chromium efficiently was probably attributed to the organic acids in the root exudates.

HPLC analysis identified that the dominating LMWOAs in the root exudates of *L. hexandra* were oxalic, tartaric, malic, lactic, maleic and citric acid (Table 2). Compared with findings in other plants, the amount of malic is the most in *L. hexandra*, which is agreed with that in rice [16] and *Zea mays L* [17]. but a little different from that in mangrove [15], which may be caused by the difference between grass and woody plants. Wang et al. [16] reported that the concentration of malic acid (3.9 μmol (g DW)$^{-1}$) was the greatest among the five organic acids assayed in rice, however, which was much lower than the amount of malic in *L. hexandra* (1031.34 μg (g DW)$^{-1}$). Aulakh et al. [18] also found that malic acid secreted by ten rice cultivars grown under soil conditions had the highest concentration followed by tartaric, succinic, citric and lactic acids. These findings suggest that malic acid secreted by roots may play an important role in adapting to the local environment. Oxalic acid is a typical LMW organic acid in plant root exudates. Since it possesses the capability of both proton donation and ion complexation, oxalate/oxalic acid has been widely used as an extractant for plant available nutrients, including phosphate, from soil [19]. Some grasses, such as rice [16], elephantgrass [20] and *Phragmites australis* [21], are known to contains oxalic acid. However, the contents of oxalic acid in these plants are lower than which in *L.hexandra*. It was interesting to note that the greatest Cr mobilization capacity of oxalic acid showed in simulation experiments. This may be due to both the greater acidity and stronger complexation capability of oxalic acid than other acids. In addition, there were significant differences in the amount of Cr-mobilization of different acids, which suggests that each organic acid plays different roles in rhizosphere chemistry and nutrient acquisition procedure. The process that was responsible for changes in rhizospheric pH involved the evolution of CO_2, root exudates, and the excretion or re-absorption of H^+ or HCO_3^-, as well as microbial production of organic acids [22]. According to the above view, in the case of mobilizing Cr from soil by *L.hexandra*, decrease pH in the rhizosphere is due to the organic acids secreted by the root. The results of mobilization experiments by the mixed acids indicated the mobilization of Cr_2O_3, $PbCrO_4$ and soil Cr by mixed organic acids were lower than that of the root exudates. Because root exudates of plants are varied in kinds and contents, In addition to the six organic acids detected in this study, the root exudates of *L. hexandra* also contain other components which may also play a role in mobilization of insoluble chromium. Thus, the components of the root exudates of *L. hexandra* should be studied further.

Although organic the role of acids in mobilization of insoluble chromium has been verified in this paper, study on the root exudates of *L.hexandra* under different culture condition has not been conducted. Therefore, other composition of the root exudates under different culture condition and their role in mobilization of Cr in soil require further study.

5. Conclusion

In summary, the root exudates of *L.hexandra* exhibited a considerable capacity in mobilizing Cr from Cr_2O_3 and $PbCrO_4$ as well as from Cr-contaminated soil. It's probably attributed to the organic acids secreted by the root of *L.hexandra*. The results of the simulation experiment which used the 6 organic acids detected in *L. hexandra* root exudates to mobilize the insoluble chromium and soil Cr provided further evidence for this hypothesis. This study demonstrated that the organic acids in the root exudates of *L.hexandra* play an important role in enhancing the bioavailability of Cr and are beneficial to improve phytoremediation efficiency.

6. Acknowledgements

The authors thank the financial supports from the National Natural Science Foundation of China (41471270) and Guangxi science and technology major project (Gui Ke AA17204047).

7. Reference

[1] Thomas C, Butler A, Larson S, Medina V and Begonia M 2014 *Int J Phytoremediat* **16** 634-640.
[2] Valentinuzzi F, Cesco S, Tomasi N and Mimmo T 2015 *Biol Fertil Soils* **51** 757-765.

[3] Montiel-Rozas M M, Madejón E and Madejón P 2016 *Environ Pollut* **216** 273-281.
[4] Bais H P, Weir TL, Perry L G, Gilroy S and Vivanco M 2006 *Annu. Rev. Plant Biol* **57** 233-266
[5] Sarma H 2011 *Journal of Environmental Science and Technology* **4** 118-138.
[6] Dakora F D and Phillips D A 2002 *Plant Soil* **245** 35-47.
[7] Lu L L, Tian S K, Yang X E, Peng H Y and Li T Q 2013 *Journal of Zhejiang University-SCIENCE B (Biomedicine & Biotechnology)* **14** 106-114.
[8] Knight B, Zhao F J, McGrath S P and Shen Z G 1997 *Plant Soil* **197** 71-78.
[9] Krämer U, Pickering I J, Prince R C, Raskin I and Salt D E 2000 *Plant Physiol* **122** 1343-53.
[10] Zhang X H, Liu J, Huang H T, Chen J, Zhu Y N and Wang D Q 2007 *Chemosphere* **67** 1138-43.
[11] Chen S M, Zhang X H, Liu J, You S H, Zhang H and Lin H 2016 *Fresen Environ Bull 25* 959-968.
[12] Zhang X H, Liu J, Wang D Q, Zhu Y N, Hu C and Sun J J 2009 *Bull Environ Contam Toxicol* **82** 358-362.
[13] Liu J, Duan C Q, Zhang X H, Zhu Y N and Lu X Y 2011 *J Hazard Mater* **188** 85-91.
[14] Yuan J, Tan X F, Ye S C, Zhou N F and Shi B 2013 *Journal of Chemical and Pharmaceutical Research* **5** 572-577.
[15] Xie X Y, Weiss D J, Weng B S, Liu J C, Lu H L and Yan C L 2013 *Environ Sci Pollut Res* **20** 97-1008.
[16] Wang X, Tam N F Y, He H D and Ye Z H 2015 *Plant Soil* **394** 301-313.
[17] Gaume A, Mächler F and Frossard E 2001 *Plant Soil* **234** 73-81.
[18] Aulakh M S, Wassmann R, Bueno C, Kreuzwieser J and Rennenberg H 2001 *Plant Biol* **3** 139-148.
[19] Fransson A M 2001 *Commun. Soil Sci. Plant Anal* **32** 2469-2484.
[20] Shen H, Wang X C, Shi W M, Cao Z H and Yan X L 2001 *J Plant Nutr* **24** 1117-1130.
[21] Rocha A C S, Almeida C M R, Basto M C P and Vasconcelos M T S D 2016 *Estuar Coast Shelf S* **171** 77-84 .
[22] Huang GY, Guo G G, Yao S Y, Zhang N and Hu H Q 2016 *Int J Phytoremediat* **18** 33-40.

IOP Conf. Series: Materials Science and Engineering **484** (2019) 012010 doi:10.1088/1757-899X/484/1/012010

Preparation of Catalyst by Phosphating Slag and Its Photocatalytic Performance on Rhodamine B and Methyl Orange

Weilin Guo[1,*], Lijun Wang[1], Lingxue Kong[3], Yongmin Qiao [2] and Suna Zhang[1]

[1]Department of Environmental and Material Engineering, Shanghai Polytechnic University, No. 2360 Jinhai Road, Pudong New Area in Shanghai, China.
Email:1225873512@qq.com
[2]Shanghai Shanshan Technology Co., Ltd. No. 3158, Jinhai Road, Pudong New Area in Shanghai, China.
[3]Frontier Materials Research Institute, Deakin University, Geelong Victoria 3220 Australia.
Email: 1225873512@qq.com

Abstract. The catalysts are synthesised by hydrothermal reaction of ball-milled phosphating slag with ammonia, ethanol, and oxalic acid respectively. The catalysts are characterized by XRD, SEM and UV-Vis absorption spectra. We studied the photocatalytic degradation efficiency of different catalysts and the amount of catalysts on the organic dye rhodamine B and methyl orange. The results shows that the catalyst prepared by the hydrothermal reaction of 1 mol/L oxalic acid with phosphating slag has a microporous structure after being modified by high temperature, the degradation efficiency of rhodamine B and methyl orange which are carried out for 120 min under the irradiation of 100 w mercury lamp can reach 78% and 81%, and the optimal amount of catalyst is 0.5 g/L. This study provides a new solution for the utilization of phosphating slag.

1. Introduction

Phosphating is a common surface treatment technology and is widely used in the casting and spraying industries [1]. During the phosphating process, a large amount of by-products are generated, its main components are $FePO_4$ and $Zn_3(PO_4)_2$, and there are some trace heavy metal elements such as Mn, Ni and Cr[2]. Phosphating slag is a kind of hazardous solid waste, but it is also a resource that can be used, phosphating slag contains a large amount of phosphate, iron and zinc, etc, to extract these useful substances from phosphating slag has a wide range of application prospects [3, 4]. At present, there have been many related studies on the utilization of phosphating slag. Fan used iron phosphate slag to prepare iron oxide red paint, which can be used in construction and chemical industry [5]. Ma used iron and calcium in the phosphating slag to prepare iron oxide and tricalcium phosphate [6]. But few studies have been published on the use of phosphating slag for the preparation of photocatalysts.This paper mainly studied on the waste phosphating slag to prepare high-performance photocatalyst through hydrothermal reaction, and studied its photocatalytic degradation ability for rhodamine B and methyl orange under ultraviolet irradiation[7,8].

IOP Conf. Series: Materials Science and Engineering **484** (2019) 012010 doi:10.1088/1757-899X/484/1/012010

2. Experimental Procedures

2.1. Ball Grinding Experiment

20 g phosphating slag and 20 mL anhydrous ethanol were added into the utensils, then 10 g, 30 g, 50 g, 70 g, 100 g, 120 g, 150 g and 180 g balls were used respectively, the phosphating slag was milled at 500 r/min for 5 hours [9], the milled-phosphating slag was mixed with ethanol and water to prepare ferric hydroxy phosphate at 200°C for 8 hours.

2.2. Preparation Of Catalysts

The milled-phosphating slag was mixed with the ammonia, anhydrous ethanol and oxalic acid respectively in the mass ratio of 1:1 to carry out hydrothermal reaction at 200 °C for 8 hours. The resulting sample were filtered, washed and dried at 100 °C, then the dried sample were baked in the nitrogen atmosphere to obtain a series of photocatalysts.

2.3. Photocatalytic Experiment

200 mL of 10 mg/L rhodamine B solution and 100 mg catalyst were added into a quartz tube, then the mixture was stirred for 30 min in the dark room to achieve adsorption equilibrium.Then turned on the power of the photochemical reactor and started the reaction under the irradiation of a 100 W mercury lamp. Took a sample every 10 minutes, the sample was centrifuged and the supernatant liquid was placed in a cuvette, and then measured the absorbance at λ=554 nm with a ultraviolet-visible spectrophotometer. Calculating the concentration of rhodamine B at different time based on the relationship between the concentration and absorbance of rhodamine B, according to the formula $D = [(C_0-C_t)/C_0] \times 100\%$, the degradation rate of rhodamine B was obtained, the D is the degradation rate, C_0 is the initial concentration of rhodamine B and C_t is the concentration of rhodamine B at different time.

3. Results and Discussion

3.1. Ball Grinding Effect of Different Quality Balls

The particle size of phosphating slag and ferric hydroxy phosphate reached to the minimal which are 1.6 um and 2.9 um when the mass of ball is 120 g.The particle size of the sample is measured by the Malvern 2000 particle size tester, which is the average particle size and it is a statistical value. As the mass of the ball increases, the phosphating slag is more fully grinded, but when the mass of the ball exceeds a certain limit, the excess ball can't be fully rotated, so the optimum amount of the ball is 120 g.

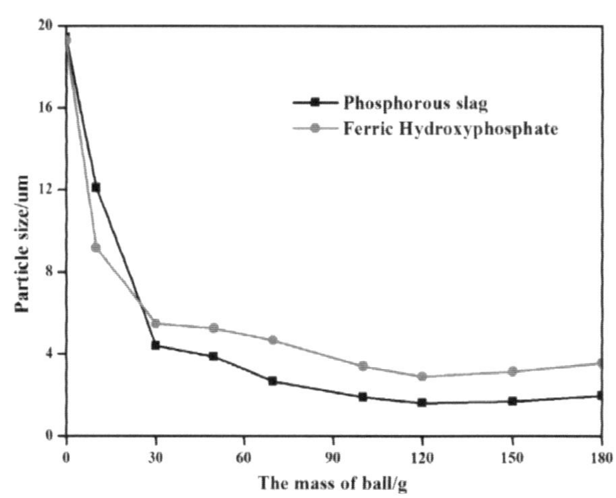

Figure 1. The particle size of phosphating slag and ferric hydroxy phosphate prepared by different ball milling ratios

3.2. XRD Characterization of Catalyst Materials

Figure 2 shows the XRD pattern of catalysts prepared by hydrothermal reaction between ball-milled phosphating slag and different solvents. Comparing the diffraction peaks of the samples with iron phosphate, iron oxide and iron hydroxy phosphate standard cards, it can be seen that the main product of hydrothermal reaction between phosphating slag and ammonia water is iron phosphate,the main product of reaction between phosphating slag with ethanol and water is iron hydroxy phosphate,the main product of hydrothermal reaction between phosphating slag with 0.5 mol/L, 1 mol/L and 3 mol/L oxalic acid are iron hydroxy phosphate,the main product of hydrothermal reaction between phosphating slag with 5 mol/L oxalic acid are iron oxide.

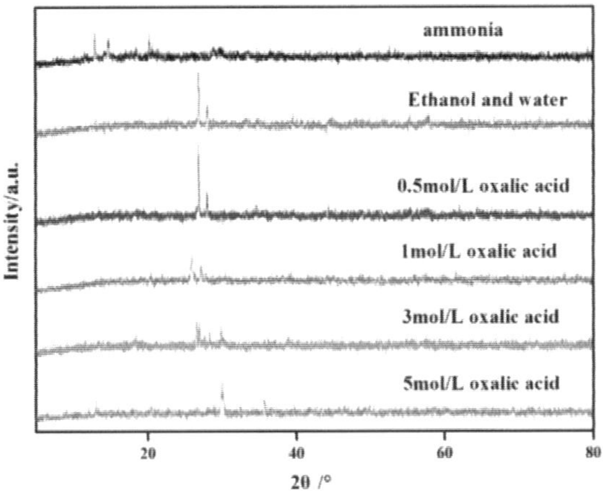

Figure 2. XRD patterns of different catalysts synthesized with different solutions

3.3. SEM Characterization of Catalyst Materials

Figure 3 shows the catalyst prepared by the ball-milled phosphating slag and different solvents. (a) and (b) are prepared by ammonia, ethanol and water with phosphating slag respectively, the particle size are approximately 1um and the surface are relatively smooth. (c), (d), (e) and (f) are prepared by different concentrations of oxalic acid and phosphating slag, the particle size are approximately 5 um. The surface are rough which has complex pore structure, and the adsorption properties is superior.

Figure 3. SEM images of different catalysts synthesized with different solutions , (a)Ammonia (b)Ethanol and water (c)0.5 mol/L Oxalic acid (d)1 mol/L Oxalic acid (e)3 mol/L Oxalic acid (f)5 mol/L Oxalic acid

3.4. UV-Vis Spectroscopy

Figure 4 shows the UV-vis absorption spectra of catalysts prepared by hydrothermal reaction of different solvents and phosphating slags. phosphating slag has a good absorption effect in the ultraviolet region, after the hydrothermal reaction with ammonia, the wavelength of ultraviolet absorption has undergone a red shift. After the hydrothermal reaction with water and ethanol, the wavelength of ultraviolet absorption shifts red more, but the absorbance in the ultraviolet range decreases. After hydrothermal reaction with oxalic acid, the catalyst has a better absorption effect in the visible light region, but the absorbance in the ultraviolet region has decreased.

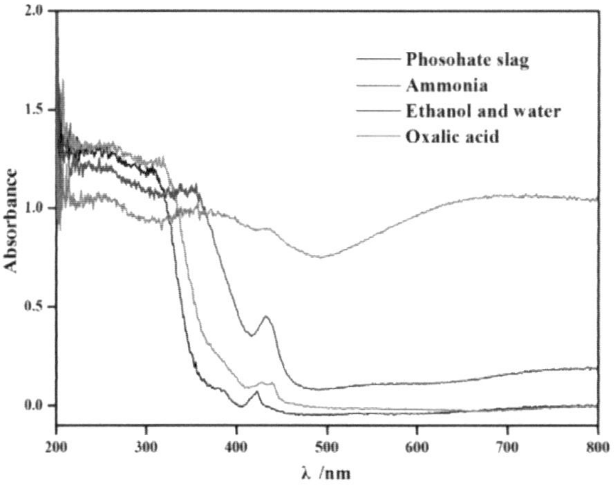

Figure 4. UV-Vis spectra of different catalysts synthesized with different solutions

3.5. Study on Photocatalytic Performance Of Catalysts

Fig. 5 shows the degradation rate of rhodamine B at different time,the degradation rate of rhodamine B is 37% which is under the irradiation of 100 W mercury lamp without catalyst. After adding different catalysts, the degradation rate of rhodamine B is significantly increased, and the catalytic effect of 1 mol/L oxalic acid is the best which reachs 78%. The specific mechanism of photocatalytic degradation of dyes is that under the irradiation of the mercury lamp, the electrons in the catalyst are excited to undergo a transition, thereby generating e^- and h^+, e^- can be captured by dissolved oxygen and generate $O\cdot_2^-$, the protonation of $O\cdot_2^-$ in the acidic medium produces ·OH, ·OH has strong oxidizing ability and can fully mineralize organic dyes [10, 11]. The speed of photocatalytic degradation of dyes is also related to the adsorption performance of the catalyst [12]. The better the adsorption performance, the faster the photocatalytic degradation[13].The catalyst prepared by hydrothermal reaction of phosphating slag and 1 mol/L oxalic acid has a rough surface, microporous structure and good adsorption performance, and its photocatalytic degradation rate is relatively good.

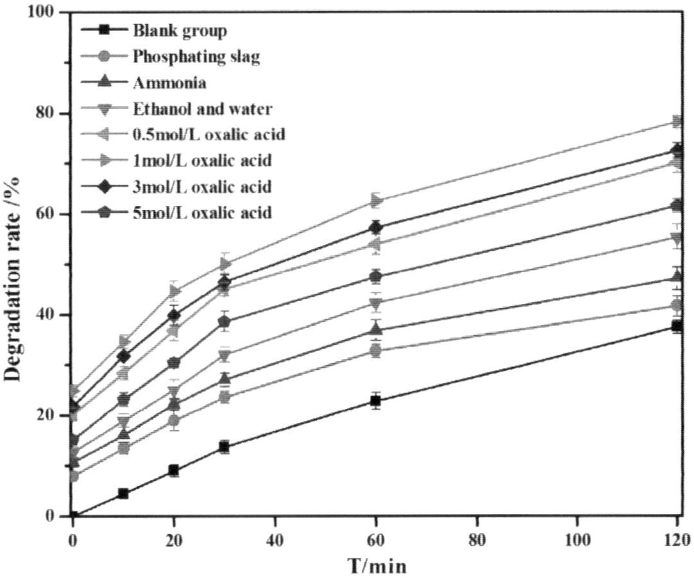

Figure 5. Degradation rate of rhodamine B at different time by different catalysts synthesized with different solutions

Figure 6 shows the degradation rate of rhodamine B and methyl orange at different time with the catalyst prepared by hydrothermal reaction of ball-milled phosphating slag and 1 mol/L oxalic acid, the catalyst is modified by high-temperature nitrogen gas. It can be seen from the figure that the catalyst has a good degradation effect on both rhodamine B and methyl orange, and the degradation effect on methyl orange is better than rhodamine B. Compared with the t-test of the degradation effect data, the value is greater than 0.05, so there is no significant difference in the degradation effect of the catalyst on the two dyes, but the degradation effect shows that the catalyst has better selectivity for methyl orange.

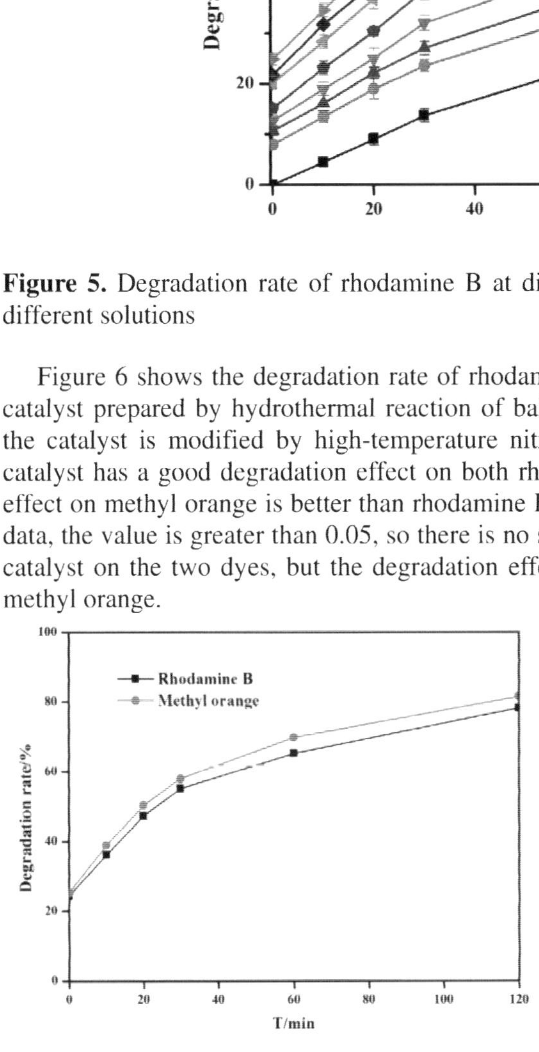

Figure 6. Effect of catalyst on the degradation rate of rhodamine B and methyl orange

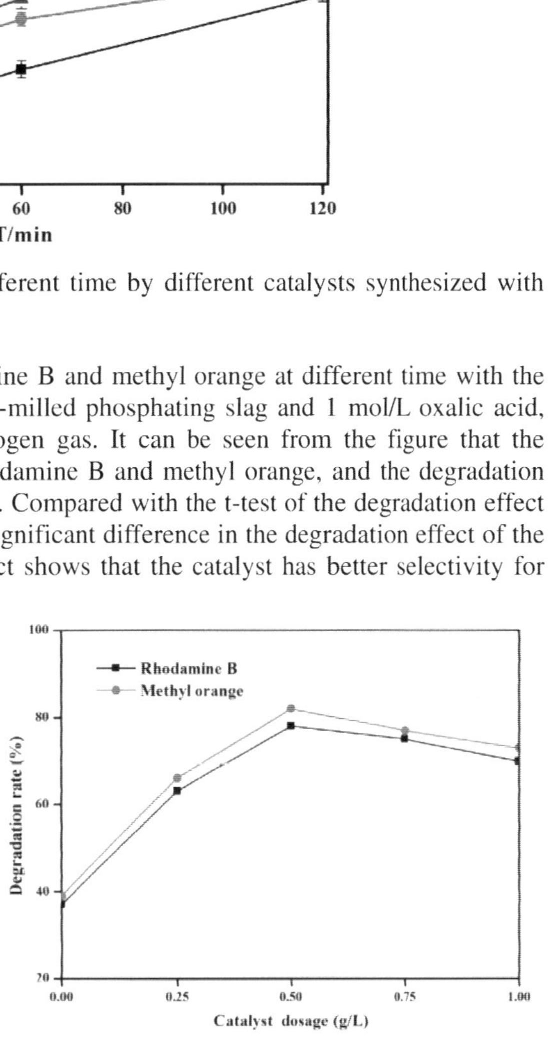

Figure 7. Effect of catalyst amount on the degradation rate of rhodamine B and methyl orange

Figure 7 shows the effect of the amount of catalyst prepared by hydrothermal preparation of 1 mol/L oxalic acid and phosphating slag on the degradation rate of rhodamine B and methyl orange. It can be clearly seen from the figure that the degradation rates of rhodamine B and methyl orange are increasing with the increase of the amount of catalyst. When the amount of catalyst is 0.5 g/L, the degradation rate of both dyes reaches the maximum. As the amount of catalyst continues to increase, the degradation rates of both rhodamine B and methyl orange have declined. This is due to the fact that when the amount of catalyst is small, the active sites provided by the catalyst are less, the dye molecules cannot fully contact with the catalyst, the photocatalytic degradation effect is poor. With the increase of the amount of catalyst, the active sites provided by the catalyst increase, the dye molecules can fully contact with the catalyst, the light source is fully utilized, and the photocatalytic degradation effect is significantly increased. However, when the amount of catalyst increases to a certain extent, the excess catalyst will shield the light source and cause light scattering, which will affect the photo catalytic effect and reduce the catalytic degradation rate of the dye.

4. Conclusion

In this study, catalysts are prepared by hydrothermal reaction of ball-milled phosphating slag with different solvents and modified by high-temperature nitrogen calcination. The catalyst prepared by 1 mol/L oxalic acid solvent and ball-milled phosphating slag has a microporous structure, which has good absorption effect of ultraviolet and visible light. Under the irradiation of 100 w mercury lamp, the degradation rate of rhodamine B reachs to the maximum, then based on the catalyst, we studied the degradation rate of methyl orange and rhodamine B and the effect of catalyst amount on degradation rate, the experimental results show that when the amount of catalyst is 0.5 g/L, the degradation rate of rhodamine B and methyl orange reach 78% and 81%. The catalyst has low preparation cost and the surface morphology of the catalyst can be regulated by controlling the concentration of oxalic acid, so that it has excellent adsorption performance. The catalyst can not only be applied to the field of photocatalysis, but also has bright application prospect in adsorption, carrier and targeted drugs [14, 15]. This study provides a new solution for the utilization of phosphate slag.

5. Acknowledgements

This work was supported by Shanghai Polytechnic University Graduate Program Foundation (EGD18YJ0057), Material Science and Engineering Key Subject of Shanghai Polytechnic University (No. XXKZD1601), and Gaoyuan Discipline of Shanghai-Environmental Science and Engineering (Resource Recycling Science and Engineering).

6. References

[1] Xiong Xinyu, Xiong Tianqing, Song Qianqian, et al. Comprehensive utilization technology of phosphorus slag for metal surface treatment [J]. *Metallurgical Power*, 2015(**2**): 43-45.

[2] Fan Hongqiang. Study on the treatment and resource utilization of solid waste phosphorus slag [D]. *Dalian University of Technology*, 2008(**1**):53-58.

[3] Kuo Y M. An alternative approach to recovering valuable metals from zinc phosphating sludge [J]. *Journal of Hazardous Materials*, 2012, 201(**1**):265-272.

[4] Wu Xiaoyan, Huang Chengcheng, Lang Qingcheng, et al. Research Progress and Prospect of Phosphating Slag Recycling[J]. *Renewable Resources and Circular Economy*, 2015(**8**): 38-41.

[5] Fan Hongqiang, Li Shuying, Jiao Yang. Study on preparation of composite antirust pigments from solid waste phosphorus slag [J]. *Equipment Environmental Engineering*, 2008, 5(**2**):83-86.

[6] Ma Fubo, Chen Wenxing, Tian Juan. Production of tricalcium phosphate using phosphate residues [J]. *Inorganic Salt Industry*, 2015, 47(**9**):54-55.

[7] Zhang Guozhen. Photocatalytic degradation of organic pollutants by cobaltate/carbon composites [D]. *Liaoning University*, 2017:41-45.

[8] Yan F C, Wang X. Treatment of dye wastewater using hydrothermally prepared nano-TiO_2, under natural light [J]. *Journal of Inorganic & Organometallic Polymers & Materials*, 2016, 26(**1**):142-146.

[9] Wu Di, Li Chen, Ji Rixin, et al. Preparation of La_2O_3 and La~(3+) doped TiO_2 by ball milling and its photocatalytic activity[J]. *Chemical Research and Application*, 2018(**5**):66-68.

[10] Zhang Guozhi. Preparation and Photocatalytic Activity of Titanium Dioxide/Carbon Nanocomposites[D]. *University of Lanzhou*, 2014:71-75.

[11] Yu Bo. Preparation of antimony trioxide in time and photocatalytic degradation of organic pollutants [J]. *Chemical Engineer*, 2018(**1**): 57-61.

[12] Wan Yan, Wei Renxing, Lu Ting, et al. Research Progress on Factors Affecting Photocatalytic Degradation of Organic Pollutants by Titanium Dioxide Photocatalysis[J].*Modern Chemicals*, 2018(**5**): 44-46.

[13] Sun Shizhen. Mechanism of BiOCl adsorption/photocatalysis for removal of typical organic pollutants in water [D]. *Zhejiang University*, 2016(**5**).56-61.

[14] Wu Ruiping. Preparation of polyhydroxyphosphate and its application in soil remediation of lead and cadmium [D]. *Central South University*, 2014(**1**).81-90.

[15] Chen Q, Wei C, Zhang Y, et al. Single-crystalline hyperbranched nanostructure of iron hydroxyl phosphate Fe5(PO4)4(OH)3·2H2O for highly selective capture of phosphopeptides[J]. *Scientific Reports*, 2014(**4**):3753.

IOP Conf. Series: Materials Science and Engineering **484** (2019) 012011 doi:10.1088/1757-899X/484/1/012011

Effect of Dietary Carbohydrate/Protein Ratios and Feeding Frequency on Carbohydrate Metabolism of Common Carp

Zhenyan Cheng, Anqi Wang, Ze FAN, Jinhui Sun, Pei Cui and Xiuting Qiao

Tianjin Key Lab of Aqua-Ecology and Aquaculture, College of Fisheries, Tianjin Agricultural University, Tianjin 300384, P.R. China.

qxt65@sohu.com

Abstract. A 8-week 3×2 two-factorial experiment was conducted to investigate effects of dietary carbohydrate to protein ratio (C5%/P32%, C10%/P30%, C20%/P28%) and feeding frequencies (fourth daily and twice daily) on the hepatic carbohydrate metabolic enzyme activities, regulatory factor contents related to glycometabolism, the insulin receptor (IR), GLUT2, hexokinase (HK), GLUT2 and SGLT1mRNA expression levels of common carp with initial average weight of 55.37±3.55 g. Three replicates with 150 common carp were randomly allotted to one treatment with combination of dietary carbohydrate level and feeding frequency in a complete randomized design. The results indicated that: (1) Dietary C/P ratio and feeding frequency significantly influenced the amylase activities of midgut ($P<0.05$). Amylase activities of midgut of fish fed fourth daily were significantly lower than those fed twice daily ($P<0.05$). (2) The phosphofructokinase (PFK) and glucose 6 phosphate dehydrogenase (G6PD) activities were significantly affected only by dietary C/P ratio. In addition, the phosphoenolpyruvate carboxykinase (PEPCK) and Glycogen synthase (GS) activities were significantly affected by dietary C/P ratio and feeding frequency, and there was a significant positive correlation between those activities and dietary C/P ratio ($P<0.05$). (3) The IR, GLUT2 and HK mRNA expression levels of hepatopancreas and GLUT2 and SGLT1 mRNA expression levels of intestine were significantly affected by dietary C/P ratio and feeding frequency, which were significantly increased with increasing feeding frequency ($P<0.05$). Furthermore, the highest values of those expression levels were observed in the C10%/P30% group and these were significantly higher than those in other groups ($P<0.05$). Above all, fish fed C10%/P30% diet was suitable with feeding frequency at fourth daily in this study.

1. Introduction

Carbohydrate to protein ratio is an important indicator of evaluating the nutritive value of fish diet. On the one hand, appropriate carbohydrate level in diet is regarded as ideal energy-providing substance, satisfying basic metabolism of fish, conducive to growth, and improving dietary utilization [1]; on the other hand, moderate carbohydrate to protein ratio can enhance protein sparing effect of carbohydrate, and reduce ammonia-nitrogen excretion to water body. Therefore, it is of great economic and ecological significance to explore appropriate carbohydrate to protein ratio in the diet [2-4]. In actual aquaculture production, aquaculture effect is closely related to feeding frequency selected according to feeding rhythms of animals cultured [5]. Rational feeding frequency not only can accelerate the growth rate of fish and influence nutritional ingredients of fish, but also increase dietary utilization [6] and reduce environmental pollution [7]. As a result, it is crucial to determine a rational feeding frequency in aquaculture management. Studies showed that with the change in nutrition composition of diet, feeding frequency suitable for cultured fish varied correspondingly [7], determining fish growth, metabolism and dietary utilization by influencing the intake and absorption of nutrient substance by fish. Therefore, in aquatic animal culture, in order to reduce culturing cost, it is significant to develop a reasonable feeding strategy and study an appropriate carbohydrate to protein ratio of fish diet and feeding frequency.

Common carp is one of the dominant freshwater aquaculture species in China. Recently, there are more and more studies of nutrition of its diet such as protein and carbohydrate and feeding strategy. Although the protein-sparing of dietary carbohydrate[8] or feed frequency[9] have been well demonstrated in common carp, most studies concerned mainly on one or other of two factors, and the interaction effect of two factors on growth, digestion and metabolism of common carp have received insufficient attention currently. Therefore, on the basis of studying the influence of interaction between carbohydrate to protein ratio of diet and feeding frequency on amylase activities, key enzyme activities of glycometabolism, and mRNA expression of glucose transporter in intestine of common carp, glycometabolism mechanism of common carp under different feeding frequencies and carbohydrate to protein ratios was explored, providing a theoretical basis for exploring an efficient and reliable feeding frequency and dietary formulation of common carp.

2. Ingredients and Methods

2.1. Experimental Diets

Taking fish meal, soybean meal, rapeseed meal, cotton meal and peanut meal as protein source, soybean oil as the lipid source, and wheat starch as carbohydrate source, three groups of equal-energy and equal-fat experimental diet with the carbohydrate/protein level of C5% / P32%, C10% / P30% and C20% / P28% (C - carbohydrate, P - protein) respectively were prepared (Table 1). Diet was prepared in Tianjin Tianxiang Aquaculture Co., Ltd. First, all dry raw materials were smashed with a pulverizer, passed through a 60-mesh screen, and were mixed evenly with a mixing machine according to the principle of step-by-step scaling up. Then soybean oil was added according to dietary formula ratio and mixed completely in the mixed machine. Next, sinking pellet diet with the diameter of 3.00 mm was made from the mixture with a feed stuff cuber (Jiangsu Muyang Group, MUZLM V4). After natural withering, the diet was preserved in a refrigerator at a temperature of - 20 °C.

Table 1. Composition and nutrient levels of experimental diets (air-dry basis) %

Ingredients (%)	Diet		
	C5P32	C10P30	C20P28
Fish meal[1]	7	5	3
Soybean meal[1]	22	22	22
Peanut meal[1]	15	15	15
Cotton meal[1]	12	12	12
Rapeseed meal[1]	13	13	13
Microcrystalline Cellulose	12	8.8	0.6
DDGS[2]	3	3	3
Soybean oil	5	5.2	5.4
Premix[3]	2	2	2
Wheat starch	5	10	20
CMC-Na	2	2	2
Monocalcium phosphate	2	2	2
Nutrient levels			
Crude protein (% DM)	31.94	30.46	28.38
Crude lipid (% DM)	6.66	6.69	6.67
Energy (MJ/kg DM)[4]	16.21	16.23	16.25

Note: [1]Fish meal: 62.3% crude protein, 12% crude lipid; soybean meal: 43.7% crude protein; peanut meal: 46.2% crude protein; cotton meal: 43.3%crude protein; rapeseed meal: 36.2% crude protein; [2]DDGS: dried distillers grains with solubles, 26% crude protein, 8% crude lipid;[3]same as cheng et al.

(2017)[4]; [4]calculated as protein: 24 kJ/g; fat: 38 kJ/g; and starch: 17 kJ/g.

2.2. Experimental Fish and Aquaculture Management

Experimental fish were purchased from Tianjin Huanxin Aquatic Product Improved Variety Station. Temporary culture and aquaculture test were finished in Tianjin Tianxiang Aquaculture Co., Ltd. After disinfection treatment, fish were cultured in a net cage temporarily (3 m × 3 m × 3 m) for 7 days of acclimatization to make them adapt to the aquaculture environment gradually. During acclimatization, commercial diets with 32% protein were fed to fish.

After 7 days of temporary culture, fish were fasted for 24 h. 900 experimental healthy carps with a uniform specification and initial weight of (55.73 ± 3.55) g were put into submersible cages with the size of 1.0 m×1.0 m×1.5 m and divided into 6 groups, with three replicates in each group. During the whole experimental period, a fixed daily feeding rate (4% BW·d^{-1}) was adopted. A 3×2 two-factorial experiment was designed. The feeding frequency of each group was twice a day (8:00 am, 17:00 pm) and four times a day (8:00 am, 11:30 am, 14:00 pm, 17:00 pm), which were named C5% /P32%/F2, C5% /P32%/F4, C10% /P30%/F2, C10% /P30%/F4, C20% /P28%/F2 and C20% /P28%/F4 respectively (C - carbohydrate, P - protein, F - feeding frequency). Each treatment had three replicates. During the experiment, weight of fish in each net cage was evaluated once every 14 days and feeding amount was adjusted according to the determined feeding rate. Water temperature was 28°C - 32°C. Dissolved oxygen was about 6.0 mg·L^{-1} and pH was 7.6 - 8.0.

2.3. Sample Collection

After 8 weeks of culture, the experimental fish were fasted for 24 hours. Then the experimental fish were counted and weighed to calculate survival rate and weight gain rate; from each cage, 20 fish were randomly selected respectively and their body lengths were measured to calculate fatness; tail venous blood was collected into centrifuge tubes and was centrifuged for 20 minutes with a benchtop high speed refrigerated centrifuge at 4 °C with the revolving speed of 4500 r·min^{-1} to obtain supernatant and prepare serum; experimental fish after blood drawing were dissected rapidly in an ice plate to obtain the hepatopancreas which were weighed to calculate liver and body weight ratio. In addition, the intestines were obtained, and samples were preserved in a refrigerator with a ultra-low temperature of - 80 °C.

2.4. Measurement of Biochemical Indexes

In the experiment, enzyme linked immunosorbent assay (ELISA) kit that was used to measure the activities of carbohydrate metabolic enzymes in the hepatopancreas, and concentrations of relevant hormone in serum (insulin, glucagon, growth hormone) and regulatory factors (insulin receptor, insulin-like growth factor) was produced by Shanghai Enzyme-linked Biotechnology Co., Ltd. Its main principle is to measure the activities of enzymes and concentrations of substances in serum or tissue-related liquid samples with a double antibody sandwich method.

2.5. Gene Expression of Insulin Receptor, Relevant Transporter and Metabolic Enzymes

After the aquaculture experiment, 24 hours of fasting was carried out and then each group was fed with respective diet. 2 hours from then, from each group, three fish were selected randomly. Their livers and intestines were obtained and ground in liquid nitrogen. Total RNA of liver tissues was extracted with Trizol method. According to the instruction of TaKaRa PrimeScript™ 1st Strand cDNA Synthesis Kit (Code: 6110A), total RNA was transcribed into cDNA reversely.

According to conserved sequences of common carp's insulin receptor (IR) (EU009571.1), glucose transporter (GLUT2) (AF247730.1), sodium-glucose cotransporter (SGLT1) (JN867793.1), hexokinase (HK) (AF119837.1) and β-actin (M24113) in Genbank, primers were designed, IR: 5' CAACTCTGGTGGTGATGG 3 (F) and 5' CGACCTGGATTATTCTGTC 3' (R), GLUT2: 5' ACTCCTGTTCGGTTTCAC 3' (F) and 5' CCATCTCAGCCTCTTCTT 3' (R), SGLT1: 5' GACTAAAGAAGAGGAGGCA 3' (F) and 5' CAGACGGTGAGGAGGATA 3' (R), HK: 5' GAGAAGAGCAAAGAGGGAC 3' (F) and 5' CCAGAGTAGCAGCAATCA 3' (R), β-actin: 5' CCGTGACATCAAGGAGAA 3' (F) and 5' GATACCGCAAGATTCCATAC 3' (R). All primers

IOP Conf. Series: Materials Science and Engineering **484** (2019) 012011 doi:10.1088/1757-899X/484/1/012011

were synthesized by Suzhou GENEWIZ Biotechnology Co. Ltd.

Real time fluorescent quantitative PCR was measured with SYBR Green-dyed under the Bio-Rad iQ5 Real Time PCR System according to the instruction of SYBR® Premix DimerEraser™ (Perfect Real Time) (Code: RR091A) kit. The data were analyzed and compared with ΔΔCt, to obtain relative expression levels of genes in each template. The calculation formula was: Relative expression level = $2^{-\Delta\Delta Ct}=2^{-[(CtProcess-CtReference)-(CtControl-CtReference)]}$. The average of three replicates was calculated and adopted.

2.6. *Data Analysis*

All data were analysed using SPSS 17.0 statistical software package for windows and are presented as the mean ± SD (n = 3). Data from each treatment were submitted to two-way analysis of variance (ANOVA). When a significant interaction was observed, data were analysed using one-way ANOVA followed by Duncan's multiple range tests to inspect all differences among the dietary treatments. *P*-values equal to or less than 0.05 indicated statistical significance.

3. Results and Analysis

3.1. *Effects of Dietary Carbohydrate to Protein Ratios and Feeding Frequency on the Amylase Activities in Intestine of Common Carp*

Amylase activities in foregut and hindgut of fish were not affected by either dietary carbohydrate to protein ratios or feeding frequency (Table 2, $P>0.05$). Amylase activities in midgut of fish maintained at a feed frequency of 4 meal/d were significantly lower than those of fish at 2 meal/d ($P<0.05$). Besides, amylase activities in midgut of fish were significantly affected by diet composition with the lowest observed in fish fed with diet C20P28 ($P<0.05$).

Table 2. Effects of dietary carbohydrate to protein ratios and feeding frequency on the amylase activities in intestine of common carp (μg/ml)

groups		Foegut	Midgut	Hindgut
C5% /P32%/F4		100.82±31.90	101.36±50.90	68.95±23.64
C10% /P30%/F4		124.28±40.70	130.16±35.90	83.66±63.16
C20% /P28%/F4		119.72±57.60	60.54±21.91	75.97±48.57
C5% /P32%/F2		128.83±48.87	149.17±16.57	121.72±16.59
C10% /P30%/F2		157.81±64.91	175.15±26.79	124.79±26.15
C20% /P28%/F2		148.85±39.94	90.63±49.44	107.11±49.70
Main effects				
C/P	C5% /P32	115.88	139.66^x	105.34
	C10% /P30%	146.26	138.25^x	104.26
	C20% /P28%	134.29	63.08^y	91.54
Frequency	4	114.94	89.02^p	86.19
	2	145.16	138.32^q	127.87
Two-way ANOVA				
Frequency		0.84	0.02	0.06
C/P		0.87	0.01	0.90
Interaction		0.34	0.46	0.87

3.2. *Effects of Dietary Carbohydrate to Protein Ratios and Feeding Frequency on the Activities of Carbohydrate Metabolic Enzymes in the Hepatopancreas of Common Carp*

As shown in Table 3, dietary C/P ratio had significant influence on the activity of PFK and G6PD of the hepatopancreas of common carps ($P < 0.05$), and had no significant influence on PK activity ($P > 0.05$); feeding frequency had no significant influence on the activity of G6PD and PK of the hepatopancreas of common carps ($P > 0.05$). Interaction between dietary C/P ratio and feeding frequency had no significant influence on key enzyme activities of glucose decomposition approach ($P > 0.05$). Under each dietary C/P ratio, the activities of PFK, G6PD and PK of hepatopancreas increased when the feeding frequency increased from twice a day to four times a day, but the difference was not significant ($P > 0.05$). In the two feeding groups, the activities of PFK, G6PD and PK of hepatopancreas decreased with the increase in dietary C/P ratio. The activities of PFK and G6PD of the C5% / P32% group were significantly higher than that of the C20% / P28% group ($P < 0.05$). In addition, there was significant difference in the activities of G6PD between the C10% / P30% / F2 group and the C20% / P28% / F2 group ($P < 0.05$). PK activity did not decrease significantly with the increase of dietary C/P ratio ($P > 0.05$).

Dietary C/P ratio and feeding frequency had significant influence on PEPCK and GS activities in hepatopancreas of common carps ($P < 0.05$), but did not have significant influence on the interaction between PEPCK and GS activities therein ($P > 0.05$). Under each dietary C/P ratio, the activities of PEPCK and GS of hepatopancreas increased significantly when the feeding frequency increased from twice a day to four times a day ($P < 0.05$). In the C10% /P30% group, PEPCK and GS activities of the F4 group were significantly higher than those of the F2 group ($P < 0.05$); in the C20% /P28% group, PEPCK activities of the F4 group were significantly higher than those of the F2 group ($P < 0.05$). In the two feeding groups, PEPCK and GS activities in hepatopancreas decreased significantly with the increase of dietary C/P ratio ($P < 0.05$). There was significant difference in PEPCK activities between the C5% /P32% group and the C20% /P28% group ($P < 0.05$); the difference in GS activities among three groups with different C/P ratios was significant ($P < 0.05$).

Table 3. Effects of dietary carbohydrate to protein ratios and feeding frequency on key enzyme activities of carbohydrate metabolic enzymes in the hepatopancreas of common carp ($U \cdot g^{-1}$)

groups		PFK	PK	G6PD	PEPCK	GS
C5% /P32%/F4		370.17±25.92	365.85±19.81	10.81±0.82	43.31±1.30	66.79±0.30
C10% /P30%/F4		304.07±17.78	329.67±21.90	8.36±0.23	34.26±1.09	59.58±0.86
C20% /P28%/F4		213.30±22.80	308.74±29.62	7.66±0.28	26.70±7.88	23.47±3.57
C5% /P32%/F2		309.56±17.93	317.61±15.65	10.45±0.37	38.78±4.64	64.79±4.64
C10% /P30%/F2		245.86±21.89	296.80±22.37	8.84±0.40	22.44±2.81	46.83±3.24
C20% /P28%/F2		208.37±21.15	290.66±21.79	7.17±0.28	17.83±1.26	19.00±2.04
			Main effects			
C/P	C5% /P32	339.86^{x}	10.63^{x}	331.32	41.05^{x}	65.79^{x}
	C10% /P30%	274.97^{y}	8.60^{y}	323.64	28.35^{y}	53.20^{y}
	C20% /P28%	210.84^{z}	7.42^{z}	299.70	22.26^{y}	21.24^{z}
Frequency	4	295.84	8.94	334.75	34.76^{p}	49.95^{p}
	2	254.59	8.82	301.69	26.35^{q}	43.54^{q}
			Two-way ANOVA			
Frequency		0.04	0.12	0.75	0.01	0.00
C/P		0.00	0.42	0.00	0.00	0.04
Interaction		0.37	0.46	0.52	0.50	0.07

3.3. Effects of Dietary Carbohydrate to Protein Ratios and Feeding Frequency on Hormones and Regulatory Factors Related To Serum Glycometabolism of Common Carp

As shown in Table 4, dietary C/P ratio and feeding frequency had significant influence on INS, GC and GH concentrations in serum ($P < 0.05$), and interaction between dietary C/P ratio and feeding frequency had significant influence on GH concentration ($P < 0.05$). Under each dietary C/P ratio, with the increase of feeding frequency, INS concentration in serum increased significantly ($P < 0.05$), but GC concentration and growth hormone did not increase significantly ($P > 0.05$). In the C20% /P28% group, INS concentration of the F4 group was significantly higher than that of the F2 group ($P < 0.05$); in the C10% /P30% group, GH concentration of the F4 group was significantly lower than that of the F2 group ($P < 0.05$). In the two feeding groups, serum INS contents increased significantly with the increase of dietary C/P ratio ($P < 0.05$), while GC and GH contents decreased significantly (P

< 0.05).

Dietary C/P ratio and feeding frequency had significant influence on serum GLU, ISR, and IGF-1 concentrations ($P < 0.05$), but interaction between dietary C/P ratio and feeding frequency did not have significant influence on above indexes ($P > 0.05$). Under each dietary C/P ratio, GLU concentration decreased significantly ($P < 0.05$) when the feeding frequency increased from twice a day to four times a day, while IST and IGF-1 concentrations increased significantly ($P < 0.05$). In the C5% /P32% group, GLU and IGF-1 concentrations of the F4 group were significantly lower than those of the F2 group ($P < 0.05$); in the C10% /P30% group, ISR and IGF-1 concentrations of the F4 group were significantly higher than those of the F2 group ($P < 0.05$); in the C20% /P28% group, GLU concentration of the F4 group decreased significantly ($P < 0.05$). In the two feeding groups, GLU concentrations increased significantly with the increase of dietary C/P ratio, and that of the C10% /P30% group and C20% /P28% group was significantly higher than that of the C5% /P32% group ($P < 0.05$); ISR concentration of the C10% /P30% group was the highest, and was significantly higher than that of the other two groups ($P < 0.05$); IGF-1 concentration of the C20% /P28% group was the lowest, and was significantly lower than that of the C5% /P32% group ($P < 0.05$).

Table 4. Effects of dietary carbohydrate to protein ratios and feeding frequency on hormone contents related to serum glycometabolism of common carp

groups		INS/ ($mmo \cdot L^{-1}$)	GC/ ($ng \cdot L^{-1}$)	GH/ ($\mu g \cdot L^{-1}$)	GLU/ ($mmol \cdot L^{-1}$)	ISR/ ($nmol \cdot L^{-1}$)	IGF-1/ ($\mu g \cdot L^{-1}$)
C5% /P32%/F4		24.70±0.35	611.20±51.27	$10.86±0.72^{d}$	6.25±0.29	86.58±2.51	27.28±1.30
C10% /P30%/F4		27.80±0.60	702.55±9.44	$14.73±0.60^{c}$	8.92±0.97	116.95±13.21	23.91±1.18
C20% /P28%/F4		31.90±1.30	728.74±9.23	$17.98±0.61^{b}$	8.79±0.84	81.87±2.59	17.39±0.31
C5% /P32%/F2		23.10±0.71	634.49±12.83	$13.33±1.23^{cd}$	8.08±0.11	75.49±2.59	23.19±0.89
C10% /P30%/F2		25.90±0.40	757.12±12.28	$17.81±0.09^{b}$	9.17±0.63	91.98±3.56	19.17±1.39
C20% /P28%/F2		28.40±0.40	777.37±7.43	$26.03±1.32^{a}$	10.66±0.36	69.35±2.98	17.32±1.53
				Main effects			
C/P	C5% /P32	23.91^{y}	12.10^{z}	622.85^{y}	7.16^{y}	81.03^{y}	25.24^{x}
	C10% /P30%	26.85^{y}	16.27^{y}	729.84^{x}	9.04^{x}	104.46^{x}	21.54^{y}
	C20% /P28%	30.15^{x}	22.01^{x}	753.09^{x}	9.73^{x}	75.61^{y}	17.36^{z}
Frequency	4	28.16^{p}	14.53^{q}	680.83^{p}	7.99^{q}	95.13^{p}	22.86^{q}
	2	25.75^{q}	19.05^{p}	722.99^{q}	9.31^{P}	78.94^{q}	19.89^{p}
				Two-way ANOVA			
Frequency		0.00	0.00	0.00	0.00	0.00	0.00
C/P		0.00	0.04	0.00	0.01	0.01	0.01
Interaction		0.38	0.76	0.01	0.25	0.47	0.09

3.4. Effects of Dietary Carbohydrate to Protein Ratios and Feeding Frequency on Glycometabolism Related Gene Expression Levels of Common Carp

As shown in Table 5, dietary C/P ratio and feeding frequency had significant influence on IR, GLUT2 and HK mRNA expression levels in hepatopancreas of common carp ($P < 0.05$). In addition, interaction between dietary C/P ratio and feeding frequency had significant influence on IR mRNA expression level in hepatopancreas ($P < 0.05$). IR, GLUT2 and HK mRNA expression levels in hepatopancreas increased significantly when the feeding frequency increased from twice a day to four times a day ($P < 0.05$). With the increase of dietary C/P ratio, IR, GLUT2 and HK mRNA expression levels in hepatopancreas of the C10% /P30% group were the highest and significantly higher than those of the other two groups ($P < 0.05$).

Dietary C/P ratio and feeding frequency had significant influence on the GLUT2 and SGLT mRNA expression levels in intestine ($P < 0.05$), but they did not had significant interactive effect on the GLUT2 and SGLT1 mRNA expression levels ($P > 0.05$). With the increase of feeding frequency, the GLUT2 and SGLT1 mRNA expression levels in intestine increased significantly ($P < 0.05$). In the C10% /P30% group, the GLUT2 and SGLT1 mRNA expression levels in the intestine of the F4 group were significantly higher than those of the F2 group ($P < 0.05$). With the increase of dietary C/P ratio, the GLUT2 and SGLT1 mRNA expression levels in intestine increased first and then decreased, and those of the C10% /P30% group were the highest and significantly higher than those of the other two groups ($P < 0.05$).

Table 5. Effects of dietary carbohydrate to protein ratios and feeding frequency on glycometabolism related gene expression levels of common carp

Groups		Hepatopancreas			Intestine	
		IR	GLUT2	HK	GLUT2	SGLT1
C5% /P32%/F4		1.00±0.00^{b}	1.00±0.00	1.00±0.00	1.00±0.00	1.00±0.00
C10% /P30%/F4		1.80±0.34^{a}	1.81±0.26	1.39±0.19	2.37±0.59	4.16±1.58
C20% /P28%/F4		0.54±0.10^{bc}	0.97±0.08	0.84±0.04	0.68±0.08	1.29±0.13
C5% /P32%/F2		0.11±0.00^{c}	0.53±0.14	0.05±0.01	0.13±0.02	0.45±0.05
C10% /P30%/F2		0.42±0.05^{c}	1.05±0.15	0.20±0.02	0.68±0.04	2.06±0.18
C20% /P28%/F2		0.20±0.03^{c}	0.47±0.09	0.08±0.02	0.21±0.08	0.33±0.06
			Main effects			
C/P	C5% /P32	0.56^{y}	0.76^{y}	0.52^{y}	0.57^{y}	0.80^{y}
	C10% /P30%	1.11^{x}	1.43^{x}	0.79^{x}	1.52^{x}	3.11^{x}
	C20% /P28%	0.34^{y}	0.71^{y}	0.46^{y}	0.44^{y}	0.73^{y}
Frequency	4	1.11^{p}	1.26^{p}	1.08^{p}	1.34^{p}	2.15^{p}
	2	0.24^{q}	0.68^{q}	0.11^{q}	0.34^{q}	0.94^{q}
			Two-way ANOVA			
Frequency		0.00	0.00	0.01	0.01	0.00
C/P		0.00	0.00	0.00	0.00	0.02
Interaction		0.03	0.57	0.09	0.12	0.30

4. Discussion

As for fish, 2-hour absorption percentage is key to dietary carbohydrate absorption and utilization[10]. In the experiment, dietary C/P ratio and feeding frequency were improved to increase 2-hour carbohydrate absorption percentage of fish. Activity of digestive enzyme is the primary factor reflecting absorption and utilization of carbohydrate in fish.

The increase of dietary C/P ratio can increase the contacting amount and time between dietary carbohydrate and amylase in intestine. But study in *Silurus meridionalis* Chen[11] and study of *Pseudosciaena crocea* R. [12]found that amylase activity did not increase with carbohydrate level, instead, it has certain conservation. However, in the study, with the increase of C/P ratio, amylase activities in intestine first increased and then decreased. Moreover, amylase in midgut was significantly affected by dietary C/P ratio, and amylase activity of the C10% /P30% group was the highest and significantly higher than that of the C20% /P28% group. This indicated that a too high concentration of carbohydrate in diet inhibits amylase activity to a certain extent, which was similar to the study in *Silurus meridionalis* Chen [11] and study of *Pseudosciaena crocea* R[12].

Feeding frequency increases 2-hour dietary carbohydrate absorption percentage of fish by reasonably controlling feeding speed of fish. However, increase in feeding frequency leads to certain dietary restriction effect, and changes in digestive enzyme affect the utilization of dietary nutrient substances by fish. The study indicated that under each dietary C/P ratio, activities of amylase in foregut, midgut and hindgut of the F4 group decreased. Cui[13] pointed out that increase in feeding frequency led to increase in amylase activity in the liver and activity of digestive enzyme in the intestine of *Acipenser gueldenstaedtii*. Difference in feeding habit of the experimental fish may be the main reason for the above difference.

In the experiment, after 24 hours of fasting, blood glucose levels of common carp showed that: C20% /P28% group > C10% /P30% group > C5% /P32% group, which reflected the digestion and absorption rate of dietary starch by fish to some extent[14]. The key to full use of starch by fish is whether glycometabolism regulation and relevant hormones regulation match changes in blood glucose. The experiment revealed that during regulating key enzymes of glucose synthesis approach, activities of key enzyme of gluconeogenesis - PEPCK and key enzyme of glycogen synthesis - GS

IOP Conf. Series: Materials Science and Engineering **484** (2019) 012011 doi:10.1088/1757-899X/484/1/012011

decreased significantly with the increase in dietary C/P ratio. This revealed that in the premise of maintaining dietary energy constant, appropriate increase in C/P ratio can inhibit gluconeogenesis, reduce the activity of key enzyme of gluconeogenesis, which conformed to findings in studies of common carp[15-16]and *Rachycentron canadum*[17]; during regulating key enzymes of glucose decomposition approach, the activities of key enzymes PFK and PK during glycolysis and the activity of key enzyme G6PD of the pentose phosphate pathway decreased significantly with the increase in dietary C/P ratio. However, Cheng[4] stated that increase in dietary corn starch level promoted the increase in PK activity in hepatopancreas of common carp. Enes[18] reported that the increase in dietary carbohydrate level induced the enhancement of G6PD activity of fish. It is inferred that difference in fish's adaptability to different carbohydrates results in the above differences. However, it is clear that besides glucose decomposition and synthesis approaches, adaptability to hormone regulation also affects the ability of glycometabolism of fish. The experiment found that concentrations of insulin and IGF-1 with the function of reducing blood glucose increased significantly with C/P ratio, which conformed to the results of study by Baños[19]. However, significant increase in concentration of glucagon that can inhibit glucose utilization shows that regulating mechanism of glycometabolism in fish is still not perfect.

Hung and Storebakken[20]reported that under continuous feeding conditions, G6PD activities in the livers of *Oncorhynchus mykiss* of the glucose and maltose groups were higher than that of the F4 group, which indicated that increase in feeding frequency can enhance fish's ability to make use of low-molecular carbohydrate to some extent. The experiment indicated that compared with the F2 group, PFK, PK and G6PD activities and insulin and IGF-1 concentrations in hepatopancreas of the F4 group were significantly higher, while glucagon and growth hormone concentrations were significantly lower. This reveals that proper increase of feeding frequency promotes fish's ability to digest and absorb polysaccharides such as starch. Therefore, after 24 hours of fasting, blood glucose level of the F4 group was lower, which proves that increasing feeding frequency, is conducive to the saving effect of dietary carbohydrate on protein.

Galgarber pointed out that SGLT1 and GLUT2 mRNA expressions in animals were influenced by nutrition supply[21]. The study indicated that amylase activities in foregut, midgut and hindgut of the C10% /P30% group were all the highest. Meanwhile, mRNA expression levels of SGLT1 and GLUT2 in the intestine of the C10% /P30% group were the highest. Hepatopancreas is the key tissue to regulate blood glucose concentration in fish. Glycolysis, gluconeogenesis and pentose phosphate cycle in glycometabolism and insulin secretion are all carried out in hepatopancreas. Therefore, glucose transporter, insulin receptor and HK mRNA expressions in hepatopancreas also determine glycometabolism ability of hepatopancreas. Moreover, the experiment also indicated that when dietary C/P ratio was C5% /P32% - C10% /P30%, GLUT2, IR and HK mRNA expression levels were positively correlated to blood glucose concentration and insulin secretion and increased significantly. Study on Oncorhynchus mykiss[22] also obtained similar findings; however, when C/P ratio was over C10% /P30%, changes in GLUT2, IR and HK mRNA expression levels were opposite to changes in blood glucose concentration and insulin secretion. Changes in above mRNA expression levels were same with the changes in amylase activity in intestine, which reveals that a too high C/P ratio restricts mRNA expression of glucose transporter. Therefore, this may be one reason why fish has the constitution of high glucose intolerance.

5. Conclusion

Above experimental results shows that increase in feeding frequency inhibits amylase activity in the intestine to some extent. However, mRNA expression levels of SGLT1 and GLUT2 in intestine, GLUT2, IR and HK in hepatopancreas, and key enzymes activities of glycolysis increased significantly with feeding frequency. This indicates that increase in feeding frequency inhibits the utilization of nutrient substance in fish, but fish makes full use of limited nutrient substance by regulating factors related to glycometabolism. In this study, diets containing 10% carbohydrate and 30% protein are suitable for common carp and its feeding frequency is fourth daily. It is necessary to further study on the regulatory mechanism by the influence of dietary carbohydrate to protein ratio and feeding frequency.

6. Acknowledgements

This study was supported by the Tianjin Innovative talent training plan for Zhenyan Cheng; the National Natural Science Foundation of China (No. 31402313); the Tianjin Research Program of Application Foundation and Advanced Technology (No. 14JCQNJC15100), the Key Technology R&D Program of Tianjin (No.13ZCZDNC00900), and the Modern Aqua-ecology and Health Aquaculture Innovation Team of Tianjin, Grant/Award Number: grant no. TD13-5089. We would also like to acknowledge Tianjin Tianxiang Aquatic Co. Ltd. for provide the farming equipment and supplements, and LI Jinhui, YIN Shuai, ZHANG Tengxian and LIN Chengli from Tianjin Agriculture University for their assistance with samlpling and index determine.

7. References

[1] Hemer G I, Mommsen T P and Krogdahl Å. Carbohydrates in fish nutrition: effects on growth, glucose metabolism and hepatic enzymes. 2002. *Aquac Nutr*, 8(3):175-194.

[2] Shiau S Y and Lin Y H. Carbohydrate utilization and its protein-sparing effect in diets for grouper (*Epinephelus malabaricus*). 2001. *Animal Science*, 73(1): 299-304.

[3] Zeng L, Lei J L, Ai C X, et al. Protein-sparing effect ofcarbohydrate in diets for juvenile turbot Scophthalmus maximus reared at different salinities. 2015. *Chinese Journal of Oceanology and Limnology*, 33(1): 57-69.

[4] Cheng Z Y, Li J H, Zhang B L, et al. Verification of protein sparing by feeding carbohydrate to common carp *Cyprinus carpio*. 2017.*Chinese Journal of Oceanology and Limnology*, 35(2): 251-257.

[5] Deng D F, Koshio S, Yokoyama S, et al. Effects of feeding rate on growth performance of white sturgeon(*Acipenser transmontanus*) larvae. 2003.*Aquaculture*, 217(1/4): 589-598.

[6] Lee S, Haller LY , Fangue NA , Fadel JG and Hung SSO . Effects of feeding rate on growth performance and nutrient partitioning of young-of-the-year white sturgeon (*Acipenser transmontanus*). 2016.*Aquac Nutr*, 22 (2) :400-409.

[7] Robinson EH , Li MH, Manning BB, Mischke CC and Bosworth BG . Effects of Dietary Protein and Feeding Rate on Channel Catfish Ictalurus punctatus Production, Composition of Gain, Processing Yield, and Water Quality. 2004 .*Journal of the World Aquaculture Society* , 35 (4) :468-477.

[8] Li J N, Xu Q Y, Wang C A et al. Effects of different dietary carbohydrates and carbohydrate levels on GH/IGF-I mRNA expression and the fish body composition of juvenile mirrorcarp (*Cyprinus carpio*). 2015. *Journal of Shanghai Ocean University* ,24(4):489-495.

[9] Deng Z W. Effects of feeding frequency on feeding and growth of juvenile FFRC. 2015. *Journal of Fujian Fisheries*, 37 (1):68-72.

[10] Hung S S O, Fynn-Aikins K F, Lutes P B, et al. Ability of juvenile white sturgeon(*Acipenser transmontanus*)to utilize different carbohydrate source. 1989.*Nutrit*,119:727-733.

[11] Gao M, Luo Y P, Cao Z D. Effect of Dietary Carbohydrate on Digestive Enzyme Activities in Southern Catfish (*Silurus meridionalis* Chen) Juveniles. 2006.*Journal of Southwest China Normal University* (Natural Science) 31(2):120-123.

[12] Wang M Q, Huang W W, Zhou P P et al. Effects of dietary protein and wheat starch levels on growthPerformance, hepatic glycolysis and gluconeogenic keyn enzymes activities in large yellow croaker (*Larimichthys crocea* Richardson). 2015. *Journal of Fisheries of China* ,39(11): 1691-1701.

[13] Cui C. Effects of light, dietary carbohydrate and feeding strategy on growth of Russian sturgeon. 2014. *East China Normal University*, Shanghai

[14] Deng D F, Refstie S, Hung S S O. Glycemic and glycosuric responses in white sturgeon *Acipenser transmontanus* after oral administration of simple and complex carbohydrates. 2001, *Aquaculture*, 199(1-2): 107-117.

[15] Fan Z, LI JH, Wang AQ et al. Effects of dietary cassava starch supplemental level on growth performance, digestive ability and glycometabolism of common carp. 2016. *Chinese Journal of Animal Nutrition*,28(12): 4044-4053.

[16] Fan Z, Qiao XT, Sun JH, et al. Effect of carbohydrate on growth performance, digestion ability of common carp (*cyprinus carpio*) at different feeding frequencies. 2018. *Advances in Applied Biotechnology*.
[17] Cui XJ. Study on mechanism of carbohydrate utilization for juvenile cobia, *Rachycentron canadum*. 2010.Zhen Jiang, Guangdong Ocean University,26.
[18] Enes P, Panserat S, Kaushik S, et al. Growth performance and metabolic utilization of diets with native and waxy maize starch by gilthead seabream (*Sparus aurata*) juveniles. 2008.*Aquaculture*, 274(1): 101-108.
[19] Banos N, Baro J, Castejon C, et al. Influence of high carbohydrate enriched diets on plasma insulin levels and insulin and IGF-1 receptors in trout. 1998.*Regul. Peptides*, 77(1-3): 55-62.
[20] Hung S S O and Strorebakken T. Carbohydrate utilization by rainbow trout affected by feeding strategy. 1994.*Nutr,* 124(2): 223-230.
[21] Galgarber O. Partial sequence and expression of the gene for and activity of the sodium glucose transporter in the small intestine of fed, starved and refed chickens. 2000.*Nutr,* 130(9): 2174-2179.
[22] Hall J R, Short C E and Druedzic W R. Sequence of Atlantic cod (*Gadus morhua*) GLUT4, GLUT2 and GPDH: Developmental stage expression, tissue expression and relationship to starvation-induced changes in blood glucose. 2006.*Exp.Bio.*, 209(Pt 22): 4490-4502.

IOP Conf. Series: Materials Science and Engineering **484** (2019) 012012 doi:10.1088/1757-899X/484/1/012012

Experiment and Mechanism Analysis of Selective Non-Catalytic Reduction with the Compound Reducing Agents in a Lab-scale CFB Reactor

Jiancong Dong[1], Degui Bi[1], Tuo Zhou[2], Juan Yu[1], Xiaojiang Wu[3] and Zhongxiao Zhang[1,*]

[1] School of Mechanical Engineering, Shanghai Jiao Tong University, 800 Dongchuan Road, Minhang District, Shanghai, China
[2] Institute of Engineering Thermophysics, Chinese Academy of Sciences, 11 Beisihuanxi Road, Haidian District, Beijing, China
[3] Shanghai Boiler Works Co., Ltd., 250 Huaning Road, Minhang District, Shanghai, China
E-mail: zhzhx222@163.com

Abstract. The conventional SNCR technology with reducing agent ammonia has been utilized in power industry extensively. In this paper, a new denitrification method with the compound reducing agents contain ammonia and reducing gas was discussed. The SNCR experiment with H_2 and CH_4 as the additives was carried out in a lab-scale CFB reactor which imitated flue gas environment in CFB boilers, meanwhile the mechanism modelling including the reaction of gas additives was adopted to simulate the experimental process. The validations of mechanism modelling and parameter optimizations were discussed by the comparisons of experimental and computational results. The compound reducing agents with H_2 and CH_4 additives can expand the adaptive temperature window of denitrification reaction and reduce the optimal reaction temperature. Hence the NOx removal efficiency of conventional SNCR can be improved significantly, and the effect of H_2 additive is better than that of CH_4 additive. The optimal reaction temperature in simulation is about 800°C with the H_2 concentration of 300 ppm, while 760°C under the concentration of 900 ppm. The minimal NOx residual concentration is estimated to approach 75 mg/m^3 at 800°C by using the H_2 compound reducing agent.

1. Introduction

In China, the nitrogen oxides (NOx) emission has been one of the major environmental concerns nowadays and the new national standard of air pollution prevention (GB13223-2011) has been already released and implemented. Thus, the control of NOx emission has become a priority aspect of environmental protection. Among various technologies, selective non-catalytic reduction (SNCR) is known to be an effective and feasible approach, especially in Circulating fluid bed (CFB) boilers, because of its low investment cost and very considerable denitrification efficiency.

Since the Thermal De-NOx was firstly patented by Lyon in 1975, various studies of SNCR technology have been carried on by researchers worldwide. One of the most important characteristics of SNCR is the narrow operating temperature window, which is usually located in the range of 1120-1380 K [1-7]. Nevertheless, this operating temperature range could be expanded or lowered by some gas additives with the common reducing agent ammonia [8-10]. It is essential to explore the influence of these compound reducing agents, ammonia with various reducing gas, on the conventional SNCR process in the CFB flue gas environment and its optimum mechanism of SNCR conditions.

For that reason, a lab-scale electricity-heated tubular flow reactor was used to simulate the CFB boiler and the compound reducing agents contain ammonia and gas additives such as H_2 or CH_4 were injected into this system for SNCR experiments. Moreover, the reaction mechanism was analysed by a business chemical dynamics software. In the comparison of simulation results and experimental data, the influence of H_2 and CH_4 addition on SNCR process in CFB boilers was given. As a result, the denitrification reaction characteristics of the conventional SNCR and the SNCR with the compound reducing agents could be investigated and the differences between H_2 and CH_4 on the reaction were analysed.

2. Experiment and Mechanism

2.1. Experiment

The experiments was performed in the experimental device as shown in figure 1, which was consists of serval parts such as pre-heater, electricity-heated tubular flow reactor, cyclone separator, simulated flue gas system, reductant injection system and gas monitoring system. The simulated flue gas was premixed by nitric oxide, nitrogen, oxygen and carbon dioxide from the gas cylinders and sequentially flowed through the pre-heater, tubular flow reactor and cyclone. The flow of flue gas was $0.5m^3/h$ and its component was set to 268 mg/m^3 NOx, 5% O_2 and 15% CO_2. The electricity-heated tubular flow reactor was made by a corundum tube with the length of 1.2m and inner diameter of 6 cm and the thermocouple was installed at the outlet of reactor to measure and adjust the flue gas temperature inside the reactor within the range of 700-1000°C. The reducing agents were delivered by the constant flow pump BT100-02 and injected at the inlet of cyclone, where the flue gas contains NOx was mixed sufficiently with these reductants due to the gas vortex inside the cyclone. At the outlet of cyclone, the compound of flue gas was measured by the Testo 340 analyser, which has an accuracy of 1 ppm or 0.1% respectively. Besides, various amounts of H_2 or CH_4 were added into the compound reducing agent to explore the effect of H_2 or CH_4 additives on denitrification efficiency and optimal reaction temperature window.

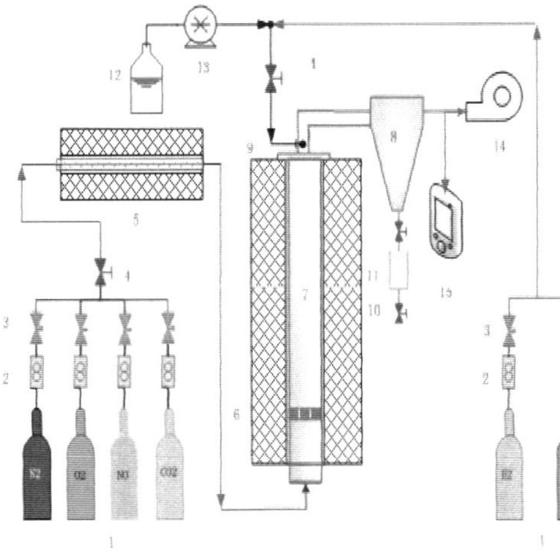

Figure 1. Schematic diagram of the experimental system

Note: 1-simulated flue gas; 2- flowmeter; 3- needle valve; 4-control valve; 5-preheater; 6-distribution plate; 7-electricity-heated tubular reactor; 8-cyclone; 9-injection;10- butterfly valve; 11- container; 12-ammonia; 13-constant flow pump; 14- fan; 15-analyzer.

2.2. Mechanism

In past decades, various reaction mechanisms and models have been implemented by researchers worldwide to investigate the SNCR process. In this work, the simulation process of SNCR was performed by a business chemical dynamics software and the Perfect Stirred Reactor (PSR) model was

IOP Conf. Series: Materials Science and Engineering **484** (2019) 012012 doi:10.1088/1757-899X/484/1/012012

adopted in the simulation, in which the adiabatic closed system and the sufficient mixture of various reactants were assumed, and heating process of gas was also ignored.

Among the reaction mechanisms described in literatures [11-14], the racial reaction mechanism of SNCR with gas additives introduced in the literature [13] was selected to simulate the SNCR process with the compound reducing agents in this study. The racial reaction mechanism includes 461 reversible reactions and 66 chemical species, including the oxidation mechanism of H_2 and CH_4 [14], and the interaction mechanism between these gas additives and nitrogen oxides.

The calculation condition of reaction simulation was set as same to the experiment. The flue gas in this simulation containing 5% O_2, 268 mg/m^3 NO (i.e. 200 ppm) and 15% CO_2 was balanced by N_2. The compound reductants were composed by 300 or 400 ppm NH_3 (i.e. NSR=1.5 or 2), 300 or 900 ppm gas additives (CH_4 or H_2). Residence time of flue gas in the reactor was set to 0.6 s and the pressure of CFB reactor was set to 1 atm. The comparison of results in mechanism simulations and experimental data was given in this paper. The characteristics of denitrification reaction in the conventional SNCR and the SNCR with different compound reductants were also analysed.

3. Results and Discussion

3.1. Validation and Parameter Optimization

The reaction temperature, flue gas residence time and normalized stoichiometric ratio (NSR) are three important indexes of SNCR process, and the optimal values for these parameters were determined according to the experimental data and calculation results. As shown in figure 2, the calculation results agree well with the experimental data, which illustrate the mechanism modelling chosen in this study is suitable to simulate the SNCR process. It is also explicit that the denitrification efficiency in the simulation is higher than the experimental data, because the ideal mixture of NH_3 and flue gas was assumed in the theoretical calculation, while it cannot occur in the actual experiment.

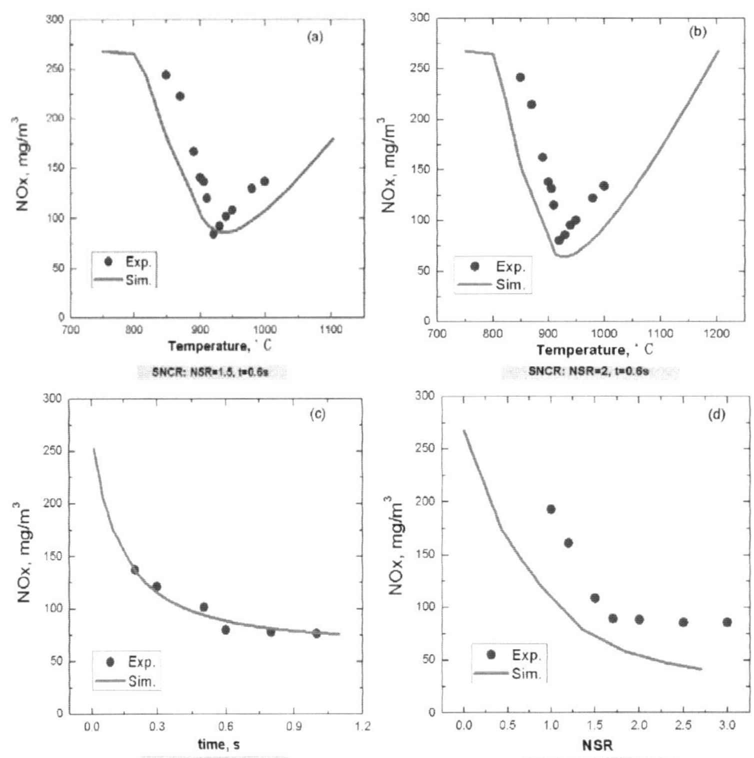

Figure 2. Comparisons of experiment data and simulation results in conventional SNCR without compound reducing agents

In the comparisons of figure 2(a) and 2(b), the optimal reaction temperature in SNCR simulation is close to the optimal temperature of 920°C observed in the experiment, and the optimal temperature window in experiment is narrower than that in simulations. As is well known the flue gas temperature in the furnace of conventional CFB Boilers is almost in the range of 850-900°C. Obviously, the optimal temperature in this study is slightly higher than the actual operational temperature of CFB boilers. However, the initial concentration of NOx emission decreases with the decreasing flue gas temperature under the CFB combustion conditions [15]. Figure 2(c) shows the relationship between the NOx residual concentration and the flue gas residence time while the reaction temperature and NSR was 920°C and 1.5 respectively. The efficiency of denitrification can reaches 50% when the residence time rises to 0.3s, and the residual concentration of NOx approaches the minimum value 70 mg/m^3 once the residence time exceeds 0.6s. Meanwhile, the simulation results agree well with the experimental data in this situation. Figure 2(d) illustrates the relation of the NOx residual concentration and NSR when the reaction temperature and residence time was 920°C and 0.6s respectively. The denitrification efficiency increases with the ascending NSR, but the growth rate of denitrification efficiency decreases significantly when the NSR exceeds 1.5. The denitrification efficiency in the experiment is lower than that in the simulation, which is attributed to the ammonia solution used in experiments and the amount of ammonia involved in actual reaction is lower than that in the theoretical calculation.

3.2. Effect of H_2 on the SNCR

Figure 3 indicates the relationship between NOx residual concentration and the reaction temperature in different concentrations of H_2 additive, and the variation trend in simulation is resemble to that in the experiment. Both in the condition of 300ppm and 900 ppm H_2 addition, the predictive results of NOx residual concentration is slightly lower than the experimental data. For example, in simulation the minimal NOx residual concentration reaches 75 mg/m^3 at the optimal temperature of 800°C, but the concentration only get 95 mg/m^3 at the actual optimal temperature of 870°C in the experiment under the H_2 concentration of 300 ppm. It is obvious in the Figure 3 that, the adaptive reaction temperature window shifts towards the lower temperature section with the increase of H_2 concentration. The optimal reaction temperature in simulation is about 800°C under the H_2 concentration of 300 ppm, while it is 760°C under the H_2 concentration of 900 ppm, and these two values are less than the temperature without the H_2 addition, 920°C. Moreover, the experimental data of NOx residual concentration in the situation of 300ppm H_2 addition deviate from the calculation results under the optimal temperature (T<870°C), while the similar deviation appears when the reaction temperature exceed the optimal value (T>750°C), as shown in figure 3(b) and (c).

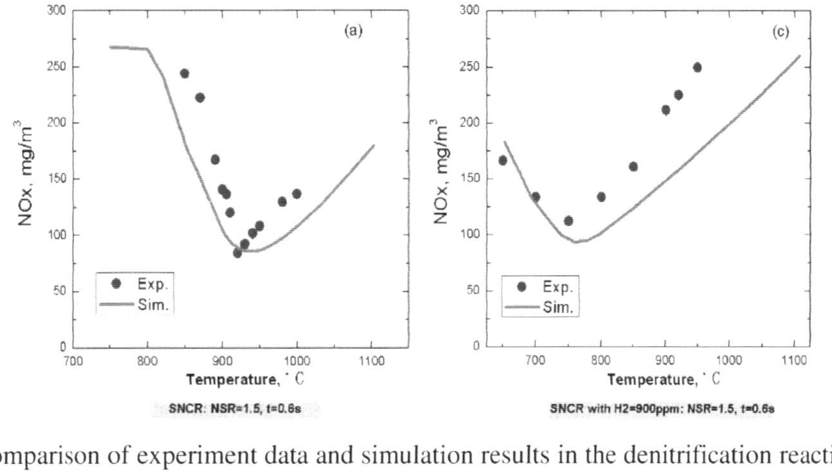

Figure 3. Comparison of experiment data and simulation results in the denitrification reaction with H2 additive

3.3. Effect of CH_4 on the SNCR

Figure 4 demonstrates the relation between NOx residual concentration and the reaction temperature in SNCR process with different amount of CH_4 addition. The impact trend of CH_4 additive on SNCR in model simulation is similar with that in experiment. In the simulation, the optimal reaction temperature with CH_4 concentration of 300 ppm is about 820°C, while 800°C for CH_4 concentration of 900 ppm. In the situation of 300 ppm CH_4 addition, as shown in Figure 4(b), the minimal NOx residual concentration is estimated to approach 80 mg/m^3 at the optimal reaction temperature of 820°C, while the experimental value reaches 105 mg/m^3 at 840°C. In general, the simulation results are better than the experimental data due to the ideal gas mixture assumed in the calculations. This phenomenon is mainly caused by the oxidation reaction of CH_4, which also generate OH and promote the reduction reaction of NO_X.

Moreover, in the comparisons of H_2 and CH_4 addition, it can be easily found that the effects of H_2 and CH_4 based compound reducing agents on the conventional SNCR process are different. H_2 additive is more effective to reduce the optimal reaction temperature of SNCR, and the denitrification efficiency of CH_4 additive is lower than that of H_2 additive. It is mainly caused by the differences of generating rate and reaction route of OH in these two conditions.

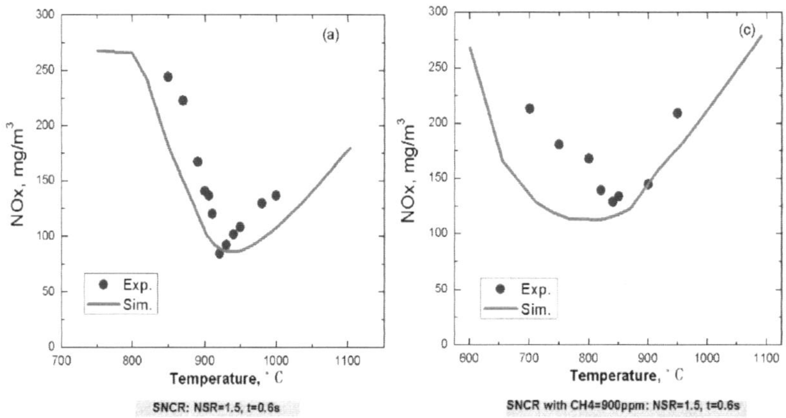

Figure 4. Comparison of experiment data and simulation results in the denitrification reaction with CH4 additive

4. Conclusions

In order to explore the denitrification reaction characteristics of conventional SNCR and SNCR with the compound reducing agents in the CFB flue gas environment, the flue gas SNCR experiment with H_2 and CH_4 as the additives was carried out in a lab-scale CFB reactor which imitated flue gas environment in CFB boilers, meanwhile the mechanism modelling including the elementary reaction of gas additives was adopted to simulate the experimental process. The comparisons of results in mechanism simulations and experimental data were discussed in this paper. By the comparisons, the mechanism model is agreeable to simulate the actual experiment. The compound reducing agents with H_2 and CH_4 additives can expand the adaptive temperature window of denitrification reaction and reduce the optimal reaction temperature. Hence the NOx removal efficiency of conventional SNCR can be improved significantly, and the effect of H_2 additive is better than that of CH_4 additive. In the simulation, the optimal reaction temperature with CH_4 concentration of 300 ppm is about 820°C, while 800°C for CH_4 concentration of 900 ppm. The optimal reaction temperature in simulation is about 800°C with the H_2 concentration of 300 ppm, while 760°C under the H_2 concentration of 900 ppm. These values are less than the temperature without the compound reducing agents. The minimal NOx residual concentration with H_2 compound reducing agent reaches 75 mg/m^3 at the optimal temperature of 800°C and the value approaches 80 mg/m^3 at the optimal reaction temperature of 820°C with CH_4 compound reducing agent.

5. Acknowledgments

The authors greatly acknowledge the financial support of this work by the National Key R & D Program of China (Grant No.: 2016YFB0600202).

6. References

[1] Lee G W, Shon B H, Yoo J G, Jung J H and Oh K J. The influence of mixing between NH_3 and NO for a De-NOx reaction in the SNCR process 2008, *J. Ind. Eng. Chem.*, **14**(4):457-467.
[2] Han Y, Gao H P, Wang H T, Li L, Wang R X and Song X G. Application of SNCR flue gas denitration technology in 330 MW CFB boiler 2013, *Clean Coal Technology*, **19**(6):85-88.
[3] Hui C, Chen D and Na W. The State-of-art and development of moderate temperature based flue gas DeNOx technology 2013, *Proceedings of the CSEE*, **33**(20):17-27.
[4] Lu Z M and Lu J D. Kinetic modeling of the NOx reduction by reburning with ammonia-containing coal gas 2011, *Proceedings of the CSEE*, **31**(26):59-64.
[5] Guo Z X, Xuan W B and Lai M G. Research and application of denitration technology in mid and mini type industrial boiler 2012, *Clean Coal Technology*, **18**(6):88-90.
[6] Loeffler G, Wartha C, Winter A and Hofbauer H. Study on NO and N_2O formation and destruction mechanisms in a laboratory-scale fluidized bed 2002, *Energy Fuels*, **16**(5):1024-1032.
[7] Lyon R.K. Thermal De-NOx controlling nitrogen oxides emissions by a non-catalytic process 1987, *Environ. Sci. Technol.*, **21**(3):231-236.
[8] Bae S W, Roh S A and Kim S D. NO removal by reducing agents and additives in the selective non-catalytic reduction (SNCR) process 2006, *Chemosphere*, **65**(1):170-175.
[9] Yang M, Zhang Z, Yu J and Li D. Denitrification mechanism analysis and experiment of selective non-catalytic reduction of flue gas of circulating fluidized bed 2014, *Combustion Science and Technology*, **20**(2):101-105.
[10] Lei J, Zhuo Y, Chen C and Xu X. Effect of methane and temperature on NO removal using by the selective non-catalytic reduction 2008, *Journal of Tsinghua University*, **48**(8):1315-1320.
[11] Rota R. Chemical kinetic analysis of thermal De-NOx process at high reactant concentration 2001, *Chem. Eng. Technol.*, **24**(5):539-541.
[12] Miller J and Glarborg P. Modelling the thermal De-NOx process: closing in on a final solution 1999, *Int. J. Chem. Kinet.*, **31**(11):757-765.
[13] Cao Q X. Effects of gaseous additives for selective non-catalytic reduction of NOx 2009, Ph.D. thesis, Harbin Institute of Technology, Harbin.
[14] Lu Z. Study on reaction mechanism and mixing characteristic of SNCR technique 2006, Ph.D. thesis, Zhejiang University, Hangzhou.
[15] Li J, Yang H, Wu Y, Lv J and Yue G. Effects of the updated national emission regulation in China on circulating fluidized bed boilers and the solutions to meet them 2013, *Environ. Sci. Technol.*, **47**(12):6681-6687.

IOP Conf. Series: Materials Science and Engineering **484** (2019) 012013 doi:10.1088/1757-899X/484/1/012013

Study on Preparation of Activated Carbon from Sludge

Mingming Du, Tao Yu, Feifei Wang and Chengtun Qu *

Shaanxi Oil and Gas Pollution Control and Reservoir Protection Key Laboratory, Xi'an Shiyou University, 710065, Xi'an

E-mail: xianquct@163.com

Abstract. Sludge, a carbon-rich material and favourable precursor of activated carbon, has stimulated new methods for the production of activated carbons for water gas treatment. Herein, various influence factors, concluding sludge resource, pyrolysis process and the type of activators, of the preparation of sludge-based activated carbon have been compiled. Moreover, the mechanism of chemical and physical activation was analyzed. Finally, Then the main research directions of preparation of activated carbons from pyrolytic sludge also were proposed.

1. Introduction

Sludge is a byproduct during the process of sewage treatment. It is difficult to handle because of its complex constituent components. Organics is an important component of sludge. In order to utilize carbon in sludge reasonably, researchers convert sludge into activated carbon with adsorption. Because the raw material of sludge-based activated carbon is cheap and easy to obtain, and its adsorption ability to some pollutants is better than that of commercial activated carbon, the application of sludge-based activated carbon prospect is appreciable [1]. Preparation of activated carbon from sludge can also achieve the goal of waste utilization. In this paper, the preparation of activated carbon from sludge is reviewed, and the development prospect of activated carbon from sludge is pointed out. The main way to prepare activated carbon from sewage sludge is pyrolysis or calcination. The adsorption capacity of activated carbon is mainly affected by factors such as sludge source, pyrolysis process and activating agent [2].

2. Influence Factors of the Preparation of Sludge-based Activated Carbon

2.1. Effect of Sludge Source

Organics content of sludge varies from sources of sludge. Organics content directly determines the surface structure and adsorption performance of sludge-based activated carbon. There are many kinds of organics in oily sludge produced by the petroleum industry, which indicates that oily sludge is a good precursor of activated carbon. The feasibility of preparing activated carbon from oily sludge was studied by Mohammad. The results showed that the activated carbon prepared from oily sludge had better adsorbability to Cd^{2+} than commercial activated carbon [3]. In addition, Deng et al.[4] had also successfully prepared activated carbon with a high surface area of more than $2000m^2/g$ from oily sludge, which illustrates that the preparation of activated carbon from oily sludge has a promising future.

At present, most of researches are to prepare activated carbon from concentrated municipal sludge. Some studies indicate that the activated carbon prepared by concentrated municipal sludge has higher specific surface area and iodine value. However, the activated carbon synthesized by dewatering sludge will be more capable of adsorbing within shorter adsorption time [5]. Pan et al. [6] prepared sludge-based activated carbon by digested sludge, chemically precipitated sludge and mixed sludge. The experiments adsorbed organics by the three activated carbon and commercial activated carbon from aqueous media showed that the activated carbon prepared by mixed sludge and pure chemical sludge had the strongest adsorption capacity for organics. In addition, many kinds of activated carbon are also synthesized by sludge produced during the water treatment process of leather making, paper making and textile industry. Kantarliet et al. [7] found that the equilibrium adsorption capacity to Cr^{6+}, methylene blue and phenol of activated carbon prepared by tannery sludge is larger than that of commercial activated carbon, which indicates that it is feasible to prepare activated carbon by sludge produced during the water treatment process of leather making. Papermaking sludge contains a large amount of cellulose and other organics, which is a good precursor for activated carbon. Vania Calisto et al.[8] used the activated carbon prepared by sludge produced during the water treatment process of papermaking to adsorb citalopram in aqueous solution. The author thinks that mixed sludge should be developed for the preparation of activated carbon in the future. We should also explore the application fields of activated carbons prepared from different sources of sludge.

2.2. *Effect of Pyrolysis Process*

Pyrolysis is a necessary process for the preparation of carbon. The adsorption properties of activated carbon prepared by different pyrolysis methods are quite different. Because the adsorption performance of activated carbon prepared by traditional one-step pyrolysis method is poor, the researchers separated the pyrolysis process from the activation process and form a two-step pyrolysis process. Yili Li et al. [9] prepared activated carbon by two-step pyrolysis process. The specific surface area of activated carbon prepared by two-step pyrolysis process is higher than that obtained by one-step pyrolysis process. The main reason is that part of ash in sludge can be removed by carbonization, and the activating agent can fully contact and react with the carbon in sludge during the process of activation, so that the pore of activated carbon is more abundant.

Microwave pyrolysis can produce heat both inside and outside the sludge, and thus sludge can be heated evenly. At the same time, the microwave heating rate is faster, so the preparation time of activated carbon can be shortened. Microwave pyrolysis includes three stages of drying, low temperature heating and high temperature carbonization [10]. Chunyan Yuan et al.[11] synthesized activated carbon from sludge by microwave pyrolysis under the condition that the microwave power is 500 W and the heating time is 8 min. The iodine adsorption value and specific surface area of the activated carbon obtained were satisfactory. This study indicates that microwave pyrolysis can apparently shorten the preparation time of activated carbon.

However, all the above pyrolysis methods require low moisture content of sludge. If the moisture content is high, the dewatering process of sludge will increase the cost of the preparation of activated carbon. In order to solve the high cost problem caused by high moisture content of sludge, the researchers put forward the idea of hydrothermal carbonization proposed by Bergius to the preparation process sludge-based activated carbon. Hydrothermal carbonization is a thermochemical conversion technology which can directly convert wet biomass into target carbon products in the reaction medium of subcritical water [12-13]. This method can effectively immobilize carbon elements in the precursor of activated carbon. Alatalo et al. [14] prepared activated carbon from digestive sludge and papermaking sludge by hydrothermal method, and the activated carbon obtained has better adsorption capacity for Pb^{2+} in aqueous solution. In summary, the hydrothermal carbonization method can transform sludge into an effective heavy metal adsorbent.

2.3. *Effect of Activators*

No matter what pyrolysis method, the total pore volume and specific surface area of the product obtained by direct carbonization are small, so it is difficult to be used as adsorbent directly. Therefore, activation of activated carbon precursors should be implemented to improve the performance index of activated carbon. There are mainly three kinds of activating agent commonly used.

2.3.1. Physical Activators. Physical activating agent, also known as gas activator, was passed into the pyrolysis furnace at high temperature. Carbon compounds in sludge react with activating agent in the pyrolysis furnace, and the gas produced by the reaction escapes from the sludge, and thus the pyrolysis products produce developed pores [15]. In the initial stage of activation, the activating gas reacts with the residual tar around the microcrystalline and between the microcrystalline, and the non-structural carbon is oxidized gradually to form a graphite-like microcrystalline surface. The reaction of microcrystals continues to expand the original pores, which forms the porous structure of activated carbon when new pores appear. The above process is that the continuous gasification of microcrystals makes the pore size of activated carbon larger. The most commonly used activating gases are water vapor, CO_2 and flue gas. The mechanism of physical activation reaction is as follows:

$$C+CO_2 \rightarrow 2CO \quad (1)$$

$$C+H_2O \rightarrow CO+H_2 \quad (2)$$

$$2CO+O_2 \rightarrow CO_2 \quad (3)$$

$$2H_2+O_2 \rightarrow H_2O \quad (4)$$

$$2KOH \rightarrow K_2O+ H_2O \quad (5)$$

$$C+H_2O \rightarrow CO+H_2 \quad (6)$$

$$CO+ H_2O \rightarrow CO_2+H_2 \quad (7)$$

$$K_2O+ CO_2 \rightarrow K_2CO_3 \quad (8)$$

$$K_2O+ H_2 \rightarrow 2K+ H_2O \quad (9)$$

$$K_2O + C \rightarrow 2K+CO \quad (10)$$

$$4KOH + \text{—}CH_2\text{—} \rightarrow K_2CO_3+K_2O+H_2O \quad (11)$$

$$8KOH + 2\text{—}CH_2\text{—} \rightarrow 2K_2CO_3+2K_2O+5H_2 \quad (12)$$

$$K_2O + C \rightarrow 2K+CO \quad (13)$$

$$2K_2CO_3+2C \rightarrow 2K+3CO \quad (14)$$

According to the above reaction, the activation process needs to absorb a lot of heat, so the physical activation is usually carried out at high temperatures of more than 800 °C. The adsorption capacity of activated carbon is related to the properties of sludge, the type of activating gas, activating temperature and activating agent flow rate. Wang et al. [16] synthesized the activated carbon through CO_2 activation, the methylene blue value of activated carbon is 245mg/g, which proves the feasibility of preparation of activated carbon by physical activation.

2.3.2. Chemical Activators. Chemical activators are usually impregnated with sludge before pyrolysis, and then the impregnation products are pyrolyzed. There are many kinds of chemical activators that include $ZnCl_2$, KOH, H_2SO_4 and others that facilitate the loss of H and O. In addition, some oxidizing agents are also used to be activators, such as KCr_2O_4 and Fenton reagent.

Taking $ZnCl_2$ and KOH as an example, the mechanism of chemical activation is analyzed. Some scholars believe that K_2O produced by the decomposition of KOH catalyze the reaction of C and H_2O and the shift reaction of CO and H_2O at 500 °C, which indicates that reaction (5), (6) and (7), take place at low temperature. When the activation temperature arrives around 800 °C, K_2O is consumed by hydrogen or carbon, the reaction (9) and (10) take place readily. The mechanism is as follows:

However, some researchers have pointed out that KOH reacts with volatile hydrocarbons in raw materials to form K_2CO_3, K_2CO_3 plays a role in activator at high temperature.

At present, the above two kinds of reaction mechanism were relatively accepted. There are two stages during pyrolysis-carbonization process when $ZnCl_2$ is used to be activator: catalytic dehydration and pore formation. In addition, $ZnCl_2$ can reduce tar production during the whole process of pyrolysis, which cause that most organics convert activated carbon. The activation principle of H_2SO_4 and other activators is similar to that of $ZnCl_2$, which facilitate deoxidation and hydrogen during pyrolysis. The macromolecule organics in the sludge can be oxidized into small molecule substances when oxidizing agent, such as $K_2Cr_2O_4$, Fenton reagent, is used to be activator. And then the small molecule substances are released from the sludge during the pyrolysis process to form rich pore structure. LI et al.[17] prepared activated carbon from sludge by using $ZnCl_2$, H_2SO_4 and KOH respectively. The three kinds of activated carbons can effectively adsorb H_2S. After activation by $ZnCl_2$ and KOH, the saturated adsorption capacity of activated carbon on H_2S is greater than that of commercial activated carbon.

2.3.3. Compound Activators. Generally speaking, the compound activator is made up of two or more than two kinds of chemical activators according to certain concentration and volume ratio. In addition, some studies combine chemical and physical activators during the preparation of activated carbon. Bingkui Yin [18] synthesized carbons with strong adsorption ability towards acid red GR through the combination of chemical and physical activators. This study revealed the combination of chemical and physical activators acts important role in the preparation of activated carbon. The author thinks that the selection of activators should be determined by the nature of sludge, the process of preparation and application area of activated carbon.

3. Summary and Prospect

It is a new idea that is preparation of activated carbon from sewage sludge, which can realize the resource utilization of sludge. Compared with the preparation of traditional activated carbon, sludge-based carbon has certain economic advantages of raw materials, but large amount of activator is needed in order to obtain adsorbents with loose structure and good performance. In addition, chemical activation gives rise to serious pollution during the preparation of activated carbon. Electro thermal furnace pyrolysis and microwave heating are still caught in laboratory stage. And thus, the development of stable and efficient pyrolysis equipment, the reducing low pollution, material structure and function control will become important research directions towards the preparation of sludge-based activated carbon in the future.

4. Reference

[1] Xiaohu Dai. *Current situation and thinking of sludge disposal in China.* Water Supply and Drainage, 2012, 38(2): 1-5.

[2] Chuan Shi, Panyue Zhang, et al. *Phosphorus adsorption characteristics of sludge biochar.* Journal of Environmental Engineering, 2016, 10(12): 7202-7208.

[3] Mohammadi S, Mirghaffari N. A preliminary study of the preparation of porous carbon from oil sludge for water treatment by simple pyrolysis or KOH activation. New Carbon Materials, 2015, 30(4): 310-318.

[4] Hao Deng, Rongsha Wang, Mingdong Zhang, et al. *Preparation of high specific surface area activated carbon from oily sludge.* Journal of Shandong University (Engineering Edition), 2014(2): 69-75.

[5] Hong Yao, Yan Shen, Xin Yuan, et al. *Characterization of physicochemical properties of activated sludge and adsorption of antibiotics.* Environmental Science and Technology, 2012, 35(2): 160-164.

[6] Pan Z H, Tian J Y, Xu G R, et al. *Characteristics of adsorbents made from biological, chemical and hybrid sludges and their effect on organics removal in wastewater treatment.* Water Research, 2011, 45(2): 819-27.

[7] Kantarli I C,Yanik J. *Use of Waste Sludge from the Tannery Industry.* Energy & Fuels, 2009, 23(6): 3126-3133.

[8] Calisto V, Ferreira C I, Santos S M, et al. *Production of adsorbents by pyrolysis of paper mill sludge and application on the removal of citalopram from water.* Bioresource Technology, 2014, 166(166): 335.

[9] Yili Li, Liping Li, Jing Yin, et al. *Optimized preparation and characterization of sludge based activated carbon by two step pyrolysis.* Journal of Beijing University of Technology, 2013, 39(12): 1887-1897.

[10] Ren X, Liang B, Liu M, et al. *Effects of pyrolysis temperature, time and leaf litter and powder coal ash addition on sludge-derived adsorbents for nitrogen oxide*. Bioresource Technology, 2012, 125(125C): 300-304.
[11] Chunyan Yuan. *Preparation and application of sludge adsorbent by microwave induced pyrolysis*. Harbin Institute of Technology, 2006.
[12] Mumme J, Eckervogt L, Pielert J, et al. *Hydrothermal carbonization of anaerobically digested maize silage*. Bioresource Technology, 2011, 102(19): 9255.
[13] Koottatep T, Fakkaew K, Tajai N,et al. *Sludge stabilization and energy recovery by hydrothermal carbonization process*. Renewable Energy, 2016, 99: 978-985.
[14] Alatalo S M, Repo E, Mäkilä E,et al. *Adsorption behavior of hydrothermally treated municipal sludge & pulp and paper industry sludge*. Bioresource Technology, 2013, 147: 71.
[15] Khalili N R, Campbell M, Sandi G, et al. *Production of micro- and mesoporous activated carbon from paper mill sludge*. Carbon, 2000, 38(14): 1905-1915.
[16] Hongliang Wang, Chungdian Si, Qingjie Guo. *Preparation and characterization of sludge derived activated carbon*. Shandong Chemical Industry, 2008, 37(2): 1-5.
[17] Li F, Yan B, Zhang Y, et al. *Effect of activator on the structure and desulphurization efficiency of sludge-activated carbon*. Environmental Technology, 2014, 35(17-20): 2575.
[18] Bingkui Yin. *Preparation of sludge activated carbon adsorbent material and its application in wastewater treatment*. Shanghai Jiao Tong University, 2007.

Investigation of Hybrid Natural Fibre Reinforced Composite for Impact Energy Absorption

Yiou Shen, Yichun Guo and Yan Li*

School of Aerospace Engineering and Applied Mechanics, Tongji University, 1239 Siping Road, Shanghai, 200092, P R China
Email: shenyiou@tongji.edu.cn

Abstract. In this paper, the low-velocity impact properties of jute/ramie fiber reinforced composites were tested by drop weight method, and the effects of mixing ratio on impact properties and energy absorption of composites were also studied. A series of impact energies ranged from 2 to 20 J were applied on five composites including jute fibre reinforced composite, ramie fibre reinforcement composite and three hybrid composites. When the jute fiber reinforced composite material is mixed with the ramie fiber reinforced composite material, energy absorption capacity of ramie fibre reinforced composites were enhanced with hybridizing this two fibre fabric together. The damage area was assessed using an ultrasound c-scan instrument. The results show that the hybrid laminate with a volume fraction of ramie fiber of 55% has better impact energy absorption and impact resistance. It was found that hybridizating different fiber together can effectively solve the problem of insufficient rigidity of jute fiber reinforced composite material and high ramie of ramie fiber reinforced composite material.

1. Introduction

In recent years, there is a greater need for weight savings and impact protection with lightweight materials. The application of cellular materials in impact energy absorbing structures has attracted more and more attention. It is important to understand the behavior of natural porous materials during impact can they be widely used in energy absorbing systems. In order to evaluate the potential of natural fibers as reinforcing materials in polymers, many studies have been carried out, and comparative studies between laminates and polymers with different types of natural fiber reinforcement have been published. [1-3]. Over the years, there has been increasing interest in the performance of hybrid composites such as those formed by combining different types of fibres or different laminating materials [4, 5]. Currently, most hybrid composites are made only between synthetic fibers, for example Kevlar and carbon, or plant fiber with synthetic fibers. Naik et al. [6] studied the impact behavior and post-impact compression properties of glass-carbon/epoxy composites in different stacking orders. Sarasini et al. [7] further tested the impact properties of woven basalt-carbon/epoxy composites. The results show that the sandwich laminate structure has better energy absorption capacity and damage tolerance, while the sandwich laminate structure has better bending resistance. Suresh Kumar et al. [8] pointed out that cannabis basalt/epoxy composites have advantages over non-hybrid composites in terms of residual flexural strength and impact properties. From the aspects of microstructure, yarn structure, fabric structure, etc., plant fibers are more different than synthetic fibers, resulting in different energy absorption and synthetic fibers. [9]. In addition, different plant fibers have different mechanical properties due to their different structures during natural growth. For example, ramie and flax fibers have better mechanical properties due to their smaller diameter and flow path size. Sisal and jute fiber have better energy absorption properties due to their multi-lumen

IOP Conf. Series: Materials Science and Engineering **484** (2019) 012014 doi:10.1088/1757-899X/484/1/012014

structures, more energy can be dissipated during their composite damage propagation. Therefore, a good composite design can combine the advantages of various fibers to improve the overall performance of the composite. In addition, the base can be prevented from cracking and the fracture toughness of the composite material can be improved.

The purpose of this study was to investigate the behaviour of plant fiber-ramie /jute hybrid laminates under low-velocity impact. The woven and jute/epoxy composites were prepared by sandwich-type hybrid structure, and the low-velocity impact was carried out by using the drop weight impact device.

2. Experimental

2.1. Materials

Ramie plain fabrics and jute plain fabrics were supplied by Hunan Huasheng Dongting Ramie Textile Co.,Ltd and Anji Zhengxing Fabric Co.,Ltd, respectively. The basic parameters of ramie and jute fabrics are listed in Table 1. Epoxy resin E51, together with MeTHPA curing agent and DMP-30 tertiary amine accelerator were supplied by Shanghai Zhongsi Industry Co., Ltd.

Table 1. The properties of ramie and jute fabrics

	Ramie fabric	Jute fabrics
Weave	Plain	Plain
Areal density (g/m^2)	255.6	116.7
Warp yarns (yarns/dm)	205	52
Weft yarns (yarns/dm)	228	58
Filament diameter (μm)	15.93	35.68

2.2. Composites Fabrication

Two types of fabric were all preimpregnated with the mixed epoxy resin system by hand stacking process. The hybrid laminates were fabricated by laying up the appropriate number of pregregs in a steel frame mould (typically 350×300 mm^2), and preheated in a hot compaction machine at the temperature of 90°C with a pressure of 1 MPa for one hour for ensuring good wetting, and cured at 120°C using a pressure of 6 MPa for two hours. After that, the mould was cool slowly to the room temperature in the press. Five types of laminates with different hybrid ratios shown in Table 2 were produced to investigate the effect of mixing ratio on the impact response of the laminates which included pure jute fiber reinforced polymer composite (JFRP), 3 kinds of ramie/jute fiber reinforced hybrid polymer composites and pure ramie fiber reinforced polymer composite (RFRP). The fiber volume percentage of ramie and jute in the composites were varying while the total fiber volume fractions of the composites were kept the same, which were approximately 50%. Fibers are layed up as a sandwich-like sequence with ramie fibre layers (skins) and jute fibre layers (core) for composites with three mixing ratios. The thickness of the laminates were 4.15±0.08 mm. Another three kinds of hybrid composites were fabricated to investigate the effect of stacking sequences on the impact energy absorb properties on hybrid laminates as shown in Table 3. The total fiber volume percentage and fiber volume fractions for each type fibers were unchanged. The total fiber volume percentage was 50%, jute fibre and ramie fibre were 23% and 27%, respectively.

Table 2. Different hybrid ratios of five types of hybrid composites.

Designation	Volume fraction ratio V_r % (Ramie/Jute)	Ply number ratio (Ramie/Jute)	Stacking sequence
JFRP	0/100	0/8	$[J]_8$
6R6J	30/70	6/6	$[R_3/J_6/R_3]$
12R4J	55/45	12/4	$[R_6/J_4/R_6]$
18R2J	80/20	16/2	$[R_9/J_2/R_9]$
RFRP	100/0	22/0	$[R]_{22}$

Table 3. Hybrid composites with different stacking sequences.

Designation	Stacking sequence
12R4J	$[R_6/J_4/R_6]$
4J12R	$[J_2/R_{12}/J_2]$
2RJ	$[R_2/J/R_2/J/R_2]_2$

Figure 1. Comparison of (a) ramie fabric and (b) jute fabric in macroscopic scale

2.3. Low-Velocity Impact Test

According to ASTM D7136 standard, a series of low-velocity impact tests were carried out on plant fiber reinforced composite laminates, and their impact properties and damage mechanisms were studied. The laminate was subjected to a low-velocity impact test using the CEAST 9350 drop hammer tower. A hemispherical indenter (steel material) is attached to a 16 nm diameter impactor, the total mass of the impactor was 2.4 kg. The sample was placed in a steel bracket with 75 x 50 mm squares sandwiched between the four corners. The anti-rebound system of the impact machine captures the crosshead and prevents it from colliding with the sample again. The load time tracking during the test was recorded by the computer. In order to evaluate the damaged area, an ultrasound c-scan (NAUT21, Japanese probe) was used when examining the impact sample.

3. Results and Discussions

3.1. The Low-velocity Impact Response of JFRP and RFRP

Typical force–time histories under impact tests of JFRP and RFRP panels with impact energy between 2 to 20 J are shown in Figure 2. It can be seen from Figures 2 (a) and (b) that some initial oscillations in the curves, this is probably owing to the ringing inside of the load cell and dynamic effects in the plate. The figures show that increasing the impact energy leads to increase in the impact force and the impact duration before being perforated. It can be seen from Figure 3 (a) for RFRP laminates, no energy can be recovered and perforated at 20 J. However, during the unloading process, the displacement of the RFRP laminate under four impact energies returned to the axial origin, indicating that it can self-recover through its own elasticity. JFRP has good compatibility with itself and can perform higher overall deformation, so it can absorb more energy.

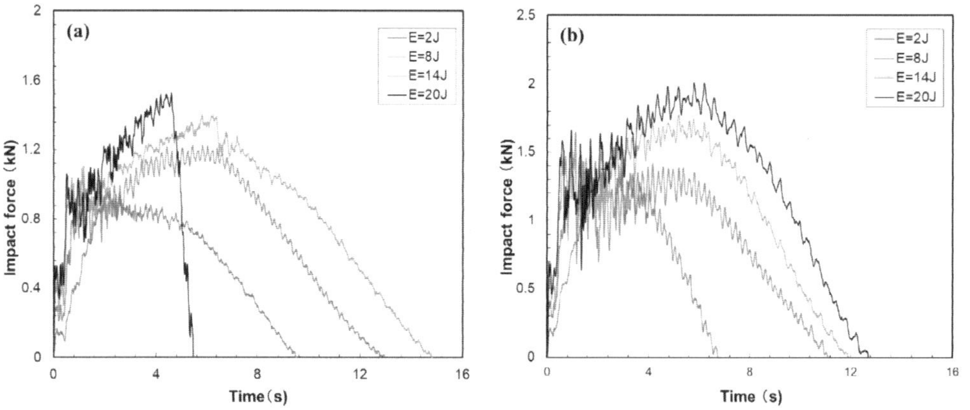

Figure 2. Typical load vs. time curves for (a) JFRP and (b) RFRP following different impact energies.

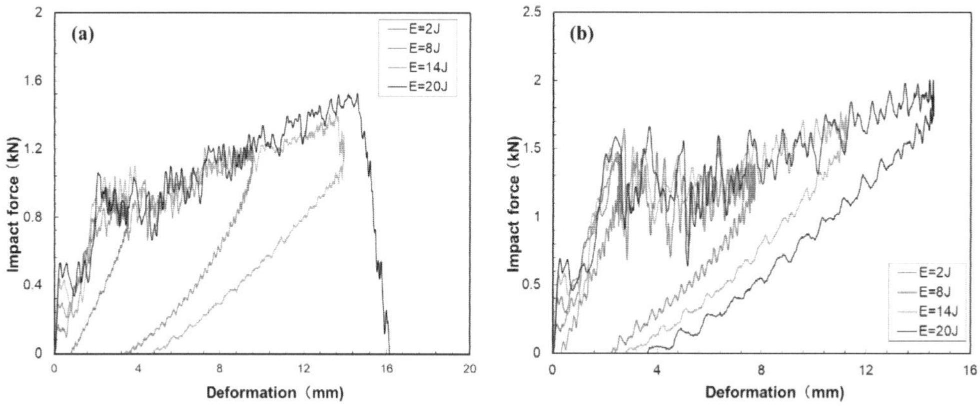

Figure 3. Typical load vs. deformation curves for (a) JFRP and (b) RFRP following different impact energies.

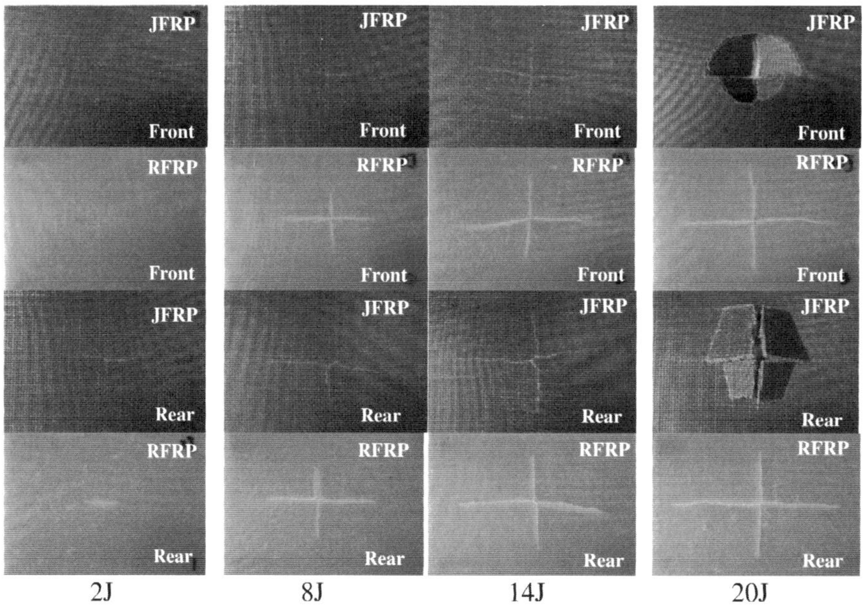

Figure 4. Photos of damage progression on front and rear faces of JFRP and RFRP.

Increase the impact energy, JFRP and RFRP laminates will produce different damage modes and corresponding damage, as shown in Figure 4. Penetration of the indentor through the laminate thickness was accompanied by cracks, fibre breakage was observed on JFRP composites following an impact test of 20 J. With same impact conditions, RFRP laminates showed a cross-shaped crack with associated matrix cracking but without evidence of penetration, this indicates that RFRP composites process superior impact resistance compare to JFRP.

3.2. Effect of Hybrid Ratios on Impact Properties of Composites

It can be seen from Figure 5 that increase the volume percentage ratio of ramie fibre up to 55% (12R4J), laminates exhibit a residual elastic response when the impact energy of laminate was 20 J. Figure 6 shows that the mixing ratio has an enhanced effect on the impact properties of the composite ramie/jute fiber. The magnitude of this maximum impact force is positively correlated with the relative volume of the ramie fiber, so the lower laminate has a higher load carrying capacity when the Vr level is higher, as shown in Figure 6(a). The RFRP laminate has a lower displacement, while the JFRP and hybrid laminates have better compliance and absorb energy through overall deformation. According to Figure 6(b), JFRP and hybrid laminates are more advantageous in terms of overall deformation, fiber breakage and splitting, so they can absorb more energy than RFPR. Three hybrid laminates show similar energy absorption capacity to the RFRP composites following impact energy up to 14 J. However, the hybrid composite show superior energy absorption ability compare to the RFRP when impact with an energy of 20 J, the absorbed energy of 6R6J (Vr % =30%) even higher than JFRP. It can be seen from Figure 7 that 6R6J laminate was also perforated when the impact energy of laminate was 20 J. This breakage of ramie fibres in hybrid composite dissipated more energy than that of in JFRP laminate. In fact, it can be seen from Figure 8 that the damaged area for 12R4J laminates is the smallest among five composites, suggesting that volume fraction ratio of ramie fiber is 55% (12J4R) is the optimized hybrid ratio for the ramie/jute hybrid composites.

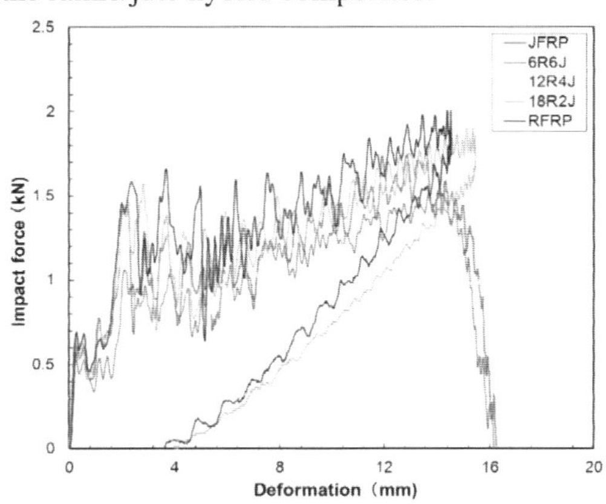

Figure 5. Force and displacement traces of five type of composites

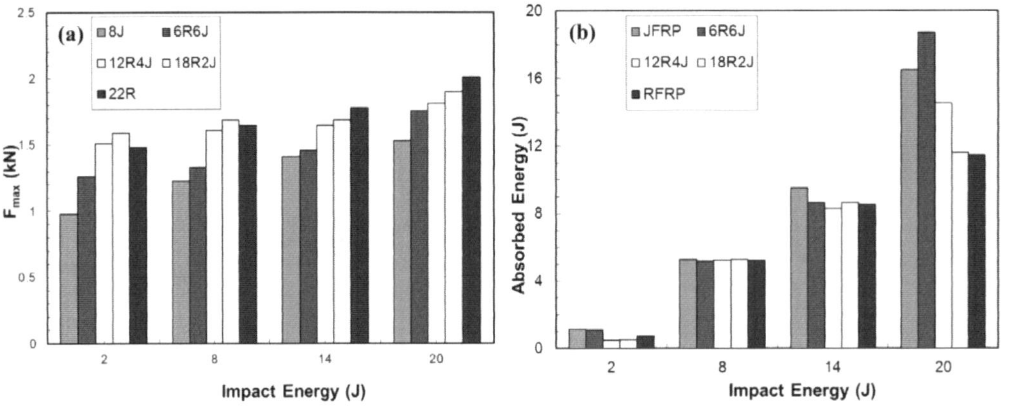

Figure 6. Bar chart of (a) Maximum force vs. Impact energy and (b) Absorbed energy vs. Impact energy.

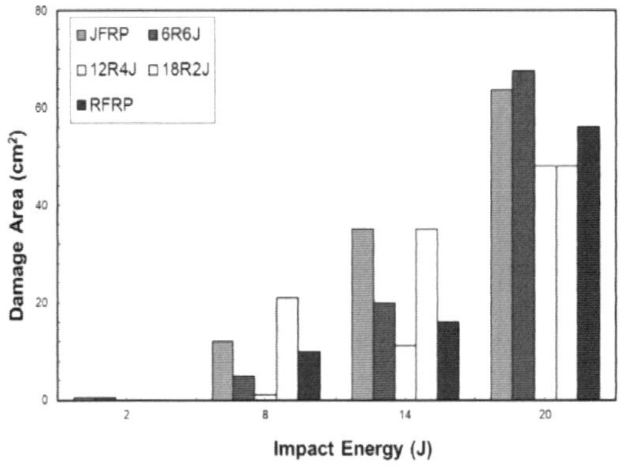

Figure 7. Damage area vs. Impact energy bar chart of five composites.

4. Conclusion

In this paper, the low-velocity impact response of ramie/jute fiber reinforced epoxy composites and the effect of ramie/jute fiber hybridization were studied. Three different hybrid ratio laminates have been prepared: the volume friction of ramie was 30%, 55% and 80%. Impact damage caused by the same energy level can lead to penetration on the jute laminate, while the mixed laminate and the ramie laminate have cross-shaped cracks on the front side, but there is no evidence of penetration. Compared to jute laminates, the hybridization of ramie and jute fibers prevents the penetration of the whole jute composite and the low ramie volume fraction mixed laminate, and the peak force is increased. Ramie/jute volume friction of 55/45 was the optimized hybrid ratio for this type of hybrid composites.

5. Acknowledgements

This research received grant from National Natural Science Foundation (11302151), the Fundamental Research Funds for the Central Universities and Shanghai Rongtai Health Technology Co., Ltd.

6. References

[1] P. Ramesh, B.D. Prasad, K. Narayana. Characterization of kenaf fiber and its composites: A review. *Journal of Reinforced Plastic and Composites*, 37(11), 2018, pp.731-737 (doi: 10.1177/0731684418760206).

[2] T.P. Sathishkumar, P. Navaneethakrishnan, S. Shankar, R. Rajasekar, N. Rajini. Characterization of natural fiber and composites – A review. *Journal of Reinforced Plastic and Composites*, 32(19), 2013, pp. 1457–1476 (doi: 10.1177/0731684413495322).

[3] J. Meredith, R. Ebsworth, S.R. Coles, M.B. Wood, K. Kirwan. Natural fibre composite energy absorption structures. Composite Science Technology, 72, 2012, pp.211–7 (doi: 10.1016/j.compscitech.2011.11.004)

[4] S. Nunna, P. Ravi Chandra, S. Shrivastava, A.K. Jalan, A review on mechanical behaviour of natural fibrebased hybrid composites, *Journal of Reinforced Plastic Composites*, **31**, 2012, pp. 759–769 (doi: 10.1177/0731684412444325).

[5] M. Jawaid, H.P.S. Abdul Khalil, Cellulosic/synthetic fibre reinforced polymer hybrid composites: a review, *Carbohydrate Polymers*, 86, 2011, pp. 1–18 (doi:10.1016/j. carbpol.2011.04.043).

[6] N.K. Naik, R. Ramasimha, H. Arya, S.V. Prabhu, N. ShamaRao, Impact response and damage tolerance characteristics of glass–carbon/epoxy hybrid composite plates. *Composite Part B*, 32, 2001, pp. 565–574 (doi: 10.1016/S1359-8368(01)00036-1).

[7] F. Sarasini, J. Tirillò, L. Ferrante, M. Valente, T. Valente, L. Lampani, P. Gaudenzi, S. Cioffi, S. Iannace, L. Sorrentino, Drop-weight impact behaviour of woven hybrid basalt–carbon/epoxy composites, *Composites: Part B*, **59**, 2014, pp. 204–220 (doi:10.1016/j. compositesb.2013.12.006).

[8] C. Suresh Kumar, V. Arumugam, H.N. Dhakal, Risil John, Effect of temperature and hybridisation on the low-velocity impactbehavior of hemp-basalt/epoxy composites, *Composite Structures*, **125**, 2015, pp. 407–416 (doi:10.1016/j. compstruct.2015.01.037).

[9] Y. Li, L. Xie, H. Ma. Permeability and mechanical properties of plant fiber reinforced hybrid composites. *Materials & Design*, 86, 2015, pp. 313-320 (doi: 10.1016/j.matdes.2015.06.164)

IOP Conf. Series: Materials Science and Engineering **484** (2019) 012015 doi:10.1088/1757-899X/484/1/012015

Dependencies between Cracking Patterns and the Physico-Mechanical Properties of Microsilica Modified Cement Matrix

M Szeląg and A Szewczak

Lublin University of Technology, Faculty of Civil Engineering and Architecture, 40 Nadbystrzycka Str., 20-618 Lublin, Poland.

Email: maciej.szelag@pollub.pl

Abstract. The article presents and evaluates the dependencies that occur between the parameters describing the geometry of cracking patterns on the sample's surface and the selected physico-mechanical properties of cement pastes with the addition of microsilica. The stereological parameters obtained with the use of image analysis were applied to the quantitative description of cracking patterns; also the compressive and tensile strength as well as the apparent density were examined. Cracks analyzed in the surface were caused by loading the samples with an elevated temperature. The relationships were calculated using the least squares method, and the quality of the fit of the model curves to the experimental data was evaluated on the basis of the diagnostic statistics. The results obtained indicated the existence of very strong relations between the geometry of cracking patterns and selected material features of the cement matrix. The analysis carried out is also of a great practical importance as it can help in assessing the degree of degradation of a cement structure exposed to elevated temperatures.

1. Introduction

As a result of the temperature gradient, the cement matrix may crack. Cracks through propagation combine or intersect to form a characteristic cracking pattern, which in the literature is also referred as the cluster cracks/thermal cracks/mapcracking [1, 2]. There are only few research papers that attempt to quantitatively analyze cracking patterns because it is difficult from a methodological point of view. However, such analyzes have a high scientific and practical value because cracks of the cement matrix increase the risk of penetration of harmful substances into the material, causing its progressive degradation. Currently, the most commonly used concept to describe cracking patterns of cementitious materials is the crack density [3-5]. However, this is a very general parameter that does not provide detailed information on, e.g., the average crack width or the average area limited by cracks.

The authors of this work, to evaluate cracking patterns use image analysis tools, the advantage of which is that they are non-destructive methods. Their potential in the technology of cementitious materials is still not fully understood. The purpose of the research conducted was to determine whether, based on the knowledge of the geometrical parameters of the cracking patterns, it is possible to estimate the physical and mechanical properties of cement pastes with the addition of microsilica (MS). So far, no similar studies have been reported in the literature. For the purpose, a statistical analysis tool – the least squares method (LSM) was used. The quality of the dependencies calculated has been verified using the diagnostic statistics, i.e., the correlation coefficient, the determination coefficient, the standard error of estimation, and the coefficient of random variation. It was determined which physical and mechanical parameters of the cement matrix can be estimated with a high accuracy based on the measurement of geometric features of the cracking patterns.

2. Materials and Methods

2.1. Preparation of Modified Cement Matrix and Thermal Loading
Two series of cement pastes were tested, in which microsilica was used as a 10% substitute for cement mass. For both series, samples were made with three *w/b* (water/binder) indicators equal to 0.4, 0.5, and 0.6, respectively. The binder is understood as the total content of cement and microsilica in the material. The series differed from each other with the class of cement used; in the first series (C42MS), the ordinary Portland cement (OPC) CEM I 42.5R was used, and in the second series (C52MS), the OPC CEM I 52.5R was used. The cements used have a very similar chemical composition, which was presented in [6]. They differ in terms of grains' size; CEM I 52.5 R has a specific surface area greater by approximately 15% compared to CEM I 42.5 R.

Samples were made as bars with dimensions 40x40x160 mm^3. They were made in accordance with EN 196-1 [7]; maturation took place in air conditions for 28 days (average relative humidity – 50%, average temperature – 22 °C). The samples were then subjected to an elevated temperature reaction based on the thermal shock. The samples were placed in a preheated oven at 250 °C for a period of 4 hours. The method of thermal loading was to extract the defects of the structure in the form of cracks, without excessive deterioration of the mechanical properties of the cement matrix. At this temperature, the cement matrix is thermally and chemically stable. The formation of cracks in the material is caused by the pressure of the saturated water vapor, which is associated with the intensive evaporation of free water. Thermal load parameters were selected based on previous studies [1, 8], in which cracking patterns were analyzed in the context of the self-assembly process of the cement matrix structure.

2.2. Properties of Modified Cement Paste
The bending tensile strength (f_{cf}) according to the EN 12390-5 [9], compressive strength (f_c) according to the EN 12390-3 [10], and the apparent density (D) according to the EN 12390-7 [11] was tested. The results of f_{cf} and D are the arithmetic means of 6 samples, in the case of f_c from 12 samples. The difference in number of samples was due to the fact that the f_c test was carried out on the halfs of the beams obtained due to the f_{cf} measurement.

To quantify the cracking patterns geometry, stereological parameters obtained from the image analysis were used, which were measured using ImageJ v. 1.50d software. The scanned surface of the samples was analyzed. A measurement procedure was applied, which was described in detail in [8]. Among the stereological parameters, the average cluster area ($\bar{A}$), the average cluster perimeter ($\bar{L}$), and the average crack width ($\bar{l}$) were measured. A cluster is understood as the area on the surface of the sample, which is limited on each side with a crack or edge of the sample. An example of a cracking pattern on the surface of a cement paste with the addition of microsilica is shown in Figure 1.

The results of the tests are presented in Table 1. Index (R) means the value obtained on the standard samples; index (T) – parameters for samples after temperature load.

Figure 1. Cracking pattern on the surface of cement paste of C42MS series with *w/b* = 0.6

Table 1. Physical and mechanical parameters of cement pastes modified with microsilica and the geometric characteristics of cracking patterns.

Series		**C42MS**			**C52MS**	
w/b	**0.4**	**0.5**	**0.6**	**0.4**	**0.5**	**0.6**
$f_{c(R)}$ [MPa]	61.97	46.88	35.47	68.13	50.99	41.01
$f_{cf(R)}$ [MPa]	5.10	4.63	3.03	6.26	3.94	3.19
$D_{(R)}$ [g/cm^3]	1.669	1.488	1.337	1.710	1.571	1.388
$f_{c(T)}$ [MPa]	41.45	25.80	14.70	52.54	41.07	24.40
$f_{cf(T)}$ [MPa]	3.26	2.55	1.36	1.94	1.79	1.27
$D_{(T)}$ [g/cm^3]	1.489	1.329	1.196	1.511	1.367	1.219
$\bar{A}$ [mm^2]	75.69	110.88	153.79	50.08	75.84	119.27
$\bar{L}$ [mm]	37.98	46.25	55.04	31.70	38.47	48.40
$\bar{I}$ [mm]	0.028	0.034	0.043	0.026	0.034	0.040

2.3. The Least Squares Method

The dependencies that occur between the geometry of cracking patterns and the physico-mechanical parameters of the modified cement pastes were calculated using the least squares method (LSM). This method consists in such a selection of the curve equation, so that the size being the sum of the squares of the distances of all empirical points from the corresponding points on the fit line was as small as possible [12]. The dependencies relate to the condition of the material after temperature load. In the previous studies, it was noted that there is a very strong correlation (0.98-0.99) between $\bar{A}$ and $\bar{L}$ variables [1, 8, 13]. The equation was calculated, which makes these parameters to be dependent from each other. Thus, knowing the value of one parameter with very high accuracy, the value of the second one can be calculated. Thus, in the present work, the description of the relationship between physico-mechanical features was limited to one of the above mentioned parameters, i.e., $\bar{A}$, and with $\bar{I}$.

In addition, the Pearson's correlation coefficient (ρ) was calculated, which statistical significance was verified by the statistical significance test of the correlation coefficient. The quality of the adjustment of the curves calculated to the empirical data was checked using the diagnostic statistics, i.e., the determination coefficient (R^2), the standard error of the estimation (S_e), and the coefficient of random variation (W).

3. Results and Discussion

Table 2 shows the calculated equations of curves together with the values of the diagnostic statistics that describe the following relationships:$f_{c(T)}(\bar{A})$,$f_{c(T)}(\bar{I})$, $f_{cf(T)}(\bar{A})$, $f_{cf(T)}(\bar{I})$, $D_{(T)}(\bar{A})$, and $D_{(T)}(\bar{I})$. Dependencies were graphically shown in Figure 2 – Figure 4. The relationships for each series were calculated separately because in this situation a better degree of matching the model curves to the experimental data was obtained; the correlation coefficients, in the criterion of division into individual series have obtained values more favorable than global coefficients. Values of ρ are negative; this means that the values of the physico-mechanical parameters of the MS modified cement paste decrease with the increase of $\bar{A}$ and $\bar{I}$. Thus, the fewer clusters on the surface of the cement matrix, the less its strength is. However, the key aspect in this case is that at the same time the crack width increases, which is the main reason for the weakening of the material structure.

Assuming the level of the statistical significance α at 0.05, all *p-values* assume a smaller value. This means that the correlations occurring between variables, both for individual series and for the whole dependency, are in each case statistically significant.

In global terms, the strongest correlation is characterized by the $f_{c(T)}(\bar{A})$ and $D_{(T)}(\bar{I})$ dependencies; ρ reach a very high value of -0.93 and -0.90, respectively. The interesting thing is that in the division into the series for $f_{c(T)}(\bar{A})$ the correlation coefficients assume lower values than in the global approach, which is a rare situation. However, this difference is very small and equal only to 0.02. In the remaining situations ρ divided into series, achieves values greater than global coefficients. The weakest global correlation occurs for dependencies with $f_{c(T)}$. This is due to the fact that the C52MS

IOP Conf. Series: Materials Science and Engineering **484** (2019) 012015 doi:10.1088/1757-899X/484/1/012015

series was characterized by lower tensile strength than C42MS, while at the same time lower values of $\bar{A}$ and $\bar{I}$. However, in the criterion for division into series, the ρ values are very high and show a very strong correlation.

The curves calculated best reflect the empirical data in a situation when the $\bar{A}$ is variable, because R^2 assumes higher values (0.84-0.91). The quality of curve fitting for $\bar{I}$ is still high (0.78-0.88), however, lower compared to the first variable. It is worth noting that for cement paste made of CEM I 52.5R, the W values in each case are smaller than for the C42MS series. This means that the estimation of material values for the paste made of cement with a higher fineness will be characterized by greater accuracy than for samples made of CEM I 42.5R.

The coefficient of random variation reaches the lowest values for the relationship with $D_{(T)}$. These values are even 6-7 times lower than the highest ones. Considering the high value of the determination coefficient, it can be concluded that the physico-mechanical properties of the cement matrix will be able to estimate most accurately on the basis of apparent density, both on the basis of $\bar{A}$ and $\bar{I}$.

Table 2. Equations describing the relationships between the cracking patterns geometry and the physico-mechanical properties of cement pastes modified with microsilica.

Relationship	Series	Equation	R^2 [-]	S_e [MPa] [g/cm³]	W [%]	ρ [-]		*p-value* (the ρ significance test) [-]	
$f_{c(T)}(\bar{A})$	C42MS	$y=5074x^{-1.130}$	0.84	4.78	17.50	-0.93	-0.91	<0.001	<0.001
	C52MS	$y=1086x^{-0.775}$	0.91	3.89	9,86		-0.91		<0.001
$f_{c(T)}(\bar{I})$	C42MS	$y=0,063x^{-1.786}$	0.79	5,50	20,15	-0.83	-0.88	<0.001	<0.001
	C52MS	$y=0.582x^{-1.228}$	0.78	5,90	14,94		-0.92		<0.001
$f_{cf(T)}(\bar{A})$	C42MS	$y=199,18x^{-0.955}$	0.87	0.32	13.47	-0.47	-0.95	0.036	<0.001
	C52MS	$y=12,81x^{-0.474}$	0.88	0.12	7,08		-0.92		<0.001
$f_{cf(T)}(\bar{I})$	C42MS	$y=0.012x^{-1.565}$	0.88	0.30	12,72	-0.71	-0.96	<0.001	<0.001
	C52MS	$y=0,103x^{-0.815}$	0.82	0.14	8,58		-0.93		<0.001
$D_{(T)}(\bar{A})$	C42MS	$y=4,427x^{-0.256}$	0.85	0.051	3.78	-0.87	-0.92	<0.001	<0.001
	C52MS	$y=3,600x^{-0.225}$	0.90	0.042	3.06		-0.90		<0.001
$D_{(T)}(\bar{I})$	C42MS	$y=0,347x^{-0.400}$	0.80	0.058	4,35	-0.90	-0.89	<0.001	<0.001
	C52MS	$y=0.360x^{-0.390}$	0.81	0.057	4,16		-0.91		<0.001

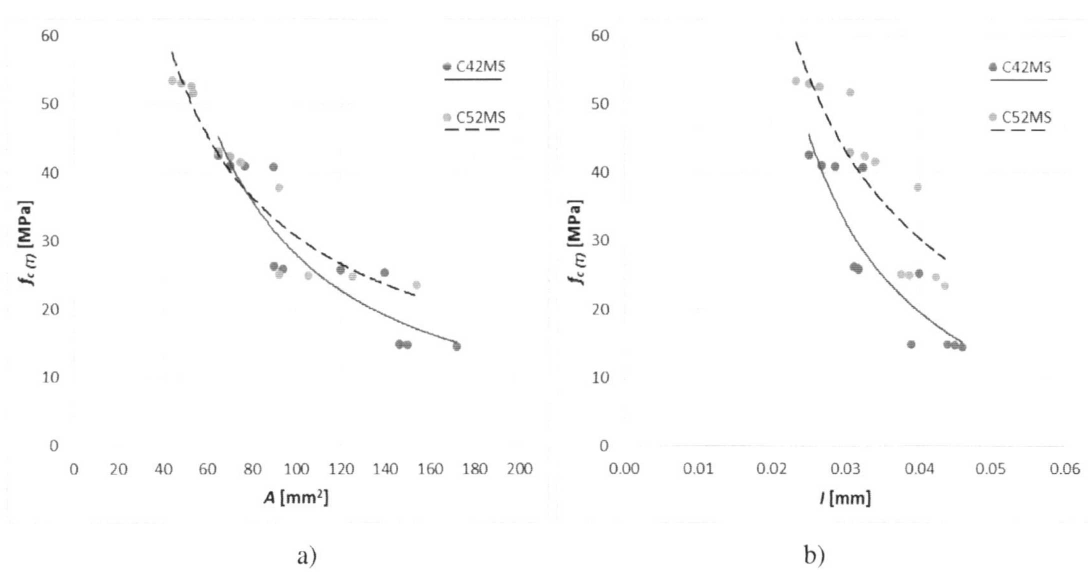

Figure 2. Relationship between $f_{c(T)}$ and: a) $\bar{A}$; b) $\bar{I}$

IOP Conf. Series: Materials Science and Engineering **484** (2019) 012015 doi:10.1088/1757-899X/484/1/012015

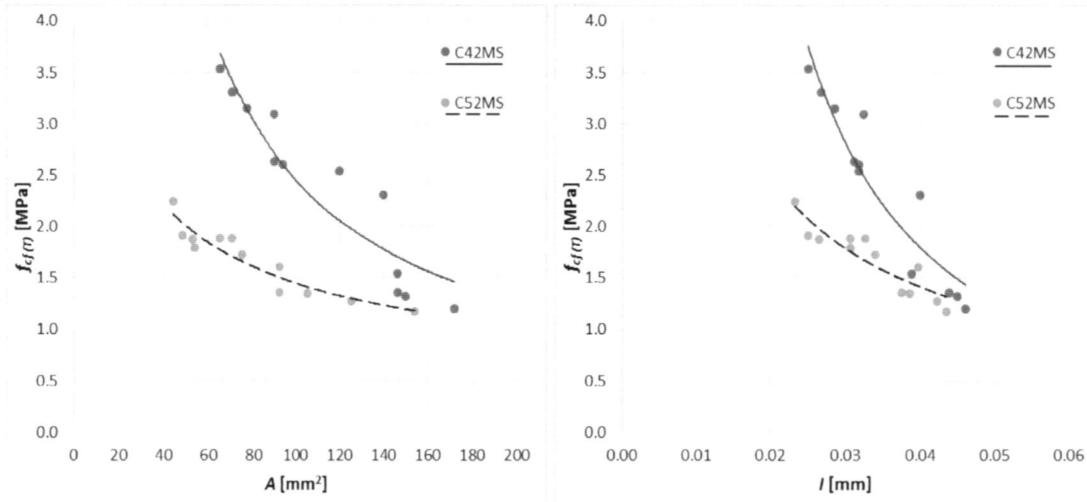

a) b)

Figure 3. Relationship between $f_{cf(T)}$ and: a) $\bar{A}$; b) $\bar{l}$

a) b)

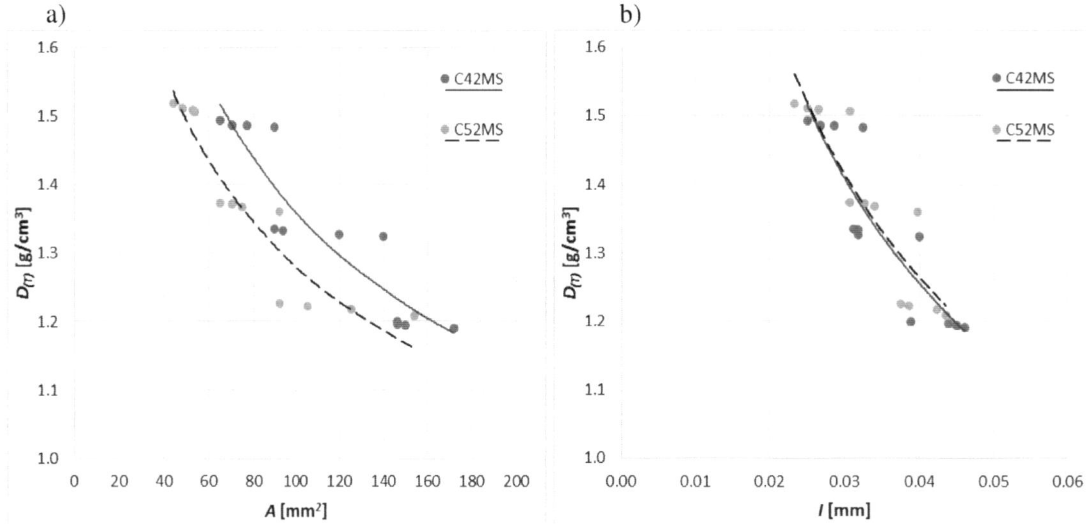

a) b)

Figure 4. Relationship between $D_{(T)}$ and: a) $\bar{A}$; b) $\bar{l}$

4. Conclusions

The article specifies the relationships between the cracking patterns geometry and selected physical and mechanical properties of cement pastes modified with microsilica. On the basis of the analyzed and interpreted research results, final conclusions were formulated:

- With the increase in the $\bar{A}$ and $\bar{l}$ values, all the examined physical and mechanical features of the cement matrix are reduced.
- The method of the least squares supplemented with the evaluation of diagnostic statistics is an effective tool in defining dependencies that occur between variables.
- The dependencies between physico-mechanical features and the cracking patterns geometry, in the criterion of division into series, are characterized by a very strong correlation (less than

-0.88). In the case of global correlation, only in the case of the relation with $f_{cf(T)}$, the average correlation was achieved (-0.47 and -0.71).

- $D_{(T)}(\bar{A})$ and $D_{(T)}(\bar{I})$ are the relationships for which the highest fit of the model curves has been achieved (R^2 from 0.80 to 0.90). In addition, the lowest value of the random coefficient of variation (3.06-4.35) indicates that the apparent density of the cement matrix modified with microsilica can be estimated with very high accuracy based on the geometric characteristics of the cracking patterns.
- The assessment of physical and mechanical parameters of cement pastes based on the measurement of parameters obtained from the image analysis is very valuable, because it is a non-destructive test. This is of great practical importance, because the tests conducted have proved that by measuring parameters of the cracking patterns such as $\bar{A}$, $\bar{L}$ and $\bar{I}$, it is possible to estimate the value of selected material properties of the cement matrix with high accuracy. It may improve the process of analysis of the degree of damage to the cement structure due to the influence of an elevated temperature.

5. Acknowledgments

This work was financially supported by the Polish Ministry of Science and Higher Education with the statutory research number S-14/B/2017, and the subsidy for the development of young scientists – 507-3-14-140.

6. References

[1] Fic S, Szelag M. Analysis of the development of cluster cracks caused by elevated temperaturesin cement paste. Construction and Building Materials. 2015; 83:223-9.
[2] Fic S, Vyrovoy VN, Dorofeev VS. Procesy samoorganizacji struktury kompozytowych materiałów budowlanych. Lublin: Politechnika Lubelska; 2013.
[3] Magalhaes M, Toledo R, Fairbairn E. Thermal stability of PVA fiber strain hardening cement based composites. Construction and Building Materials. 2015; 94:437-47.
[4] Mobhaser B, Stang H, Shah S. Microcracking in fiber reinforced-concrete. Cement and Concrete Research. 1990; 20(5):665-676.
[5] Xu Y, Wong Y, Poon C, Anson M. Influence of PFA on cracking of concrete and cement paste after exposure to high temperatures. Cement and Concrete Research. 2003; 33(12):2009-2016.
[6] Szelag M. The Influence of Metakaolinite on the Development of Thermal Cracks in a Cement Matrix. Materials. 2018; 11(4):520.
[7] EN 196-1. Methods of testing cement – Part 1: Determination of strength.
[8] Szeląg M. Wpływ składu kompozytów cementowych na geometrię ich spękań termicznych. Lublin: Lublin University of Technology; 2017.
[9] EN 12390–5. Testing hardened concrete – Part 5: Flexural strength of test specimens.
[10] EN 12390-3. Testing Hardened Concrete - Part 3: Compressive Strength of Test Specimens.
[11] EN 12390-7. Testing hardened concrete - Part 7: Density of hardened concrete.
[12] Wolberg J. Data Analysis Using the Method of Least Squares: Extracting the Most Information From Experiments. Berlin: Springer; 2006.
[13] Szeląg M, Fic S. Analiza rozwoju spękań klastrowych w zaczynie cementowym modyfikowanym mikrokrzemionką. Budownictwo i Architektura. 2015; 14(4):117-27.

IOP Conf. Series: Materials Science and Engineering **484** (2019) 012016 doi:10.1088/1757-899X/484/1/012016

Research on New Materials in Civil and Construction Engineering

Peng Jiang[1, 2] and Junhai Hu[2]

[1]School of Civil Engineering, Lanzhou Jiaotong University, Lanzhou, Gansu, China.
[2]School of Materials Science and Engineering, Wuhan, Hubei, China.
Email: pengjiang61787688@163.com

Abstract. Along with the development of industry and economy all around the world, the traditional materials have made great achievements. But traditional materials have many disadvantages such as high cost, difficult con-struction, etc. Besides, we have to face the following problems such as air pollution, water pollution, depletion of natural resources, etc., so new materials which have better functions will play an important role in the field of civil engineering. This article is through the introduction to new materials for civil engineering to do an overview.

1. Introduction

New material industry has penetrated into all fields of national economy, national defence construction and social life. It plays an important role in the development of national economy. The main developed countries attach great importance to the investment and development of new materials industry. With the development of our country, people start to pay more and more attention to the living environment. The state also put priority on the development of new functional materials, advanced structural materials and composite materials. Civil engineering construction is such an important part of national development, so the new materials in this field have broad prospects.

2. The Definition of New Materials

New materials are referred to materials that have excellent performance and special features which traditional materials do not have newly emerging or developing. In addition, the traditional materials whose performance have been significantly improved or have produced new features also belong to the new material. It is generally believed that some key materials to meet the needs of the development of high-tech industries are also the category of new materials. New materials have been widely used in various industries. As the basic and pillar strategic industry, new materials are the foundation and the forerunner of modern high technology and industry.

3. Thermal Insulation and Energy Saving Materials

3.1. Foam Concrete

Foam coagulation is transferring the vesicant into full foam by mechanical means through the foaming system of foaming machine and uniform mixing the foam and cement slurry. Then the pumping system of the foaming machine is used to carry out the pouring construction or the mold forming, and a new type of lightweight thermal insulation material which contains a large number of closed pores formed after the natural curing. Foamed concrete is a new kind of building energy-saving materials which is waste re-cycled, environmental friendly, energy saving, low-cost and non-burning. As a new type of energy-saving and environmental protection building materials, domestic and foreign scholars have done a lot of research, so that it is widely used in wall materials. The advantages are: small density, light weight, heat preservation, sound insulation, earthquake resistance and other properties. But it also has some shortcomings, such as low strength, cracking, water absorption and so on. So it is necessary to do further re-search on foaming agent, mix ratio, process flow and equipment to extend its application field.

3.2. Estazolam Plate

Estazolam plate is a new type of wall material which can be used to replace the brick for load-bearing walls, non-bearing walls, floor, balcony, etc. It has many advantages such as expanding the use area of buildings (10%-15%), heat insulation, sound insulation, seismic resistance, waterproof, construction simple, etc. At the same time, greatly reduce the cost of the project. It is made of cold drawn low carbon steel wire which is welded together that has the characteristics of strong joints, reasonable structure, uniform mesh, etc. It is a kind of special mesh used in the construction of floor heating.

3.3. Autoclaved Aerated Concrete

The concept of aerated concrete is light porous silicate product with calcareous and siliceous materials as the main raw material and foaming agent (aluminium) as adulterating agent. The technological pro-cess includes batching, mixing, pouring, pre raising, cutting, autoclaving, maintenance, etc. it contains a large number of uniform and small pores. The performance advantages are: light weight, fire proofing, sound insulation, heat preservation, anti-seepage, an-ti-seismic, green, lasting long, etc. Because of these advantages, it is commonly used in high storey frame architecture, seismic area building, building in icy cold area, soft ground construction, etc. Aerated concrete production is rich in raw materials, especially the use of fly ash as raw materials which has two good effects. First it is good for the integrated utilization of the industrial residue, control of the environmental pollution, protect of the arable land. Second it also can create a good social benefit and economic benefit. Aerated concrete is one kind of ideal wall material to replace the traditional solid clay brick which is strongly supported by the nation-al wall reform policy, tax policy and the environmental protection policy over the years. It has broad marketing space and developing prospect. The title is set 17 point Times Bold, flush left, unjustified. The first letter of the title should be capitalized with the rest in lower case. It should not be indented. Leave 28 mm of space above the title and 10 mm after the title.

4. Roofing Energy Saving Materials

4.1. Vermiculite and Vermiculite Brick

Vermiculite is a natural non-toxic mineral which expands under high temperature. It is a relatively rare mineral and belongs to the silicate. The use of expanded vermiculite is very extensive, but it is mainly used as building material. According to the American consumption structure in 1986, expansion vermiculite accounted for 52% among all the materials used as mortar and cement pre mixture and light-weight aggregate concrete. In Britain expansion vermiculite accounted for 40% among all the different materials used as concrete, mud painted on the wall and cement coagulant .It is applied in many different fields, including building, forestry, metallurgy, etc.

Vermiculite brick is one kind of heat insulation products with regular shape. Vermiculite is the raw materials added with binder. There are vermiculite cement products, asphalt vermiculite products, etc. The density of vermiculite brick is 0.4~0.5g/cm3, room temperature thermal conductivity is 293 to 502.4 J/ (m·h·°C) (70~120cal/ (m·h·°C), compressive strength is under 0.5MPa and water absorption rate is around 6%. Vermiculite brick belongs to the thin-wall hollow structure products. The strength and the melting point are low. The refractoriness and high temperature properties were lower than other light-weight refractory materials. It should not be used for load-bearing parts and should not be used as high temperature heat insulation material; the temperature it can withstand is 800 to 900°C. Vermiculite brick is widely used in roof treatment, noise protection, etc.

4.2. Polystyrene Foam Board
Polystyrene foam board -- also known as EPS foam board, EPS board is white object composed of expandable polystyrene beads containing volatile liquid foaming agent which is heated and formed in the mold after preheating. Its structure feature is the subtle obturator.

EPS foam board has a lot of advantages including heat preservation, sound-absorbing, self-extinguishing, antiaging, etc. So it is mainly used for building wall, roof insulation, composite board insulation, cold storage, air conditioning, floor heating, decorative carving, etc.

5. Steel Architecture Materials
Polyurethane board is made of PU, or polyurethane sandwich panel formed by PU and color steel plate. It is wildly used in energy saving industrial plants, civil construction, dwelling house, etc.

The characteristics are: good fire resistance, non-toxic and tasteless, waterproof, moisture proof, low thermal conductivity, good thermal insulation performance, etc. So it is a kind of widely used high efficiency energy-saving building envelope material. It has great potential and a new type of energy saving plate which is advocated and promoted by the national Ministry of Construction.

6. Summary
New materials for civil engineering occupy an important position for the use of resources and the environmental impact. The output, energy consumption, environmental protection and other aspects are big part of the national economy. New materials have the features of multi - function and high performance, to be specific, light weight, high strength, good technology and excellent durability. All of these features can meet the needs of social development and make people have a more comfortable living environment. At present, the national policy guidance on development of new materials is a huge opportunity.

7. References
[1] Yaming Liu. 2013. Application of new materials in Civil Engineering. Build World (23):132.
[2] Xiaoyu Xie.2015. The new development of civil engineering materials in City Construction. Consume Guide (6):275.
[3] Liqin Wu.2014. Application of new concrete materials in the field of Civil Engineering. Guangdong Science & Technology (8):135-136.
[4] YujiaWang.2014. The application analysis of new concrete materials in civil engineering. Urbanism and Architecture (14):257.
[5] Wei Qian. 2009. The application of new materials and new technology in Civil Engineering. Zhengzhou: the Yellow River Water Conservancy Press.
[6] Danping Shi&Jingfeng Fu.2015. The characteristics and development of new building materials. Jiangxi Building Materials (16): 304.
[7] Shengyi Ren&Ling Lai.2015. Civil engineering materials. Beijing: China Building Materials Industry Press.

IOP Conf. Series: Materials Science and Engineering **484** (2019) 012017 doi:10.1088/1757-899X/484/1/012017

Chip Deformation Mechanism during Drilling of Nickel-based Superalloy FGH97

Yang Qiao[a], Peiquan Guo[b,*], Hongtang Chen[c], Shouren Wang[d] and Hui Wang[e]

Department of Mechanical Engineering, University of Jinan, 336 Nanxinzhuangxi Road, Jinan, Shandong 250022, P.R. China.
Email: [a]me_qiaoy@ujn.edu.cn, [b]oss_guopq@ujn.edu.cn, [c]me_chenht@ujn.edu.cn, [d]me_wangsr@ujn.edu.cn, [e]wanghui@163.com

Abstract. Nickel-based superalloy is widely used in aviation, aerospace, shipping and petrochemical industry. An experimental investigation was conducted to analyze the effects of cutting parameters on chip deformation during drilling process of nickel-based superalloy FGH97. Drilling experiments have been carried by means of a machining center YCM-V116B and TiN/TiAlN coated carbide drill bit SECO made SD203-8.0-27-8R1. The chip root is obtained by means of ourselves-made device. The chip deformation decreases with the increasing of cutting speed during drilling process of Ni-based superalloy. The larger the feed rate, the less the chip deformation is. When cutting speed is from 6 m/min to 12 m/min, the shear angle of drilling changes from 14.8 degrees to 19.9 degrees. When feed rate speed is from 0.03mm per tooth to 0.06mm per tooth, the shear angle of drilling changes from 13.7 degrees to 17.4 degrees.

1. Introduction

Nickel-based powder metallurgy superalloy materials are widely used in aviation, aerospace, shipping and petrochemical industries because of their excellent high temperature strength, thermal stability and thermal fatigue properties [1]. In the early 1960s, U.S. began to develop and apply Ni-based powder metallurgy superalloys firstly. U.S.S.R. began to study Ni-based powder metallurgy superalloys in the late 1960s. P.R. China began to develop Ni-based powder metallurgy superalloys until 1970s. Superalloy FGH95 is characterized as high strength at working temperature of 650 degrees Centigrade. Superalloy FGH96 can be used at temperatures up to 700-750 degrees Centigrade. It is a key material for preparing high performance engine turbine disk. Superalloy FGH97 prepared by plasma rotating electrode pulverizing and hot isostatic pressing has the advantages of high tensile strength and high creep resistance, and its crack propagation rate is one order of magnitude lower than that of FGH95 or FGH96 [2-4]. Ozcelik and Bagci [5] proposed a new method for measuring drilling temperature and measured the drilling temperatures of AISI-1040 and AI7050-T651. Sato, et al. [6] found that cutting thickness is the greatest factor affecting drill wear, drill wear of processing blind-hole is more serious than that of processing through -hole as drilling the same depth of hole because the drilling temperature near the bottom of blind-hole material increases sharply. The higher the drilling temperature, the faster the wear rate of drill is. Du [7, 8] investigated the microstructure of white layer appeared in the high speed machining of FGH95. Some research about cutting force [9-11], machined surface integrity [12, 13], bit wear and tool life [14, 15] during drilling process have been done.

The chip deformation in cutting process can reflect the essence of cutting phenomena. Nickel-based superalloy is of high strength, high hardness, low thermal conductivity and porous structure. It is more difficult to drill Ni-based superalloy than to turn or to mill because the heat generated in the drilling

IOP Conf. Series: Materials Science and Engineering **484** (2019) 012017 doi:10.1088/1757-899X/484/1/012017

process cannot be dispersed in time and the temperature in the drilling area is very high. The shear angle can characteristic the chip deformation degree in drilling process. It reflects the essence of physical phenomena such as drilling force, drilling temperature, bit wear and machined surface quality. It is very important to study chip deformation for optimizing the drilling parameters of Ni-based superalloy.

2. FGH97 Drilling Test

2.1. Properties of FGH97

Table 1 shows the composition of FGH97. The alloy elements dissolve a large number of solid solutions in Ni and Co resulting in the strengthening phase γ'. The content of strengthening phase γ' in FGH97 is about 64% of the alloy mass fraction. Its tensile strength is 1510 MPa, yield strength is 1080 MPa and solid solution temperature is 1180-1190 degrees Centigrade. Therefore, the physical properties of FGH97 seriously restrict its machinability. Because of the special manufacturing process of materials, thermal-induced pores are formed in the material during the manufacturing process. When the tool cuts the induced pores, the tool will be impacted. The continuous impact has a great impact on the tool performance. Moreover, the existence of thermal-induced pores is not conducive to the heat dissipation during the cutting process. The thermal induced porosity also severely restricts the machinability of the material. So, FGH97 is very difficult to be machined.

Table 1. Composition of FGH97 (wt %).

C	Cr	Co	Al	Ti	W	Nb	Hf	B	Zr	Mg	Ce	Ni
0.04	9.0	15.5	4.9	1.8	5.5	2.6	0.3	≤0.015	≤0.015	≤0.02	≤0.01	bal

2.2. Experiment and Test Setting

Drilling experiments have been carried by means of the machining center YCM-V116B as shown in Figure 1. TiN/TiAlN coated carbide drill bit SECO made SD203-8.0-27-8R1 is used to drill FGH97. The value setting of cutting speed, feed rate are shown in Table 2. Figure 2 shows the photograph of drilling experiment. The chip root is obtained by means of our self-made device as shown in Figure 2. The specimen is placed in the holder supported by two bearings and fixed by a positioning pin. When the positioning pin is pulled out, the specimen rotates with the drill bit to ensure that the chip and the specimen rotate with the drill bit together, so chip root can remain in the base materail of specimen.

Figure 1. Machining center YCM-V116B.

IOP Conf. Series: Materials Science and Engineering **484** (2019) 012017 doi:10.1088/1757-899X/484/1/012017

Table 2. Cutting parameters.

Factor	Level			
	1	2	3	4
Diameter of drill bit (mm)	8	8	8	8
Cutting speed v_c (m/min)	6	8	10	12
Feed per teeth f_z (mm/z)	0.03	0.04	0.05	0.06

Figure 2. Device for drilling chip root obtaining.

3. Test Results and Discussion

3.1. Test Results

Shear slip occurs in the first deformation zone, grain elongation, and the angle between the direction of cutting speed and the shear plane is the shear angle φ as shown in Figure 3. The value of shear angle is opposite to the extent of chip deformation, that is, the less the shear angle, the greater the chip deformation is. The combinations of cutting speed, feed rate and the measured values of shear angle are shown in Table 3.

Table 3. Variables and test results.

No.	Cutting speed v_c (m/min)	Feed per tooth f_z (mm/z)	Shear angle φ (°)	No.	Cutting speed v_c (m/min)	Feed per tooth f_z (mm/z)	Shear angle φ (°)
1	6	0.05	14.8	5	10	0.03	13.7
2	8	0.05	15.9	6	10	0.04	13.9
3	10	0.05	16.1	7	10	0.05	16.1
4	12	0.05	19.8	8	10	0.06	17.4

3.2. Test Result Analysis and Discussion

As the diameter of drill bit is 8mm and feed rate is 0.05mm per tooth, the shear angles corresponding to cutting speed of 6m/min, 8m/min, 10m/min and 12m/min are shown in Figure 4. Figure 4 shows that the shear angle during drilling process of Ni-based superalloy increases, that is, chip deformation decreases with the increasing of cutting speed.

Figure 5 shows shear angles of drilling nickel-based superally FGH97 corresponding to feed rate of 0.03mm per tooth, 0.04mm per tooth, 0.05mm per tooth and 0.06mm per tooth as drill bit diameter is 8mm and cutting speed is 10mm/min. The shear angle of drilling nickel-based superally FGH97 increases with the increasing of feed rate. On the other word, drilling chip deformation of nickel-based superalloy will decrease with the increasing of feed rate.

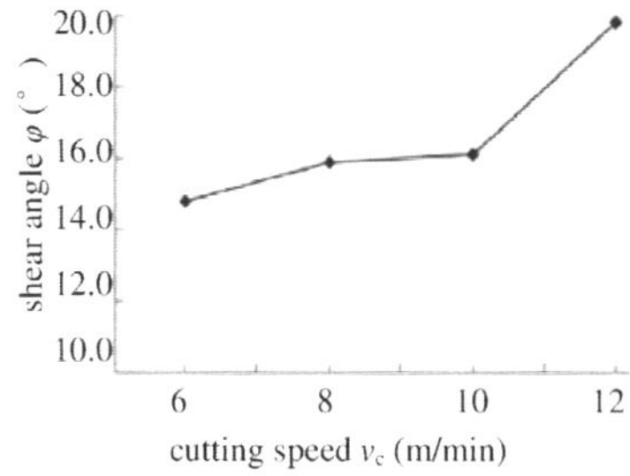

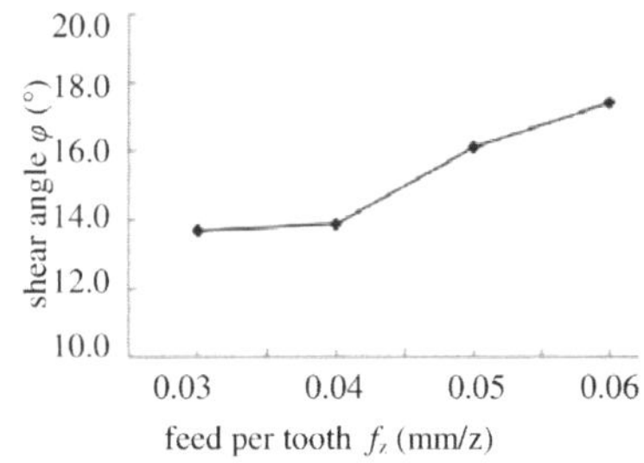

Figure 3. Shear angle. **Figure 4.** Effect of v_c on shear angle. **Figure 5.** Effect of f_z on shear angle.

4. Conclusion

The chip roots in drilling nickel-base superalloy were obtained by using our self-made device, and the drilling deformation mechanism was studied. As the drilling speed increases, chip deformation decreases, and chip deformation becomes smaller with the increase of feed rate. When cutting speed is from 6 m/min to 12 m/min, the shear angle of drilling changes from 14.8 to 19.9 degrees. When feed rate speed is from 0.03mm per tooth to 0.06mm per tooth, the shear angle of drilling changes from 13.7 degrees to 17.4 degrees

5. Acknowledgments

This work was supported by Shandong Natural Science Foundation of China (No. ZR2017LEE015), Key R & D projects (civil military integration) of Shandong, China.

6. References

[1] Thellaputta G R, Chandra P S and Rao C S P 2017 *Materials Today Proceedings 4* 3712
[2] Zhang M, Li F, Yuan Z, Li J and Wang S 2013 *Materials & Design 49* 705
[3] Zhang M J, Li F G, Wang S Y and Liu C Y 2011 *Materials Science and Eng.: A 528* 4030
[4] Zhong B, Wang Y, Wei D, Zhang K and Wang J 2018 *Int. J. of Fatigue* 109 26
[5] Ozcelik B and Bagci E 2006 *Materials & Design 27* 920
[6] Sato M , Aoki T, Tanaka H and Takeda S 2013 *Int. J Machine Tool & Manufacturing* 68 40
[7] Du J, Zhang J and Wang L 2018 *Rare Metal Materials and Engineering 47* 2275
[8] Du J, Liu Z and Lv S 2014 *Applied Surface Dcience 292* 197
[9] Arif R, Fromentin G and Rossi F 2018 *Procedia CIRP* 77 425
[10] Uçak Nand Çiçek 2018 *J.of Manufacturing Processes* 31 622
[11] Hamade R F, Seif C Y and Ismail F 2006 *Int. J Machine Tool & Manufacturing* 46 387
[12] Wang X, Yang L and Qiao Y 2018 *Procedia CIRP* 77 370
[13] Venkatesan K 2017 *J. of Advanced Research* 8 407
[14] Wang X, Huang C and Zou B 2013 *Int. J. of Advanced Manufacturing Technology* 64 41
[15] Varote N and Joshi SS 2017 *J. Materials Eng. and Perf.* 26 4391

IOP Conf. Series: Materials Science and Engineering **484** (2019) 012018 doi:10.1088/1757-899X/484/1/012018

Temperature Effect on Dislocation Slip Mechanism of Nanotwinned Mg with Void Defect at the Twin Boundary

Kaixin Huang, Junping Yao, Qiyao Hu, Letian Shao and Zhong Sun

School of Aeronautical Manufacture Engineering, Nanchang Hangkong University, Nanchang 330063 China

Email: h13647096714@outlook.com

Abstract. In this paper, a molecular dynamics model is used to simulate the uniaxial tensile loading process at different temperatures of nano-twinned Mg with void defect at the twin boundaries. The dislocation slip mechanism during the stage of plastic deformation for different temperatures was studied. The results show that the initial edge dislocations formed from lattice misfit. It is found that the propagations of the slip bands emitted and the stacking faults also generated from Shockley partial dislocation.

1. Introduction

In recent years, many achievements have been made in the research on macroscopic defects of materials and breakthroughs have been made in microscopic defects. For example; Stella Brach [1] *et al.* used MD to study the effect of pore shape on the strength properties of nano-porous materials. The results show that the strength properties of nano-porous materials depend on all three Hagrid- Sigyard stress invariants. Xu[2] *et al.* studied the contraction of the spherical nano hole under the uniaxial compression load by using MD simulation, and found that the original defect of the model may shrink during the compression process. Tang [3] *et al.* used MD tosimulate the expansion and fracture mechanism of single-crystal γ-Ti Alnanopores. The results show that the generation of dislocation nuclei and the propagation of shear rings expand the pores. Zhang [4] *et al.* simulated the effect of hole density of nickel single crystal on hole contraction mechanism based on molecular dynamics. The results showed that whole density could affect yield strain of nickel single crystal. However, they are mainly concentrated on the crack propagation of nano-single crystals, but there are few studies on the mechanism of the dislocation slip of twin boundaries (TB). Therefore, it is necessary to study the dislocation slip mechanism of the defect in the grain boundary of the TB. In addition, with the continuous development of classical MD and computer science and technology, we have been able to provide theoretical and technical support.

In summary, MD simulations are carried out with the Largescale Atomic Molecular Massively Parallel Simulator (LAMMPS) [5], and software OVITO [6] was used for visual processing to study the dislocation slip mechanism of the TB hole defects, to view the evolution of partial dislocations and slip bands during the tensile deformation, the Dislocation analysis (DXA) is adopted.

2. Model and Theoretical Method

In order to better observe the change process of dislocation slip mechanism, the model established should be of appropriate size. The size of the model is 320×240×12.8*nm*, and it is divided into the upper part and the lower part, which are respectively 320×240×12.8 *nm*. The diameter of hole defect at the TB is R=16*nm*, and the total number of atoms is 84360.Periodic boundary conditions were applied in all three directions, and the EAM potential function developed by Sun [7] *et al.* was used to describe the atomic potential of magnesium. The energy is minimized and 10,000 time step relaxation is performed before the uniaxial tensile load is applied, so that the total energy of the system reaches the equilibrium state. The uniaxial tensile load was then applied along the Y direction; the uniaxial tensile loading is applied with a constant strain rate of$2\times10^{9}s^{-1}$, with a MD time step of 0.001 *ps*, the simulations are performed with NPT ensemble to erase the stresses of other two directions that are perpendicular to the loading direction at different temperature.

3. Results and Discussion

In order to explore the effect of temperature on the dislocation slip mechanism of void defects at the TB, three different temperatures 200 K, 300 K and 500 K, were set up respectively. As can be seen from Fig. 1(a), stacking faults (dark red cylinders in the Fig. 1(a)) and Shockley partial dislocations (yellow brown cylinders in the Fig. 1(a)) occur at the TB and the upper and lower boundaries of the model around the void defect. The reason for this is the Shockley partial dislocation caused by lattice misfit, and the partial dislocation will inevitably lead to stacking fault, that is, the partial dislocation line is the boundary line of stacking fault. In Fig. 1(b), due to the occurrence of plastic deformation, a large number of slip bands appear in the whole model. At the same time, the number of partial dislocation and stacking fault also increased. Fig. 1(c) as the plastic deformation increases, the TB of the upper and lower parts fuse with each other, resulting in small void defects at the left and right boundaries of the model. Fig. 1(d) it can be seen that the slip bands become denser and longer, and there is a slip network characterized by multiple slips. The number of partial dislocations and stacking faults increases continuously, indicating that the plastic deformation is very serious at this time.

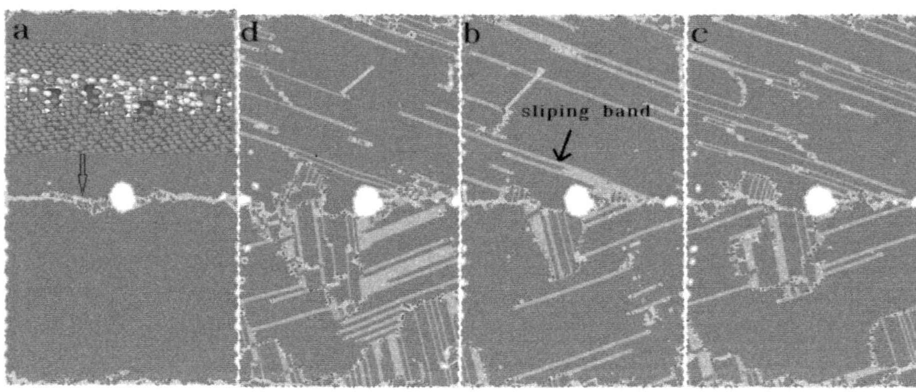

Figure 1. Dislocation analysis of the deformed atomic configurations at 200 K
(a) ε=0.02 (b) ε=0.05 (c) ε=0.08 (d) ε=0.12

As can be seen from Fig. 2(a), the distribution of partial dislocations and stacking faults in Shockley is similar to that in Fig. 1(a), which can be proved that temperature has no effect on partial dislocations and stacking faults. Fig. 2(b) compared with Fig. 1(b), the number of slip bands decreases but the number of twins increases. Fig. 2(c) shows that the grain boundary around the defect begins to break, indicating that the yield strength decreases with increasing temperature. Fig. 2(d) partial dislocations and stacking faults accumulate with the increase of deformation and form twin nucleation on the left side of the void defect, which hinder the fracture of the material.

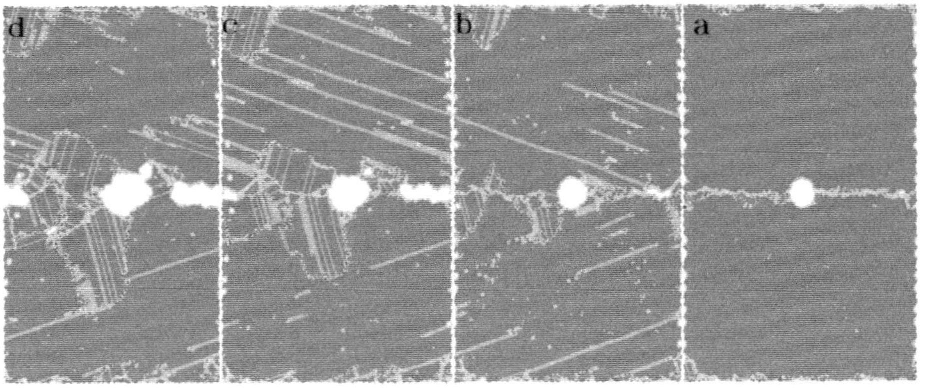

Figure 2. Dislocation analysis of the deformed atomic configurations at 300 K
(a) ε=0.02 (b) ε=0.05 (c) ε=0.08 (d) ε=0.12

When the temperature is 500 K, the same distribution as that of Fig. 1 (a) and Fig. 2 (a) is observed in Fig. 3(a), which proves that the temperature has no obvious effect on the mechanical properties of the material during the elastic stage. It is shown from Fig. 3(b), 3(c) and 3(d) that the critical shear stress of slip decreases sharply with the increase of temperature, and the probability of slip initiation increases greatly, resulting in the increase of slip band and the decrease of twins. At the same time, the increase of temperature makes the thermal motion of atoms and the energy of atoms intensify, which leads to the increase of plasticity of materials and is not easy to fracture.

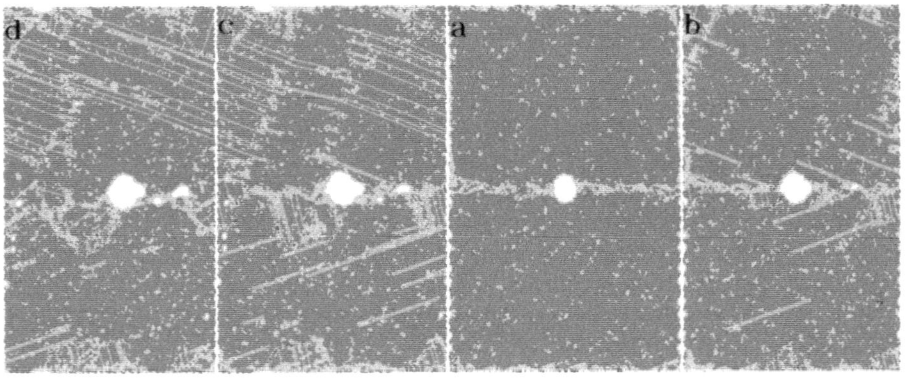

Figure 3. Dislocation analysis of the deformed atomic configurations at 500 K
(a) ε=0.02 (b) ε=0.05 (c) ε=0.08 (d) ε=0.12

4. Conclusions

In this paper, a molecular dynamics model is used to simulate the uniaxial tensile loading process at different temperatures of nano-twinned Mg with void defect at the TB. The following conclusions are drawn:

(1)Temperature does not have a significant effect on the mechanical properties of the elastic deformation stage of the material, but it has a significant effect on the dislocation slip mechanism in the plastic deformation stage;

(2) The number of partial dislocations and stacking faults is proportional to the material deformation, and then they accumulate to form twin nucleation, hindering the fracture of the material.

5. Acknowledgements

This work was financially supported by the National Natural Science Foundation of China (51164027, 51661024) and Key Science and Technology Project of Jiangxi Provincial Department of Education

(GJJ14502).

6. References

[1] StellaB StefanoCDjimedoK2017*J*. Void-shape effects on strength properties of nanoporous materials *Mech .Res. Commu*86:11-17
[2] XuX TTangF LXueH T2015*J*Molecular dynamics simulations of void shrinkage in γ-TiAl single crystal*Comput.Mater.Sci*107: 58-65
[3] Tang F LCaiH MH WBao2014*J*Molecular dynamics simulations of void growth in γ-TiAl single crystal*ComputMater.Sci.*84:232-237
[4] Zhang Y QJiangS YZhu X M2017*J*Influence of void density on dislocation mechanisms of void shringkage in nickel single crystal based on molecular dynamics simulation*Phy.E*90:90-97
[5] Plimpton S J1995*J*Fast parallel algorithms for short-range molecular dynamics *JCompuPhys* 117(1):1-19
[6] Stukowsk A2010 *J*Visualization and analysis of atomistic simulation data with OVITO-the open Visualization Tool *Modell. Simul .Mater .Sci. Eng*18 (6):2154-2162

IOP Conf. Series: Materials Science and Engineering **484** (2019) 012019 doi:10.1088/1757-899X/484/1/012019

Experimental Study on Preparation and Mechanical Properties of CF Knitted Fabric Composites for Nut Lining

Bin Zhao, Zhipei Tong, Qiang Zhang, Peng Zhang, Yongqi Cheng and Yousong Sun

Material and energy institute, Guangdong University of Technology, Guangzhou, Guangdong 510006, China
Corresponding author e-mail: zp1990@sina.com

Abstract. To solve the problem of preparing composite screw nut lining, a preparation method of lining by the CF knitting stretch fabric was discussed. The effects of vacuum treatment and secondary reinforcement of chopped fiber on the mechanical properties of CF woven fabric and knitted fabric composite were studied. The results showed that the compressive strength of composites increased by 8.86% and 8.08% after vacuum treatment, and the impact strength increased by 15.55% and 9.43% respectively. The secondary reinforcing effect of chopped fibers was obvious, and the compressive strength increased by 9.91% and 22.52% respectively. The actual application showed the composite nut lining prepared by using the CF tubular knitted stretch fabric could better solve the defects of insufficient strength and partial peeling.

1. Introduction

Transmission screw pairs are important functional elements for machineries. In heavy load and low velocity conditions, the lubrication will be worse and wearing of the friction pairs become serious, reducing transmission efficiency and shortening service life. A new type of transmission screw nut has been developed with the structure of SB/CF-EP lining. This nut is made of steel backing (SB) with carbon fiber reinforced plastic (CFRP) composite lining composed by epoxy resin (EP) as matrix and some self-lubricants as additives. This structure has both advantages of heavy bearing capacity and excellent tribological behaviors, increasing transmission efficiency by 35-50% [1]. This nut lining structure is usually prepared by the CF fabric laying process, and the spiral surface is not easily formed, resulting in the composite nut being not widely used in engineering practice: The nut lining structure prepared by using the chopped fiber as the reinforcing body and by injection molding is another way of preparation [2], but this nut has poor mechanical strength and affects the nut transmission efficiency. In this study, a spiral single-layer CF knitted fabric was used to prepare the lining structure. The lining layer had a complete thread structure without lap joints, which could effectively solve the problem of strength discontinuity in the fiber layup process. At the same time, the CF knitted fabric had a grid-like skeleton support structure, which could compensate for the defects such as insufficient strength of the chopped fiber injection molding lining and easy local peeling. In this paper, the mechanical properties and preparation processes were studied.

2. Fabrication of Samples

2.1. Experimental Materials

Carbon fiber (CF) reinforcement material: plain woven fabric, T300-1K, produced by Toray Co. Japan, the detailed specifications are shown in Table 1 and Table 2; Resin and additives: Epoxy resin E51, industrial grade, produced by Yueyang Baling Petrochemical Co., Ltd.; Curing materials: 4-4' diamino Diphenylmethane (DDM), analytical grade, Light yellow crystalline flake at room temperature, produced by Aladdin Reagent Co.; Other materials: Epoxy reactive diluent 501; deputy phthalate; Solid lubricating components: graphite, MoS2, analytical grade, produced by Tianjin Damao Chemical Reagent Factory; Talc powder, commercially available. According to previous studies, a formula for CF/PTFE composite was shown in Table 3.

Table 1. The essential property of T300-1K carbon fiber.

Type	Tow	Tensile strength(Gpa)	Elongation	Linear density (g/km)	Monofilament diameter (μm)
T300-1K	1000	3530	1.5%	66	5~7

Table 2. The product specification of Japan Toray 1K carbon fabric cloth.

Warp yarn	Weft yarn	Weaving type	Thickness	Width(cm)	Density(g/m^2)
T300-1K	T300-1K	Plain	0.15	100	119

Table 3. Composite matrix material formula (%).

NO.	Epoxy	DDM	Active diluent	Plasticizer	Graphite	MoS_2	French chalk	Chopped Fiber
No.1	50	12.5	2.5	5	6	9	15	/
No.2	50	12.5	2.5	5	3.25	5	8.25	13.5

2.2. Fabrication of Samples

2.2.1. Preparation of the CF Knitted Stretch Fabrics. The basic sample of the CF knitted stretch fabrics was manually knitted by the flat knitting machine. The knitting parameters were as follows: the elastic scale value was 4, the machine number was 7 stitches per inch, and the T300-1000/3000 type fiber bundle was used. Fabric thickness was about 1.2mm; Stretch rate of the CF knitted fabrics is 69.1%.

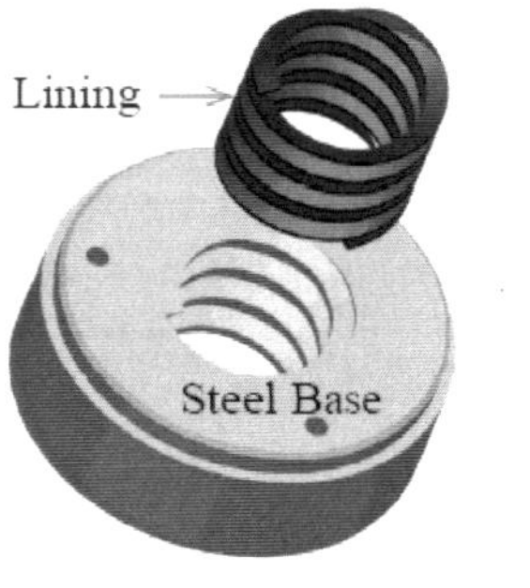

Figure 1. SB/CF-EP nut structure.

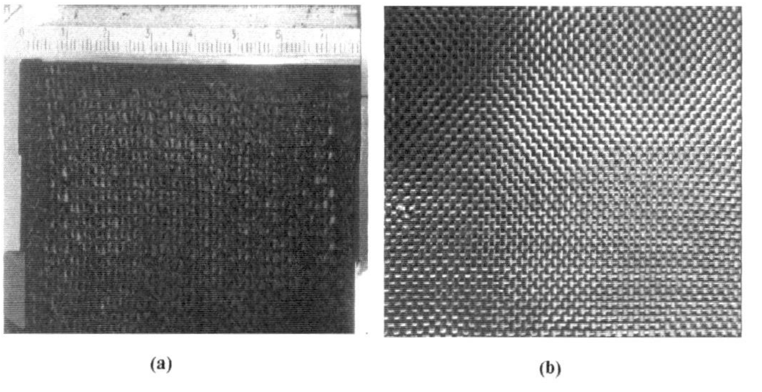

Figure 2. Photos of CF knitted fabric (a) and woven fabric(b).

2.2.2. Preparation of resin and Additives Colloid. Before being used, Weighed the EP, curing agent, modified components, etc. according to the specific formula ratio, The EP and the modified component were mixed and stirred uniformly at room temperature, and the DDM curing agent was heated to 120°C to melt, in order to prevent the liquid DDM from cold crystallization again, the other components were pre-baked at low temperature for 20 min, and then DDM was evenly mixed. Finally, the vacuum defoaming treatment was carried out at room temperature for 30 min.

2.2.3. Preparation of CFRP Composite Samples. The prepared knitted fabric or purchased woven fabric was cut into a certain size, and pre-treated according to the sequence of: acetone cleaning, nitric acid heating oxidative, coupling treatment. Using hand lay-up method, uniformly coated the glue on the surface of the fabric, stacked and placed it in the mold. After been pressurized to the required thickness, the mold was placed in an oven and heated in accordance with the curing step of the resin body. After cured, it was naturally cooled, demolded, processed and polished.

2.2.4. Fabrication of SB/CF-EP Nuts. The SB/CF-EP nut molding procedures were as follows: Prepared the CF tubular knitted fabric and resin, Then coated the screw mandrel with wax of about 0.12 mm thickness, set the knitted CF tube on the screw mandrel, tightened the tube on grooves of the thread with thin CF band, dipped CF tube into resin matrix, then vacuumed, rotated the mandrel with CF tube into steel baking part and assembly the set. Replenished resin matrix through the hole of the steel backing, cured, demolded, post processed. Figure 3 showed the key processes and the sample screw nut of the standard trapezoid screw Tr32x8 [3].

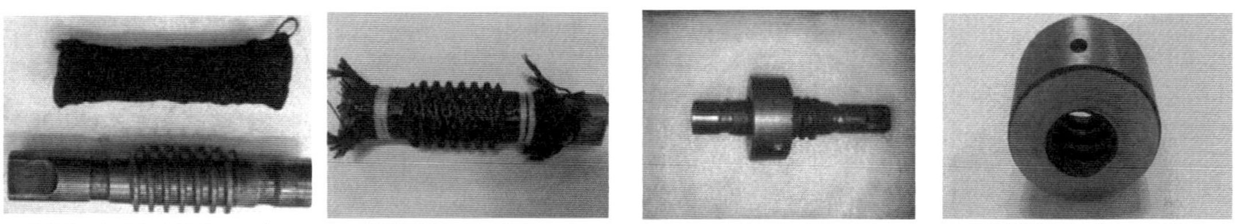

(a) the knitted Tube and Thread mandrel (b)CF knitted Tube wrapped on screw mandrel (c)molding set after curing (d) SB/CF-EP nut after demoulding

Figure 3. Key photos of SB/CF-EP Nut molding procedures.

2.3. Experimental Equipment

CMT5504 Microcomputer Control Electronic Universal Tensile Machine, Shenzhen Xinsansi Material Testing Co., Ltd.; Hot pressing mold, homemade; S-3400N scanning electron microscope, Japan Hitachi Co.; Vacuum drying oven, Shenzhen Kuaike Manufacturing Technology Co., Ltd.; MR5000 inverted metallographic microscope, Nanjing Jiangnan Yongxin Optical Co., Ltd.

2.4. Test Methods

2.4.1. Impact Strength. The impact strength test of the CFRP composites was carried out according to the national standard "GB/T 1843-2008 Plastic cantilever beam impact strength measurement". The sample size was 80×10×4mm, and the unnotched impact sample was adopted. Support plane of sample was attached to the support block of the tester; the impact edge was aligned with the center of the sample.

2.4.2. Compressive Strength. The compressive strength test of CFRP composite was carried out according to the national standard "GB/T 1448-2005 FRP compression performance test method". To avoid the instability, the sample size was 10×10×10mm, and the error range was 0.5mm. The upper and lower end faces of the sample were required to be parallel to each other, and the non-parallelism was less than 0.1% of the height value. Loading speed was 2 mm/min. The direction of compression was perpendicular to the lamination direction of the fabric, and the maximum load at the moment of destruction of the specimen was recorded.

2.4.3. Resin Mass Percentage of Composite. The experiment was carried out according to the national standard "GB/T 3855-1983 CFRP resin content test method", the content of CFRP resin was determined by sulfuric acid digestion method, the method was suitable for the CFRP which could completely decompose the resin colloid under certain conditions and did not excessively corrode the fiber.

3. Experiment results and Analysis

3.1. Effect of Vacuum Treatment Process on Mechanical Properties

Table 4 showed the mechanical properties in compression and impact strength and the resin mass percentage prepared by vacuum or non-vacuum treated to the CF plain woven and knitted fabric sample. Compared to non-vacuum process, It could be seen that the compression performance of the vacuum-treated CF plain woven and knitted fabric was improved by 8.9% and 6.1%, and the impact performance was improved by 15.6% and 9.6%, respectively. The addition of chopped CF (No.2) increased the compressive strength of CF plain woven and knitted fabric sample by 9.91% and 22.52%, respectively. However, the impact strength of the two showed a different degree of decline, which decreased by 29.34% and 17.97%, respectively. The resin mass percentages of the CFRP were changed by the vacuum treatment. The resin mass percentage of the CFRP vacuum treated of CF plain woven fabric and the CF knitted stretch fabric was 19.9% and 21.8% higher than the non-vacuum treated.

Table 4. The test date of CFRP Sample.

CFRP Sample type	Impact strength (KJ/m^{-2})	Compression strength (Mpa)	Resin content (%)
woven fabric /non-vacuum	55.3	453.3	29.42
woven fabric /vacuum	63.9	493.5	35.27
woven fabric-chopped /vacuum	45.2	542.4	/
knitted fabric / non-vacuum	21.8	248	59.68
knitted fabric /vacuum	23.9	263.1	72.69
knitted fabric-chopped / vacuum	19.6	293.4	/

Figure 4(a, b) showed that the CF knitted stretch fabric had a void percentage of 5.6% when not vacuum treated, and the void percentage was 1.0% after 60min vacuum treatment. The percentage of porosity before and after plain woven fabric treatment was 16.2% and 2.3%, respectively. It was indicated that vacuum treatment could effectively fill the resin between the reinforcements, greatly reducing the porosity (Figure 4(c, d)). It could be seen that the gas generated by the reaction between the air in the fabric and the resin was sufficiently discharged before the matrix resin was solidified under vacuum, The fiber bundle of the fabric was also completely wetted by the colloid resulting in a tight bond, There was almost no gap in the resin matrix, fabric layers and layers or grids, which greatly improved the bond strength and transferred load between the fabric reinforcement layers, and improved the bearing capacity [4]. The mechanical strength of the fabric was also improved.

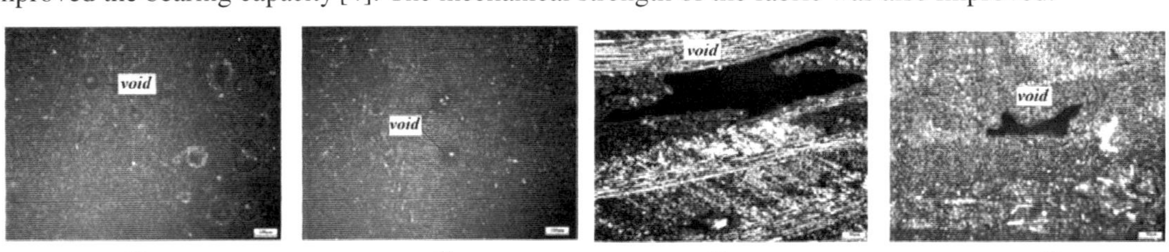

Figure 4. The metallographic micrograph of CF knitted fabric composite.

3.2. Analysis of Fracture Morphology

It could be seen from the shape of the impact fracture sample as shown in Figure 5. In Figure 5(a), the CF plain fabric has a distinct impact fracture, and the fracture morphology was roughly a "V" shaped section. And near the surface of both sides of the sample, there was obvious delamination. It was

IOP Conf. Series: Materials Science and Engineering **484** (2019) 012019 doi:10.1088/1757-899X/484/1/012019

indicated that the bond strength between the layers of the multi-layer carbon plain fabric was not good. Secondly, during the impact process, the fibers in the middle part of the test piece first debond, and the transverse shear fracture breaks and spreads to both sides. In Figure 5(b), the knitted fabric has a relatively flat fracture, and some of the carbon filaments were pulled out. The cross-section fiber material was relatively less than the resin colloid. It was subjected to the impact load, and the fiber bundle acts as a discontinuous reinforcement effect, which was a typical brittle fracture phenomenon [5].

(a) (b)

Figure 5. Photos of impact fracture morphology.

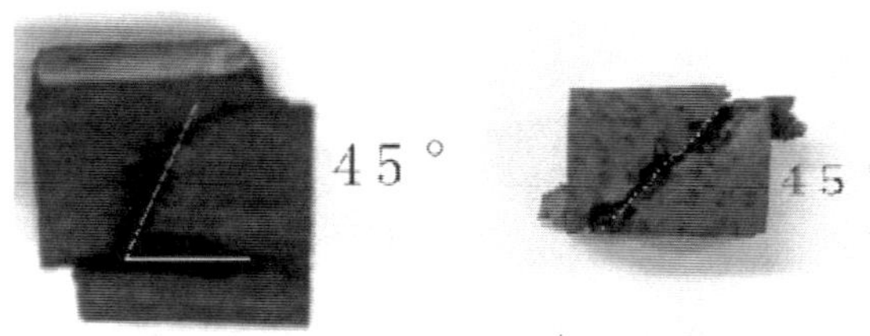

(a) (b)

Figure 6. Photos of compression fracture.

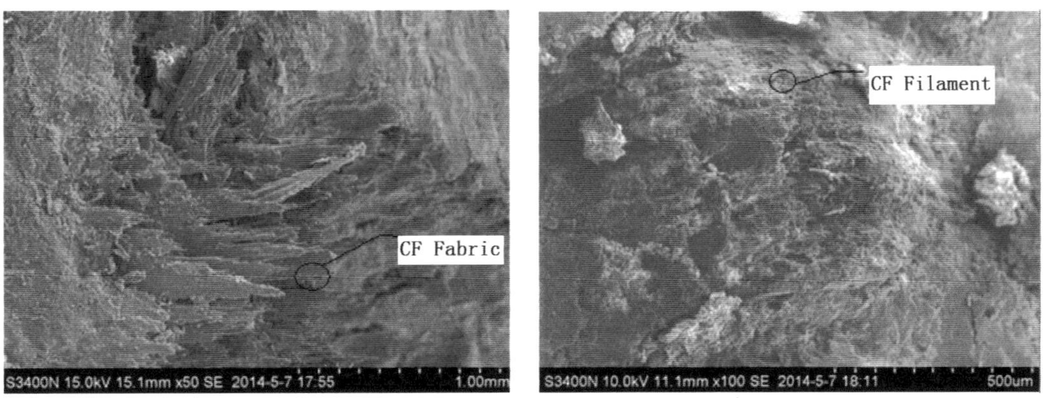

(a) (b)

Figure 7. SEM photo of compression section.

3.3. Analysis of Compression Morphology

Figures 7 and 8 showed the macro topography of the compressed sample and SEM photo of the compressed section. It could be seen from Figure 6 that, after the compression failure, the fiber fracture of the samples were relatively flat along the width of the test piece, and the fracture in the thickness direction was about 45°, which was a typical shear failure [6, 7]. It could be seen from the Figure 7(a) that the fiber bundles of sample by the woven fabrics were arranged very closely, and the fiber breakage at the fracture was clearly visible, that was, the fiber was fractured in a certain direction, indicated that the fiber was the main carrier during compression. The stress concentration between the layers caused the fibers to break in turn and spread the cracks to the next layer. The initial crack location was usually located where the fiber bond strength was weak, which was the inter-layer resin-poor zone. It could be seen from the Figure 7(b) that the main carrier of sample by the knitted fabric was an intermittent CF tow and resin matrix filler. The resin matrix had a lower load carrying capacity than the fiber. The CF fibers on the cross section of the knitted sample were discontinuous and the CF ratio was small, so the mechanical strength of the knitted sample was weaker than that of the plain woven structure. During the compression test, the load-displacement curve showed a linear change before the fracture, which was a typical brittle fracture. However, the compression curve of the knitted sample had several fluctuations before the crack was generated to the final fracture, and the plain sample had a steep drop after the force reaches the maximum value, that was, a sudden crack phenomenon. It was indicated that the knitted fabric sample had small fiber content, and the load was mainly carried by the resin matrix. When the crack extends to the periphery of the fiber bundle, the matrix and the fiber were debonded, and then the fiber was broken by load.

4. Conclusion

Vacuum treatment causes the fibers of the CF plain woven fabric and knitted fabric to be completely wetted by the colloid, resulting in tight adhesion of the fibers. The composite compressive strength was increased by 8.86% and 8.08% compared with non-vacuum conditions, and the impact resistance was increased by 15.55% and 9.43%.

The chopped CF had a significant enhancement effect on the compressive strength of CF plain woven fabric and knitted fabric composites, which played a secondary strengthening role, and its value was increased by 9.91% and 22.52%, respectively, but the impact resistance was reduced to some extent.

The single-layer CF tubular knitted stretch fabric was integrally formed on the screw nut, the chopped CF was used as the secondary reinforcement to prepare the nut lining by vacuum treatment, the problem of inter-layer peeling in the CF fabric laying process could be avoided, and the defect that the strength of the chopped fiber injection molding liner was insufficient was also improved. The composite nut of this structure could basically meet the mechanical properties and preparation process requirements of the transmission nut under heavy duty and low velocity.

5. Acknowledgment

This research was got the financial support by Guangdong Science and Technology Plan Project (2014A010105042, 2017A010102021). The authors express their sincere thanks for these.

6. References

[1] Sun Y S, Cheng Y Q, Zhang P, Zhang Z R, Dong Z X, Zhang J J and Zhou X H 2017 A novel type of transmission screw nuts with structure of steel backing/self-lubricant composite lining. *Journal of Mechanical Engineering*, 53(13) p150-158.

[2] Sun Y S, Li C M, Cheng Y Q and Wei L M 2013 Effective transmission screw nuts of steel backing/self-lubricating composite lining fabricated by injection molding. *J. of Mech. Sci. and Tech.* 27 (1) p77-83.

[3] Zhang Q 2016 *Study on SB/CF-EP self-lubricating composite lining transmission nut preparation and performance* (Guangzhou: Guangdong University of Technology) p21-28.

[4] Pan L J, Liu W P, Chen P and Qi F F 2012 Change in preform thickness throughout the vacuum assisted resin infusion process and the post-filling control. *Acta Materiae Campmitae Sinica*, 29(5) p244-248.

[5] Huang F R and Zhou Y 2008 *Advanced Resin Matrix Composites* (Beijing: Chemical Industry Press).

[6] Meng S H, Tian X X, Du S Y, Li J , We L M and Xu C H 2012 Mechanical properties testing and failure mechanism analysis of 3D woven carbon/epoxy composites. *Acta Materiae Campmitae Sinica*, (02) p115-120.

[7] Zheng Z C, Sun S X, Chen Y, Wang Q and Zhang Y Z 2008 Preparation and axial compressive properties of thin-wall pipe of carbon fiber composite. *Engineering Plastics Application*, (04) p44-47.

Mode I Elastic Solution of Cracked Composite Plate

Yongyong Suo, Purong Jia*, Qiang Wang and Leilei Zeng

School of Mechanics and Civil-Engineering & Architecture, Northwestern Polytechnical University, Xi'an 710129, PR China.
*Email: prjia@nwpu.edu.cn

Abstract. The basic equation of composite mechanics has been established according to the orthotropic material properties. The analytic function of the complex variable including the material parameter is analyzed fully. Boundary problems of composite plate with Mode I crack are solved by using the complex variable method. The formulae for stress fields and strain fields are derived. The derivation of elastic displacement fields is carried out, and the whole displacement expression is determined by the complex variable function.

1. Introduction

Fracture mechanics of non-homogeneous materials or anisotropic materials has more applications in new engineering structure materials. Prediction of crack initiation and propagation must be based on the fracture mechanics. Fiber-reinforced polymer matrix materials are the most typical composites and usually modeled as anisotropic materials at the macroscopic level [1]. The orthotropic plate may have been the base of composites in common use. So this paper focuses on the fracture of orthotropic materials. Mode I stress fields and displacement fields in the cracked plate are mainly analyzed.

Linear elastic fracture mechanics (LEFM) has been found to be a very useful tool for design purposes and investigated in great detail for many engineering materials, whether isotropic or anisotropic [2]. The analysis of stresses near the crack tip holds an essential part of LEFM. The methods of elasticity are used to obtain stresses and displacements in cracked bodies [3]. The only viable method to solve stress-field problems in anisotropic composites is using complex analytic function theory, and the results have been reported [4~6]. But the general solution may have some weakness. So it is necessary to make up a new solution, and this is the paper purpose.

2. Basic Equations and Stress Function

The plane stress state of composite sheets is common and very importance for the application. It is the key point to solve stress-field problems in orthotropic materials. In plane elasticity, the equilibrium equations of three stress components are (body forces are absent):

$$\frac{\partial \sigma_x}{\partial x}+\frac{\partial \tau_{xy}}{\partial y}=0 \quad , \quad \frac{\partial \tau_{xy}}{\partial x}+\frac{\partial \sigma_y}{\partial y}=0 \tag{1}$$

For two-dimensional problems, three strain components can be expressed by the partial derivatives of two displacement components u and v, namely:

$$\varepsilon_x=\frac{\partial u}{\partial x} \quad , \quad \varepsilon_y=\frac{\partial v}{\partial y} \quad , \quad \gamma_{xy}=\frac{\partial v}{\partial x}+\frac{\partial u}{\partial y} \tag{2}$$

IOP Conf. Series: Materials Science and Engineering **484** (2019) 012020 doi:10.1088/1757-899X/484/1/012020

Suppose the principal elastic directions of the plate coincide with the coordinate directions, and let the directions 1, 2 parallel to the axes x, y, respectively. Now consider the linear elastic strain-stress relations, the constitutive equations for the orthotropic materials are given as (plane stress state):

$$\varepsilon_x = \frac{\sigma_x}{E_1} - \frac{v_{12}\sigma_y}{E_1} \ , \ \varepsilon_y = \frac{\sigma_y}{E_2} - \frac{v_{12}\sigma_x}{E_1} \ , \ \gamma_{xy} = \frac{\tau_{xy}}{G_{12}} \tag{3}$$

It is well known that the stress components can be expressed by a real stress function $F=F(x, y)$, which is defined as follows:

$$\sigma_x = \frac{\partial^2 F}{\partial y^2} \ , \ \sigma_y = \frac{\partial^2 F}{\partial x^2} \ , \ \tau_{xy} = -\frac{\partial^2 F}{\partial x \partial y} \tag{4}$$

The equilibrium equations in Eqs (1) can be satisfied naturally. For the plane stress problems, the stress function F must be selected reasonably in terms of the boundary conditions.

3. General Complex Function

In the consideration of cracked orthotropic materials, and also in the construction of suitable stress functions, it is very advantageous to use complex variables. It is well known that two real numbers x and y form the complex number $z = x + iy$ conventionally ($i = \sqrt{-1}$). And the conjugate complex number is expressed by $\bar{z} = x - iy$. For the convenience of general investigation, now we introduce another complex variable (w), that is:

$$w = x + ihy \ , \ \ \overline{w} = x - ihy \tag{5}$$

Where h is a real constant. And we suppose the constant h to be positive, $h > 0$, also it can be called tensile or compressive ratio for the coordinate system. The derivative relation must be given as:

$$\frac{\partial w}{\partial x} = \frac{\partial \overline{w}}{\partial x} = 1 \ , \ \ \frac{\partial w}{\partial y} = -\frac{\partial \overline{w}}{\partial y} = ih \tag{6}$$

For the crack problem, rectangular and polar coordinates are shown in Figure 1. The polar coordinate system centered at the crack tip may be adequate for local stress analysis. In terms of the polar coordinates, the complex variables can be written as:

$$\begin{aligned} w &= x + ihy = a + r\cos\theta + ihr\sin\theta \\ \overline{w} &= x - ihy = a + r\cos\theta - ihr\sin\theta \end{aligned} \tag{7}$$

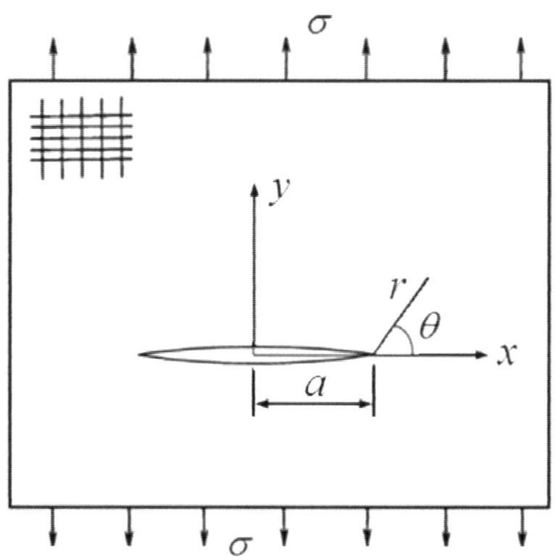

Figure 1. Opening mode plate

Consider an analytic function of complex variable $\Psi = \Psi(w)$ and its conjugate, $\overline{\Psi} = \overline{\Psi}(\overline{w})$. And they can be expressed as:

$$\Psi(w) = \mathrm{Re}\,\Psi + i\,\mathrm{Im}\,\Psi \quad , \quad \overline{\Psi}(\overline{w}) = \mathrm{Re}\,\overline{\Psi} + i\,\mathrm{Im}\,\overline{\Psi} \tag{8}$$

Furthermore, another analytic function Φ and its conjugate $\overline{\Phi}$ are introduced as below:

$$\Phi = \Phi(w) = \frac{d\Psi}{dw} = \Psi' \quad , \quad \overline{\Phi} = \overline{\Phi}(\overline{w}) = \frac{d\overline{\Psi}}{d\overline{w}} = \overline{\Psi}' \tag{9}$$

The partial derivatives of $\Psi(w)$ with respect to x and y should be given by:

$$\begin{aligned} \frac{\partial \Psi}{\partial x} &= \frac{d\Psi}{dw}\frac{\partial w}{\partial x} = \Phi = \mathrm{Re}\,\Phi + i\,\mathrm{Im}\,\Phi = \frac{\partial \mathrm{Re}\,\Psi}{\partial x} + i\frac{\partial \mathrm{Im}\,\Psi}{\partial x} \\ \frac{1}{h}\frac{\partial \Psi}{\partial y} &= \frac{1}{h}\frac{d\Psi}{dw}\frac{\partial w}{\partial y} = i\Phi = i\,\mathrm{Re}\,\Phi - \mathrm{Im}\,\Phi = \frac{1}{h}\left(\frac{\partial \mathrm{Re}\,\Psi}{\partial y} + i\frac{\partial \mathrm{Im}\,\Psi}{\partial y}\right) \end{aligned} \tag{10}$$

Obviously, it is easy to derive the following differential equation:

$$\mathrm{Re}\,\Phi = \frac{\partial \mathrm{Re}\,\Psi}{\partial x} = \frac{1}{h}\frac{\partial \mathrm{Im}\,\Psi}{\partial y} \quad , \quad \mathrm{Im}\,\Phi = \frac{\partial \mathrm{Im}\,\Psi}{\partial x} = -\frac{1}{h}\frac{\partial \mathrm{Re}\,\Psi}{\partial y} \tag{11}$$

Once more, they can be written as follows:

$$\frac{\partial^2 \mathrm{Re}\,\Psi}{\partial x^2} + \frac{1}{h^2}\frac{\partial^2 \mathrm{Re}\,\Psi}{\partial y^2} = 0 \quad , \quad \frac{\partial^2 \mathrm{Im}\,\Psi}{\partial x^2} + \frac{1}{h^2}\frac{\partial^2 \mathrm{Im}\,\Psi}{\partial y^2} = 0 \tag{12}$$

Similarly, the partial derivatives of $\Phi(w)$ ought to be given by:

$$\begin{aligned} \mathrm{Re}\,\Phi' &= \frac{\partial \mathrm{Re}\,\Phi}{\partial x} = \frac{1}{h}\frac{\partial \mathrm{Im}\,\Phi}{\partial y} \quad , \quad \mathrm{Im}\,\Phi' = \frac{\partial \mathrm{Im}\,\Phi}{\partial x} = -\frac{1}{h}\frac{\partial \mathrm{Re}\,\Phi}{\partial y} \\ &\frac{\partial^2 \mathrm{Re}\,\Phi}{\partial x^2} + \frac{1}{h^2}\frac{\partial^2 \mathrm{Re}\,\Phi}{\partial y^2} = 0 \quad , \quad \frac{\partial^2 \mathrm{Im}\,\Phi}{\partial x^2} + \frac{1}{h^2}\frac{\partial^2 \mathrm{Im}\,\Phi}{\partial y^2} = 0 \end{aligned} \tag{13}$$

4. Stress Function and Solution

The Airy stress function U can be expressed by the complex variable functions. Therefore, we consider an infinite plane with the central crack of 2a-length along the x-axis shown in Figure 1, and consider the problem of Mode I crack (Opening Mode) in tension loading along the y-direction. The derivation of elastic stress field equations will be limited to Mode I load because it is the predominant mode in many practical cases. Once the derivation is completed, it is possible to obtain a number of useful expressions for stresses and displacements in the crack tip region. In terms of the stress intensity factor and the *T*-stress in references [7~9], and by experience, the stress function can be determined by the below form:

$$F = A\,\mathrm{Re}\,\Psi + By\,\mathrm{Im}\,\Phi + \frac{T}{2}y^2 \tag{14}$$

To take advantage of the preceding equations, the partial derivatives of $F(x, y)$ with respect to x or y can be given as follows:

$$\begin{cases} \dfrac{\partial F}{\partial x} = A\,\mathrm{Re}\,\Phi + By\dfrac{\partial\,\mathrm{Im}\,\Phi}{\partial x} \\ \dfrac{\partial F}{\partial y} = (B - hA)\,\mathrm{Im}\,\Phi + By\dfrac{\partial\,\mathrm{Im}\,\Phi}{\partial y} + Ty \end{cases} \tag{15}$$

$$\begin{cases} \dfrac{\partial^2 F}{\partial x^2} = A\dfrac{\partial\,\mathrm{Re}\,\Phi}{\partial x} + By\dfrac{\partial^2\,\mathrm{Im}\,\Phi}{\partial x^2} \\ \dfrac{\partial^2 F}{\partial y^2} = (2B - hA)\dfrac{\partial\,\mathrm{Im}\,\Phi}{\partial y} + By\dfrac{\partial^2\,\mathrm{Im}\,\Phi}{\partial y^2} + T \\ \dfrac{\partial^2 F}{\partial x\partial y} = (B - hA)\dfrac{\partial\,\mathrm{Im}\,\Phi}{\partial x} + By\dfrac{\partial^2\,\mathrm{Im}\,\Phi}{\partial x\partial y} \end{cases} \tag{16}$$

By substituting above formulae into the expression (4), the stress components can be expressed by

$$\begin{aligned} \sigma_x &= (2B - hA)h\,\mathrm{Re}\,\Phi' - h^2 By\,\mathrm{Im}\,\Phi'' + T \\ \sigma_y &= A\,\mathrm{Re}\,\Phi' + By\,\mathrm{Im}\,\Phi'' \\ \tau_{xy} &= -(B - hA)\,\mathrm{Im}\,\Phi' - hBy\,\mathrm{Re}\,\Phi'' \end{aligned} \tag{17}$$

Evidently, the analytic function Φ can be as a new usual stress function. For Mode I crack problem, the plane with a line crack is subjected to symmetric loading σ at infinity along y direction, and the stress boundary conditions are:

$$\sigma_y = \tau_{xy} = 0 \quad \text{at } |x| < a \text{ and } y = 0 \quad \text{(free crack surfaces)} \tag{18}$$

$$\sigma_y = \sigma,\ \sigma_x = \tau_{xy} = 0 \quad \text{at } |x| \to \infty \text{ and } y = 0 \tag{19}$$

In order for the stress function to meet the preceding boundary value problem, the complex function can be selected as:

$$\Phi(w) = w \cdot \sqrt{1 - \frac{a^2}{w^2}} \tag{20}$$

Then the derivatives of $\Phi(w)$ must be given by:

$$\Phi' = \sqrt{\frac{w^2}{w^2 - a^2}} \quad , \quad \Phi'' = \frac{-a^2 w}{(w^2 - a^2)^2}\sqrt{1 - \frac{a^2}{w^2}} = -\frac{a^2}{w^3}\left(\frac{w^2}{w^2 - a^2}\right)^{3/2}$$

Substituting the expressions into Equation (17), and to make the stress components to meet the stress boundary conditions, then the coefficients can be determined as:

$$A = \sigma \quad , \quad B = hA = h\sigma \quad , \quad T = -\sigma h^2$$

Therefore, the stress components can be expressed as follows:

$$\begin{aligned}
\frac{\sigma_x}{\sigma} &= h^2 \operatorname{Re}\sqrt{\frac{w^2}{w^2 - a^2}} + h^3 \operatorname{Im}\left[\frac{a^2 y}{w^3}\sqrt{\left(\frac{w^2}{w^2 - a^2}\right)^3}\right] - h^2 \\
\frac{\sigma_y}{\sigma} &= \operatorname{Re}\sqrt{\frac{w^2}{w^2 - a^2}} - h \operatorname{Im}\left[\frac{a^2 y}{w^3}\sqrt{\left(\frac{w^2}{w^2 - a^2}\right)^3}\right] \\
\frac{\tau_{xy}}{\sigma} &= h^2 \operatorname{Re}\left[\frac{a^2 y}{w^3}\sqrt{\left(\frac{w^2}{w^2 - a^2}\right)^3}\right]
\end{aligned} \tag{21}$$

5. Deformation Analysis

For discussing the deformation of above elastic body as shown in Figure 1, let u and v express the displacements of any point in the x an y directions. Now suppose the displacement components can be given by the complex function as follows:

$$\begin{cases} u = A_1 \operatorname{Re}\Phi + A_2 y \operatorname{Im}\Phi' + A_3 x \\ v = B_1 \operatorname{Im}\Phi + B_2 y \operatorname{Re}\Phi' + B_3 y \end{cases} \tag{22}$$

Substituting the expressions into equation (2), the strain components can be determined as:

$$\begin{cases} \varepsilon_x = A_1 \operatorname{Re}\Phi' + A_2 y \operatorname{Im}\Phi'' + A_3 \\ \varepsilon_y = (B_2 + hB_1)\operatorname{Re}\Phi' - B_2 hy \operatorname{Im}\Phi'' + B_3 \\ \gamma_{xy} = (B_1 - A_1 h + A_2)\operatorname{Im}\Phi' + (B_2 + A_2 h) y \operatorname{Re}\Phi'' \end{cases} \tag{23}$$

And again, in terms of the constitutive equation (3) and the stress equation (21), the identical relation of the strain components can be built. On the basis of the strain equivalence, the constants can be solved and given as follows:

$$A_1 = \frac{\sigma}{E_1}(h^2 - \nu_{12}) \quad , \quad A_2 = -\frac{\sigma h}{E_1}(h^2 + \nu_{12}) \quad , \quad A_3 = -\frac{\sigma}{E_1}h^2$$

$$B_1 = \frac{2\sigma}{hE_2} \quad , \quad B_2 = -\frac{\sigma}{E_1}\left(\frac{E_1}{E_2} + \nu_{12}h^2\right) \quad , \quad B_3 = \frac{\sigma}{E_1}\nu_{12}h^2$$

$$h^2 = \frac{E_1}{2G_{12}} - \nu_{12} = \sqrt{\frac{E_1}{E_2}} \tag{24}$$

It is evident that the constant h is relative to the elasticity of the orthotropic material. Up to now, all the constants have been determined. Therefore, the displacement components can be given by:

$$\begin{cases} u = \dfrac{\sigma}{E_1}(h^2 - \nu_{12})\,\mathrm{Re}[w\cdot\sqrt{1-\dfrac{a^2}{w^2}}] - \dfrac{\sigma h}{E_1}(h^2+\nu_{12})\,y\,\mathrm{Im}\sqrt{\dfrac{w^2}{w^2-a^2}} - \dfrac{\sigma}{E_1}h^2 x \\ v = \dfrac{2\sigma}{hE_2}\mathrm{Im}[w\cdot\sqrt{1-\dfrac{a^2}{w^2}}] - \dfrac{\sigma}{E_1}(\dfrac{E_1}{E_2}+\nu_{12}h^2)\,y\,\mathrm{Re}\sqrt{\dfrac{w^2}{w^2-a^2}} + \dfrac{\sigma}{E_1}\nu_{12}h^2 y \end{cases} \tag{25}$$

Obviously, the displacement fields for the orthotropic material are quite complex. For the isotropic materials ($h=1$), $E_1 = E_2 = E = 2G(1+\nu)$, $\nu_{12} = \nu$, $G_{12} = G$ and $w = z = x + iy$. And so the expressions of the displacements can be reduced to:

$$\begin{cases} u = \dfrac{\sigma}{E}(1-\nu)\,\mathrm{Re}[z\cdot\sqrt{1-\dfrac{a^2}{z^2}}] - \dfrac{\sigma}{E}(1+\nu)\,y\,\mathrm{Im}\sqrt{\dfrac{z^2}{z^2-a^2}} - \dfrac{\sigma}{E}x \\ v = \dfrac{2\sigma}{E}\mathrm{Im}[z\cdot\sqrt{1-\dfrac{a^2}{z^2}}] - \dfrac{\sigma}{E}(1+\nu)\,y\,\mathrm{Re}\sqrt{\dfrac{z^2}{z^2-a^2}} + \dfrac{\sigma\nu}{E}y \end{cases} \tag{26}$$

Obviously, the expressions are conventional displacement fields of isotropic materials.

6. Acknowledgments

This research was financially supported by the Natural Science Foundation of China (NSFC Grant No.51275422, 51475372).

7. References

[1] Klaus Friedrich: Application of fracture mechanics to composite materials. Elsevier Sci. Pub. B.V., (1989)
[2] G.C. Sih: Mechanics of fracture initiation and propagation. Kluwer Academic Publishers, (1991)
[3] C.T. Sun: Fracture mechanics. Elsevier Inc., (2012)
[4] M.H. Sadd: Elasticity, theory, applications and numerics. Elsevier Inc., (2014)
[5] M. Fakoor, N.M. Khansari. Mixed mode I/II fracture criterion for orthotropic materials based on damage zone properties. Eng Fract Mech 2016; 153: 407-420
[6] M. Fakoor. Augmented strain energy release rate (ASER): A novel approach for investigation of mixed-mode I/II fracture of composite materials. Eng Fract Mech 2017; 179: 177-189
[7] M. Myatollahi, M. Pavier, D. Smith. Mode I cracks subjected to large T-stresses. Int J Fract 2002; 117, 159-174
[8] T. Nakmura, D.M. Parks. Determination of elastic T-stress along three-dimensional crack fronts using an interaction integral. Int J Solids Struct 1992; 29: 1597-1611
[9] K. Lu, T. Meshii. A systematic investigation of T-stresses for a variety of center-cracked tension specimens. Theor Appl Fract Mech 2015; 77: 74-81.

IOP Conf. Series: Materials Science and Engineering **484** (2019) 012021 doi:10.1088/1757-899X/484/1/012021

The Price Linkage between Oil, Gold, Stock and Exchange Rate Based on Vine Copula

Huan Sun

School of computer and information, Hohhot Vocational College, Inner Mongolia, 010051, China
E-mail: sunhhvc@163.com

Abstract. With the development of China's economy and the deepening of the reform of the socialist market economic system, the relationship between different markets has increased and the relationship is becoming more complex, which poses new challenges to the portfolio and risk management of the market. Therefore, this paper empirically analyses the relationship between the price of China's oil market, the gold market, the stock market and the foreign exchange market by using the correlation measure based on Vine copula. The results show that the oil market occupies the dominant position in the four markets, and the price changes in other markets have a driving effect, and the lower tail correlation among the markets is generally higher than that of the upper and tail correlation, showing asymmetric correlation and thick tail. These results show that the risk price linkage among markets is more closely and the risk is more contagious in the extreme cases.

1. Introduction

The Copula function has certain advantages in the characterization of correlations, and in particular it can accurately describe the nonlinearity and thick-tailedness in the correlation. Therefore, it is widely used in the financial field. Li and Yang [1], Nguyen et al. [2] used the Copula theory to analyze the correlation between two financial markets. However, due to its dimensionality constraints, the binary Copula function cannot measure the correlation between multiple variables. Joe [3] defined the Vine Copula function based on the Copula function proposed by Sklar[4] to describe correlation between multivariables. Bedford and Cooke [5, 6] further refined Vine Copula's related theories. The Vine Copula model has both Vine structure and Copula model characteristics. The Vine structure is used to incorporate multiple variables under a unified research framework. Then the node relationships in the Vine structure are characterized by the binary Copula function to achieve a unified measurement of multivariable relationships.

Researches on Vine Copula function has gradually become a hot topic in academic circles in recent years. Aas and Czado [7] used the Vine Copula model to study the multivariate correlation structure; Mendes [8] used the Vine Copula model to study the correlation between multiple markets. Righi [9] used a new Pair Copula structure to characterize the correlation between several financial markets in Brazil and calculated the VaR of the portfolio.

Therefore, this article intends to empirically analyze the price linkage relationship between petroleum, gold, stocks and foreign exchange markets based on the correlation measure of Vine Copula, with a view to provide valuable participation for investment decision-making and risk management in the China market.

2. Correlation Measurement based on Vine Copula

2.1. Vine Copula Function

Vine Copula has obvious advantages when modeling the marginal distribution of multidimensional random variables and related structures. It does not require that the marginal distribution of variables has the same pattern. From the decomposition process of the Pair-Copula function, we know that an n-dimensional Copula function can be decomposed into $\frac{p(p-1)}{2}$ binary Copula functions, and there are Pair-Copula construction methods. In order to simplify the number of models, Cooke and Bedford (2001) proposed the theory of "Rule Vine". Canonical vine (C-vine, C-Vine for short) is the most commonly used rattan structure. Therefore, this paper uses C-Vine Copula function to fit the data in the vine structure separately so as to consider the difference of the vine structure. The C-Vine Copula function is constructed as follows:

Step 1: Verify orders of n variables in T_1, and rearrange the order and denote $(u_1, u_2, \dots, u_n)$.

Step 2: Because there are $n-1$ edges between the n variables in the vine structure, $n-1$ binary Copula functions can be used to link the sorted adjacent variables, creating the first layer of Pair-Copula$\left(C_{1,2}, C_{1,3}, \dots, C_{1,n}\right)$

Step 3: Take the n-1 dimensional variable $\left(C_{1,2}, C_{1,3}, \dots, C_{1,n}\right)$ in the second step as a random variable in T_2 and continue with the second step. The calculation step calculates the second layer Pair-Copula$\left(C_{2,3|1}, C_{2,4|1}, \dots, C_{2,n|1}\right)$.

Step 4: Repeat steps 2 and 3 until the end of the calculation.

2.2. Correlation measure

Joe (1997) gives the definition of the tail correlation coefficient, which is used to express the probability that when one variable's implementation value is extremum, another variable also has extremum, it is a measure widely used in extremum theory. First, the definition of quantile correlation measure of relevance is given. If $u^* \in [0,1]$, define

$$\lambda(u^*) \equiv P[U > u^* | V > u^*] = \frac{\overline{C}(u^*,u^*)}{1-u^*} \tag{1}$$

as a quantile correlation measure of relevance, whereas $\overline{C}(u,v) = 1 - u - v - C(u,v)$.

Then, the definition of the tail correlation coefficient is given. If X and Y are two consecutive random variables, their marginal distribution is $F(\bullet)$, $G(\bullet)$, and Copula function is $C(\bullet,\bullet)$. Let

$$\tau^U \equiv \lim_{u^* \to 1} P[Y > G^{-1}(u^*) | X > F^{-1}(u^*)] = \lim_{u^* \to 1} \frac{\hat{C}(1-u^*,1-u^*)}{1-u^*}, \tag{2}$$

$$\tau^L \equiv \lim_{u^* \to 0} P[Y < G^{-1}(u^*) | X < F^{-1}(u^*)] = \lim_{u^* \to 0} \frac{C(u^*,u^*)}{u^*}, \tag{3}$$

Respectively represent upper and lower tail correlation coefficients, and $\hat{C}(1-u, 1-v) = \overline{C}(u,v)$.
If the upper tail relationship τ^U (or the lower tail correlation coefficient τ^U) exists and is within the interval (0,1], the upper tail (or lower tail) is related; if the upper tail relationτ^U (or If the lower tail correlation coefficient τ^U) is zero, the variables are independent of each other, which shows that the Copula function has a great advantage in measuring the tail correlation.

2.3. Parameter Estimation

The parameters of Copula model can use maximum likelihood estimation or moment estimation. In this paper, the maximum likelihood method is used to estimate the parameters. The likelihood function can be written as

$$\mathcal{L}(\xi; x) = \sum_{j=1}^{T} \left(\sum_{i=1}^{p} \log\left(f_i\left(x_{i,t}; \phi_i\right)\right) - \log\left(c\left(F_1\left(x_{1,t}\right), \dots, F_p\left(x_{p,t}\right); \theta\right)\right)\right),$$

where vector $\xi = (\phi, \theta)$ include marginal distribution parameter $\phi = \left(\phi_1, \ldots, \phi_p\right)$ and Copula parameter θ.

If the number of parameters to be estimated is too large, a two-stage estimation method is usually used to simplify the estimation process. The first step is to estimate the marginal distribution parameters and the second step is to estimate the Copula parameters. The maximum likelihood estimation of marginal parameters is as follows:

$$m\mathcal{L}(\phi; x) = \sum_{i=1}^{p} mill_i = \sum_{i=1}^{p} \sum_{j=1}^{T} \log\left(f_i\left(x_{i,t}; \phi_i\right)\right). \tag{4}$$

The maximum likelihood estimation of the Copula parameter is as follows:

$$c\mathcal{L}(\theta; u, \phi) = \log\left(c\left(F_1\left(x_{1,t}\right), \ldots, F_p\left(x_{p,t}\right); \theta\right)\right), \tag{5}$$

where $u = \left(F_1(x_1), \ldots, F_p\left(x_p\right)\right)$.

3. Empirical Analysis

3.1. Data Sources and Descriptive Statistics

The indexes are selected for the oil and gold markets are the FUL9 Fuel Index in the Shanghai Commodity Futures Market and the AUL9 Gold Index, and the Shanghai-Shenzhen 300 Index for the stock market index. These figures are derived from CITIC Securities. The foreign exchange market index is selected from the People's Bank of China's RMB exchange rate against the US dollar (RMB/100 US\$). The data comes from the Bank of China. Since gold futures were formally traded on the Shanghai Futures Exchange in 2008, in order to ensure consistency in the timing of various market indicators, the daily data from 2008 to 2016 was selected and a total of 1,892 valid data were obtained. First, the logarithmic returns of the four financial time series are calculated. The logarithmic daily return rate is shown in Figure 1. The four figures are from left to right, from top to bottom, the gold index, the fuel index, the CSI 300 Index, and the exchange rate.

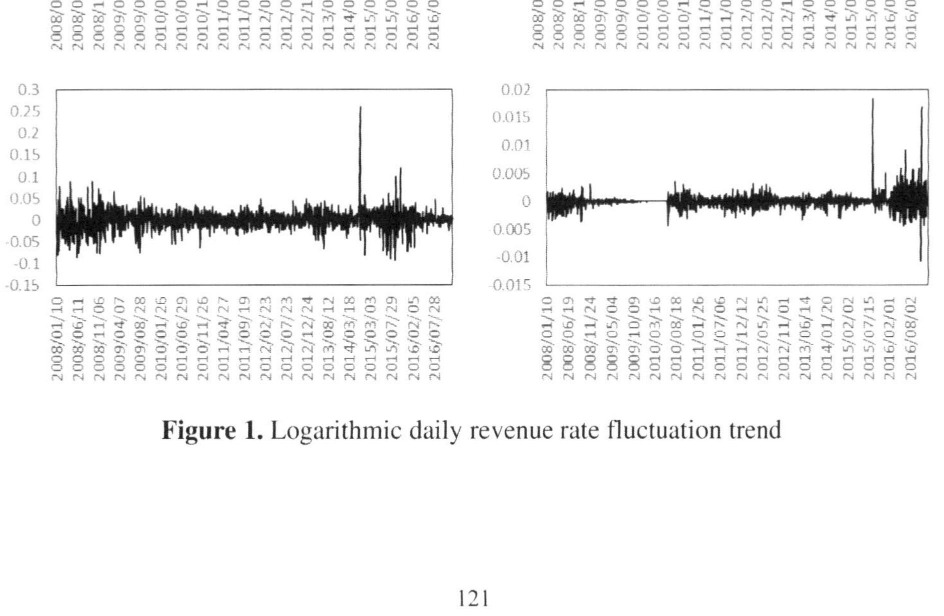

Figure 1. Logarithmic daily revenue rate fluctuation trend

From Figure 1, it can see the fluctuation of the rate of returns of four markets. Except for the extreme value, the exchange rate benchmark price has the smallest volatility, and the CSI 300 index has the largest volatility. The volatility of the gold index and the fuel index is medium and the degree of volatility is similar. The further statistical analysis of the yield data are shown in Table 1.

Table 1. Descriptive statistics of the daily return rate of data index

Index	Gold Index Yield	Fuel Index Yield	CSI 300 Yield	Exchange rate yield
minimum	-0.0788	-0.0945	-0.0915	-0.0108
maximum	0.0567	0.0719	0.2599	0.0184
average	1.02E-04	3.64E-05	-2.79E-04	-2.45E-05
Standard deviation	0.0130	0.0160	0.0209	0.0013
Skewness	-0.48	-0.29	0.75	3.57
Kurtosis	3.92	3.55	15.28	47.54
J-B statistics	1273.63	1009.48	18470.66	181141.80
ADF statistics	-44.17	-27.43	-43.10	-39.90

Note: JB statistics and ADF statistics are significant at the 5% level.

According to the statistical results in Table 3-1, the average return rate of the fuel index and the gold index is positive, and the average return rate of the CSI 300 index and the exchange rate benchmark price is negative, indicating that the oil market and the gold market show an upward trend and the stock market and the foreign exchange market show an downward trend. The standard deviation of the CSI 300 index is the largest, indicating that the stock market has the highest price volatility, followed by the volatility of the fuel index and the gold index, and the difference is not significant, indicating that the volatility of the oil market and the gold market are similar. The volatility of the exchange rate benchmark price is the smallest in the four indicators. From the perspective of skewness, the four markets are biased. The skewness of the fuel index and the gold index is negative and the performance is left deviation. The CSI 300 and the benchmark exchange rate are positive, showing a right deviation. Judging from the kurtosis value, the four markets are all presented with "spikes" and the performance of the exchange rate market is particularly obvious.

In addition to the basic descriptive statistics, the J-B statistics (J-B) and ADF unit root tests (ADF) were also performed for each index's rate of return. From the J-B statistics, each index refuses to obey the Gaussian normal distribution at a significant level of 5%, and accepts the assumption that each indicator's time series is stable at a significant level of 5%. Therefore, the Copula function can be used to mathematically modeling the first relations between markets.

3.2. Modeling Analysis of Price Interaction Based on Binary Copula

In the binary Copula model, the Clayton Copula function with asymmetry can quickly capture the relevant changes in the lower tail, and its representing parameters the Kendall rank correlation coefficient τ. The SJC-Copula is a function that allows the asymmetry of the left tail and right tail, and its two parameters to be evaluated τ^Uandτ^U, which represent the upper tail correlation coefficient λ^{up} and the lower tail correlation coefficient λ^{lo}.

Therefore, the Kendall rank correlation coefficient τ is calculated using the Clayton Copula function, and the upper tail correlation coefficient λ^{up} and the lower tail correlation coefficient λ^{lo} of the SJC-Copula function. The parameter estimation results are shown in Table 2.

Table 2. Non-conditional tail correlation coefficient estimation results

combination	Kendall's tau	upper tail dependence	lower tail dependence
Gold Index - Fuel Index	0.1594 (0.014)	0.0682 (0.031)	0.1797 (0.031)
Gold Index - CSI 300 Index	0.0623 (0.014)	0.0000 (0.000)	0.0394 (0.028)
Gold Index - Exchange Rate Benchmark Price	0.0062 (0.012)	0.0000 (0.036)	0.0002 (0.042)
Fuel Index-CSI 300	0.1165 (0.014)	0.0386 (0.026)	0.1026 (0.030)
Fuel Index-Exchange Rate Base Price	0.0097 (0.012)	0.0000 (0.056)	0.0003 (0.079)
CSI 300-Exchange Rate Benchmark Price	0.0096 (0.013)	0.0005 (0.003)	0.0000 (0.000)

Note: The value in brackets is the standard deviation.

From Table 2, we can see that the tail correlation coefficient estimated by the binary Copula model is positive and the relationship is weak. Among them, the tail correlation between the oil market and the gold market is most obvious among the six groups of related relations, and the tail market correlation between the oil market and the stock market is second. At the same time, in the two correlations, the values of lower tail correlation coefficient are significantly greater than the values of upper tail correlation coefficient, indicating that the tail price linkage between the oil market and the gold market and between the oil market and the stock market is strong. Especially when the market price plummets, the risk of infection is even more pronounced.

Under the condition of binary correlation (non-conditional relationship), a portfolio that can be related to other markets with weaker correlation, such as foreign exchange, can be constructed. When the entire market is in a bear market, or when price fluctuations are large, risks can be effectively avoided. However, the binary investment portfolio does not apply to China's complex financial market system. It should also incorporate the selected samples into a holistic model and comprehensively analyze the correlations among various markets.

3.3. Modeling Analysis of Multivariate Price Linkage Based on Vine Copula

For ease of description, in the modeling process below, 1 is used to represent the gold index, 2 is the fuel index, 3 is the CSI 300 index, and 4 is the exchange rate base price.

C-Vine copula model is a model with pilot node. According to the empirical research, the first layer consists of 2-1, 2-2, 2-4, the second layer consists of 1-3|2, 1-4|2 and the third layer consists of 3-4|1,2, with a 2-1-3-4 order. Hence, the sequence based on the relative importance of the four markets is oil market, gold market, stock market and exchange rates market. And it is clear that the oil market is the most dominant market among them.

Similar to the analysis of the binary correlation, the Kendall rank correlation coefficient was calculated using the Clayton Copula function, and the tail relationship and the tail-tail correlation coefficient were used on the SJC-Copula function. The parameter estimation results are shown in Table 3.

Table 3. C-Vine Copula Correlation Coefficient Estimate Results

Decomposition	Kendall's tau(τ)	upper tail dependence	lower tail dependence
2-1	0.1557	0.0715	0.1730
2-3	0.0939	0.0219	0.0757
2-4	0.0017	0.0000	0.0000
1-3|2	0.0635	0.0000	0.0411
1-4|2	0.0119	0.0000	0.0000
3-4|1,2	0.0105	0.0011	0.0000

The analysis of the first-level tree shows that the Kendall rank correlation coefficient between the oil market and the gold market is the highest, and it is most likely to rise or fall in the same direction; At the same time, the tail correlation coefficient between the oil market and the gold market is also the highest, indicating the possibility that oil market and the gold market soar or plunge simultaneously is greatest.

The Kendall rank correlation coefficient between the oil market and the stock market is relatively high, indicating that the overall correlation between them is also strong. Although the tail correlation is not as significant as the tail market correlation between the oil market and the gold market, it still has a certain tail correlation. However, the correlation coefficient between the two tails in the two groups of relations is significantly greater than the correlation coefficient of the last tail, indicating that the collapse of oil prices is more likely to cause the price of gold and stock markets to plummet.

The possibility of skyrocketing gold prices and stock prices caused by skyrocketing oil prices is not as obvious as when the oil market was in a bear market. The Kendall rank correlation coefficient between the oil market and the foreign exchange market is less than 0.002, and the overall correlation is the weakest.

The analysis of the second-tier tree shows that when taking the oil market's yield as a known condition, taking the gold market as the leading node and measuring the correlation between the gold market and the stock market and the foreign exchange market, the overall correlation between the gold market and the stock market is significantly larger than the overall correlation between the gold market and the foreign exchange market. And there is a certain degree of correlation between the gold market and the stock market, and the price plunge in the gold market is likely to cause a bear market in the stock market. However, the soaring price of gold is less likely to bring a bull stock market.

The analysis of the third-level tree shows that the overall correlation between the stock market and the foreign exchange market is small, compared with the result of the binary Copula model, when the oil market yield and the gold market yield are known conditions. However, there has been some improvement. The value of the Kendall rank correlation coefficient exceeds 0.01, and the tail-tail correlation between the two markets begins to show.

Overall, in these four market portfolios, the oil market dominates, and price fluctuations have the greatest impact. In extreme circumstances, especially the decline in the overall trend of the financial market brought about by the collapse of oil prices has the most significant impact. When there is more the known yield information, the conditional correlation between the two markets is less obvious. The correlation between the foreign exchange market and the tails of the other three markets has approached zero, indicating that even in extreme circumstances, the risk transmission and price linkage effects between the foreign exchange market and the other three markets are not significant.

4. Conclusion

This paper constructs a correlation measurement based on Vine copula to study the price linkage relationship about petroleum, gold, stocks and foreign exchange markets. Firstly, the correlation measurement model based on Vine Copula function and the parameter estimation method of its model are given. Then the correlation measurement models of binary Copula and Vine Copula functions are used to empirically analyze price linkage relationship of the oil, gold, stock and foreign exchange markets.

The empirical results show that there is a positive correlation among the oil market, the gold market, the stock market and the exchange rate market, and the correlation between the oil market and the gold market is the strongest, and the correlation between the exchange rate market and the other three markets is relatively weak. The correlation between the two markets is significantly stronger than the correlation between the tails, indicating that there is a certain degree of risk contagion among the markets, especially in extreme cases; the price linkage relationship is more significant. And the oil market dominated the four sample markets, leading to price linkages in sample markets.

According to the comparison between the binary Copula model and the multivariate Vine Copula model, Vine Copula has more advantages than binary Copula in terms of intra-sample fitting and data analysis. Therefore, in practice, Vine Copula-based models can be used for risk analysis in the process of portfolio construction, and then it can improve the accuracy of risk measurement results.

5. References

[1] Li M, Yang L. Modeling the volatility of futures return in rubber and oil—A Copula-based GARCH model approach[J]. Economic Modelling, 2013, 35:576-581.

[2] Nguyen C, Bhatti M I, Komorníková M, et al. Gold price and stock markets nexus under mixed-copulas[J]. Economic Modelling, 2016, 58:283-292.

[3] Joe H. Multivariate models and multivariate dependence concepts[M]. CRC Press, 1997.

[4] Sklar M. Fonctions de répartition à n dimensions et leurs marges[M]. Université Paris 8, 1959.

[5] Bedford T, Cooke R M. Probability density decomposition for conditionally dependent random variables modeled by vines[J]. Annals of Mathematics and Artificial intelligence, 2001, 32(1-4):245-268.

[6] Bedford T, Cooke R M. Vines: A new graphical model for dependent random variables[J]. Annals of Statistics, 2002:1031-1068.

[7] Aas, K., Czado, C., Frigessi, A., & Bakken, H. (2009). Pair-copula constructions of multiple dependence. Insurance Mathematics & Economics, 44(2), 182-198.

[8] Mendes, B. V. M., & Costa Lopes, S. R. (2012). Dynamic copulas and long range dependence. Social Science Electronic Publishing, 8, 89-111.

[9] Righi, M. B., & Ceretta, P. S. (2013). Analyzing the dependence structure of various sectors in the brazilian market: a pair copula construction approach. Economic Modelling, 35(5), 199-206.

IOP Conf. Series: Materials Science and Engineering **484** (2019) 012022 doi:10.1088/1757-899X/484/1/012022

Experimental Investigation of NOx Reduction by Urea Solution Injection in Fuel-rich Zone of Coal Combustion

Degui Bi[1], Zhongxiao Zhang[1, *], Zhixiang Zhu[1], Jiancong Dong[1] and Tuo Zhou[2]

[1]School of Mechanical Engineering, Shanghai JiaoTong University Shanghai, 200240, PR China
[2]Institute of Engineering Thermophysics, Chinese Academy of Sciences, Beijing 100190, China
Email: bidegui1103@163.com

Abstract. Experiments were carried out in an electrically heated vertical tandem-type drop-tube furnace system to evaluate the influence of stoichiometric ratio (SR), normalized stoichiometric ratio (NSR) and temperature (T) on NOX reduction by urea solution in coal combustion. The results indicated that, SR was the deciding factor for NOX reduction by ammonia under high temperatures. The reduction efficiency was inproved with the increase of NSR. There was a significant reduction in NOX emission with increase in temperature under reducing conditions. However, ammonia had the tendency to get oxidized to NO, in presence of excess O2 in the combustion zone at high temperature.

1. Introduction

NOx emitted from utility boilers makes contributions to environmental pollution problems. Many countries have lowered their NO_X emission limits in flue gases. For example, in the European Union the permissible NO_X emission has been reduced to below 200 mg of NO_2/m^3 (6% O_2), after 2016[1]. In China, by 2014, the permissible NO_X emission was controlled to less than 100 mg/Nm^3 at 6%O_2 for all power plants [2]. There are numerous well-established methods, such as air staging, staged fuel combustion, selective non-catalytic reduction (SNCR), and selective catalytic reduction (SCR). SNCR is a conceptually simple process for reducing NO_X emissions and has therefore been the most widely adopted method in coal-fired power plants [3-4]. However, in SNCR technology, a serious limitation of the SNCR for NO_X reduction is the requirement of a narrow temperature window [5], developing new NOx control technologies is quite necessary, especially the technology of removing NOx in coal combustion process.

NH_3 is formed during staged combustion of pulverized coal [6], which indicates itself that combustion modification could be an effective strategy for NOx abatement. Hence, Adding NH_3 reagent(such as ammonia, urea solution) into the fuel rich zone may be to furher reduce NO in staged combustion, which have received little attention in the present literature.Wang [7] reported the NO_X reduction by urea solution,which showed that under hypoxic conditions(O_2%=0), as the temperature increased, the NO_X reduction efficiency increased rapidly. The maximum value of the optimal temperature was higher than in excess of O_2(1~8%). In other words, NH_3 could react with NO_X in fuel-rich conditions, even at the higher temperatures. Fan and William [8-9] also reported this phenomenon in their recent experimental and modeling studies, which were carried out without the addition of any catalytic. According to Javed [10], the rate of NO destruction by ammonia would just be balanced by its formation by oxidation at the temperature of around 1230°C, in excess of O_2. Therefore, it can be believed that the temperature balance is increased, if the oxygen concentration is decreased. Taniguchi [11] conducted a study to investigate the index for modeling a NO_X reaction mechanism of combustion of pulverized coal, under various burning conditions. Cremer et al [12-15] found that when ammonia reagent was introduced into the reduction region of the primary combustion zone in a cyclone-barrel-fired boiler, the NO_X reduction efficiency improved significantly. Most of the researches focused on the features of NO_X reduction at relatively lower temperatures, in excess of oxygen. However, there are only few reports, which explain the characteristics of NH_3 reaction with NO at high-temperatures, especially in a pulverized coal furnace.

In the present study, the characteristics of NH_3 reaction with NO in flue gas for typically high-volatile bituminous coal was investigated experimentally and concentrates in the evaluation of the NO_X emission in reducing region from a novel vertical drop-tube furnace system.

2. Experamental Section

The experimental bench system is shown in Fig.1. The experimental setup consisted of an electric furnace reactor system, reducing agent injection device, pulverized coal feeder, flue gas analyzer, and other affiliated components. The reactor was an corundum tube which had an inner diameter of 50 mm and the length of 2100mm, and reaction zone is 1200 mm.. The pulverized coal was fed through a micro-screw pulverized coal feeder, and coal feed rate was0.2- 0.5 kg/h for experiments. Pulverized coal and the air of supporting combustion was mixed in the nozzle which used a water-cooled probe to prevent pyrolysis of coal particles before injection.. Urea solution, with mass concentration of 5%, was injected into the reactor by a micro screw peristaltic pump.

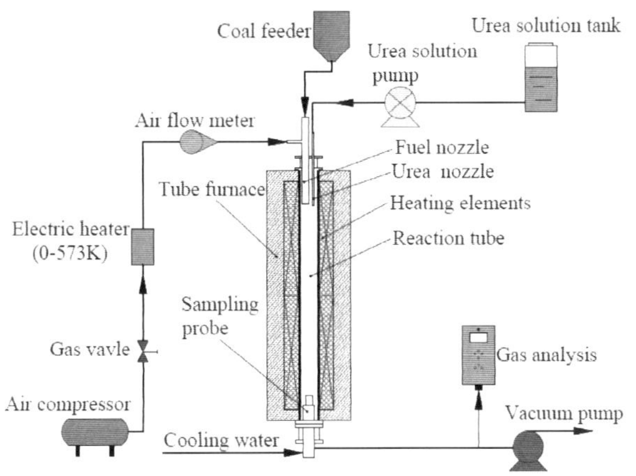

Figure 1. Schematic diagram of the experimental system

The coal used in the tests was high-volatile Shenhua bituminous coal from China. The characteristics of this coal are presented in Table 1. The coal sample was grounded and sieved to a size of 75-90 μm and air -dried at 101°C for 10 h.

Table 1. Properties of pulverized coal used in the investigation

Coal	*Ultimate analysis, wt%*					*Approximate analysis, wt%*			
	Cad	Had	Oad	Nad	Sad	Mad	Aad	Vad	FCad
bituminous coal	62.74	3.82	7.78	0.7	0.5	1.03	23.43	26.88	48.66

The concentrations of NO, O_2, CO, and NO_2 were analyzed by a AFRISO-M60 flue gas analyzer, which maximum relative errors for the measured species were ±5% (>100ppm). The on-line data of NO concentration measured was in parts per million [ppm]. For the purpose of comparison, the NO_X

concentration was converted into concentration of NO_2 (mg/m^3, dry 6%O_2). The formula for conversion is

$$[NO_X, mg/m^3] = [NO, ppm] \bullet \frac{46.0}{22.4} \bullet \frac{21\% - 6\%}{21\% - [O_2, vol\%]} \quad (1)$$

3. Results and Discussion

3.1. Influence of SR on NOX Reduction

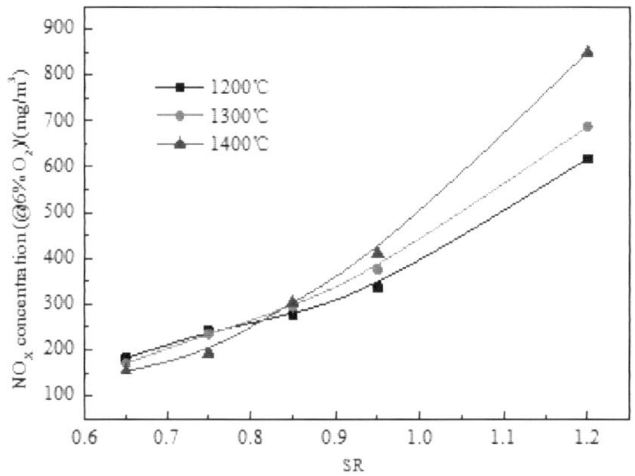

Figure2. Influence of SR on NO_X emission witout urea solution

Figure2 shows the influence of SR on NOX emission, when no urea solution was injected into the furnace. The results showes that the effect of the temperature on NOX generation exhibited different characteristics, depending upon the SR. The concentration of NOX emission decreased radically with SR at different temperatures. In an oxidizing atmosphere (SR>1.0), higher temperature favored the formation of a larger amount of NOX. However, under a reducing atmosphere (SR<1.0), the tendency was reversed. It could also be seen that at a higher temperature in the reducing zone, there was a sharper decline in the NOX emission as SR decreased.

In order to investigate the effect of SR on NO_X reduction by urea solution, the experiments were carried out in the SR range of 0.65 to 1.2. Figure 3 shows the influence of the SR in the combustion zone at different NSR and temperatures.

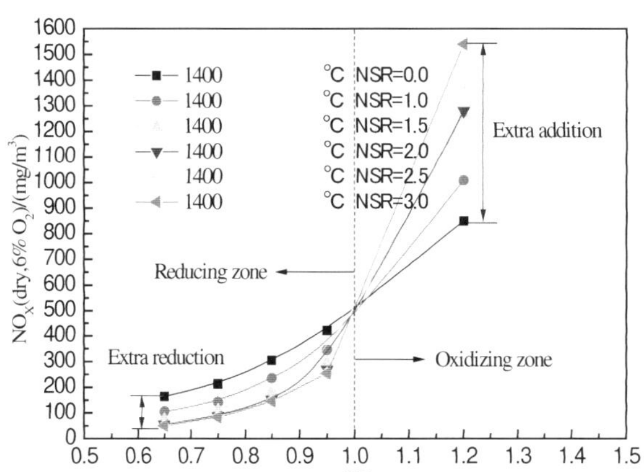

Figure 3. Influence of SR on NO_X emission with injecting urea solution in reaction zone

IOP Conf. Series: Materials Science and Engineering **484** (2019) 012022 doi:10.1088/1757-899X/484/1/012022

As is shown, with decrease of SR in reaction zone, the atmosphere turned into a reducing atmosphere (SR<1.0). The concentration of NO_X emission had a obvious decrease .The lower the SR is,The higher of the NO_X reduction efficiency when the NSR and temperatures were constant. This was due to the fact that when SR of combustion zone decreased, the concentration of O radicals decreased and the concentration CO increased. A small number of O radicals could be prevented from getting oxidized to NO_X. The regenerated OH radicals continued to promote the conversion of NH_3 to NH_2, which reduce NO primarily The concentration of NO_X emissions could be reduced to 50~100mg/m³in highly reduced atmosphere, with NSR=3 and SR=0.65. There are few studies that reported of SR on NO_X reduction by ammonia at high temperature.

3.2. Influence of NSR on NOX Reduction with Injecting Urea Solution

In order to investigate the influence of NSR on NO_X emission at high temperature, the experiments were carried out at different NSR values of 0, 1.0, 1.5, 2.0, 2.5, and 3.0.

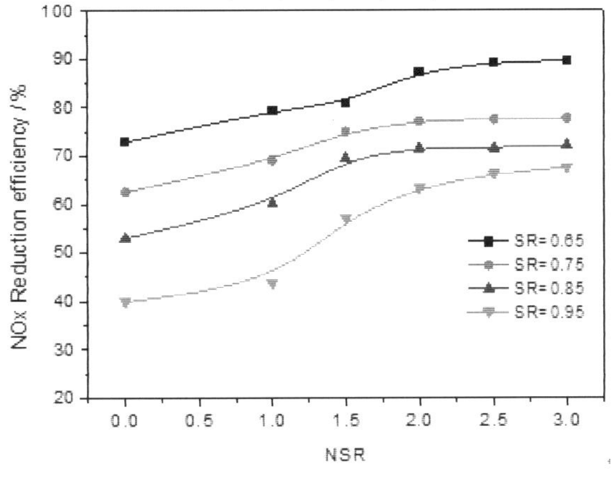

Figure4. Influence of NSR on NO_X reduction by ammonia (t=1300°C.)

Figure.4 illustrates the differences in NO_X reduction efficiency with variations of NSR, in which baseline conditions for concentration of NO_X emission corresponded to the different SR values when temperature is at 1300°C. . The efficiency of NO_X reduction increased rapidly when NSR increased. However, when NSR>2, the rate of NO_X reduction become slowed down. This was due to the fact that the concentrations of NO_X and other reactive groups decreased as ammonia content was constantly increased. Thus, the reaction rate gradually slowed down.

3.3. Influence of Temperature on NOX Reduction with Injecting Urea Solution

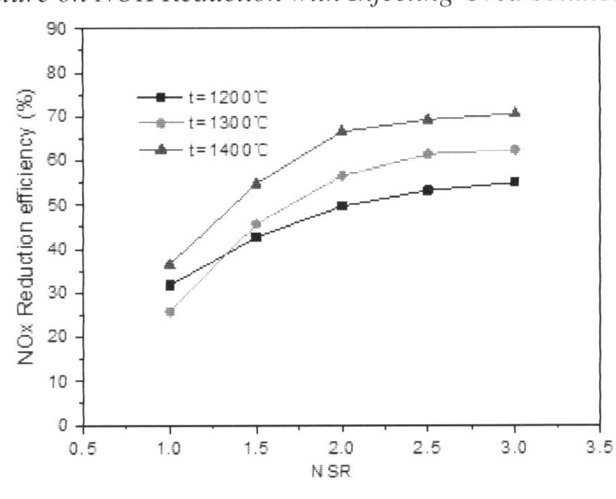

Figure 5 Influence of temperature on NOx reduction by urea solution at SR= 0.65

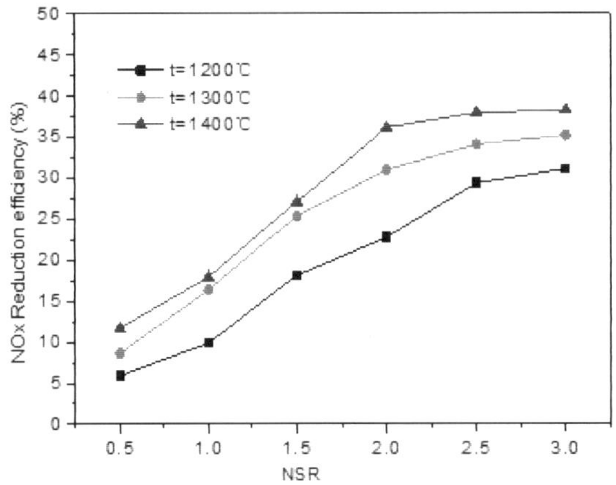

Figure 6 Influence of temperature on NOx reduction by urea solution at SR= 0.95

The effect of temperature on NO_X reduction of efficiency while SR=0.65, 0.95 were shown in Figures5 and Figures6. It can be observedminimum NO_X emission occurred at 1400°C, and the NO_X emission value was lower at SR=0.65 than at SR=0.95. This could be attributed to the fact that the higher temperature in the reducing zone lowered the NO_X emission with decrease in SR. Those results suggested when O_2 concentration was low in the fuel rich zone; temperature had great effect on the removal efficiency of NO_X. When SR=0.95, O radicals were relatively high in number, which implied that reactions: NH_2 + O = NH + OH, NH + O = NO + H, HNO + OH = H_2O +NO, which were the crucial steps that would decrease the reduction efficiency of NO_X. At higher SR value, as shown in Figure 6(b), NOx concentration increased rapidly as temperature increased at different NSR.

4. Conclusion

An experimental investigation was conducted on a vertical drop-tube furnace system, and the conclusion is summarized as follows: SR had an obvious effect on the NO_X emissions for pulverized coal combustion. The decrease of SR led to a highly reducing atmosphere and increased the reduction efficiency of NO_X when urea solution participated in the reaction, and vice versa. NO_X reduction efficiency increased more with increased in NSR, under reducing atmosphere, when the NSR>2, the rate of NO_X reduction increased is not obvious. Increase of temperature was crucial factor for promoting the NO_X reduction efficiency in a reducing atmosphere, and reversed results were obtained in a oxidizing atmosphere.

5. Acknowledgements

The authors gratefully acknowledge the financial support from National Key R&D Program of China, Grant No. 2016YFB0600202.

6. Reference

[1] European union (Large combustion plants) regulations 2012. S.I. No. 566 of 2012.
[2] GB13223-2011, Emission standard of air pollutants for thermal power plants[S], 2011.
[3] TayyebJaved M, NaseemIrfan, Gibbs B M. *Control of combustion -generated nitrogen oxides by selective non-catalytic reduction*. Journal of environmental management 2007, 83: 251-89.
[4] Daood S S, Javed M T, Gibbs B M, Nimmo W. *NO_X control in coal combustion by combining biomass co-firing, oxygen enrichment and SNCR*. Fuel 2013, 105: 283-92.
[5] Norbert Modinski. Numerical simulation of SNCR (Selective non-catalytic reduction) process in coal fired grate boiler. Energy 2015, 92: 67-76.
[6] Bose AC. *Pulverized coal combustion: Fuel nitrogen mechanisms in the rich post-flame*[C] // Symposium. 1989:1127-1134.
[7] Quan Wang, Qingcheng Wang, JianjunCai. *Experimental study of urea on SNCR removal of NOx*.Journal of chemical and pharma ceutical research 2014, 6(7): 2541-2546.

[8] Fan W, Lin Z, Kuang J, et al. *Impact of air staging along furnace height on NOx emissions from pulverized coal combustion* [J]. Fuel Processing Technology, 2010, 91(6):625-634.

[9] William H Sun, Penelope Stamatakis, John E Hofmann. *Reaction kinetics of selective Non-catalytic NOx reduction with urea.* https://web.anl.gov/PCS/acsfuel/preprint%20archive/Files/38_2_DENVER_03-93_0734.pdf.

[10] Javed M T, Irfan N, Gibbs B M. *Control of combustion-generated nitrogen oxides by selective non-catalytic reduction.*[J]. Journal of Environmental Management, 2007, 83(3):251-289.

[11] Taniguchi M, Kamikawa Y, Okazaki T, et al. *A role of hydrocarbon reaction for NO x, formation and reduction in fuel-rich pulverized coal combustion* [J]. Combustion & Flame, 2010, 157(8):1456-1466.

[12] Marc A Cremer, Huafeng D Wang, David E Boll, Edmund Schindler, Edmundo Vasquez. *Improved rich reagent injection (RRI) performance for NOx control in coal fired utility boilers.* U.S. DOE Conference on SCR and SNCR for NOx control 2003.

[13] Marc A Cremer, Bradley R Adams, David C O'Connor, Venkata Bhamidipati, R Gifford Broderick. *Design and Demonstration of rich reagent injection (RRI) for NOx reduction at Conectiv's B.L. England station.* http://citeseerx.ist.psu.edu/viewdoc/download?doi=10.1.1.497.5961&rep=rep1&type=pdf

[14] Cyclone boiler field testing of advanced layered NOx control technology in Sioux Unit 1. Final technical report. Reaction Engineering International. Sept. 2006. http://www.osti.gov/scitech/servlets/purl/927774

[15] NOx control options and integration for US coal fired boilers. Final report. Reaction Engineering Interntional, Jun 2006. http://www.osti.gov/scitech/biblio/927773-BjKBpO/

IOP Conf. Series: Materials Science and Engineering **484** (2019) 012023 doi:10.1088/1757-899X/484/1/012023

Experimental Research on NOx Emission Characteristics During Coal Combustion in the Stoker-fired Boiler

Xinwei Guo[1,a], Zhongxiao Zhang[1,b*], Juan Yu[2], Jian Zhang[1], Degui Bi[2], Jiancong Dong[2], Hao Bai[1] and Junjie Fan[1]

[1]Key Laboratory of High-Efficiency Combustion and Ultra-Low Emission of Coal (Gas) in Mechanical Industry, University of Shanghai for Science and Technology, Yangpu District, Shanghai 200093, China.

[2]School of Mechanical Engineering,Shanghai Jiaotong University, Minhang District, Shanghai 200240, China.

Email: [a]guoxinwei1991@126.com,[b*]zhzhx222@163.com

Abstract. In this paper, the combustion stability and nitrogen oxides (NOx) emissions test were made in two coal-fired chain grate boilers with hot water supplies of 29MW and 46MW. The investigations aim to clarify the key characteristics of NOx emission, e.g.instability of NOx emission, temperature and gaseous components profiles and the unsteady variation of NOx emission with power level changing. Main conclusions include: 1) In the process of stable combustion, NOx emissions still have±5% fluctuations, it is not stable. 2) The furnace temperature distribution show double temperature peaks. One peak (~1500°C) is located on the burning solid-fuel surface on the chain grate, and the other (~1050°C) is ~5.6m above the arch. 3) With increasing the thermal power, NOx emission shows a dramatic increase, higher around 1.25 times, as due to the rapid adjustment of coal feeding in grate furnace. Those above show the NOx control difficulty and the proper injector location of SNCR (selective non-catalytic reaction), which is key data collected for further NOx reduction retrofits.

1. Introduction

Coal accounts for a large proportion of China's energy structure. Nitrogen oxides emitted from coal-fired power station boilers are one of the key control targets for air pollution. In the northern part of China, the central heating in winter uses a large number of coal-fired boilers. The furnace type in the north is characterized by large number, small single-unit capacity, uneven distribution, low coal burn-up rate, and large pollutant discharge. It has great pressure on environmental pollution and urgently needs to be solved. The nitrogen oxides emitted by industrial boilers account for 15%-20% of the total nitrogen oxide emissions in China[1], while the layered combustion chain furnace is the main component of industrial boilers, and the nitrogen oxides emitted by it cannot be ignored. In this regard, the state has also increased its control to reduce the emission of nitrogen oxides from coal-fired boilers. The stratified combustion of coal is divided into four sections: preheating drying stage, volatile precipitation stage, coke combustion stage, and burnout stage. The fuel is first cracked under high temperature conditions, and the volatile matter of coal are precipitated, followed by coke combustion, and finally the ash is discharged out of the furnace [2]. The combustion of coal is mainly composed of two processes: volatile matter precipitation combustion and coke combustion. Therefore, fuel N is divided into volatile matter N and coke N in different precipitation stages [3]. The generation and emission of NOx are related to factors such as temperature, combustion atmosphere, and oxygen concentration. More than 90% of the NOx in the stratified furnace belongs to fuel type NOx[4]. For the NOx in the chain furnace, Zhao Laifu used industrial experiments and unit furnace experiments to verify the effectiveness of the unit furnace method in the study of the NOx precipitation law of the coal seam in the layer-burning boiler, proposed to use anaerobic pyrolysis combustion method to select appropriate coal seam thickness to reduce NOx emissions[5]. Song Shunxin et al. carried out a transformation experiment on a SNCR (selective non-catalytic reduction) process for a 35 t/h layer furnace, the denitration rate of the chain furnace was between 13% and 32%, which preliminarily verified the use of SNCR technology on small and medium industrial chain furnaces is feasible[6].

In regard of the above research and industrial transformation of NOx control in chain furnaces, the research on combustion and NOx emission characteristics in layer furnaces is not specific, and no multi-load and versatility studies have been conducted. Therefore, this paper systematically studies the instability of NOx emissions from boilers, the flame temperature/component distribution in the furnace, and the changes in NOx emissions during load changes in large-grain coal stratified combustion, expects to provide detailed key data support and guidance for the low NOx combustion retrofit in the furnace and the rational position design of SNCR ammonia.

2. Brief Introduction and Test Method of Tesearch Layer Burning Boiler

The test was conducted on two layer hot water boilers, with powers of 29 MW and 46 MW respectively. The design and operation parameters of the two boilers are shown in Table 1, the analysis data of the burning coal quality is shown in Table 2.

Table 1. The major parameters of chain grate boiler

	Boiler-1	Boiler-2
Rated Heating Capacity, MW	29	46
Nominal Water Pressure, MPa	1.6	1.6
Number of Drums	Single	Double
Design Efficiency of Boiler, %	82.30	82.50
Return Water Temperature, °C	70	70
Nominal Hot Water Temperature, °C	100	100
Grate Length/Width, m	11.595/5.82	14.54/12.16

Table 2. The industrial and elemental analysis of coal

M,ar/%	M,ad/%	A,ad/%	V,ad/%	FC, ad/%	Q,gr ad/kJ/kg	Q,net ad/kJ/kg
9.1	3.91	16.3	30.19	49.6	26.199	23.672

During the test, the flame temperature, gas composition, NOx emissions, etc. were tested mainly at the fire hole of the boilers. Oxygen concentration and NOx emissions were tested at the outlet of the boiler air preheater. The measurable position in the furnace is shown in Figure 1. There are 5 fire holes in the vicinity of the layer burning furnace, and the fire hole size is about 40cm. The data was tested at the left and right points of a fire hole, and the data of 9 points was obtained (points 1 to 9 in Fig. 1). In addition, there are 5 measuring points at different heights of the furnace (points 10 to 14 in Figure 1). The heights of the five points from the layer burning are 1.35m, 3.4m, 5.56m and 9.2m respectively (where the measuring point 11 and 12 are at the same horizontal position, the two measuring points are 0.93m apart). Due to the limitations of the conditions, the temperatures of these five locations were measured.

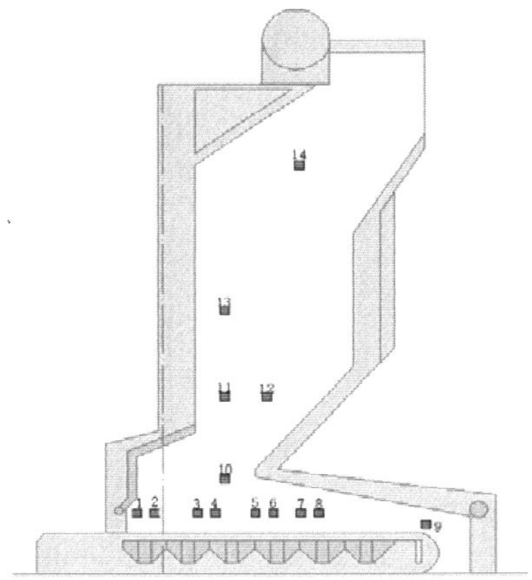

Figure 1. The location of the station in boiler one

During the test, the data was collected by inserting the thermocouple measuring point and the high temperature resistant steel pipe to the center of the boiler and using the "small fan exhaust-cooling-flue gas analyzer test" method. The results below are the average values, and the test data fluctuates within ±5% of the test error. The measured NOx value is converted by the following formula:

$$[\mathrm{NOx, mg/Nm^3}] = [\mathrm{NOx, ppm}] \cdot \frac{46.0}{22.4} \cdot \frac{21\% - 9\%}{21\% - [\mathrm{O_2, vol\%}]}$$

In the formula, the ppm unit of NOx is converted into the concentration corresponding to NO_2, and the oxygen concentration in the flue gas is converted to 9%. Refer to the national standard [GB13271-2014] [7].

Table 3. The test list of condition

	Boiler	Experimental Purpose	Actual Heating Load	Percentage of Full Load	Steady State or Unsteady State
1	Boiler-1	NOx release rule under stable working condition	~25.4MW	87.5%	Steady State
2	Boiler-2	Law of NOx release in process of instantaneous load increase	34-44MW	73.8%~95%	Unsteady State
3	Boiler-1	Measurement of temperature field and gas composition field on grate	~25.4MW	87.5%	Steady State
4	Boiler-1	Test of temperature components in lower part of throat mouth of boiler	~25.4MW	87.5%	Steady State

3. Experimental Results and Discussion

3.1. NOx Emission Test in Smooth Conditions

Six sets of data tests were carried out for the 1# boiler (load 29MW), as shown in Figure 2, where the load in Figure 2(a) is DCS dial data and the temperature measurement point is measure point 14 in Figure 1. The load of the layer-fired boiler in this half hour is basically maintained at about 25.4 MW, which is only slightly fluctuating. The flame in the furnace is close to the right wall, which is mainly affected by the distribution of the coal in the furnace and the size of the coal. The oxygen concentration in the flue gas is very stable, but the NOx concentration fluctuation range is about ± 5%, the lowest is 390 mg/m^3, the highest is 425 mg/m^3, and the average value is 410 mg/m^3. Under steady conditions, the fluctuation range of NOx concentration is large. It can be seen that the large fluctuation of NOx emission data is a major feature of the layer furnace.

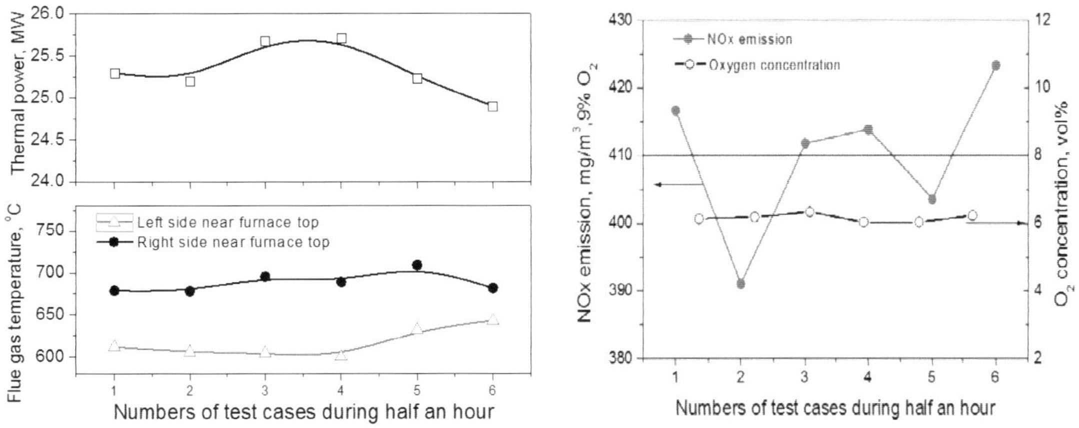

(a) Boiler load and temperature test of two sidewalls (b) Oxygen concentration and NOx concentration in flue gas

Figure 2. Test of NOx and oxygen concentration in flue gas under steady burning condition

3.2. Changes in NOx Emissions during a Sudden Increase in Boiler Load

As shown in Figure 3, this figure describes in detail a typical load change process for 2# boiler (load 46 MW). Figure 3 (a-b) shows the operation change process of coal volume and blower air volume, each quantity will be increased and then reduced, and the whole process will be completed in one and a half hours. During the lifting process, the thickness of the coal seam is certain, and the advance speed of the layer burning is increased to increase the amount of coal burned to the boiler. At the same time, increase the air volume of the blower, give the boiler more air volume, and increase the air volume of the draft fan to ensure the safe operation of the micro-negative pressure in the boiler furnace.

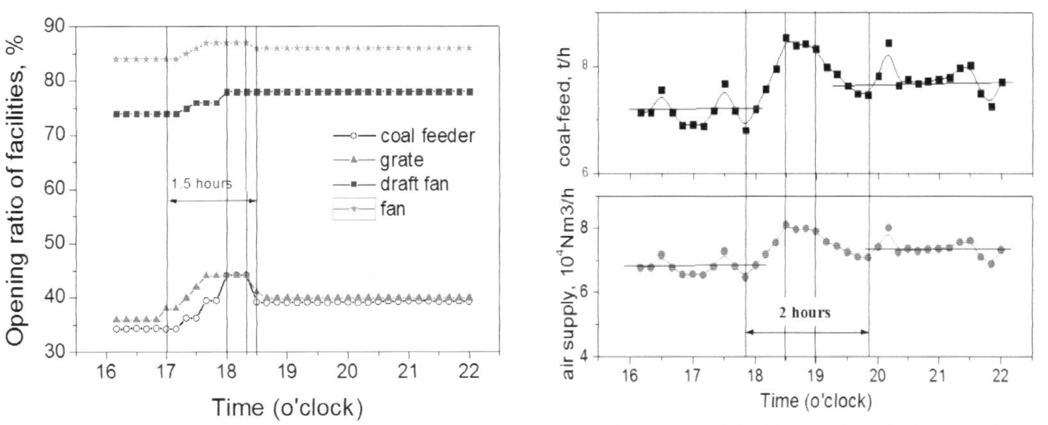

(a) Commissioning data on various equipment of the boiler (b) Changes of boiler coal and air quantity

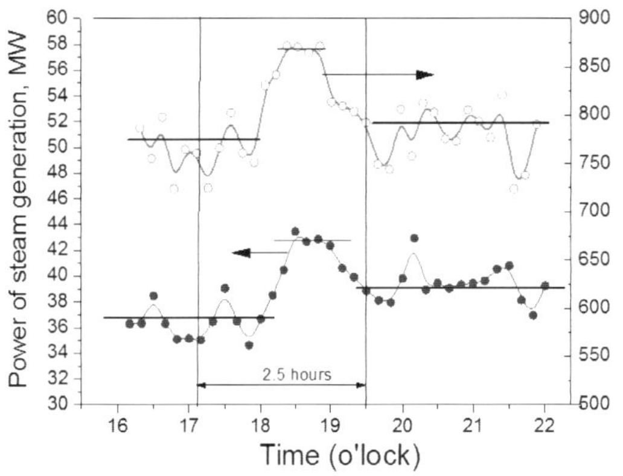

(c) Changes of boiler load and top flue gas temperature

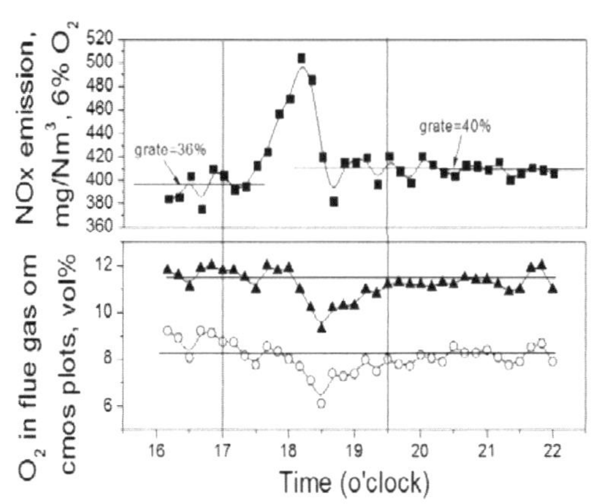

(d) Changes of oxygen and NOx concentrations in flue gas

Figure 3. Boiler one changes in each parameter during load change

Figure 3(c) shows the change in boiler load and flue gas temperature with the boiler operation (temperature measurement point is at the top of the boiler). Figure (d) depicts the process of concentration changes of nitrogen oxides (NOx) and oxygen in the flue gas throughout the load change. When the load is 37MW, the amount of O_2 and the amount of NOx released in the flue gas are both stable and remain basically unchanged. In the process of lifting the load, it takes 2.5 hours for the boiler load to stabilize. That is to say, the stabilization process of the stratified furnace has a significant time delay characteristic compared to the operation. This situation is the lag problem in the control of the layer burner, which is caused by the burning of large particles of coal. At this time, the oxygen concentration in the front and rear stable sections of the flue gas remains the same. During the load change, NOx peaked and its peak value (510 mg/m^3) was increased (24%) from the original stable value (395 mg/m^3). The boiler load suddenly rises and the NOx value jumps. The increase in NOx emissions data is due to the fact that the boiler accumulates instantaneously more coal ash on the grate, the release of N in the coal on the grate increases, and the fuel nitrogen increases, causing a sudden increase in pollutant emissions. When the load in the furnace reaches 39 MW, the NOx content in the flue gas is at a steady state, but the NOx is also increased by about 2.5% compared with the previous stable load. It indicates that the amount of oxygen, coal, air volume remains unchanged, and when the load increases, the NOx content in the flue gas will also increase. It can be seen that the increase of load in the layer furnace will bring more NOx emissions and increase the difficulty of boiler pollutant control.

3.3. Temperature Field and Component Field Test on the Grate

Data shown in Figure 4, Figure 5 is measured from the fire hole measurement point shown in Figure 1, the height of the fire hole is 0.54m away from the grate, and each view hole is tested with two sets of data, the test position is close to both sides of the fire hole, the distance is 40cm. The surface temperature of the fuel is not much different from the temperature on both sides of each fire hole, and only one set of data is measured. In the area of 2m to 3.3m, the surface temperature of the coal seam exceeds 1350°C, and there are many thermal NOx. A temperature peak (1500°C) appeared at 2.3 m.

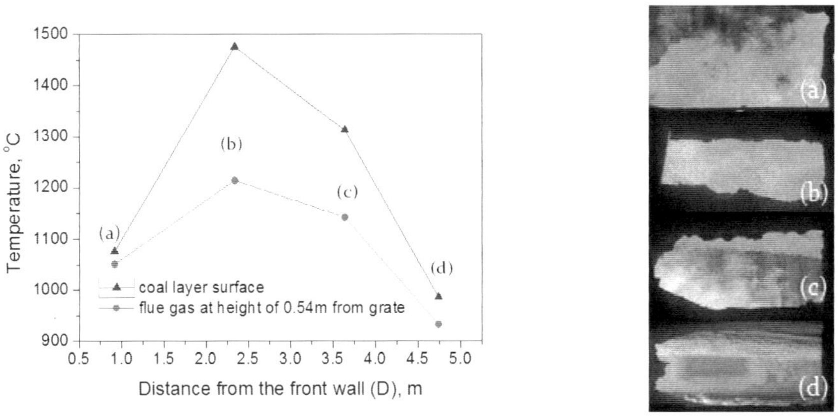

Figure 4. Temperature test (including coal surface and space smoke) in the direction of the furnace.

In combination with Fig. 4 and Fig. 5, it can be seen that in the initial stage of combustion, due to the poor heat transfer performance of large granular coal, the heat generated by the combustion of the upper coal has not yet passed to the bottom, and the coal seam is mainly in a preheated and dry state. The smoke content in this interval is mainly CO_2 and O_2, with a CO_2 concentration of 4% and an oxygen concentration of 17%. After 2.2m along the grate direction, CO_2, CO formation and O_2 consumption in this stage indicate that the upper coal seam in this section is heated and volatilized to precipitate and partially burn. The N element in the fuel is released in the form of HCN, NH3, etc. with the precipitation of volatiles, after volatilization precipitation, it burns inside the coal seam and on the surface of the coal seam. HCN is oxidized to NCO in an oxidizing atmosphere and further oxidized to form NO. NH_3 can react with O_2 to form NO or react with NO to form N_2, depending on the reaction conditions. Since the oxygen concentration is high at the beginning, a large amount of volatile matter N is converted into NOx. The heat released by the combustion of volatile matter causes the coke to burn. The O_2 concentration is rapidly reduced to 0% and maintained at around 0% in the 2.2m-3.9m range, while CO_2 peaks. The lack of oxygen in the coal seam leads to incomplete combustion of the coal seam to produce a large amount of CO, and the CO concentration reaches the highest around 3m from the front wall. The formation of a large amount of reducing gas forms a reducing atmosphere in the coal seam, and the oxygen-reducing strong reducing atmosphere suppresses the formation of NOx. As a reducing gas, CO can reduce the generated NOx. The NOx concentration gradually decreases from 2.7m. After 4m, the O_2 concentration began to rise, and part of it was oxidized to NOx. Due to the coke burning in this area, the coke particles in the coal seam are in a hot and highly active state. The coke layer has a catalytic strengthening effect on CO reduction of NOx [8,9].

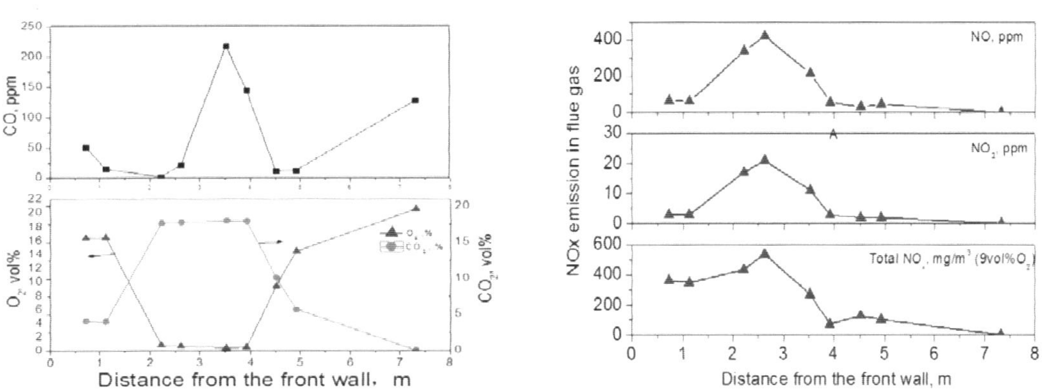

Figure 5. Changes in the composition of the gas in the direction of the furnace

During the diffusion of volatile matter and N in the coke between the coal particles and the coal layer, under the catalytic and reduction of nitrogen oxides by the coke layer and the reducing gas, a large amount is converted into N2. Therefore, although the combustion in this region is very intense, the amount of NOx generated is decreasing. There is more thermal NOx on the grate fuel, and more fuel type NOx is released in the volatile matter burnout position. It can be seen that in the upper part of the grate, the amount of NOx released during the process of volatiles matter precipitation is relatively large, this is the emission reduction position needs to be focused on. When using the low-nitrogen technology in the furnace such as flue gas recirculation or stratified combustion in the stratified furnace, this position should be more focused on for NOx control.

3.4. Temperature and Smoke Composition Test near the Throat

The left graph of Figure 6 shows the temperature profile at vertical height (measurement points 10-14, excluding measurement points 12), and the right graph shows the compositional concentration comparison of points 11 and 12. It can be seen from Fig. 6 that the flame in the vertical direction of the furnace also has a temperature peak (~1050°C), the position is about 5.6m from layer burning height, and when the height is 3.4m, there is a large amount of CO, to 5.6m CO burns out and the temperature reaches its maximum. Therefore, there is another temperature peak in the CO burnout zone. Figure 6 is a diagram showing the composition and temperature of the flue gas at the upper points 11, 12 of the throat, above the rear arch, there is a row of exhausted air vents, the purpose of which is to supply oxygen to burn off carbon and CO in the flue gas, reduce incomplete combustion heat loss, It can be clearly seen from the figure that the measuring point 12 has high oxygen amount than point 11 due to it is opposite to the burnout wind vent, while the CO, CO_2 and temperature are lower than those of measuring point 11, and the NOx is slightly lower, which indicates that the burn-in wind can be added in this position to improve the burnout rate, and the local high temperature can be reduced, thereby reducing the generation of a part of the thermal NOx.

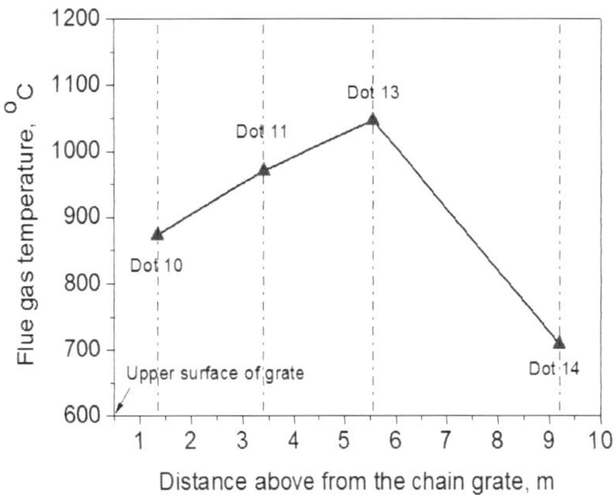

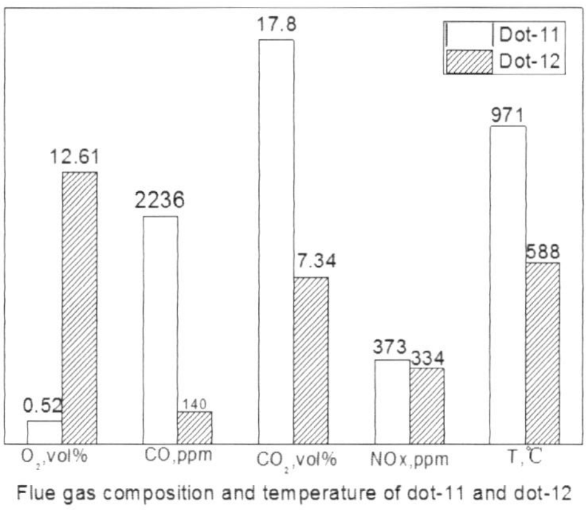

Figure 6. Temperature and composition change in the vertical direction of arch

The selective non-catalytic reduction (SNCR) system is a key method for controlling the layer furnace, it means that the reducing agent ammonia or urea is injected into the furnace without adding a catalyst to reduce NOx to N2[10]. However, SNCR has severe requirements for flue gas temperature, when the reaction temperature is too high, ammonia may directly react with O_2 to form NO, which leads to a decrease in the efficiency of NOx reduction, and also increases NOx emissions. When the temperature is too low, the denitration reaction rate is low, which causes a large amount of NH3 to escape. Therefore, it is necessary to spray an amino solution to a position of about 1000°C, and the temperature window of the SNCR is between 850 and 1100 °C. This experiment provides a preliminary study for the SNCR transformation. Through the research on the combustion and NOx emission characteristics of the layer-burning boiler, we have a deep understanding of the NOx emissions of the layer furnace. At the same time, it also provides a theoretical basis for the installation position of the SNCR, it can be seen from the figure that the optimal installation position of the SNCR of the layer-fired grate boiler in this experiment is between 4.6-5.5 m from the height of the grate.

4. Conclusions

This experiment was carried out on two layer-fired boilers (29 MW and 46 MW), focusing on the temperature field, composition field, and NOx release law in the furnace. The purpose is to reduce NOx emissions from the layer furnace, such as flue gas recycling methods, staged combustion methods, SNCR methods, etc., providing critical boiler data to make it play a denitration effect reasonably. The main conclusions are as follows:

1) In the stable combustion process of the layer furnace, because the furnace is a confined space, the NOx emission still has ±5% fluctuation, which is very unstable;

2) When the boiler needs to change the load, the rapid boiler combustion adjustment will bring about an instantaneous increase in NOx, which is about 24% higher;

3) The temperature distribution in the furnace shows a double peak phenomenon, that is, there is a temperature peak (~1500°C) near the fuel on the grate, and another temperature peak (~1050°C) in the CO burnout area ~5.6m above the inner throat; There is a certain thermal NOx release on the grate, and the release of fuel-type NOx is mainly distributed during the release of volatiles matter. The temperature peak of the space is caused by the combustion of CO, and its position is 3.4m from the grate. The use of a controlled flame method to control NOx should focus on flue gas recirculation and staged combustion techniques at the volatile matter precipitation location, which is beneficial for overall NOx reduction.

4) A reasonable SNCR injection zone, that is, the best installation location is between 4.6-5.5 m from the grate height, at which point the flue gas temperature is within the optimum temperature window temperature (900-1100°C) of the SNCR.

5. Acknowledgment

This research was supported by the National Key R&D Program of China, Grant No. 2016YFB0600202 , all support is gratefully acknowledged.

6. References

[1] Guo Xinwei, Zhang Jian 2017. Test Study of New Flue Gas Re-circulation Techniques Working on NOx Removal in Chain-stoked Boiler. *Energy Engineering*, **4** 48-52.

[2] Yu Z, Fan X 2017. NOx Reduction in the Iron Ore Sintering Process with Flue Gas Re-circulation. *JOM*, **69**(9) 77-83.

[3] Du Shi, Fan Junjie 2018. Multilayer Flue Gas Recirculation Techniques on NOx Removal in Chain-stoked Boiler. *Clean Coal Technology*, **3** 103-107.

[4] Eric A., Nemoda S. and Dakic D. 2016. Experimental and Numerical Study on Combustion of Baled Biomass in Cigar Burners and Effects of Flue Gas Recirculation. *Thermal Science,* **20** 151-165.

[5] Zhao Laifu 2011. Study on NOx Formation Characteristics of Burning Boiler Coal Fired Boiler. Harbin: Harbin Institute of Technology Paper.

[6] Song Shunxin, Zhang Guangguang 2014. Experimental Study on Denitrification Characteristics of SNCR in 35t/h Layer Furnace. *Power System Engineering*, **4** 36-38.

[7] [National Standards] GB13271-2014 Boiler Air Pollutant Emission Standard.

[8] Yang Dong, Lu Chunmei 2007. Experimental Study on NOx Emission Characteristics of Different Types of Pulverized Coal Combustion. *Proceedings of the CSEE*, **27**(5) 18-21.

[9] Zhang Juwei 2009. Multilayer Flue Gas Recirculation Techniques on NOx Removal in Chain-stoked Boiler. Harbin Institute of Technology Paper.

[10] Li Guangzhao 2013. Analysis of SNCR Denitration Technology in Coal-fired Power Plants. *Resources Economization & Environmental Protection*, **12** 32.

IOP Conf. Series: Materials Science and Engineering **484** (2019) 012024 doi:10.1088/1757-899X/484/1/012024

Analysis on Characteristics of Structural Impact Response of Airborne Armored Vehicle in Landing Process

Xinhua Fu

Airborne Training Center, Guilin, Guangxi, 541003, China
E-mail: fuxinhua_003@163.com

Abstract. For such problems as high cost and complex implementation in analyzing the characteristics of the structural impact response of airborne armored vehicle in landing process by real equipment airdrop test, finite element method(FEM) was taken to build the finite element(FE) model of airborne armored vehicle and airbag system to simulate the landing impact process of airborne armored vehicle and that under different typical airdrop conditions of airborne armored vehicle, including different altitude and different vertical landing speed. In addition, comprehensive quantitative analysis was conducted on the characteristics of the structural impact response of airborne armored vehicle to provide theoretical guidance for the design, development, operational use and maintenance of airborne armored vehicle.

1. Introduction

The impact load in the landing process of airborne armored vehicle is one of the major factors causing vehicle structure damage. The structural dynamic response in the landing buffering process of airborne armored vehicle is an important characteristic and is an important indicator to evaluate the stiffness and strength of equipment structure. Real equipment airdrop test and numerical simulation are currently the most common methods to analyze the characteristics of structural impact response of airborne armored vehicle under landing impact conditions. The landing conditions of airborne armored vehicle vary in a large range and are difficult to predict due to the influence of climate and ground environmental conditions. The implementation of the real equipment airdrop test under multiple conditions needs enormous human, material and financial resources. At the beginning of the development and approval test of airborne armored vehicle, the real equipment airdrop test was conducted on airborne armored vehicle. However, limited test data were insufficient to study structural impact response characteristics [1-2]. The development of computer technology and finite element theory has provided a new method for studying the characteristics of structural impact response of airborne armored vehicle in landing process in recent years. FEM was taken in this paper to build the FE model of airborne armored vehicle and airbag system to simulate the landing impact process of airborne armored vehicle and analyze structural impact response characteristics so as to provide theoretical guidance for the design, development, operational use and maintenance of airborne armored vehicle.

2. Analysis Process and Build of Finite Element Analysis (FEA) Model

2.1. Analysis Process

The analysis process of the characteristics of structural impact response is as shown in Fig. 1. The whole analysis process is based on Ansys Workbench, LS-DYNA, Mechanical and so on. First of all, the FE model of airborne armored vehicle and airbag system is built by such steps as geometric model building, material definition, meshing, setting contact relationship, setting initial conditions, setting constraints, solution setup, outputting K files and editing and modifying K files. Next, the landing buffering process under typical airdrop conditions (including different altitude and different vertical landing speed) of airborne armored vehicle was simulated on the basis of the FE model of airborne armored vehicle and airbag system built to obtain such simulation results as stress distribution cloud map, stress change curve and plastic strain field cloud map and conduct comprehensive quantitative analysis on the characteristics of the structural impact response of airborne armored vehicle.

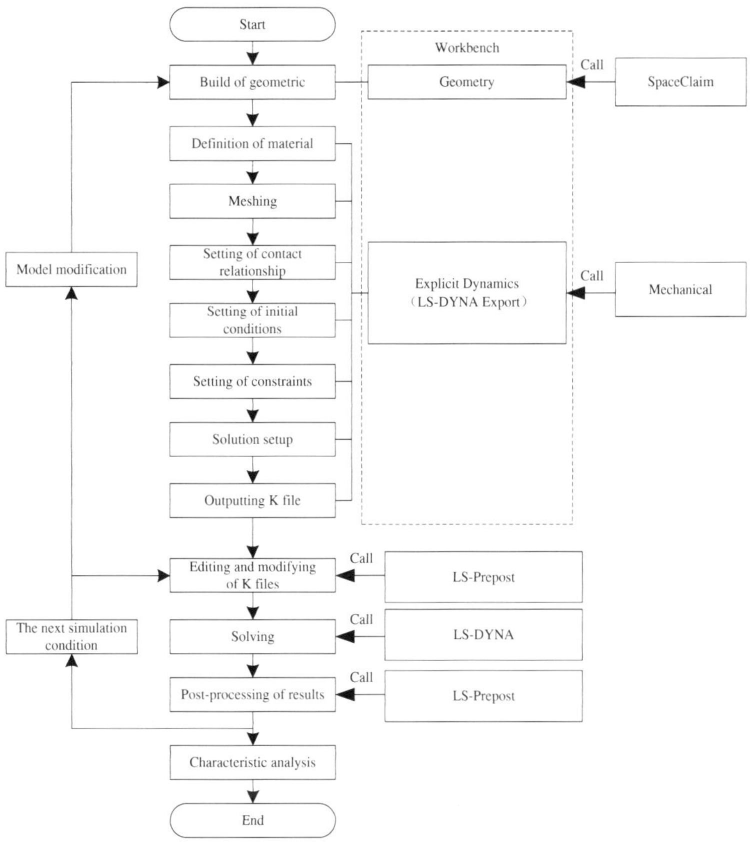

Figure 1. Analysis Process of Characteristics of Structural Impact Response

2.2. FE Model of Airborne Armored Vehicle and Airbag System

In order to improve the efficiency of simulation, when building the FE model of the vehicle body, the structure is necessarily simplified by omitting non-load-bearing components, removing tiny structural characteristics, smoothing the surface of the components and configuring the non-load-bearing components with large mass in the form of mass points on the premise of ensuring the accuracy of calculation [1]. Since the armor plate of the vehicle body of airborne armored vehicle is thin, it is meshed by shell element. Some solid structures, such as turret base ring and support, are meshed by body element. The main body support structure, engine support and other components are identified as the weak parts of the vehicle body structure according to the results of preliminary simulation. Therefore, mesh refinement is conducted on such components.

The buffering airbag system of the airborne armored vehicle is composed of 8 independent airbags in parallel. Since only the landing buffering process is studied, the folding of airbag before landing process and the unfolding of airbag during landing process are not considered. Therefore, the airbag model is in an unfolded state. The airbag is meshed by shell element. Control volume method is taken to establish buffering airbag model on the basis of ideal gas homogeneous pressure model.

The airbag will generate significant compression deformation in the landing buffering process. Both the self-contact model of airbag and the contact model of the airbag with the ground, the vehicle body and the airbag plate are solved by penalty function. The FE model of airborne armored vehicle and airbag system built is as shown in Fig. 2.

Figure 2. FE model of airborne armored vehicle and airbag system

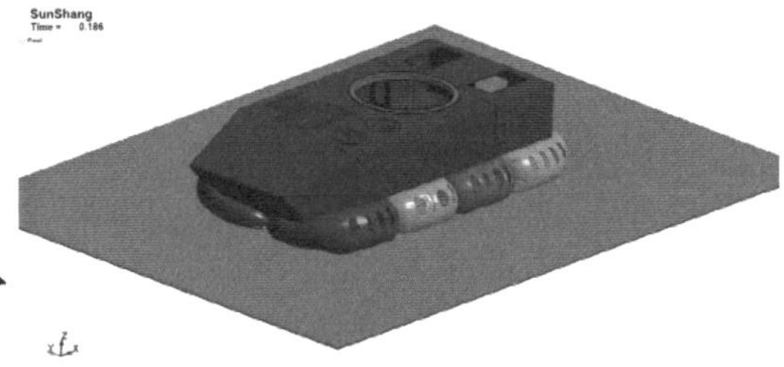

Figure 3. Landing buffering state of vehicle at V=8m/s and the max stress

2.3. Constitutive Material Model

Johnson-Cook model is used to construct the constitutive relations of structural materials of airborne armored vehicle. Johnson-Cook model can reflect the effects of strain hardening, strain rate hardening and temperature softening. Its flow stress-plastic strain expression is [3]:

$$\sigma = \left(A + B\bar{\varepsilon}^{p^n}\right)\left(1 + C\ln\dot{\varepsilon}^*\right)\left(1 - T^{*m}\right)$$

Where, σ is flow stress; *A, B, n, C and m* are constitutive parameters: *A* is yield strength of material, *B* is hardening modulus, *n* is plasticity hardening index, and *C* is strain rate coefficient; $\bar{\varepsilon}^p$ is equivalent plastic strain; and $\dot{\varepsilon}^*$ is dimensionless strain rate.

$\dot{\varepsilon}^* = \frac{\dot{\bar{\varepsilon}}^p}{\dot{\varepsilon}_0}$, $\dot{\bar{\varepsilon}}^p$ is actual strain rate, $\dot{\varepsilon}_0$ is reference strain rate which usually is $1s^{-1}$.

$T^* = \frac{T - T_{room}}{T_{melt} - T_{room}}$, T is material temperature (test temperature), T_{room} is reference temperature which usually is the ambient temperature, and T_{melt} is the melting temperature of material.

3. Simulation of Characteristics of Structural Impact Response

3.1. Simulation Results

The landing buffering process under typical airdrop conditions of airborne armored vehicle was simulated. The structural impact response of vehicle at different vertical landing speed in plain area (the altitude is approximately 0m) is as shown in Table 1 and that upon landing at different altitude as shown in Table 2. The italic numbers in the table means that the element number corresponding to the max stress value of certain part is different from that corresponding to the max plastic strain.

Table 1. Structural Impact Response of Vehicle at Different Vertical Landing Speed (Plain Area)

Vertical landing speed (m/s)	Typical part	Specific element number	Max stress value (MPa)	Max plastic stress (%)
V=8	Armor plate	361583	216.88	0
	Front left column bottom	202515	304.95	0
	Front right column bottom	202183	348.21	0
	Back left column bottom	201906	320.58	0
	Back right column bottom	202330	361.42	0
	Bottom of engine back support	265729	137.14	0
	Bottom of engine front left support	265730	185.02	0
	Bottom of engine front right support	268979	177.47	0
V=9	Armor plate	320806	487.50	0.46
	Front left column bottom	143924	513.52	1.50
	Front right column bottom	202183	518.64	0.92
	Back left column bottom	201906	522.67	0.97
	Back right column bottom	202330	525.42	1.01
	Bottom of engine back support	265729	459.91	0.06
	Bottom of engine front left support	265730	497.84	0.61
	Bottom of engine front right support	268979	513.66	1.43
V=10	Armor plate	288854	***559.34***	2.40
		289279	532.10	***2.58***
	Front left column bottom	143962	865.41	6.68
	Front right column bottom	137082	738.75	4.50
	Back left column bottom	137731	***910.83***	7.47
		137730	910.32	***7.54***
	Back right column bottom	150185	701.82	3.94
	Bottom of engine back support	265729	514.47	1.65
	Bottom of engine front left support	265730	***528.66***	1.14
		265734	512.80	***1.34***
	Bottom of engine front right support	268979	702.03	6.55

Table 2. Structural Impact Response of Vehicle at Different Landing Ground Altitude (V=8m/s)

Landing Ground Altitude (m)	Typical part	Specific element number	Max stress value (MPa)	Max plastic stress (%)
H=1000	Armor plate	361583	***436.30***	0.00
		289279	326.74	***0.05***
	Front left column bottom	202515	469.59	0.19
	Front right column bottom	202183	482.62	0.38
	Back left column bottom	201906	475.42	0.27
	Back right column bottom	202330	481.11	0.36
	Bottom of engine back support	265729	428.13	0.00
	Bottom of engine front left support	265730	459.27	0.03
	Bottom of engine front right support	268979	458.49	0.05

Table 3. Structural Impact Response of Vehicle at Different Landing Ground Altitude (V=8m/s)(Cont.)

Landing Ground Altitude (m)	Typical part	Specific element number	Max stress value (MPa)	Max plastic stress (%)
H=2000	Armor plate	320806	***490.04***	0.50
		289279	454.79	***0.75***
	Front left column bottom	143924	596.74	3.95
	Front right column bottom	202183	538.22	1.23
	Back left column bottom	201906	553.54	1.69
	Back right column bottom	202330	542.64	1.27
	Bottom of engine back support	265729	458.61	0.02
	Bottom of engine front left support	265730	499.78	0.64
	Bottom of engine front right support	268979	536.59	1.79
H=3000	Armor plate	288854	***501.79***	1.30
		289279	497.92	***1.74***
	Front left column bottom	143962	727.91	5.52
	Front right column bottom	202183	573.35	2.11
	Back left column bottom	137731	750.00	5.36
	Back right column bottom	202330	588.90	2.21
	Bottom of engine back support	263659	***458.95***	0.02
		265725	458.29	***0.04***
	Bottom of engine front left support	267234	514.54	0.88
	Bottom of engine front right support	268979	593.50	3.56
H=4000	Armor plate	288854	658.66	4.12
	Front left column bottom	143962	901.88	7.75
	Front right column bottom	137082	743.91	4.54
	Back left column bottom	137731	945.54	8.82
	Back right column bottom	150185	764.22	4.66
	Bottom of engine back support	265729	515.85	1.80
	Bottom of engine front left support	265734	548.46	2.45
	Bottom of engine front right support	268979	764.44	8.46

3.2. Result Analysis

(1) Without loss of generality, characteristic analysis is conducted on the basis of the structural impact response of airborne armored vehicle in plain area and at vertical landing speed of V=8m/s.

1) As shown in Fig. 3, in the landing buffering process of airborne armored vehicle, at 0.186s, the armor plate of airborne armored vehicle contacts the ground (closest to the ground), the vehicle body structure undergoes the max impact overload and the whole vehicle generates the max stress.

2) The max von Mises stress distribution cloud map of the whole vehicle is as shown in Fig. 4. The max comprehensive stress is located at the root of the bottom of the back right column of the main support of the vehicle body, and the max stress is 361.42MPa. The stress of the whole vehicle is mainly distributed on the frame, especially at the bottom of the 4 columns of the fighting cabin and on the girder of the bottom armor plate. The stress on the vehicle armor plate is significantly smaller than that on the main support, indicating that the frame has played a very good enhancement function to the vehicle body structure and has absorbed most of the impact energy. In the landing process of the airborne armored vehicle, the time and position of generation of the max stress, the max first principal stress and the max shear stress may be different.

3) The stress change curve at the max stress position is as shown in Fig. 5. The generation time of the max stress of each component/part of the vehicle is different during landing buffering. Before the bottom armor plate of vehicle contacts the ground, the stress at all positions of vehicle body gets greater with the buffering process and increases quickly to the max value in the moment of contacting the ground, after which the stress oscillates at a low frequency. The stress oscillation frequency and amplitude of body components at different positions are different.

4) When the airborne armored vehicle is under the effect of normal landing process impact, i.e. the airborne armored vehicle realizes landing process in plain area at a vertical landing speed below 8m/s, according to the first and fourth strength theory of material mechanics, since the structural stress of the vehicle body does not exceed the yield limit of the material, the vehicle body will not produce plastic strain, indicating that the strength of the vehicle body meets airdrop requirements.

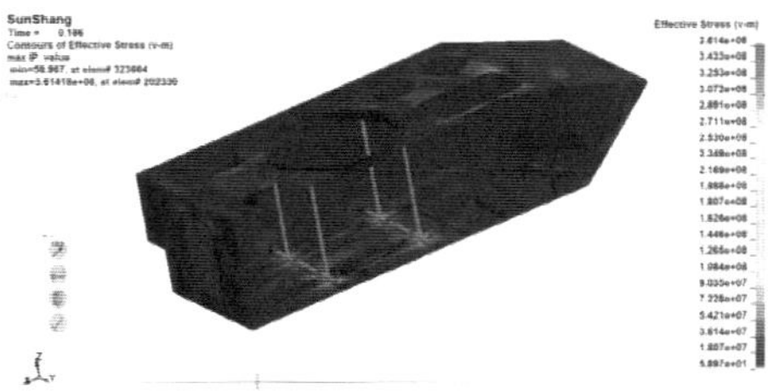

Figure 4. The max von Mises stress distribution cloud map at V=8m/s

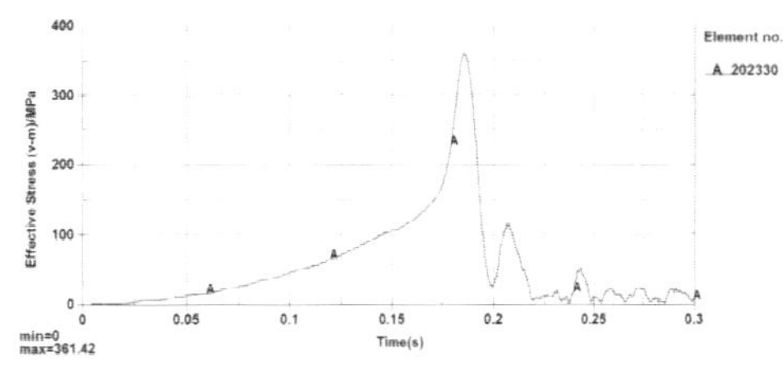

Figure 5. Stress change curve at the max stress position at V=8m/s

(2) When the vertical landing speed of vehicle V=9m/s, the vehicle body structure shows plastic strain, indicating that landing impact overload on the vehicle body structure increases with the increase of the vertical landing speed. Besides, the structural stress of the vehicle body increases accordingly and exceeds the yield limit of the material. The vehicle body structure shows deformation. When the vertical landing speed of vehicle V=10m/s, the vehicle body structure shows great plastic strain (as shown in Fig. 6) which is almost close to the plastic strain corresponding to the damage threshold of the material, indicating that the ultimate vertical landing speed of the airborne armored vehicle allowed by single airdrop in plain area is 10m/s.

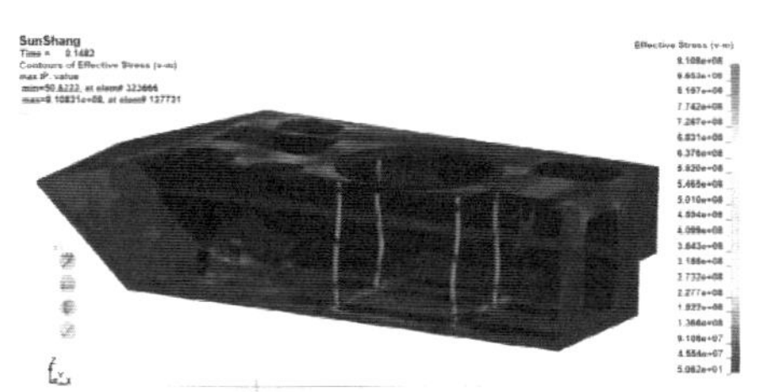

Figure 6. The max von Mises stress distribution cloud map at V=10m/s

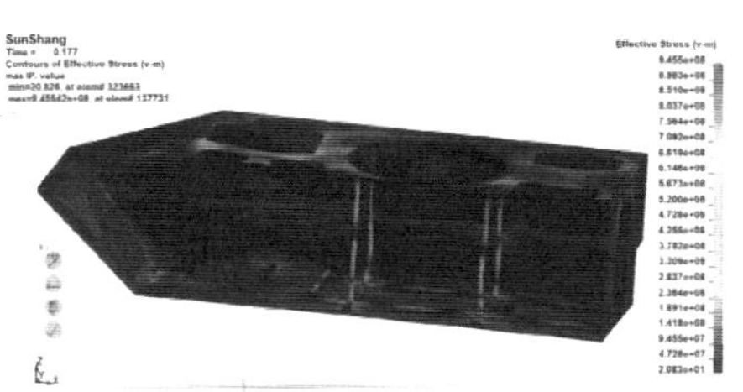

Figure 7. The max von Mises stress distribution cloud map at H=4,000m

(3) Under different vertical landing speed, i.e. under different airdrop conditions, the time and position of the max stress of the vehicle body structure may be different, and the position of the max stress value may be different from the max plastic strain. The oscillation frequency and amplitude of the stress change curve at the max stress position increase with the increase of the vertical landing speed.

(4) When the landing ground altitude H=0m, the vehicle body structure shows no plastic strain. When the landing ground altitude H=1000m, the vehicle body structure shows plastic strain, indicating that landing impact overload on the vehicle body structure increases with the increase of the landing ground altitude. Besides, the structural stress of the vehicle body increases accordingly and exceeds the yield limit of the material. The vehicle body structure shows deformation.

(5) With the gradual increase of landing ground altitude, the max stress and the max plastic strain also increase gradually. When the landing ground altitude H=4,000m, the vehicle body structure shows great plastic strain. That at some positions has exceeded the plastic strain corresponding to the damage threshold of the material (as shown in Fig. 7), indicating that the airborne armored vehicle can realize landing process only in areas below the altitude of 4,000m when the vertical landing speed is 8m/s. If the landing ground altitude is higher than this altitude, the airborne armored vehicle will have

more serious damage and exceed the damage threshold of the material, making the mission unsuccessful.

(6) Under different landing ground altitude, i.e. under different landing process conditions, the time and position of the max stress of the vehicle body structure may be different; and the position of the max stress value may also be different from the max plastic strain. The oscillation frequency and amplitude of the stress change curve at the max stress position increase with the increase of the landing ground altitude.

4. Conclusions

FEM was applied to build the FE model of airborne armored vehicle and airbag system to simulate the landing impact process of airborne armored vehicle, analyze its structural impact response characteristics and simulate the landing impact process under typical airdrop conditions of airborne armored vehicle. The following conclusions are drawn: 1) when the airborne armored vehicle realizes landing process in plain area at a vertical landing speed below 8m/s, since the structural stress of the vehicle body does not exceed the yield limit of the material, the vehicle body will not have plastic strain and the strength of the vehicle body meets airdrop requirements; 2) in order to ensure safe airdrop of the airborne armored vehicle, the ultimate vertical landing speed of the airborne armored vehicle allowed by single airdrop in plain area is 10m/s. When the vertical landing speed is 8m/s, the airborne armored vehicle can land only in areas below the altitude of 4,000m.

5. References

[1] LI Jian-yang, WANG Hong-yan, RUI Qiang , HONG Huang-jie , ZHANG Fang. Research on Cumulative Damage Assessment Method for Airborne Vehicle at Landing Based on Finite Element Analysis [J]. Journal of System Simulation, 2014.1, 26(1): 208-214.

[2] LI Jian-yang, WANG Hong-yan, HAO Gui-xiang. Simulation of Landing Process Based on Explicit Finite Element Method [J]. Journal of Academy of Armored Force Engineering, 2010.6, 24(3): 25-28.

[3] LI Yun-fei, ZENG Xiang-guo, LIAO Yi. Thermal-viscoplastic constitutive relation of Ti-6A1-4V alloy and numerical simulation by modified Johnson-Cook modal [J]. The Chinese Journal of Nonferrous Metals, 2017.7, 27(7): 1419-1425.

IOP Conf. Series: Materials Science and Engineering **484** (2019) 012025 doi:10.1088/1757-899X/484/1/012025

Investigation on Fusion Bridge of Au Electrical Contacts Material under Different Voltage

Jiheng Fang, Di Wu, Jinghong Chen, Xi Liu, Ming Xie*, Song Chen, Jieqiong Hu, Yongtai Chen, Youcai Yang, Hongwei Ma, Jiming Zhang and Manmen Liu

Department of Materials Science and Engineering, Kunming Institute of Precious Metals, No. 988 science and technology road, Kunming, Yunnan, China.
Email: fangjiheng@ipm.com.cn

Abstract. The fusion bridge phenomenon caused by Au contacts was investigated with the high speed vedio-electrical contact testing system built by ourselves. The microscopic topography of Au contacts surface was observed after the survey of fusion bridge size. The results in this paper revealed some novel phenomena and rules. The size of fusion bridges which generated during electrical contact process of Au contacts is micron scale when the load is DC 10 A (6 V to 15 V); The shape of fusion bridges has three types which are the cylindrical, dumbbell and frustum; With the increase of the voltage, the size of the diameter exhibit the characteristic of S type, and the change in length also has such a rule; the molten bridge behaviour can erode the Au contact surface and present the erosion surface with the characteristics of the molten pool, spray particles and convex hill, the droplet spray is the main form of the loss of Au contact material in the process of electrical contact.

1. Introduction

Electrical contact issues have received more and more attention due to their wide application in switching and conducting [1, 2]. Au was widely used in the fields of electrical, electronic, aviation and aerospace with the superior ductility and conductivity, the contact resistance of Au is low and stable and the oxidation resistance is strong [3]. In the microsecond time interval, the fusion bridge is produced in the breaking gap of contacts within micon scale [4-7]. The limit of contacting life and materials transfer can be attributed to the generation of fusion bridges [8-11]. The development of computer technology and numerical simulation technology provides an effective theoretical analysis means for the study of electrical contact phenomena. However, the research on the numerical simulation of Fusion Bridge and the establishment of related models are mostly based on the simplification and neglect of the shape and other factors from the formation to the disconnection process because of the lack of sufficient evidence about the evolution of fusion bridges and other related experimental [12-14]. Therefore, the research on the behaviour of the molten bridge is of great significance [15]. Observing fusion bridges directly, many difficulties will arise because of fusion bridges' dimension (millimetre to micron size) and duration time (millisecond to microsecond interval) is too small [16].

To explore the fusion bridges phenomenon of Au contacts generated in dynamic processing, we built a high speed video- electrical contact testing system. With the Scanning Electron Microscope, the microstructure of contacts surface was studied after the electrical contact testing to further explore the fusion bridges' evolution behaviour.

2. Experimental

The appearance of Au contact object studied in the experimental was cylindrical with the diameter of 2.0 mm. The contact surface was the bottom of the cylinder. The contact surface of the upper contact was processed into cambered by grinding and polishing while the contact surface of the lower contact is processed into plane. Experiments were carried out at DC 10 A (6 V, 8 V, 10 V, 12 V, 14 V, respectively). Thereinto, the anode contact was the upper one while the cathode contact was the lower one.

The high speed video-electrical contact testing system was mainly composed of four parts: the illuminant equipment, the testing machine of electrical contact, the system of scanning and the computer, illustrated in Fig.1. The adjustment of illuminant equipment and scanning system device were performed before the starting of experimental. Next, installed the video to high-speed mode, and opened the testing machine of electrical contact for experimental research. In the mode of break only and the DC load, the Au contacts were studied by the testing machine of electrical contact. The experimental parameters involved were breaking speed ($v = 20.0\ mm{\cdot}s^{-1}$) and contact force ($f = 0.50$ N). The image capture rate of the high-speed video was 480 fps ($n = 480$ fps) in this study, and the resolution of the captured picture was 224 × 160. The computer received video data from the high-speed video. The data was analysed by EDIUS6.0 software to select meaningful and representative images. Taking the diameter (2.0 mm) of the lower contact as the scale, the size of both the length and diameter of fusion bridges were measured with the Digimizer software. At last, the SEM was used to study the microscopic morphology of Au contacts surface after the testing of electrical contact.

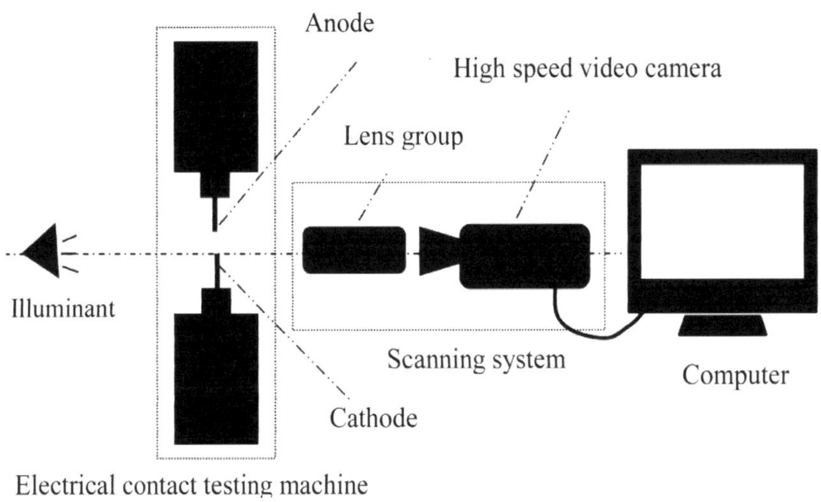

Figure 1. The high speed video-electrical contact testing system

3. Results and Discussion

3.1. High Speed Video Research on Fusion Bridges

The dynamic process trajectory of fusion bridges behaviour is illustrated in Fig.2 under the load of DC 10 A and 6 V. The time interval between each tow pictures is 2.08 ms. The breaking action of the anode shown in Fig.2a has begun, but the Au contact pairs are still not completely separated. As the breaking action of the anode contact continues, a fusion bridge appears between the gaps as shown in Fig.2b. The shape of the fusion bridge can be observed to be dumbbell when the picture is enlarged. The shape of the fusion bridge can be observed to be dumbbell when the picture is enlarged. With the increasing separation of the anode, the fusion bridge disappears so that the Au contact pairs are completely broken. Based on the cathode diameter value of 2.0 mm in Fig.2a, the contact area of Au contact pairs before disconnection is 600.406 μm in diameter. In above method, the fusion bridge can be 81.093 μm in diameter and 66.989μm in length as illustrated in Fig.2b.

IOP Conf. Series: Materials Science and Engineering **484** (2019) 012025 doi:10.1088/1757-899X/484/1/012025

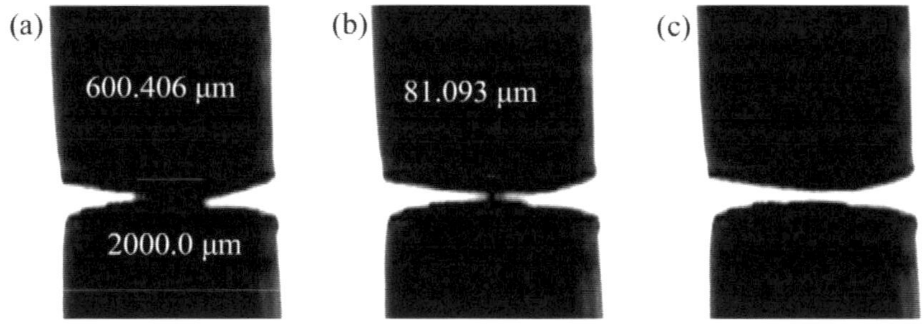

Figure 2. The dynamic process trajectory of fusion bridges under the load of DC 10 A, 6 V

The behaviour characteristics of fusion bridges under the load of DC 10 A and 8 V are similar to that under DC 10 A and 6 V, as shown in Fig.3. It can be seen from Fig.3b that the shape of the fusion bridge looks like a dumbbell. The contact area of Au contact pairs in Fig.3a has a diameter of 445.358 μm. The size fusion bridge occurred in Fig.3b can be 68.460 μm in diameter and 63.194μm in length.

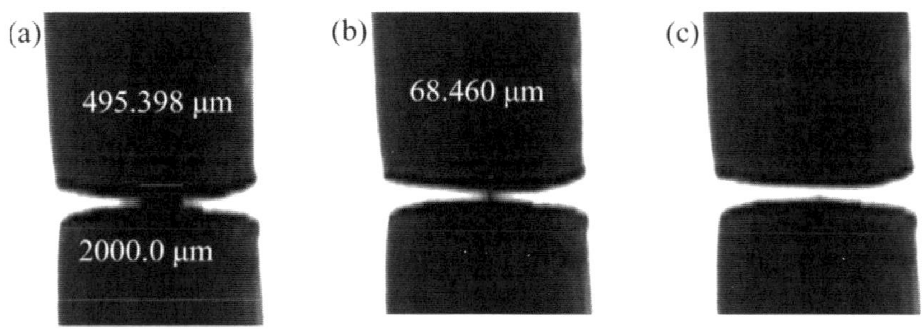

Figure 3. The dynamic process trajectory of fusion bridges under the load of DC 10 A, 8 V

Under the condition of DC 10 A and 10 V, there is a cylindrical fusion bridge in the gap of Au contact pairs as illustrated in Fig.4b. The contact area of Au contact pairs in Fig.4a has a diameter of 632.860 μm. The size fusion bridge occurred in Fig.4b can be 267.245 μm in diameter and 59.161μm in length. The contact area of Au contact pairs in Fig.4a has a diameter of 632.860 μm.

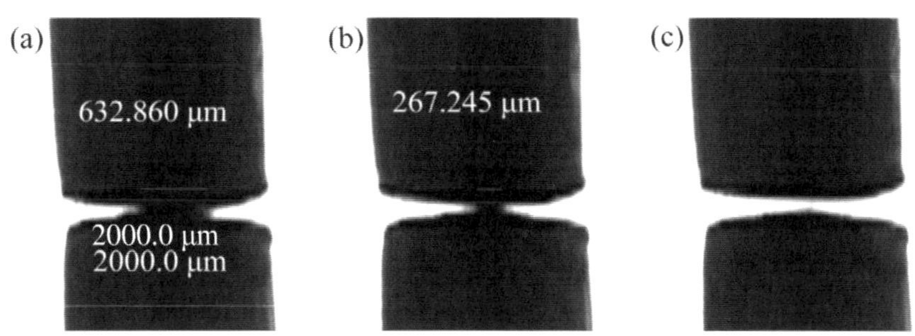

Figure 4. The dynamic process trajectory of fusion bridges under the load of DC 10 A, 10 V

Figure 5b illustrates that the fusion bridge of frustum appearance in the gap of Au contact pairs under the load of DC 10 A and 12 V. With the continue breaking of the cathode, the fusion bridge is broken and then falling back to the surface of cathode. A small residual melt can be observed in Fig.5c which is retained on the surface of the cathode. The size fusion bridge occurred in Fig.5b can be 477.876 μm in diameter and 106.195 μm in length.

IOP Conf. Series: Materials Science and Engineering **484** (2019) 012025 doi:10.1088/1757-899X/484/1/012025

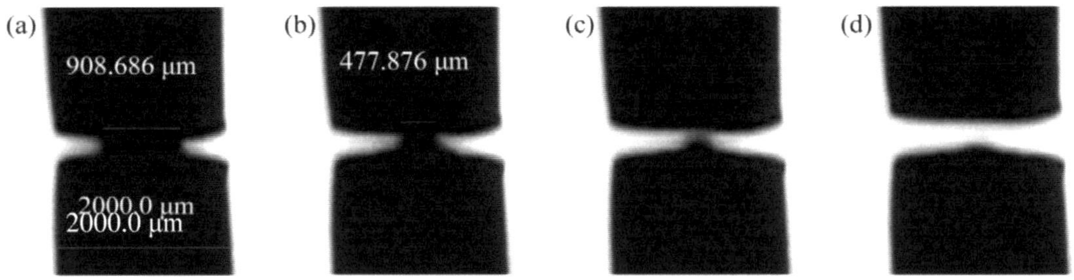

Figure 5. The dynamic process trajectory of fusion bridges under the load of DC 10 A, 12 V

Figure 6 shows the dynamic process trajectory of the fusion bridge under the load of DC 10 A and 14 V which are similar to that under DC 10 A and 12 V. A fusion bridge can be seen in Fig.6b which looks like a dumbbell appearance. The size fusion bridge occurred in Fig.6b can be 493.927 μm in diameter and 133.603 μm in length.

Under the load of DC 10 A and 15 V, the dynamic process trajectory characteristics of fusion bridge are similar to that under DC 10 A and 12 V, as shown in Fig.7. A fusion bridge can be seen in Fig.7b which looks like a frustum appearance. The size fusion bridge occurred in Fig.7b can be 347.242 μm in diameter and 121.186 μm in length.

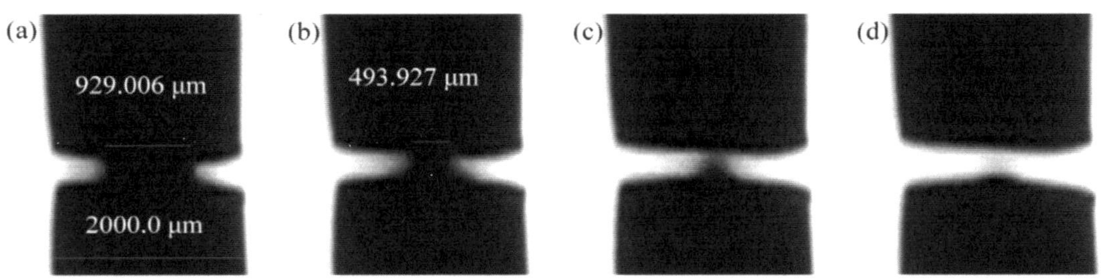

Figure 6. The dynamic process trajectory of fusion bridges under the load of DC 10 A, 14 V

3.2. SEM Morphology on Au Contacts Surface

The microscopic topography of the Au anode surface after the testing of electrical contact is illustrated in Fig.8. An approximately circular erosion area is seen on the surface of Au contact pairs as shown in Fig.8a, which is formed by lots of overlapping fusion pools. The diameter of the erosion area is between 910 μm and 974 μm, which is consistent with the contact area diameter of the high speed video-electrical contact testing experiments. The typical fusion pool morphology is illustrated in Fig.8b. The fusion pool is seen like a circular shape and the size is about 50 μm which belongs to the same order of magnitude as the fusion bridge diameter of high speed video-electrical contact experiments. The surface of the molten pool is relatively flat and a small amount of spray particles can be observed in the vicinity of the molten pool edge.

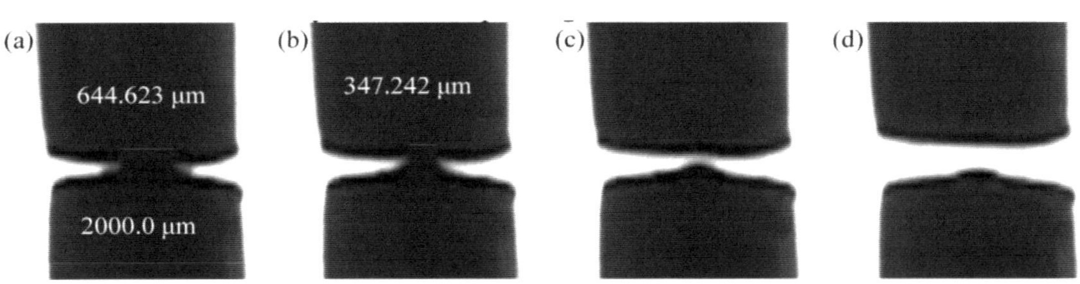

Figure 7. The dynamic process trajectory of fusion bridges under the load of DC 10 A, 15 V

Figure 9 gives the microscopic topography of the Au cathode surface after the testing of electrical contact. The basic features of the erosion area of the contact surface can be observed from Fig.9a, including the shape of the circular, which consists of a number of molten spot and convex hills. The morphology characteristics of the convex hill are shown in Fig.9b. After the broken of molten bridge, the convex hill may be formed by the solidification of the melt down to the contact surface. There are a lot of molten spot and splash particles around the convex hill.

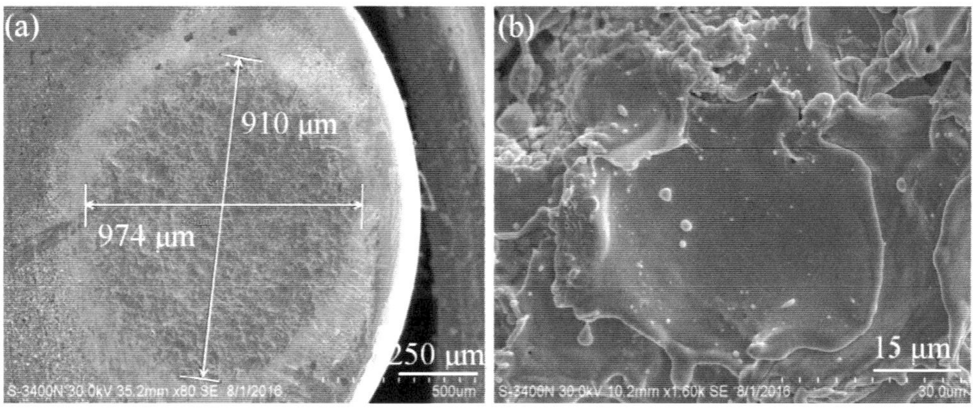

Figure 8. SEM morphology of Au anode contact surface

3.3. Discussion on the Experimental Results

The results of high speed video data on fusion bridges show that the duration time of the molten bridge from the appearance to the fracture is no more than 2.08 ms when the contact voltage is less than 10 V. While the contact voltage is higher than 10V, the duration time of molten bridge will be between 2.08 ms and 4.16 ms.The size of fusion bridges formed during electrical contact of Au contact pairs is varied from 68.460 μm to 493.927 μm in diameter, and 59.161 μm and 133.603 μm in length. The existing report also gives the results that the size of precious fusion bridges is micron grade, which is similar to the results in this paper [17]. The size of fusion bridges is much smaller in diameter than that of the contact area. This result fully demonstrates that the total area of all conductive spots is only a small fraction of the area of the contact area. H Ishida reported the consistent research conclusion with our results [18]. It can be found from the experimental that the appearance of fusion bridges has three types which are the cylindrical, dumbbell and frustum. The results in this paper also prove the conclusion from P M Davidson that the appearance of fusion bridges has three types such as pier, cylindrical and dumbbell [19].

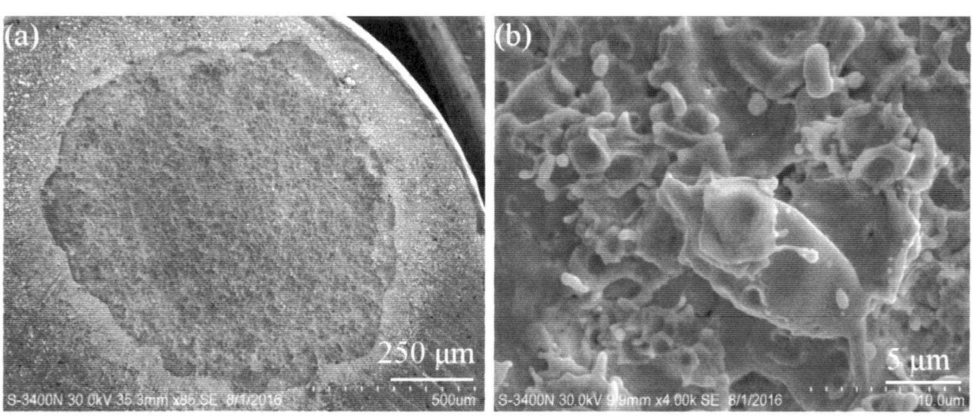

Figure 9. SEM morphology of Au cathode contact surface

Figure 10 illustrates the rules of diameter and length of fusion bridges versus the load of voltage. Under the condition that the voltage is less than 8 V, the diameter of fusion bridges is nearly not affected by the voltage change. When the voltage is more than 8 V, the diameter increases rapidly with the increase of voltage. When the voltage is up to 12 V, the diameter stops increasing and shows a decreasing trend. When the voltage is lower than 10 V, the length of fusion bridges decreases because of the voltage increase. After the voltage is greater than 10 V, the length increases sharply with the increase of voltage. When the load of voltage reaches to 14 V, the fusion bridges length decreases with the voltage increase. Taken as a whole, both the length and diameter of fusion bridges have the growth characteristics of the S type with the voltage increase. The total tendency of experimental results and data analysis give in this research shows difference from the conclusion from H Ishida [20] on Pd contact pairs under low current.The main reason for the emergence of these different phenomena may be that the properties of these two materials are different. With the increase of voltage, the value of both the diameter and length increase while the diameter has a larger growth gradient. Therefore, fusion bridges will show the characteristics of stub type. When the voltage is less than 8 V, the Au contacts are not easy to generate fusion bridges during the electrical contact process.

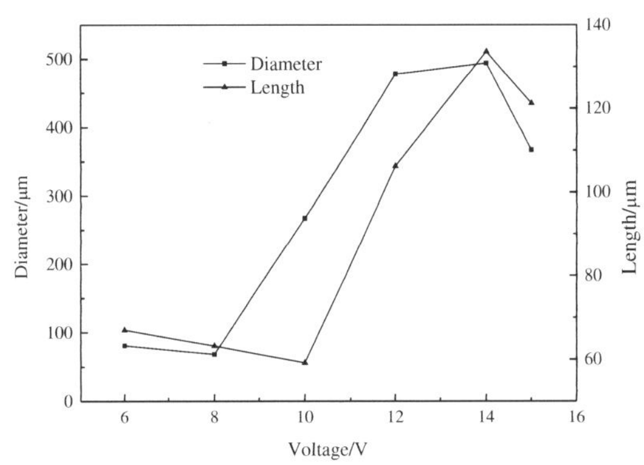

Figure 10. Rules of diameter and length of fusion bridges versus the load of voltage

The behaviour of electrical contact can affect the microscopic topography on the surface of Au contact pairs. Both the surface characterization of Au contacts in the paper of W B Ren [21-23] under the (5 V, 15 V and 25 V)/1 A load and our research conclusions in this paper have a high degree of consistency. During the electrical contact behaviour, fusion bridges will make the contacts surface to form an erosion surface with characteristics of fusion pool, spray particles and convex hill. The shape of fusion pool is circular indicates that the shape of fusion bridge observed in experimental has three types which are the cylindrical, dumbbell and frustum are correct. A large number of spray particles exist on contacts surface shows clearly that the droplet spray is the main form of the loss of Au contact material in the process of electrical contact.

4. Conclusion

The experimental results are summarized as following: under the condition of DC 10 A (6 V to 15 V), the shape of fusion bridges has three types which are the cylindrical, dumbbell and frustum. The dimension of fusion bridges which generate during electrical contact process of Au contacts is micron grade. The diameter and length of fusion bridges both illustrate the growth characteristics of the S type with the voltage increase. The duration time of voltage increase from the appearance to the fracture is no more than 2.08 ms when the contact voltage is less than 10 V. While the contact voltage is higher than 10V, the duration time of fusion bridges will be between 2.08 ms and 4.16 ms. when the voltage is less than 8 V; the Au contacts are not easy to generate fusion bridges during the electrical contact process. The fusion bridges behaviour can erode the Au contact surface and present the erosion surface with the characteristics of the fusion pool, spray particles and the convex hill, the droplet spray is the main form of the loss of Au contact material in the process of electrical contact.

As discussed above, various experimental which are still vague are remained, one is the build of theory to reveal fusion bridges phenomena and the other is the lack of accurate experimental data. Those things are our future works.

5. Acknowledgments

This work was financially supported by the National Natural Science Foundation of China (Project No.U1602271 U1302272 and U1302272) and the National Key Research and Development Program of China (Project No. 2016YFB0305700)

6. References

[1] Gao Q W, Wang X H, Zou J T, Yang X H and Liang S H 2017 Microstructure and properties of Ag-TiB_2 electrical contact materials prepared by hot pressing *J. Rare Metal Mat Eng* 2 503
[2] Cinar I, Aslan O B, Gokce A, Dincer O,Karakas V, Stipe B, Katine J A, Aktas G and Ozatay O 2015 Three dimensional finite element modeling and characterization of intermediate states in single active layer phase change memory devices *J. J Appl Phys* 21 214302
[3] Xie X L, Zhang L, Xiao J K, Qian Z Y, Zhang T and Zhou K C 2015 Sliding electrical contact behavior of AuAgCu brush on Au plating. *J. Trans. Nonferrous Met. Soc. China* 9 3029
[4] Pompanon F, Fouvry S and Alquier O 2018 Influence of humidity on the endurance of silver-plated electrical contacts subjected to fretting wear *Surf Coat Tech* 354 246
[5] Hetzmannseder E and Rieder W F 1996 Make-and-break erosion of Ag/MeO contact materials *J. IEEE T Compon Pack A* 3 397
[6] Dejan P, Mitjan K and Ludvik K 2018 Influence of contact parameters on the tribological behaviour of various graphite/graphite sliding electrical contacts *Wear* 406 75
[7] Kim S K 2008 Process development of low-cost AgNi and AgFe contact materials with no range of solid solubility *J. Adv. Mater. Res.* 47 873
[8] Cao Z F, Xia Y Q, Liu L H and Feng X 2018 Study on the conductive and tribological properties of copper sliding electrical contacts lubricated by ionic liquids *Tribol Intl* 08 033
[9] Chen Z K and Witter G J 2004 Electrical Contacts for Automotive Applications A Review *J. IEICE T Electron* 8 1248
[10] Fourdrinier L, LePoche H, Chevalier N, Mariolle D and Rouviere E 2008 Electrical properties measurements on individual carbon nanofibers by scanning spreading resistance microscopy *J. J. Appl. Phys* 11 114305
[11] Miyanaga K, Kayano Y, Komakine T, Inoue H and Takagi T 2011 Effect of heat conductivity on bridge break at different material contact pairs *J. IEICE T Electron* 9 1431
[12] Wu C P, Zhao Q, Li N N, Wang H S, Yi D Q and Wen W 2018 Influence of fabrication technology on arc erosion of Ag/10SnO2 electrical contact materials *J Alloy Compd* 06 317
[13] Lindner T, Paasch G and Scheinert S 2007 Operation and properties of ambipolar organic heterostructure field-effect transistors. *J. J. Appl. Phys* 1 014502
[14] Schluckner C, Subotić V, Preißl S and Hochenauer C 2018 Numerical analysis of flow configurations and electrical contact positions in SOFC single cells and their impact on local effects *Int J Hydrogen Energy* 11 132
[15] Tan Z L, Guan W M, Guo J M, Chen S, and Wang C J 2013 Mechanical, electrical and thermal coupled-field simulation of a molten metal bridge during contact separation *J. IEEE T Compon Pack* 6 960
[16] Tolpygo S K and Amparo D 2008 Electrical stress effect on Josephson tunneling through ultrathin AlO xbarrier in Nb/Al/AlO x/Nb junctions. *J. J. Appl. Phys* 104 063904
[17] Borkowski P and Walczuk E 2011 computerized measurement stands for testing static and dynamic electrical contact welding *J. Measurement.* 9 1618
[18] Ishida H, Taniguchi M, Sone H, Inoue H and Takagi T 2005 Relationship between length and diameter of contact bridge formed under thermal equilibrium condition *J. IEICE T Electron* 8 1566
[19] Davidson P M 1954 the growth of the liquid bridge in an electrical contact *J. Br J Appl Phys* 5 189

[20] Ishida H, Suzuki S, Hideaki S, Sone H, Inoue H, Taniguchi M and Takagi T 2006 Formation mechanism of dark bridge between contacts with very slow opening speed *J. IEICE T Electron* 8 1136
[21] Ren W B, Chang C, Chen Y 2016 Formation process of intermittent molten bridge between Au-Plated contacts at super low breaking velocity *J. Plasma Sci. Technol.* 3 236
[22] Ren W B, Chang C, Chen Y, Xue S J and Coutu R 2015 Investigation of the surface adhesion phenomena and mechanism of gold-plated contacts at superlow making/breaking speed *J. IEEE T Compon Pack* 6 771
[23] Ren W B, Wang H, Chang C, Xue S J and Chen Y 2016 Experimental investigation and understanding of intermittent molten bridge phenomena and mechanism of contacts with super low opening speed *J. IEEE T Compon Pack* 3 418

Numerical Analysis of Influencing Factors and Capability for Line Laser Scanning Thermography Nondestructive Testing Technology in Chemicals Corrosion Defect Detection

Zhengwei Yang[a,b]*, Xingyu Xie[a], Yin Li[a] and Gan Tian[a]

[a]Xi'an Research Inst. of Hi-Tech, Hongqing Town, Xi'an 710025, PR China
[b]School of Mechanical Engineering, Xi'an Jiaotong University, Xi'an 710049, PR China

Abstract. A new Infrared Thermal Wave Nondestructive Testing (IR TWNDT) technology—line laser scanning thermography nondestructive testing (LLST NDT) technology is applied to detect the defect that caused by chemicals corrosion. The heat conduction process is derived and simulated by the numerical method. Parameters of LLST NDT as the maximum surface temperature difference (ΔT_{max}), the best detection time (t_{best}), and the temperature difference holding time ($\tau_{\Delta T\geq 0.1}$) are discussed as the targets. Based on these parameters, factors influencing the detection results of materials, defect characters (size and depth), and thermal excitation source parameters (line laser scanning speed and power) are analyzed. Simulation results show that LLST NDT is a rapid and effective technique to detect the inside defect of the specimen. The relationship exists between ΔT_{max} with the parameters are described. This work lays a theoretical foundation for further application of the technology and paves the way for optimal design of LLST NDT system.

1. Introduction

Corrosion defect is a common event in the storage of chemicals, especially in petroleum chemical engineering and aerospace areas. Corrosion decreases wall thickness and strength, and serious corrosion will lead to leakage, which is dangerous for not only bringing economic loss, but also environmental pollution and personnel incidents. Thus, it's very necessary to detect the corrosion defect to make sure the safety of chemicals. Infrared Thermal Wave Nondestructive Testing Technology (IR TWNDT), as a great way for defect detection, has been widely used, due to its large test area, rapid test speed, non-contact, easy to operate, etc0-. In active IR TWNDT, the surface area of target is heated by the thermal excitation source; as a result, the thermal wave is produced and conducted into the specimen. When the thermal wave is conducted to the defect interface, part of the thermal wave is reflected to reheat the surface, which could cause the surface temperature of the defect area is higher than that of the normal area as shown in Fig.10. In this case, by applying an infrared camera to monitor the surface temperature fields, the information of defect inside the specimen can be obtained.

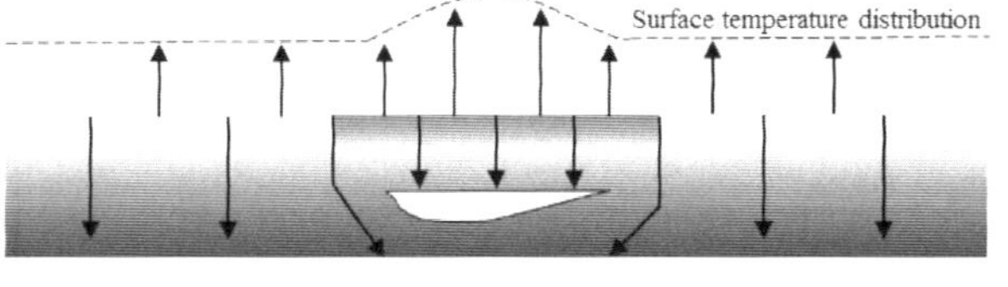

Figure1. Basic principle of active IR TWNDT.

At present, the thermal excitation source mainly includes high-energy flash lamp, ultrasonic, eddy current, etc. Previous scholars stated that the detection results is closely affected by the adopted thermal excitation source and each excitation source has its specific advantages, e.g. the high-energy flash lamp excitation source is suitable for detecting the large area defects with large thermal impedance0, the ultrasonic excitation source is suitable for detecting the kissing defects such as cracks0, while the eddy current excitation source is suitable for detecting the deep surface defects0. However, any one of the excitation source cannot be applied to all kinds of defects. Thus, it is vital to explore new excitation ways to meet different testing needs, which is always one of the research hotspots in IR TWNDT.

In this work we present a new Thermal Wave Nondestructive Testing Technology——Line Laser Scanning Thermography Nondestructive Testing Technology (LLST NDT), which uses line laser to scan the surface of specimen to implement the thermal excitation as shown in Fig. 2.

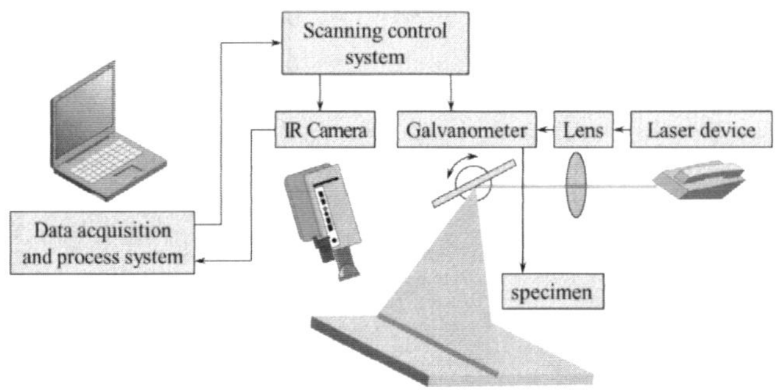

Figure 2. Line laser scanning thermography nondestructive testing technology system diagram.

When it works, a beam of high power laser is shaped by lens into a line uniform and controlled by galvanometer to irradiate the target surface, forming a heat wave to propagate in the target. Once the surface temperature of specimen changes caused by the defects, it can be caught by infrared camera and the defects can be detected by analyzing the infrared image sequence. Its unique excitation way makes this method with the advantages of high controllability and excellent detection accuracy, and the most appropriate detection parameters can be set according to the characteristics of materials and defects, which also can be viewed as an effective supplement to cover the limitations of main thermal excitation sources, including the heating inhomogeneous and tail-light in the high-energy flash lamp excitation, the coupling problem between ultrasonic-gun and specimen in the ultrasonic excitation, and eddy current excitation only suitable for detecting ferromagnetic materials0-.

Form the above analysis, it can be known that the LLST NDT is a promising thermal wave nondestructive testing technology. But the influence factors in the LLST NDT are barely involved, which limits its applications. Thus, in this work, we derive the heat conduction theory and use numerical simulation method to calculate the heat transfer process in LLST NDT. On this basis, parameters that can influence the detection results are analyzed, and the quantitative relationship between defect detection parameters is established. This work is of importance as it paves the way for optimal design of LLST NDT system, which is the future study.

2. Modeling and Computation

2.1. Simulation Model

Flat bottomed holes are often used to stand for the real defects in simulation 0, . Following this way, the model used in this paper is shown in Fig.3, where Δh the depth of defect and D is is the diameter.

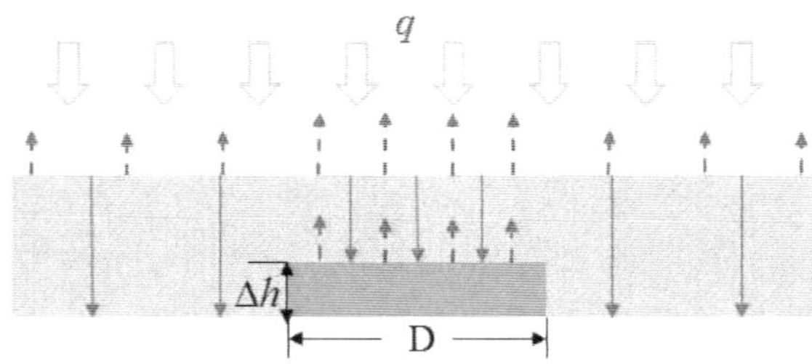

Figure 3. Cylindrical bottomed holes model.

In this research, we assume that the ambient temperature $T_e = 22\,°\text{C}$ and the convective heat transfer coefficient of environment $h_c = 20\text{W}/m^2 \cdot °\text{C}$ remains unchanged, there is no internal heat source, and the convection and radiation at the side is ignored. Selected grid cell is SOLID87 3D 10-node tetrahedral thermal solid element, the curvature meshing method is used in this work, and the grid at the defect is much denser to ensure that the calculation results with high accuracy.

2.2. *Parameters of the Detection Capability*

Generally, the detection capability of IR TWNDT is characterized by such three parameters as 0, : the maximum surface temperature difference ΔT_{max}, the best detection time t_{best} and the temperature difference holding time τ . IR TWNDT is actually based on the surface temperature difference between defect area and normal area to achieve the detection, the bigger the temperature difference, the easier the defect is to be detected, so ΔT_{max} is the most important parameter and it's very necessary to determine the relationship between ΔT_{max} and each influence factors. t_{best} is the time when the surface temperature difference reaches the maximum and the detection effect is best. The defects can be effectively evaluated by analyzing relationships of the two parameters with the detection results. Besides, due to the limitation of image acquisition frequency of the thermal camera (currently, the acquisition frequency is a few Hz to tens of Hz), so the temperature difference holding time should be longer than the time of the camera sensitivity to ensure that the camera has recorded the temperature difference information. In this work, the thermal sensitivity of the camera is set to 0.1°C, so $\tau_{\Delta T \geq 0.1}$ is the parameter that we care about.

2.3. *Surface Temperature Field Distribution*

As shown in Fig.4.(a) ,a specimen with the dimension of 0.25 × 0.15 × 0.005 *m* is established in ANSYS 18.0, and it includes all simulation models. The first row corresponds to the model 1, while the second row corresponds to the model 2. The calculated parameters of the models are shown in Table 1. In the following analysis, all simulations are calculated based on the model 1, except the simulation to calculate the depth of defect which is done in the model 2. All the data are collected from the outside surface center over the 20mm diameter defect. In addition, the heat conduction is two-dimensional heat conduction, so the temperature of the defect center will rise before scanned by the laser. Therefore, the data extraction shall start at the first three seconds before the laser scanning reaches the center of the defect, so as to show the whole process of heat conduction.

Table 1. Model parameters

Parameter	Model 1				Model 2			
Δh (mm)	4	4	4	4	2	3	4	4.5
D(mm)	30	20	15	10	20	20	20	20

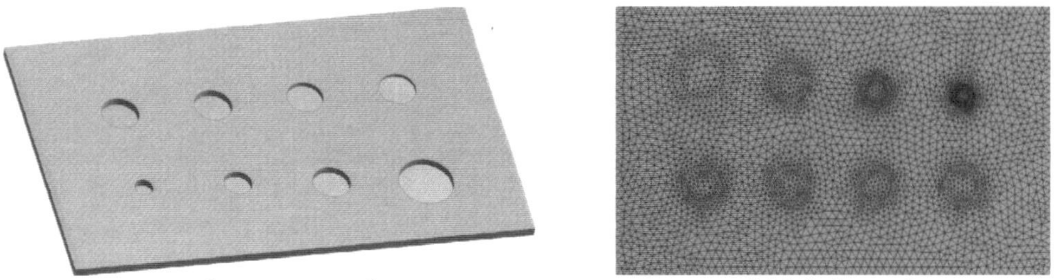

(a) Geometry result (b) Mesh result

Figure 4. The diagram of the specimen.

Simulation results of surface temperature field distribution for structural steel are shown in Fig.5 (also shown in video file: simulation result. avi). It can be seen that at t=5.6s, the line laser has not yet reached the defect, and the defects can not been seen on the surface of the specimen. At t=12.5, the line laser scanned at the center of defect with D=30mm and Δh=2mm, the defect can be observed through hot spots on the surface, but due to Δh=2mm is too deep, the temperature difference between the defect and the normal area disappears very quickly. At t=18.4s, the defect with D=30mm can be observed clearly but the defect with Δh=2mm cannot be observed already. At t=25s, the line laser scanned at the center of defect with D=20mm and Δh=3mm, because of the de decrease of depth, the defect with Δh=3mm can be observed much more clearly than Δh=2mm, but still cannot be completely observed as the defect with the same diameter but a lower depth in the second row at t=28s. As the depth becomes more and more deep, the defect in the first row become more and more obvious, but the defect in the second row becomes more and more unclear due to the decrease of diameter. Finally, at t=53.7s, only the defect with Δh=4.5mm can be observed on the surface. And at t=57.2s, there is nothing on the surface of the specimen. Through the simulation, we can draw a simple conclusion that the bigger the diameter and the deeper the defect is, the easier it will be detected. And the defect can only be detected for a short time. Therefore, study the parameters influence on defect detection is of great significant to improve the detection ability of LLST NDT.

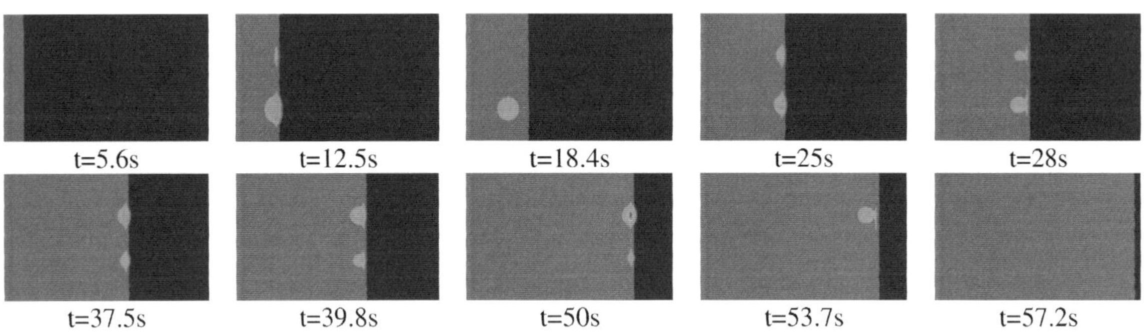

Figure 5. Thermal image sequences of temperature distribution on the surface

3. Discussion and Conclusion

In this work, the heat conduction theory is analyzed and the calculation model is established. On this basis, the effects of parameters on detection results are comprehensively investigated. The conclusions are listed as follows:

From the simulated detection results, the ΔT_{max} is generally greater than 0.1 °C, which can fully meet the demands of sensitivity of infrared camera, and proves LLST NDT is an effective detection technique to detect the inside defect of the specimen. Additionally, the discussion on the different line laser scanning speed shows that LLST NDT is also an efficient detection technique, usually only takes a few seconds.

IOP Conf. Series: Materials Science and Engineering **484** (2019) 012026 doi:10.1088/1757-899X/484/1/012026

4. Acknowledgments
This work was supported by National Natural Science Foundation of China (Grant nos. 51575516 and 51605481), Xi'an science and technology project (Grant no. 2017089CG/RC052 (HJKC001)).

5. Reference
[1] Zhou Y, Cai J. Infrared thermal wave nondestructive testing technology and its application. International conference on computer engineering, information science & application technology 2016.
[2] Heriansyah R, Abu-Bakar S A R. Defect detection in thermal image for nondestructive evaluation of petrochemical equipments. NDT & E International 2009; 42(8):729-40.
[3] Palumbo D, Tamborrino R, Galietti U, et al. Ultrasonic analysis and lock-in thermography for debonding evaluation of composite adhesive joints. NDT&E International 2016; 78:1-9.
[4] Iino M, Carr D J. Pulse thermography applied on a complex structure sample: comparison and analysis of numerical and experimental results. Proc Conferencia Panamericana De End. ELSEVIER 2007; 69–72.
[5] Maldague X P V. Introduction to NDT by active infrared thermography. Materials evaluation 2002; 60(9):1060-73.
[6] Sakagami T, Kubo S. Applications of pulse heating thermography and lock-in thermography to quantitative nondestructive evaluations. Infrared Physics & Technology 2002; 43(3–5):211-218.
[7] Chen D, Zhang C, Feng L, et al. Ultrasonic infrared thermal wave technology and its applications in nondestructive evaluation. Laser & Infrared, 2008; 457(1):1011-30. (in Chinese)
[8] Yan H P, Yang Z W, Tian G, et al. Analysis of influencing factors of geometry size in crack inspection using eddy current thermography. Chinese Journal of Scientific Instrument 2016; 37(7):1610-17(in Chinese).
[9] Abidin I Z, Tian G Y, Wilson J, et al. Quantitative evaluation of angular defects by pulsed eddy current thermography. NDT & E International 2010; 43(7):537-46.
[10] Cheng L, Tian G Y. Surface crack detection for carbon fiber reinforced plastic (CFRP) materials using pulsed eddy current thermography. IEEE Sensors Journal 2011; 11(12):3261-68.
[11] Jin G F, Wei Z, Yang Z W, et al. Application of Ultrasonic Infrared Thermal Wave Technique in Detection and Recognition of Interface Kissing Damage. Journal of Sichuan University 2013; 45(2):167-75 (in Chinese).
[12] Zhang H J. Non-quasi-steady analysis of heat conduction from a moving heat source (III). Journal of Engineering Thermophysics 1991; 112(3):777-79.
[13] Louaayou M, Naït-Saïd N, Louai F Z. 2D finite element method study of the stimulation induction heating in synchronic thermography NDT. NDT & E International 2008; 41(8):577-81.
[14] Wang D, Tian Z, Shen L, et al. Numerical simulation of temperature field of laser remelting MCrAlY coating prepared by plasma spraying on titanium alloy. Applied Laser 2007; 27(6):444-49.
[15] Gui J B. Studies of high-power laser heat effect simulation and laser beam description. Master thesis: Kunming University of Science and Technology 2004. (in Chinese)
[16] Guofeng J, Wei Z, Jun S, et al. Numerical analysis of influencing factors and capability for thermal wave NDT in liquid propellant tank corrosion damage detection. Measurement Science Review 2013; 13(4):214-22.
[17] Dodd C V, Pate J R, Deeds W E. Eddy-current inversion of flaw data from flat-bottomed holes. NDT & E International 1989; 30(3):305-12.
[18] Liu J M. Simple technique for measurements of pulsed Gaussian-beam spot sizes. Optics Letters 1982; 7(5):196.
[19] Jönsson M, Rendahl B, Annergren I. The use of infrared thermography in the corrosion science area. Materials & Corrosion 2015; 61(11):961-65.
[20] Yang Z W, Zhang W, Tian G, et al. Numerical simulation of influencing factors for thermal wave NDT in composite materials. Journal of System Simulation 2009; 21(13):3918-21.
[21] Dai, G.H.. Heat Transfer, 2nd ed. Beijing: Higher Education Press 1999.

IOP Conf. Series: Materials Science and Engineering **484** (2019) 012027 doi:10.1088/1757-899X/484/1/012027

The Research and Application of Industrial Robot in Steel Deep Processing Field

Xiaodong Hao, Yan Bian, Lu Bai, Lihua Li and Changjie Sun

China Iron and Steel Research Institute Group, No.76 Xueyuan South Road, Haidian District, Beijing, China.
Email: yinzihao163@163.com

Abstract. The paper describes technical research results of four categories of robots in steel deep processing field, including integrated code reading and temperature measuring robot for hot rolling production line, code spraying robot, slag-dredging robot for galvanizing line and coiling machine sleeve mounting robot for color coating line and proposes the development trend of industrial machinery in the steel industry.

1. Foreword

In recent years, steel industry of our country has gradually realized the digitization and automation, and is moving toward informatization and intellectualization. To further implement the "Made in China 2025" plan and the national "13th Five-Year" strategic planning, with the support of China Association for Science and Technology under the subject of "Interdisciplinary Collaborative Innovation in Intelligent Manufacturing in Steel Industry", Automation Research and Design Institute of Metallurgical Industry (hereinafter referred to as ARIM) and other units jointly launch a research work concerning development status and demand of intelligent manufacturing in Chinese iron and steel enterprises. The results show that it is an important trend for the future development of the steel industry to integrate networked and intelligent tools such as big data, Internet of things, cloud computing and robots with the process links such as research and development, production, management and logistics to effectively optimize the structure of the manufacturing process and improve the full-process intelligence level of the steel industry [1] [2]. Among them, the large-scale application of industrial robots in the steel industry is an important foundation, and the application fields include: 1) process before iron-making, iron-making process: more than 20 categories of robots for stacker unmanned, molten iron sampling and transfer, automatic sampling of coke for analysis, automatic mud adding of clay gun etc; 2) steel making process: more than 10 categories of robots for temperature measuring and sampling, tundish operation, bale and long nozzle operation, internal defect detection of continuous casting slab, continuous casting slab grading, automatic heating protection of crystallizer etc; 3) deep processing such as steel rolling and galvanizing: more than 20 categories of robots for automatic bar sampling, automatic welding plate printing for bar, spraying and handling, automatic baling and code-spraying for hot rolled coils, zinc slag removing, roll changing, rolled steel cooling bed tracing, cold rolling stripping etc.

This paper focuses on application technology of industrial robot in steel deep processing field and describes technical research results of four categories of robots in steel deep processing field, including integrated code reading and temperature measuring robot for hot rolling production line, code spraying robot, slag-dredging robot for galvanizing line and coiling machine sleeve mounting robot for color coating line, contributing to the application of robotics in steel industry.

2. Integrated Code Reading and Temperature Measuring Robot for Hot Rolling Production Line

At present, most steel enterprises do not realize full- process MES system control, and thus it is necessary to manually read the billet code and enter into control system before the billet enters the heating furnace. Therefore, according to the spraying imaging characteristics, process requirements and operation processes, ARIM has developed an integrated code reading and temperature measuring robot for hot rolling production line, which can effectively measure the temperature of five surfaces of the billet and enter all the temperature into control system automatically while reading code. The robot is composed of network camera, infrared thermometer, robot hardware, image identification software based on machine vision with independent intellectual property rights. The robot control system takes pictures of the billet, analyses the pictures and identifies the number; and it verifies the identification results after comparing the number issued by secondary system. At the same time, infrared temperature measurement system is used to measure the temperature of the billet at three points, and the temperature is transmitted to the secondary system (see figure 1). Integrated code reading and temperature measuring robot for hot rolling production line can systematically identify and record the entire billet information, provide data support for the subsequent production, quality control and sales of the billet; in this way, error-prone problems in manual copying and input coding can be effectively avoided, and whole control of recording process can be realized.

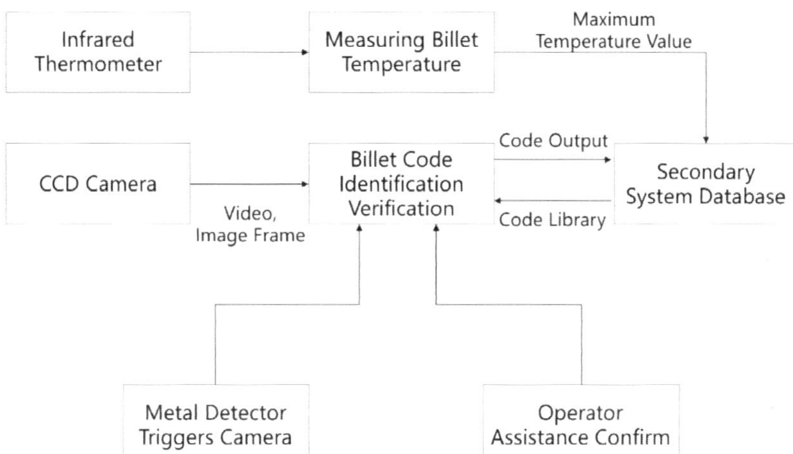

Figure 1. System Structure Diagram of Code Reading and Temperature Measuring Robot for Hot Rolling Production Line

3. Code Spraying Robot

Since hot rolled coil coding process involves high temperature of steel coil, relative movement of work position, close working distance of personnel, heavy workload and other problems, meaning the process has a high risk coefficient and runs in a harsh working environment, the code spraying robot has a wide market prospect. The robot consists of laser coding machine, ranging system, robot body and robot control system. The robot control system receives the coil number, coil weight and other information from the secondary system and waits for the steel coil to be ready. After the coil being in place, the robot detects the distance between the coding machine and the steel coil through with ranging system, and controls the laser coding machine to move to the appropriate distance to complete the code spraying operation (see figure 2).

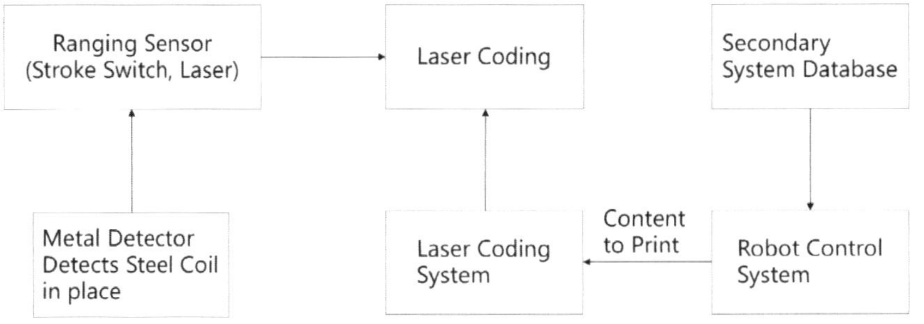

Figure 2. System Structure Diagram of Code Spraying Robot

4. Slag-Dredging Robot For Galvanizing Line

Traditional slag-dredging mode is manual slag-dredging and it has the following problems: 1) safety hazards. During the operation, the worker is only 0.2m away from the high-temperature plating liquid; 2) Harsh working conditions affect health. Position liquid pot reaches 460 °C, ambient temperature at working position reaches approx. 80 °C, air knife noise is as high as 120 db and accompanied by a large amount of dust; 3) slag-dredging process is difficult to control and affects product quality; 4) enterprise cost is difficult to control. The robot consists of CCD camera, image visual servo system, robot body and robot control system. The zinc-liquid surface image is obtained using industrial camera, the zinc-liquid surface slag amount and distribution are obtained through image processing, and slag dredging instructions are output to control the slag-dredging robot for operation (see figure 3 and figure 4). Software identification shows that the calculated proportion of slag is consistent with the actual estimation, the accuracy of zinc slag distribution recognition is more than 90%, the surface of zinc liquid is stable, and galvanized sheet enjoys good surface quality [3] [4].

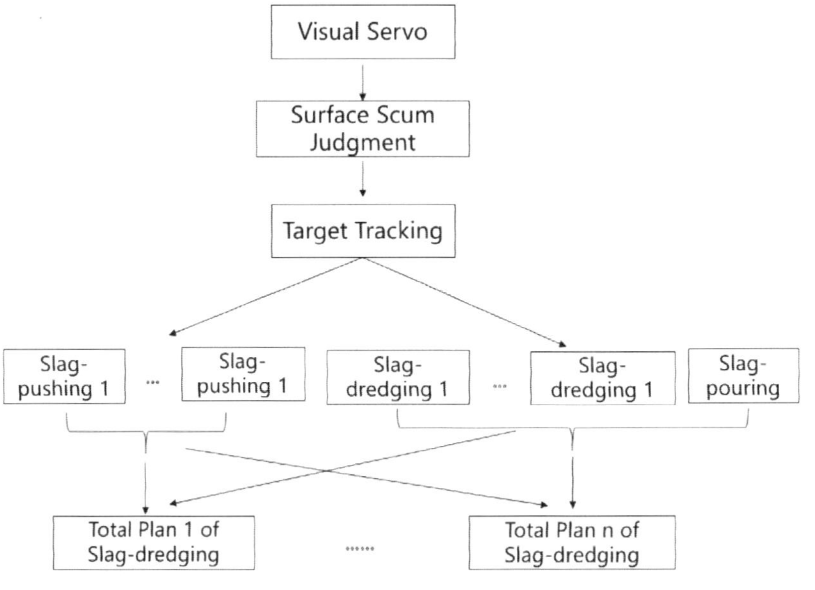

Figure 3. Slag-Dredging Plan of Slag-dredging Robot for Galvanizing Line

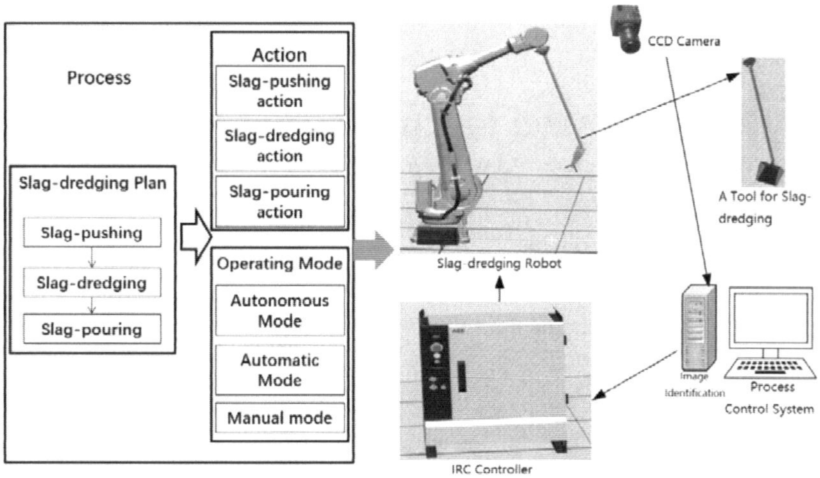

Figure 4. Design Outline of Slag-dredging Robot for Galvanizing Line

5. Coiling Machine Sleeve Mounting Robot for Color Coating Line

Generally, coiling machine sleeve at outlet of color coating line is installed under joint efforts of two operators, presenting problems such as high labor intensity, tight rhythm and small working space, so it is difficult to guarantee the safety of personnel. The robot consists of CCD camera, self-adaptive flexible sleeve grasping equipment, laser measuring instrument, and robot body and robot control system. Coiling machine sleeve mounting robot for color coating line can install paper sleeve of different specifications onto crankshaft to effectively solve coil collapse problem. At the same time, the system realizes the self-determination flexible sleeve measurement and calibration, and the judgement accuracy reaches 100%. The self-adaptive buffering sleeve grasping mechanism is studied and designed and robot is used as a substitute during sleeve mounting process.

6. Conclusion

To sum up, it is an inevitable trend for the future development of the steel industry to integrate the robot with informatization and big data and incorporate it into the full-process control system to form large-scale "machine replacement of human". ARIM focuses on the design and application of robot system under high temperature, high pressure, high humidity, low temperature, poisonous gas, heavy physical strength conditions, and it modifies robot based on existing technology using intelligent equipment such as robots and sensors through the research of application technology to form a low cost, high efficiency production mode which basically replaces existing workers to improve labor efficiency and precision and provide important means for the transformation and upgrading of iron and steel enterprises.

7. References

[1] Lu P and Chen M 2018 A brief analysis of the application of industrial robots in intelligent manufacturing Metallurgical Industry Automation vol 42 pp 134–5
[2] Meng M, Zhou C, Chen L, Feng M and Miao C 2016 Industrial robot development and application overview J. Shanghai Jiaotong University vol 50 p 98
[3] Merluzzi A and Brunetti G 2017 Metals industry: Road to digitalization 2017 40th Int. Conv. on Inf. and Com. Tec., Ele. And Microelectronics vol 40 (Croatia: Opatija) pp 967–73
[4] Dai X and Wei Z 2017 Application research on dross removing equipment in hot dip galvanizing pot Metallurgical Industry Automation vol 41 pp 10–14

The Relationship between Morphology of Milling Surface and Its Corrosion Resistance for Nickel base Superalloy FGH97

Jinping Shi[a], Peiquan Guo[b,*], Yang Qiao[c], Shouren Wang[d] and Tonghui Liu[e]

Department of Mechanical Engineering, University of Jinan, 336 Nanxinzhuangxi Road, Jinan, Shandong 250022, P.R. China.
Email: [a]me_shijp@ujn.edu.cn, [b]oss_guopq@ujn.edu.cn, [c]me_qiaoy@ujn.edu.cn, [d]me_wangsr@ujn.edu.cn, [e]tonghui_liu@163.com

Abstract. The relationship between morphology of milling surface and its corrosion resistance for nickel base superalloy FGH97 have been investigated based on the milling experiment corresponding to different cutting parameters. Motif area ratio parameter is proposed in order to visually characterize the surface morphology. Motif area ratio is defined as the ratio of the surface area that can meet the threshold condition to the machined surface. The morphology of the workpiece surface can be characterized by the amplitude parameter and Motif area ratio method. The relationship model between the milling parameters and the corrosion rate of the workpiece is established by taking surface morphology, residual stress and other surface characteristics as the intermediate variables. Experimental results show that the corrosion rate of the workpiece will be smaller and smaller as surface morphology improved. The higher the residual stress on the workpiece surface, the larger the corrosion rate of the workpiece will be.

1. Introduction

Powder superalloy is of excellent mechanical properties, thermal properties, thermal fatigue properties and high reliability. It is the preferred material for the key parts of modern aeroengine, such as turbine disc, guide vane, ship gas turbine, rocket engine and nuclear reactor. In the early 1970s, the P&WA Company of the United States first developed the IN100 powder superalloy, which is used in the turbine discs of F15 and F16 engines and other components. In order to improve the engine performance, GE successfully developed René88DT turbine disc by argon atomization (AA) and hot extrusion and isothermal forging in 1988. The turbine disc is widely used in civil and military aircraft. The outstanding characteristic of the second generation powder superalloys represented by René88DT is that the crack propagation rate is obviously reduced by 50% compared with the first generation superalloys, the maximum service temperature is up to 750 degrees Centigrade. [1, 2]. In the 1990s, the United States has developed the third generation powder superalloys, such as Alloy10, René104 and RR1000 [3, 4]. Superalloy FGH95, FGH96 and FGH97 alloys have been developed successively in the late 1970s in China. Superalloy FGH95 is characterized as high strength at working temperature of 650 degrees Centigrade. Superalloy FGH96 can be used at temperatures up to 700-750 degrees Centigrade. It is a key material for preparing high performance engine turbine disk. Superalloy FGH97 prepared by plasma rotating electrode pulverizing and hot isostatic pressing has the advantages of high tensile strength and high creep resistance, and its crack propagation rate is one order of magnitude lower than that of FGH95 or FGH96 [5-7]. In October 1971, an aircraft of VICHERS Company crashed due to corrosion-induced breakage of pressurized compartment. In August 1981, a B737-200 of Taiwan Far East Airlines crashed due to corrosion-induced large area perforation of fuselage skin. In August 1985, a Japanese B747 aircraft crashed due to corrosion-induced fatigue fracture of pressurized compartment. In May 2005, a Boeing 747 of Taiwan crashed into the South China Sea due to corrosion-induced cracks in the tail wing. There are so many catastrophic accidents caused by corrosion that it is very important to study the effect of milling surface morphology on corrosion resistance of superalloy FGH97. A lot of research works about machining of nickel-based alloy have been done such as cutting forces and wear [8, 9], integrity of machined surface [10-13], machining induced residual stresses [14-16]. There is lack of research on its corrosion resistance.

2. FGH97 Milling Test

2.1. Properties of FGH97

The composition of FGH97 is shown in Table 1. The alloy elements dissolve a large number of solid solutions in Co and Ni, i.e. the strengthening phase γ'. In order to ensure good strengthening effect, the content of strengthening phase γ' in FGH97 is about 64% of the alloy mass fraction. Its tensile strength is 1510 MPa, yield strength is 1080 MPa and solid solution temperature is 1180-1190 degrees Centigrade. Therefore, the physical properties of FGH97 seriously restrict its machinability. Because of the special manufacturing process of materials, thermal-induced pores are formed in the material during the manufacturing process. When the tool cuts the induced pores, the tool will be impacted. The continuous impact has a great impact on the tool performance. Moreover, the existence of thermal-induced pores is not conducive to the heat dissipation during the cutting process. The thermal induced porosity also severely restricts the machinability of the material. So, FGH97 is very difficult to be machined.

Table 1. Composition of FGH97 (wt %).

C	Cr	Co	Al	Ti	W	Nb	Hf	B	Zr	Mg	Ce	Ni
0.04	9.0	15.5	4.9	1.8	5.5	2.6	0.3	≤0.015	≤0.015	≤0.02	≤0.01	bal

2.2. Test Condition

Milling experiments were conducted on a machining center YCM-V116B as shown in Figure 1. TiAlN-TiN coated carbide insert SECO made SEEX1204AFTN-MD18 is used to mill FGH97. Mill cutter consists of a cutter head R220.53-0100-15-5A and inserts mentioned above. Inserts are fixed on cutter head. Cutter's geometric parameters are rake angle of 10°, clearance angle of 12°, cutting edge angle of 30°, distortion angle of 3°, tool cutting edge inclination angle of -10°.

Figure 1. Machining center and cutter.

3. Morphology Characterization of Milling Surface

3.1. Amplitude Parameter Characterization Method

Under the condition of same feed per tooth and depth of cut, the influence of cutting speed on 3D surface topography parameters is shown in Figure 2. When the speed is in range of 30-70m/min, the value of kurtosis S_{ku} and partial slope S_{sk} is higher, however, the other parameter values are relatively small. This suggests that the surface morphology is better. When the cutting speed is 90m/min, the parameter values are generally higher, especially S_v and S_z, which shows that there are relatively large and deep valleys on the surface. This is because when the cutting speed is 90m/min, for machining of nickel base powder metallurgy superalloy, it has been in range of high-speed cutting. In such case, cutting temperature and tool rotational speed are very high; the vibration of machine tool system is relatively large. For this reason, it eventually makes surface topography bad.

Under the condition of same cutting speed and depth of cut, the influence of feed per tooth on 3D surface topography parameters is shown in Figure 3. As shown in Figure 3, in addition to kurtosis S_{ku} and partial slope S_{sk}, with the increase of feed per tooth other amplitude parameters also subsequently increased. When the feed per tooth is 0.05mm/z, S_{sk}<0 and S_{ku}>3. This shows that there are relatively steep valleys which affect the quality of 3D surface topography. When the feed per tooth is 0.10mm/z, the value of each magnitude parameter is relatively small, in addition, the value of slope S_{sk} is close to zero and the kurtosis S_{ku} <3. This shows that the surface is smooth and no larger peaks and valleys exist. Therefore, when the feed per tooth is 0.10mm/z, the 3D surface topography of workpiece is the best.

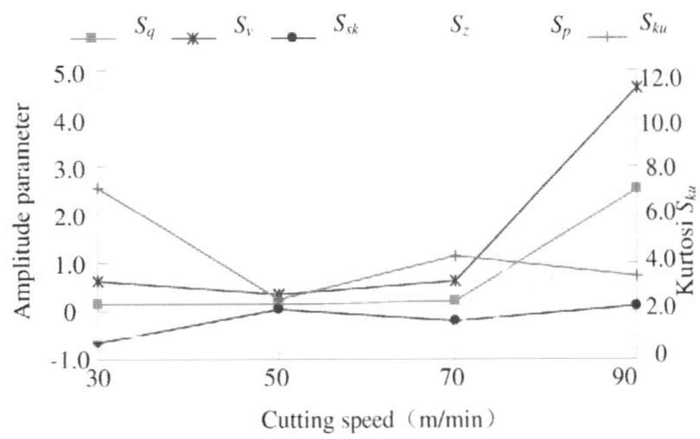

Figure 2. Influence of v on surface topography.

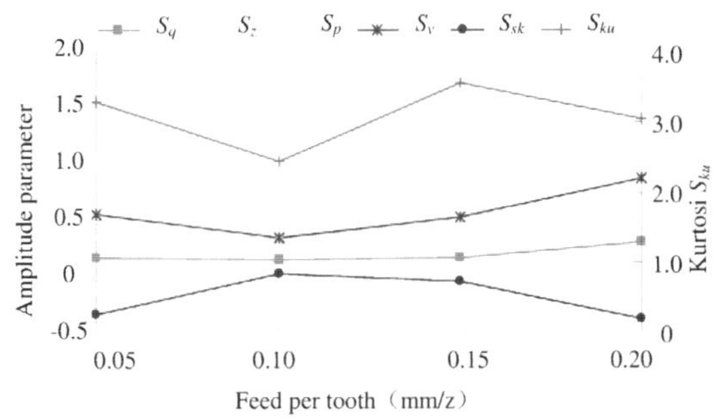

Figure 3. Influence of f_z on surface topography.

Under the condition of same cutting speed and feed per tooth, the influence of depth of cut on 3D surface topography parameters is shown in Figure 4. When the depth of cut is 0.15mm, after the completion of the cutting each amplitude parameters are relatively large, in addition, S_{ku}>3 and S_{sk}>0. This shows that there distribute relatively steep peaks on the surface, which affect the quality of the surface topography.

3.2. Motif Area Ratio Characterization Method

In order to be more intuitive to characterize the surface morphology of different milling parameters and evaluate the degree of corrosion, this paper defines Motif area ratio parameter. It is defined as the ratio of the surface area that can meet the threshold condition to the machined surface. The formula is shown as Equation (1). The selection of threshold need to consider the actual situation that does not affect the subsequent experimental results. As shown in Figure 5, the black area represents the Motif area that meet the threshold condition. In this experiment, the selection of threshold area does not affect the result of surface morphology on corrosion resistance of the workpiece.

$$P_S = \frac{\sum_{i=1}^{n} S_i}{S} \times 100\% \tag{1}$$

Where P_s is the Motif area ratio, S_i is the surface area that meets the threshold condition; S is the whole surface area.

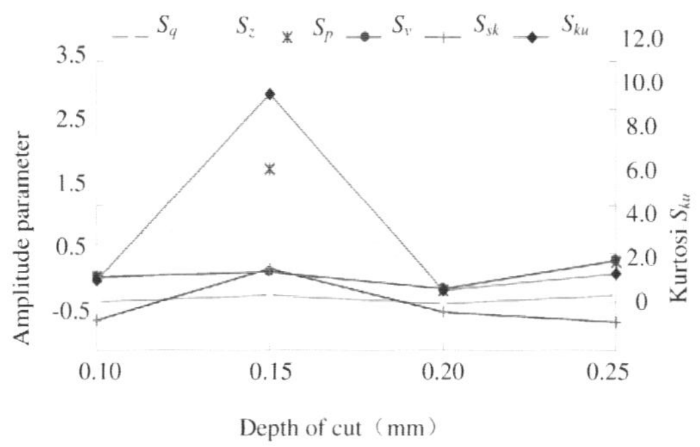

Figure 4. Influence of a_p on surface topography.

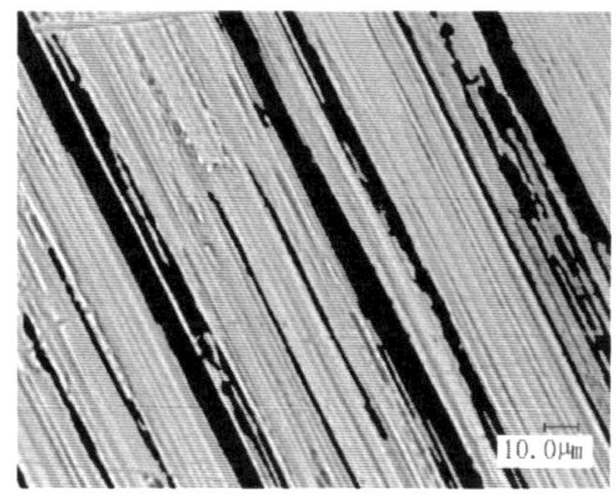

Figure 5. Motif area distribution.

With the increase of cutting speed, the Motif area ratio starts smoothly and then increases rapidly. From Figure 2 and Figure 6, we can find that the result of Motif area ratio method is close to the amplitude parameter characterization method. As shown in Figure 7, with the increase of feed per tooth, Motif area ratio increases gradually. In the process of cutting, with the increase of feed per tooth, cutting force and plastic deformation become larger. Therefore, the scallop height on workpiece increases so as to increases the Motif area ratio.

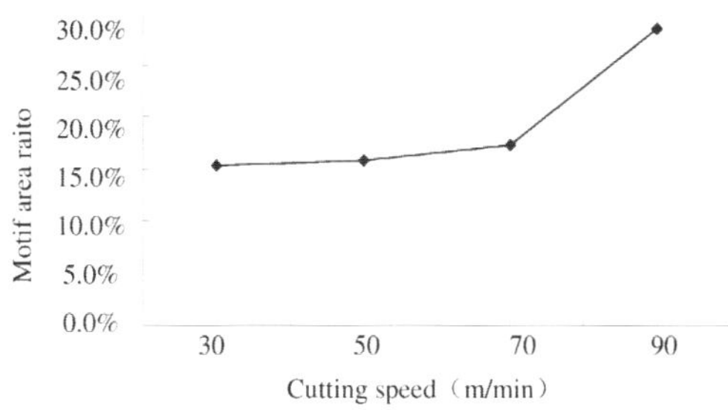

Figure 6. Relationship between cutting speed and Motif area ratio.

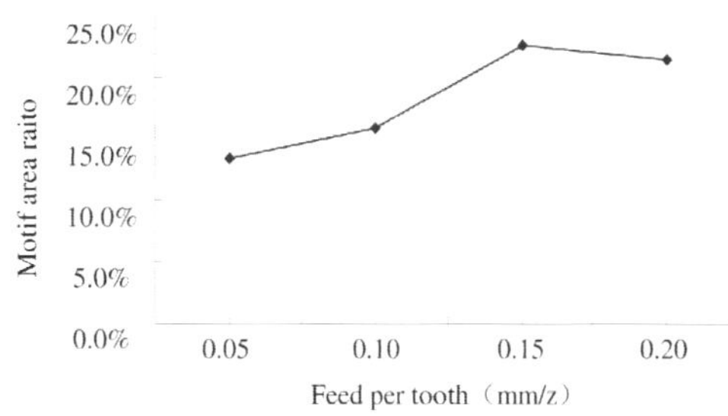

Figure 7. Relationship between feeding per tooth and Motif area ratio.

It can be seen from Figure 8 that the influence of depth of cut on surface Motif area ratio is not obvious, but because of the Motif area ratio value itself is at a relatively high position, mean value at around 20%. Therefore, we can conclude that the depth of cut has a significant influence on Motif area ratio.

4. Relationship between Surface Morphology and Corrosion

4.1. Immersion Corrosion Test

Immersion corrosion test has been done in order to simulate marine environment. All surfaces of the specimen are painted except for machined surface in order to avoid corrode. The test time is 1200h and the experimental process reference JB/T 7901-1999 (metallic materials laboratory uniform immersion corrosion test method). Average corrosion rate can be used to evaluate test result as shown in Equation (2).

$$V_a = \frac{kW}{At\rho} \tag{2}$$

Where V_a is average corrosion rate, k is a constant, W is corrosion weight loss, A is sample surface area, t is test time, ρ is the density of the material.

IOP Conf. Series: Materials Science and Engineering **484** (2019) 012028 doi:10.1088/1757-899X/484/1/012028

4.2. Influence of Cutting Parameters on Corrosion

The surface roughness of the workpiece is extracted from the 3D surface topography information, and the variation law of the average corrosion rate and the surface roughness is drawn into curves as shown in Figure 9, 10 and 11. They respectively indicate the influence of cutting speed, feed of per tooth or depth of cut on corrosion.

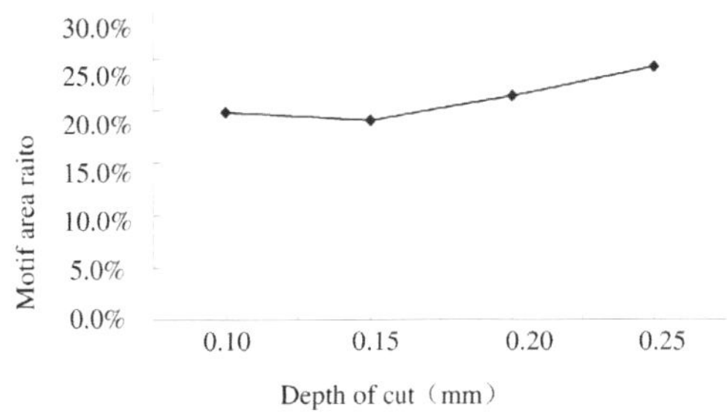

Figure 8. Relationship between depth of cut and Motif area ratio.

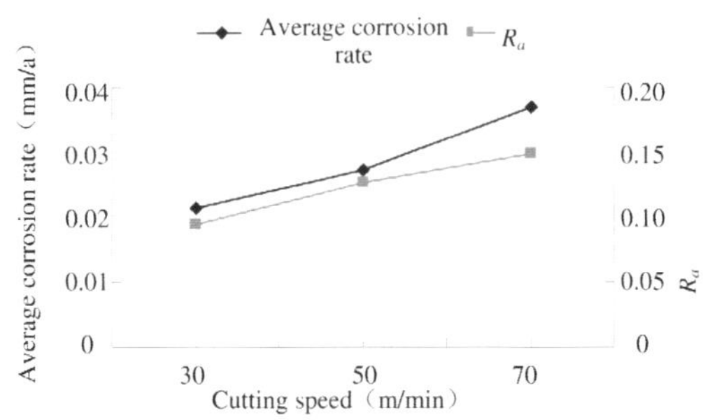

Figure 9. Influence of cutting speed on surface roughness and corrosion rate.

Compared to the average corrosion rate curve and the surface roughness curve can be found that the change tendency of both is basically the same, when the value of the surface roughness increases, the average corrosion rate becomes larger; when the value of surface roughness is small, the average corrosion rate is smaller. That is to say average corrosion rate of workpiece varies with the change of surface roughness. When the cutting speed is 90m/min, there occurs tool tipping, so in this paper we omit the corrosion result out of Figure 9.

When the feed per tooth is 0.05mm/z, from the figure can be found that the value of average corrosion rate reaches the maximum. However the surface roughness is very small, which is due to measurement error that led to this result. Thus, we omit this data point out of Figure 10. When the feed per tooth is in range of 0.10-0.20mm/z, average corrosion rate increases when surface roughness becomes larger as shown in Figure 10. As shown in Figure 11, it can be found that along with the change of axial cutting depth, change low of average corrosion rate is consistent with the change of surface roughness.

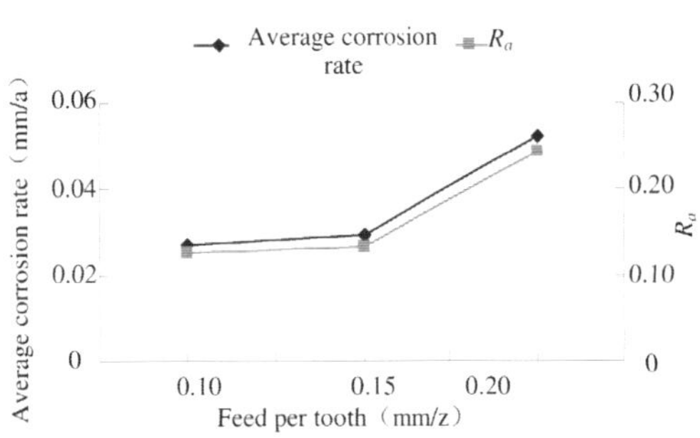

Figure 10. Influence of feeding per tooth on surface roughness and corrosion rate.

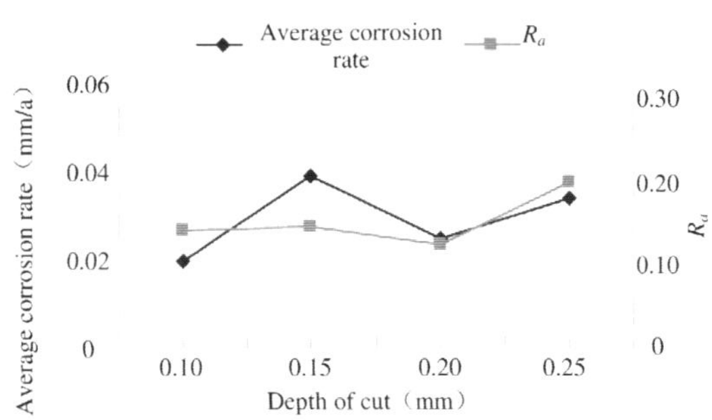

Figure 11. Influence of depth of cut on surface roughness and corrosion rate.

5. Conclusion

In order to be more intuitive to characterize the surface morphology of different milling parameters and evaluate the degree of corrosion, this paper defines Motif area ratio parameter. It is defined as the ratio of the surface area that can meet the threshold condition to the machined surface.

Through the experiment found that average corrosion rate varies with the change of surface roughness. The larger the surface roughness, the faster the average corrosion rate. Conversely the smaller the surface roughness, the slower the average corrosion rate. Based on the theoretical analysis, machined surface topography contains a large amount of information, including the surface roughness. If the workpiece surface topography is poorer, the workpiece surface roughness value is relatively large, therefore, the workpiece surface is easy to store large amounts of corrosive medium, combined with the increased surface defects on the workpiece, all this reasons eventually result in that average corrosion rate increased. Therefore, there is a correlation of average corrosion rate and surface roughness.

6. Acknowledgments

This work was supported by Shandong Natural Science Foundation of China (No. ZR2017LEE015), Key R & D projects (civil military integration) of Shandong, China.

7. References

[1] Ezugwu E O, Bonney J and Yamane Y 2003 *J. Materials Proc. Tech.* 134 233
[2] Herbert C R J, Axinte D A, Hardy M C and Brown P D 2011 *Procedia Engineering 19* 138
[3] Liu Y, Ning Y, Yao Z, Li H and Miao X 2016 *J. of Alloys and Compounds 675* 73
[4] Liu Y, Yao Z, Ning Y, Li H and Nan Y 2016 *J. of Alloys and Compounds 691* 554
[5] Zhang M, Li F, Yuan Z, Li J and Wang S 2013 *Materials & Design 49* 705
[6] Zhang M J, Li F G, Wang S Y and Liu C Y 2011 *Materials Science and Eng.: A 528* 4030
[7] Zhong B, Wang Y, Wei D, Zhang K and Wang J 2018 *Int. J. of Fatigue* 109 26
[8] Devillez A, Schneider F, Dominiak S, Dudzinski D and Larrouquere D 2007 *Wear* 262 931
[9] Hao Z P, Fan Y H, Lin J Q and Yu Z X 2015 *Int. J. Adv. Manuf. Tech.* 78 1329
[10] Hardy M C, Herbert C R J, Kwong J, Li W, Axinte D A, Sharman A R C Encinas-Oropesa A and Withers P J 2014 *Procedia CIRP* 13 411
[11] Ginting A and Nouari M 2009 *Int. J Machine Tool & Manufacturing 49* 325
[12] Liu J, Sun J and Chen W 2017 *Int. J. Adv. Manuf. Tech.* 92 1
[13] Varote N and Joshi SS 2017 *J. Materials Eng. and Perf.* 26 4391
[14] Arrazola P J, Kortabarria A and Madariaga A 2014 *Simu. Mod. Prac. and Theo.* 41 87
[15] Özel T and Ulutan D 2012 *CIRP Annals* 61 547
[16] Madariaga A, Arrazola PJ, Esnaola JA, Ruiz-Hervias J and Muñoz P 2014 *Proc. CIRP* 13 175

Stress and Strain of Cracked Orthotropic Sheet

Purong Jia, Gang Wang, Long Zhang and Yongyong Suo

Department of Engineering Mechanics, Northwestern Polytechnical University, Xi'an 710129, PR China.
Email: prjia@nwpu.edu.cn

Abstract. The analytic function of complex variable including a material parameter is analyzed fully. The new stress function has been formed by the complex function method. Typical shear mode II crack states are considered to orthotropic materials. The boundary shear loading problem of cracked orthotropic sheet is studied and the formulae for stress fields are derived. The singular strain fields are also discussed. And the expressions of the singular stress and strain in the crack-tip are determined with inclusion of the material elasticity coefficients.

1. Introduction

Linear elastic fracture mechanics (LEFM) has been found to be a very useful tool for design purposes and investigated in great detail for many engineering materials, whether isotropic or anisotropic [1]. Nowadays the fracture mechanics of anisotropic materials may be relative to many structural aspects in engineering evaluation. The prediction of crack initiation and propagation must be based on the fracture mechanics method [2]. Fiber-reinforced polymer matrix materials are as the most typical composites and usually modeled as anisotropic materials at the macroscopic level. And the orthotropic plates have been for the base of composites in common use [3, 4]. The plane fracture mechanics of composite materials may be more important, particularly the study for the shear mode II crack problem is very necessary.

The investigation of singular stress or strain fields near the crack tip holds an essential part of LEFM. Generally the method of elastic mechanics can be used to obtain stresses and displacements in cracked bodies [5]. The valid method to solve crack tip field problems in anisotropic composites may be in using complex analytic function theory, and the results have been reported [6]. But the general solution may have some weakness. So it is necessary to make up a new solution for the singular stresses and strains of cracked composite materials.

2. General Complex Function

Complex variable theory has been a very powerful tool for the solution of many engineering mechanical problems. The technique is also useful for anisotropic materials. It is well known that the basic complex variable (z) and its conjugate ($\bar{z}$) are defined as:

$$z = x + iy \ , \quad \bar{z} = x - iy \qquad (i = \sqrt{-1})$$

For the convenience of general investigation, it is truly necessary to introduce another complex variable (w) and its conjugate ($\bar{w}$), which are defined as follows:

$$w = x + ihy \ , \quad \bar{w} = x - ihy \tag{1}$$

Where, the parameter h is real. Now we suppose the constant h to be positive ($h>0$), and it can be called tensile or compressive ratio for the coordinate system. The derivative relation of $w(x,y)$ must be given as follows:

$$\frac{\partial w}{\partial x}=\frac{\partial \overline{w}}{\partial x}=\frac{\partial z}{\partial x}=\frac{\partial \overline{z}}{\partial x}=1\ ,\qquad \frac{\partial w}{\partial y}=-\frac{\partial \overline{w}}{\partial y}=ih \tag{2}$$

For the mode II crack problem, rectangular and polar coordinates are shown in Figure 1. The polar coordinate system centered at the crack tip may be adequate for local stress analysis. In terms of the polar coordinates, the complex variables can be written as:

$$\begin{aligned} z &= x+iy=a+r\cos\theta+ir\sin\theta \\ w &= x+ihy=a+r\cos\theta+ihr\sin\theta \end{aligned} \tag{3}$$

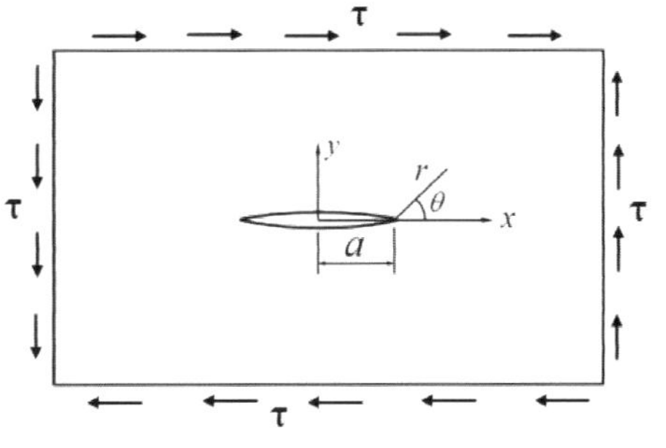

Figure 1. Basic model of the composite with mode II crack

Consider the analytic functions of complex variable. $\Psi=\Psi(w)$, $\Phi=\Phi(w)$. The analytic functions can be expressed as follows:

$$\begin{aligned} &\Psi(w)=\operatorname{Re}\Psi+i\operatorname{Im}\Psi,\qquad \overline{\Psi}=\overline{\Psi}(\overline{w})=\operatorname{Re}\Psi-i\operatorname{Im}\Psi \\ &\frac{d\Psi}{dw}=\Phi(w)=\operatorname{Re}\Phi+i\operatorname{Im}\Phi \\ &\frac{d\overline{\Psi}}{d\overline{w}}=\overline{\Phi}(\overline{w})=\operatorname{Re}\Phi-i\operatorname{Im}\Phi \end{aligned} \tag{4}$$

The derivatives of Ψ are given by

$$\begin{aligned} &\frac{\partial\Psi}{\partial x}=\frac{d\Psi}{dw}\frac{\partial w}{\partial x}=\frac{d\Psi}{dw}=\Psi'=\Phi(w)=\operatorname{Re}\Phi+i\operatorname{Im}\Phi=\frac{\partial\operatorname{Re}\Psi}{\partial x}+i\frac{\partial\operatorname{Im}\Psi}{\partial x} \\ &\frac{1}{h}\frac{\partial\Psi}{\partial y}=\frac{1}{h}\frac{d\Psi}{dw}\frac{\partial w}{\partial y}=i\frac{d\Psi}{dw}=i\Psi'=i\Phi=i\operatorname{Re}\Phi-\operatorname{Im}\Phi=\frac{1}{h}(\frac{\partial\operatorname{Re}\Psi}{\partial y}+i\frac{\partial\operatorname{Im}\Psi}{\partial y}) \end{aligned} \tag{5}$$

Obviously, it is easy to derive the following differential equations:

$$\begin{aligned} \operatorname{Re}\Psi'=\operatorname{Re}\Phi=\frac{\partial\operatorname{Re}\Psi}{\partial x}=\frac{1}{h}\frac{\partial\operatorname{Im}\Psi}{\partial y} \\ \operatorname{Im}\Psi'=\operatorname{Im}\Phi=\frac{\partial\operatorname{Im}\Psi}{\partial x}=-\frac{1}{h}\frac{\partial\operatorname{Re}\Psi}{\partial y} \end{aligned} \tag{6}$$

3. Basic Equations and Stress Function

The plane stress state of composite sheets is common and very importance for the application. It is the key point to solve the stress-field problems in orthotropic materials. In plane elasticity, the equilibrium equations are (body forces are absent):

$$\frac{\partial \sigma_x}{\partial x}+\frac{\partial \tau_{xy}}{\partial y}=0 \quad , \quad \frac{\partial \tau_{xy}}{\partial x}+\frac{\partial \sigma_y}{\partial y}=0 \tag{7}$$

Suppose the principal elastic directions of the plate coincide with the coordinate directions, and let the directions 1, 2 parallel to the axes x, y, respectively. Now consider the linear elastic strain-stress relations, the constitutive equations for the orthotropic materials are given as (plane stress state):

$$\varepsilon_x=\frac{\sigma_x}{E_1}-\frac{\nu_{12}\sigma_y}{E_1} \quad , \quad \varepsilon_y=\frac{\sigma_y}{E_2}-\frac{\nu_{12}\sigma_x}{E_1} \quad , \quad \gamma_{xy}=\frac{\tau_{xy}}{G_{12}} \tag{8}$$

The compatibility condition of strains must be satisfied as follows:

$$\frac{\partial^2 \varepsilon_x}{\partial y^2}+\frac{\partial^2 \varepsilon_y}{\partial x^2}=\frac{\partial^2 \gamma_{xy}}{\partial x \partial y} \tag{9}$$

It is well known that the Airy stress function $U=U(x,y)$ is defined by:

$$\sigma_x=\frac{\partial^2 U}{\partial y^2} \quad , \quad \sigma_y=\frac{\partial^2 U}{\partial x^2} \quad , \quad \tau_{xy}=-\frac{\partial^2 U}{\partial x \partial y} \tag{10}$$

The equilibrium equations in Eqs (7) can be satisfied. Next by using above relations, the governing equation of the compatibility condition can be expressed by the Airy stress function, that is:

$$\frac{\partial^4 U}{\partial y^4}+(\frac{E_1}{G_{12}}-2\nu_{12})\frac{\partial^4 U}{\partial x^2 \partial y^2}+\frac{E_1}{E_2}\frac{\partial^4 U}{\partial x^4}=0 \tag{11}$$

The Airy stress function U can be expressed by the complex variable function. Consider an infinite plane with the crack along the x-axis shown in Figure 1, and also consider the problem of Mode II shear loading. The Airy stress function can be determined by the form:

$$U=y\frac{\Psi+\overline{\Psi}}{2}=y\,\mathrm{Re}\,\Psi \tag{12}$$

The derivatives of U with respect to x or y are given by:

$$\frac{\partial U}{\partial x}=y\frac{\partial \mathrm{Re}\,\Psi}{\partial x}=y\,\mathrm{Re}\,\Psi'=y\,\mathrm{Re}\,\Phi \ , \quad \frac{\partial U}{\partial y}=\mathrm{Re}\,\Psi+y\frac{\partial \mathrm{Re}\,\Psi}{\partial y}=\mathrm{Re}\,\Psi-hy\,\mathrm{Im}\,\Phi$$

$$\frac{\partial^2 U}{\partial x^2}=y\,\mathrm{Re}\,\Psi''=y\,\mathrm{Re}\,\Phi' \ , \quad \frac{\partial^2 U}{\partial y^2}=-2h\,\mathrm{Im}\,\Phi-h^2 y\,\mathrm{Re}\,\Phi' \ , \quad \frac{\partial^2 U}{\partial x \partial y}=\mathrm{Re}\,\Phi-hy\,\mathrm{Im}\,\Phi'$$

On the basis of above relations, the governing equation (11) becomes:

$$4(\frac{E_1}{2G_{12}}-\nu_{12}-h^2)\frac{\partial^3 \mathrm{Re}\,\Psi}{\partial x^2 \partial y}+[\frac{E_1}{E_2}-h^2(\frac{E_1}{G_{12}}-2\nu_{12})+h^4]y\frac{\partial^4 \mathrm{Re}\,\Psi}{\partial x^4}=0 \tag{13}$$

Evidently, the solution must be followed by the characteristic equations, and also reduced to:

$$\begin{aligned} &\frac{E_1}{2G_{12}} - \nu_{12} - h^2 = 0 \\ &\frac{E_1}{E_2} - 2h^2(\frac{E_1}{2G_{12}} - \nu_{12}) + h^4 = 0 \end{aligned} \tag{14}$$

The solution of the characteristic equation is given as:

$$h^2 = \frac{E_1}{2G_{12}} - \nu_{12}, \qquad h^4 = \frac{E_1}{E_2}$$

Because the real constant h is positive, so the positive real root must be as:

$$h = \sqrt{\frac{E_1}{2G_{12}} - \nu_{12}} = \sqrt[4]{\frac{E_1}{E_2}} \tag{15}$$

The stresses are considered again. By substituting the above derivative relations of U into stress expressions (10), the stresses can be expressed as:

$$\begin{aligned} \sigma_x &= -2h \operatorname{Im}\Phi - h^2 y \operatorname{Re}\Phi' \\ \sigma_y &= y \operatorname{Re}\Phi' \\ \tau_{xy} &= -\operatorname{Re}\Phi + hy \operatorname{Im}\Phi' \end{aligned} \tag{16}$$

The stress expressions of stresses are similar to the traditional stress formulations that have been given in the books of fracture mechanics for the isotropic materials.

4. Solution of Stress and Strain

For Mode II crack problem, the sheet with a line crack is subjected to uniform shear loading τ at remote edges. Consider the following stress boundary conditions:

$$\sigma_y = \tau_{xy} = 0 \quad \text{at } |x| < a \text{ and } y = 0 \quad \text{(free crack surfaces)}$$

$$\sigma_x = \sigma_y = 0, \quad \tau_{xy} = \tau \quad \text{at } x^2 + y^2 \to \infty$$

In order for the stress function to meet the preceding boundary value problem, the complex function can be selected as:

$$\Psi = Cw\sqrt{1 - \frac{a^2}{w^2}} \tag{17}$$

Then the derivatives of the complex functions $\Psi(w)$, $\Phi(w)$ are given as:

$$\begin{aligned} \Psi' &= \Phi = C\sqrt{\frac{w^2}{w^2 - a^2}} \\ \Phi' &= -C\frac{a^2}{w}\sqrt{\frac{w^2}{(w^2 - a^2)^3}} \end{aligned} \tag{18}$$

By substituting the stress functions Φ and Φ' into Eq. (16), and also using the remote boundary conditions, the coefficient C can be derived. It is given out: $C = -\tau$. So the stress expressions can be determined as follows:

$$\begin{aligned}
&\frac{\sigma_x}{\tau} = 2h\,\mathrm{Im}\sqrt{\frac{w^2}{w^2-a^2}} - h^2\,\mathrm{Re}[\frac{a^2 y}{w}\sqrt{\frac{w^2}{(w^2-a^2)^3}}] \\
&\frac{\sigma_y}{\tau} = \mathrm{Re}[\frac{a^2 y}{w}\sqrt{\frac{w^2}{(w^2-a^2)^3}}] \\
&\frac{\tau_{xy}}{\tau} = \mathrm{Re}\sqrt{\frac{w^2}{w^2-a^2}} + h\,\mathrm{Im}[\frac{a^2 y}{w}\sqrt{\frac{w^2}{(w^2-a^2)^3}}]
\end{aligned} \tag{19}$$

In the vicinity region of the crack tip ($r << a$), as shown in Figure 1, the functions can be simplified by the polar coordinate system:

$$x = a + r\cos\theta\ , \qquad y = r\sin\theta\ , \qquad w = a + r\cos\theta + ihr\sin\theta = a + r\eta$$

$$w + a \approx 2a\ , \quad w^2 - a^2 \approx 2ar(\cos\theta + ih\sin\theta) = 2ar\eta\ , \quad \eta = \cos\theta + ih\sin\theta$$

Then the stresses at near crack-tip can be expressed as:

$$\begin{aligned}
&\frac{\sigma_x}{\tau} = \sqrt{\frac{a}{2r}}(2h\,\mathrm{Im}\frac{1}{\sqrt{\eta}} - h^2\frac{\sin\theta}{2}\mathrm{Re}\frac{1}{\sqrt{\eta^3}}) \\
&\frac{\sigma_y}{\tau} = \sqrt{\frac{a}{2r}}\frac{\sin\theta}{2}\mathrm{Re}\frac{1}{\sqrt{\eta^3}} \\
&\frac{\tau_{xy}}{\tau} = \sqrt{\frac{a}{2r}}(\mathrm{Re}\frac{1}{\sqrt{\eta}} + h\frac{\sin\theta}{2}\mathrm{Im}\frac{1}{\sqrt{\eta^3}})
\end{aligned} \tag{20}$$

These expressions show that the stress components tend to infinity at the crack tip, this is so-called stress singularity. By substituting the stress expressions into the constitutive equations (8), the strain equations can be determined as follows:

$$\begin{aligned}
&\varepsilon_x = \frac{\tau}{E_1}\sqrt{\frac{a}{2r}}[2h\,\mathrm{Im}\frac{1}{\sqrt{\eta}} - (h^2 + \nu_{12})\frac{\sin\theta}{2}\mathrm{Re}\frac{1}{\sqrt{\eta^3}}] \\
&\varepsilon_y = \frac{\tau}{E_1}\sqrt{\frac{a}{2r}}[-2\nu_{12}h\,\mathrm{Im}\frac{1}{\sqrt{\eta}} + (\nu_{12}h^2 + \frac{E_1}{E_2})\frac{\sin\theta}{2}\mathrm{Re}\frac{1}{\sqrt{\eta^3}}] \\
&\gamma_{xy} = \frac{\tau}{G_{12}}\sqrt{\frac{a}{2r}}(\mathrm{Re}\frac{1}{\sqrt{\eta}} + h\frac{\sin\theta}{2}\mathrm{Im}\frac{1}{\sqrt{\eta^3}})
\end{aligned} \tag{21}$$

It is evidently shown that the strain components tend to infinity at the crack tip, and also the material elasticity must be closely related with the strain fields.

For the isotropic materials ($h = 1$), the complex variable η is as: $\eta = e^{i\theta} = \cos\theta + i\sin\theta$. So the expressions of stresses and strains at the crack-tip region can be reduced to:

$$\begin{aligned}
&\frac{\sigma_x}{\tau} = -\sqrt{\frac{a}{2r}}\sin\frac{\theta}{2}(2 + \cos\frac{\theta}{2}\cos\frac{3\theta}{2}) \\
&\frac{\sigma_y}{\tau} = \sqrt{\frac{a}{2r}}\sin\frac{\theta}{2}\cos\frac{\theta}{2}\cos\frac{3\theta}{2} \\
&\frac{\tau_{xy}}{\tau} = \sqrt{\frac{a}{2r}}\cos\frac{\theta}{2}(1 - \sin\frac{\theta}{2}\sin\frac{3\theta}{2})
\end{aligned} \tag{22}$$

And

$$\begin{aligned}
\varepsilon_x &= -\frac{\tau}{E}\sqrt{\frac{a}{2r}}\sin\frac{\theta}{2}[2+(1+\nu)\cos\frac{\theta}{2}\cos\frac{3\theta}{2}] \\
\varepsilon_y &= \frac{\tau}{E}\sqrt{\frac{a}{2r}}\frac{\sin\theta}{2}[2\nu+(1+\nu)\cos\frac{\theta}{2}\cos\frac{3\theta}{2})] \\
\gamma_{xy} &= \frac{\tau}{G}\sqrt{\frac{a}{2r}}\cos\frac{\theta}{2}(1-\sin\frac{\theta}{2}\sin\frac{3\theta}{2})
\end{aligned} \tag{23}$$

Obviously, the expressions are conventional stress and strain fields of isotropic materials.

5. Acknowledgments

The authors have a wish to acknowledge the financial support of the Natural Science Foundation of China (NSFC Grant No: 51275422, 51475372).

6. References

[1] Sun C T. Fracture mechanics. Elsevier Inc., (2012)
[2] Friedrich K. Application of fracture mechanics to composite materials. Elsevier Sci. Pub. B.V., (1989)
[3] Sih G C. Mechanics of fracture initiation and propagation. Kluwer Academic Publishers, (1991)
[4] Fakoor M, Khansari N M. Mixed mode I/II fracture criterion for orthotropic materials based on damage zone properties. Eng Fract Mech 2016; 153: 407-420
[5] Sadd M H. Elasticity, theory, applications and numerics. Elsevier Inc., (2014)
[6] Fakoor M. Augmented strain energy release rate (ASER): A novel approach for investigation of mixed-mode I/II fracture of composite materials. Eng Fract Mech 2017; 179: 177-189.

IOP Conf. Series: Materials Science and Engineering **484** (2019) 012030 doi:10.1088/1757-899X/484/1/012030

A Simple and High-performance Platform for Refractive Index Sensing based on Plasmonic Metal Disks on a Metal Mirror

Xianchao Liu[1, 2], Rui Ma[2, 3], Duncan Sutherland[2] and Jun Wang[1,*]

[1]School of Optoelectronic Information, University of Electronic Science and Technology of China, No. 4, Section 2, North Jianshe Road, Chengdu 610054, China.
[2]iNANO Center, Aarhus University, Gustav Wieds vej 14, Aarhus 8000, Denmark.
[3]Beijing University of Posts and Telecommunications, Haidian district, Xitucheng Roads 10th, Beijing 100876, China.
Email: wjun@uestc.edu.cn

Abstract. Numerical simulations are used to study light interactions with a platform composed of arrays of stacked metal and spacer dielectric disks on a continuous Au film. Tunable and high absorption is obtained near 1.6 μm by adjusting the thickness of the dielectric disks. Nearly perfect-absorption remains for a wide range of disk diameters. The near-perfect absorption is attributed to a magnetic resonance within the dielectric disks (combined with electric resonance at the metal disks) inducing the platform response to incident light. In addition, the resonance response to environment is more sensitive than that of metal disks on dielectric/metal films. Furthermore, when the diameter of the dielectric disks is reduced, the local electromagnetic field is more exposed and the refractive index sensitivity (RIS) increases to 1133nm/RIU (refractive index unit). The method is simple and effective to improve RIS of sandwich plasmonic structures and will be adopted in various plasmonic devices.

1. Introduction

Micro/nano-scale metal particles support optical carrier resonances called localized surface plasmon resonances (LSPRs) [1,2], which are paid much attention in various fields, such as light harvest[3], gas sensing [4], molecular sensing [5-7] and surface enhanced Raman scattering [8]. One area of strong interest concerns optical refractive index sensing based on spectra shifts of resonances in response to refractivity change of the surrounding environment. Refractive index sensors are widely used in testing the amount of specific materials in solution and identification of types of gas and solutions [4, 9, 10]. It is a vital method for the sea economy and chemical industries. There are several strategies for refractive index sensing, such as utilizing propagate surface plasmon polariton resonances (PSPR), plasmon lattice resonance (PLR), Fano resonance or Fabry–Perot (FP) cavities to achieve high-sensitivity sensing [11]. Recently, the Bin Ren group suggested a unique two-dimensional gold plasmonic nanohole array, which presented a PSPR with full width at half maximum (FWHM) of 8nm and figure of merit (FOM) of 730 [12]. Jianfang Wang's team fabricated plasmonic gold mushroom arrays, whose RIS approaches 1010nm/RIU and FOM is 108 [13]. Hatice Altug's group demonstrated a Fano resonance asymmetric metamaterial, which can be used to identify molecular monolayers [14]. However, sensors based on PSPR, PLR and Fano resonance usually need well-ordered micro/nano-structure arrays, which is a challenge for manufacturing technology due to high cost. In addition, simple nanostructures and disordered or short range ordered nanostructures also show high-performance sensing and are easy to fabricate. For examples, monolayer metal disks array fabricated by microsphere lithography shows resonance linewidth of 10.5nm and FOM of 15.3 [15]. When metal disks are separated by dielectric disks, placed on glass substrates, it shows improved RIS of 800nm/RIU [16]. When wet etching is implemented to the supporting dielectric pillars of these kinds of structures, the obtained sensor can expose more area to local environment, resulting in further improved sensing performances [7]. Such methods via reducing substrate effect on sensing performance of metal disks and increasing exposed area show potential for more sensors. In addition, we find metal/dielectric/metal structure can easily obtain high absorption, via FP-like cavities [17, 18].

To design a high-absorption and high-performance RI sensor, a platform composed of Au/SiO_2 disks placed on a continuous reflecting Au layer is suggested. By optimizing the thickness of the SiO_2 disk, tunable prefect absorption is achieved. The RIS of the suggested structure is two times higher than that of Au disks placed on continuous SiO_2/Au film [19]. When the diameter of spacer SiO_2 disk is reduced, the RIS can reach 1133nm/RIU. These results provide a new design that can be utilized in high sensitivity detection.

2. Design and Numerical Study

A schematic representation of the suggested platform is shown in figure 1. The platform is composed of Au/SiO_2 disks on top of a reflective Au layer. CST microwave studio software was utilized to calculate the electromagnetic distribution and absorption. The Au/SiO_2 disk and bottom Au layer were placed at the center of coordinate space and the z=0 for the interface between Au disk and SiO_2 disk. Their geometric center was set at the z axis. The boundary of the Au square was parallel to the x and y axis. As the platform is symmetrical about xoz and yoz planes, symmetry planes were set to reduce mesh and calculation time. In addition, there is no PSPR and PLR existing for the structure, so a single disk placed on Au mirror is feasible as a design. In the simulation, the environment index was 1.33 or 1.33~1.36. The side length of the Au mirror was fixed at 550nm. The thickness of the top Au disks was fixed at 20nm, while other parameters such as the radius of the Au/SiO_2 disks, the thickness of the SiO_2 disks, and the radius of SiO_2 disks were adjusted to find a better set of parameters which promise a better sensing performance of the structure. The Au mirror thickness was 100nm, where no light from visible light to near-infrared light can transmit through it. The incident plane wave transmits along the z axis, with electric vector along the x axis. Optical constants were taken from Ref. [20].

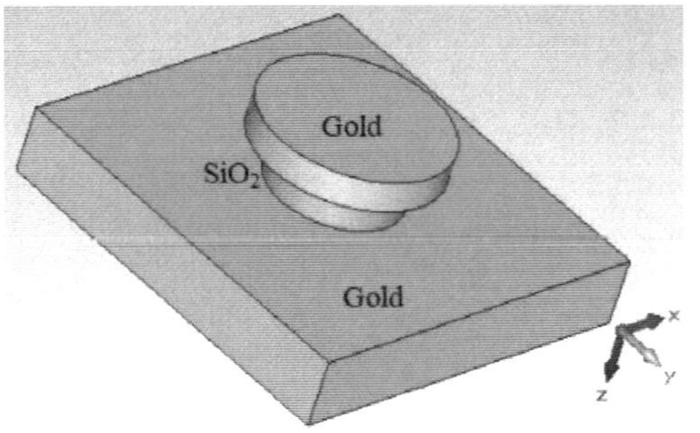

Figure 1. Scheme of the simulation unit.

2.1. Thickness of Spacer Layer Effect on the Absorption Spectra

To achieve high absorption, the thickness of SiO_2 was varied. The initial value of disk radius was 140nm. The absorption spectra varied with the thickness of the spacer SiO_2 disks layer as shown in figure 2. With the increase of the SiO_2 thickness, the resonance shifted to shorter wavelength. The amplitude of absorption initially increased, went through a maximum, and then decreased. The process is adjusting of FP-like cavity mode (magnetic resonance) among SiO_2 disks. A proper thickness of SiO_2 disk will meet perfect absorption.

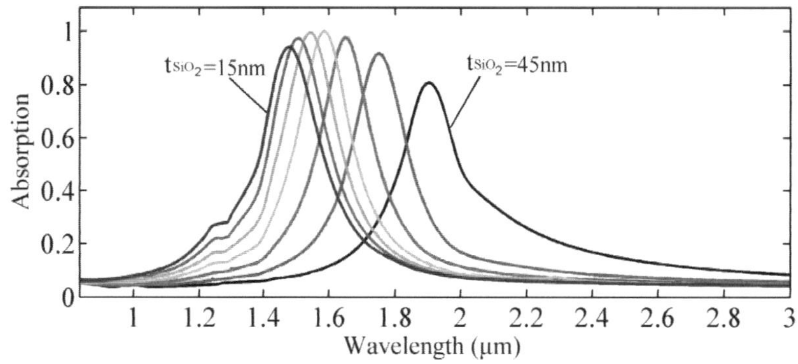

Figure 2. The absorption spectra varying with thickness of spacer SiO_2 disks layer.

2.2. Radius of Disk Effect on the Absorption Spectra

Based on the results seen in figure 1, we set the thickness of the SiO_2 disk to 30nm to keep a high absorption in following calculations. Here, the radius of both the Au/SiO_2 disks was varied from 125nm to 160nm, with a step of 5nm. The peak position shifted to longer wavelength quickly, and the FWHM of peaks in wavelength increased concomittedly with the increased radius, as shown in figure 3 as expected for lager metal particles. We also noticed that the perfect-absorption does not alter during this tunability. The high absorption is attributed to a magnetic resonance located in spacer SiO_2 disks, which dominates the platform's response to incident light.

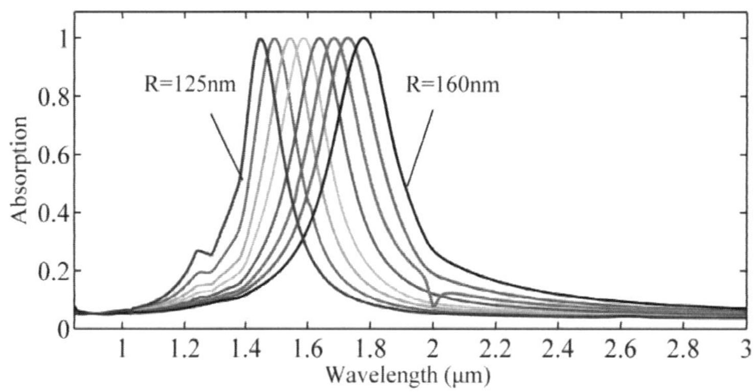

Figure 3. The absorption spectra varying with Au/SiO_2 disks radius.

2.3. Spectra Response to Surrounding Environment

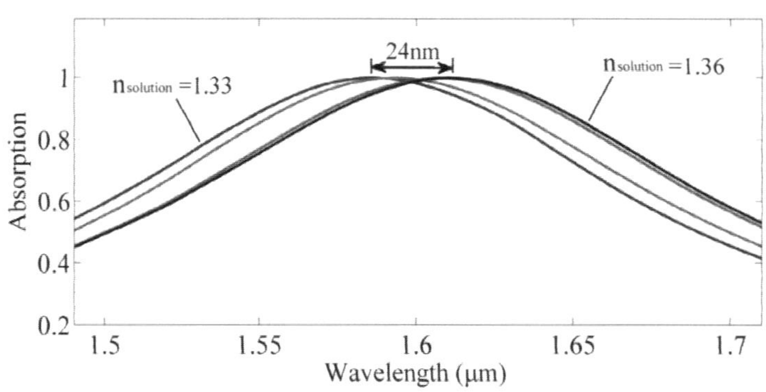

Figure 4. Spectral response to environment.

IOP Conf. Series: Materials Science and Engineering **484** (2019) 012030 doi:10.1088/1757-899X/484/1/012030

To compare with results in Ref. [17], we selected radius of the Au and SiO_2 disks of 140nm, whose absorption peak was near 1.6 μm. The resonance peak shift is present in figure 4. The environment index was varied from 1.33 to 1.36 and the peak-shift was 24nm corresponding to a RIS of 800nm/RIU, which is twice as large as the RIS result reported in Ref. [17].

3. Improving Spectra Sensitivity to Environment by Increasing Exposing Area

To expose more local electromagnetic field, here the spectra response to change in the environment refractive index with reducing diameter of the SiO_2 disk was studied. The resonance peak shift is shown in figure 5(a) and (b). The RIS was 1000nm/RIU and 1133nm/RIU, respectively for SiO_2 disks diameter of 80nm and 60nm. These results validate that increasing the exposed area is effective for metal/dielectric/metal microstructure to improve sensing performance. Figure5 (c) and (d) presents electric field and magnetic field distribution of *XOZ* plane (*y*=0nm) at the resonance wavelength observed in figure5 (b) with bulk refractive index of 1.33, which indicates that the high absorption results from the magnetic resonance located in the SiO_2 disk [17]. The local field distribution revealed near the SiO_2 disk is the reason for the improved RIS.

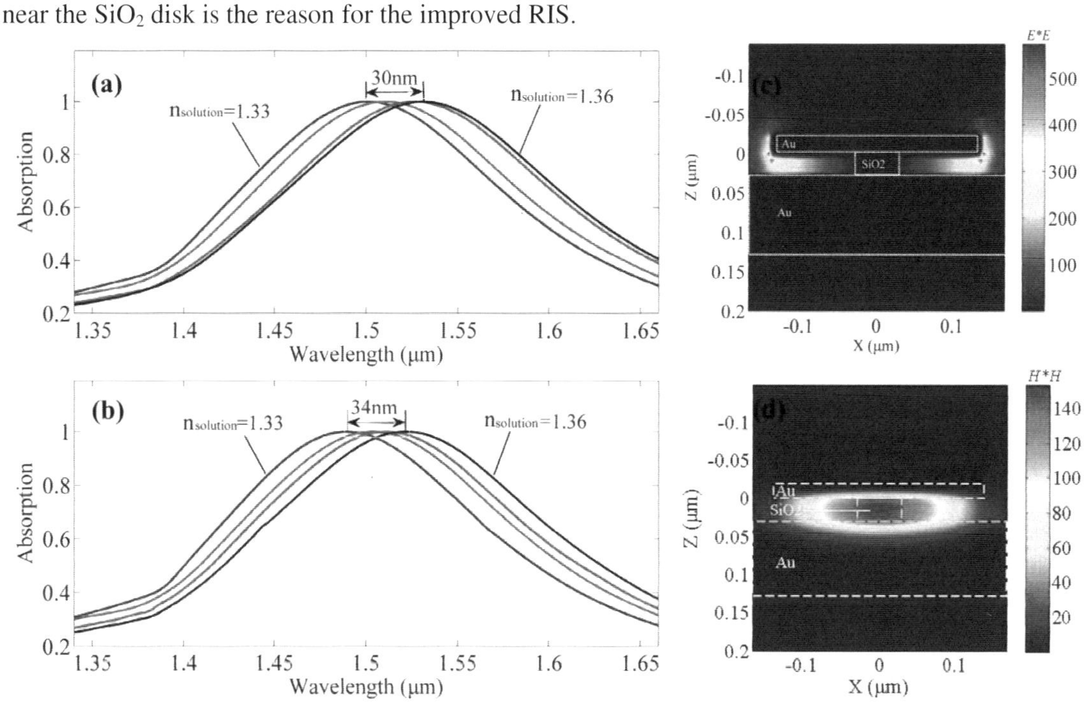

Figure 5. (a) and (b): Spectral response to change in the local dielectric environment, the radius of SiO_2 disks is 40nm and 30nm, respectively. (c) and (d): electric field and magnetic field images of *XOZ* plane(y=0nm), related to (b). The *z* value is inverted in (c) and (d).

4. Conclusion

A simple and high-performance sandwich structure plasmon refractive index sensor is suggested. The platform is with near-perfect absorption at a resonance peak due to a hugely enhanced magnetic resonance inducing response to incident light. Improved RIS of the sensor is validated via increasing expose area of local magnetic resonance. In detail, by replacing spacer dielectric layer with a dielectric disk, and further reducing the diameter of the dielectric disk, the RIS of 800 nm/RIU, 1000 nm/RIU and 1133 nm/RIU are obtained, respectively. The high-performance sensor can find application in RI test and solution identification.

5. Acknowledgments

This work was supported by Science Fund for Creative Research Groups of the National Natural Science Foundation of China (No. 61421002), Sichuan Province Science and Technology Innovation Talent Project (No. 2018066) and the China Scholarship Council Fund.

6. References

[1] Jensen T R, Malinsky M D, Haynes C L and Van Duyne R P 2000 *J. Phys. Chem. B* 104 10549-10556
[2] Willets K A and Van Duyne R P 2007 *Annu. Rev. Phys. Chem.* 58 267-297
[3] Aubry A, Dang Y, Fernández-Domínguez A I, Sonnefraud Y, Maier S A and Pendry J B 2010 *Nano lett.* 10 2574-2579
[4] Kreno L E, Hupp J T and Van Duyne R P 2010 *Anal. Chem.* 2 8042-8046
[5] Larsson E M, Alegret J, Käll M and Sutherland D S 2007 *Nano lett.* 7 1256-1263
[6] Chen K, Dao T D, Ishii S, Aono M and Nagao T 2015 *Adv. Funct. Mater.* 25 6637-6643
[7] Aćimović S S, Šípová H, Emilsson G, Dahlin A B, Antosiewicz T J and Käll M 2017 *Light Sci. Appl.* 6 e17042
[8] Le F, Brandl D W, Urzhumov Y A, Wang H, Kundu J, Halas N J, Aizpurua J and Nordlander P 2008 *ACS nano* 2 707-718
[9] Häfele V, Trügler A, Hohenester U, Hohenau A, Leitner A and Krenn J R 2015 *Opt. express* 23 10293-10300
[10] Chen Y, Liu X, Chen W, Xie Z, Huang Y and Li L 2017 *Chinese Physics B* 26 017807
[11] Li J, Chen C, Lagae L and Van Dorpe P 2015 *J. Phys. Chem. C* 119 29116-29122
[12] Liu B, Chen S, Zhang J, Yao X, Zhong J, Lin H, Huang T, Yang Z, Zhu J, Liu S, Lienau C, Wang L and Ren B 2018 *Adv. Mater.* 30 1706031
[13] Yang S, Zhou J, Liu T, Tao Y, Jiang R, Liu M, Xiao G, Zhu J, Zhou Z, Wang X, Jin C and Wang J 2013 *Nat. Commun.* 4 2381
[14] Wu C, Khanikaev A B, Adato R, Arju N, Yanik A A, Altug H and Shvets G 2012 *Nat. mater.* 11 69
[15] Chang Y C, Chung H C, Lu S C and Guo T F 2013 *Nanotechnology* 24 095302
[16] Dmitriev A, Hägglund C, Chen S, Fredriksson H, Pakizeh T, Käll M, and Sutherland D S 2008 *Nano lett.* 8 3893-3898
[17] Chen K, Dao T D, Ishii S, Aono M and Nagao T 2015 *Adv. Funct. Mater.* 25 6637-6643
[18] Liu X, Wang J, Gou J, Ji C and Cui G 2018 *Nanoscale Res. Lett.* 13 335
[19] Palik E D 1998 Handbook of optical constants of solids (Academic Press, New York)
[20] Liu N, Mesch M, Weiss T, Hentschel M and G Harald 2010 *Nano lett.* 10 2342-2348

IOP Conf. Series: Materials Science and Engineering **484** (2019) 012031 doi:10.1088/1757-899X/484/1/012031

Application of Lanthanum Oxide and Cerium Oxide in E4303 Electrode

Enhuai Li[1, a, *], Nengwei Wang[2, b, *] and Shuquan Wan[3, c, *]

College of Vanadium and Titanium, Panzhihua University, 617000 Panzhihua, Sichuan Province, People's Republic of China.

[a] 2695567907@qq.com, [b]wnw_2008@163.com, [c]48182833@qq.com

Abstract. In view of the existing problems of E4303 electrode in the process of use, the welding rod formula was optimized in this test. The total content of rare earth oxide was added to the coating of E4303 electrode, and the experiment was conducted according to the following proportions: 0% lanthanum oxide +10% cerium oxide, 2% lanthanum oxide +8% cerium oxide, 4% lanthanum oxide+6% cerium oxide, 6% lanthanum oxide+4% cerium oxide, 8% lanthanum oxide +2% cerium oxide and 10% lanthanum oxide +0% cerium oxide. The welding rod was made by hand according to the formula, and then the welding rod was used to perform double-side flat welding on the Q235 steel plate to test the properties of weld deposit metal. According to the experiment, when the optimal amount of lanthanum oxide and cerium oxide was 6%, the optimal amount of cerium oxide was 4%. At this point, the porosity of weld deposit metal is good, forming effect is good, splash rate is 3.89%, deposition efficiency is 74.62%, welding rod efficiency is 53.13%, melting coefficient is 0.124, hardness is 136.60HV, and tensile strength is 492.25Mpa.

1. Introduction

The electrode for type E4303 is the most widely used in China, the production of which accounted for more than 80% of the total electrode production [1-5]. The basic formula of the E4303 electrode is mainly composed of titanium dioxide, aluminum silicate, carbonate, ferroalloy and organics. Due to the high content of Al2O3 in the coating of type E4303 welding electrode, it is often difficult to remove slag and poor stability of arc during welding metallurgy [6-7], which limits the usage amount of type E4303 welding electrode. In terms of cost, the type E4303 welding rod still has a reduced space. In order to reduce the production cost of type E4303 welding rod, it is one of the hot topics for the wide range of welding rod researchers to improve the welding rod comprehensive performance.

Rare earth elements are very chemically active and can react with almost all elements. The rare earth elements can be used as reducing agent, desulfurizer and deoxidizer in electrode, which can improve the toughness, high-temperature resistance and thermal processing of welding seam. The addition of rare earth elements in the electrode will promote the dispersion distribution of non-metallic inclusions, reduce the grain size of austenite and homogenize the grain size of austenite. Rare earth elements have the effect of changing the shape, quantity and distribution of inclusions (sulphur inclusions of iron and manganese, oxides, silicon inclusions, oxygen and sulfur inclusions, etc.) in welding electrode [8-10]. It has been proved that during welding, the oxidation reaction of rare earth oxide can be reduced, and the effect on the mechanical properties of welding seams is equivalent to that of pure rare earth elements [11-12]. In addition, the production cost of rare earth elements and rare earth alloys is much higher than that of rare earth oxides. Therefore, the effects of lanthanum oxide and cerium oxide as composite electrode additives on the comprehensive properties of the electrode could be analyzed [13-16].

2. The Experimental Process

2.1. The Experimental Method

In order to study the effect of lanthanum oxide and cerium oxide on the electrode properties, lanthanum oxide and cerium oxide with a total mass fraction of 10% in the electrode coating were added. The skin formula of the experimental welding rod was listed in table 1.

A diameter of 2.5 mm H08A welding core was used in experiment, the electrode was made by hand rub, which was dried at 200°C for 24 hours. A zx7-255l inverter of DC welding machine was used to perform double-sided welding on the Q235 steel plate (140mm×40mm×3mm). The welding current was 80A+/-10A, and the welding voltage was 20V+/-5V.

Table 1. Weight percentage of testing electrode coating

The serial number	Lanthanum oxide	Cerium oxide	Reduction ilmenite	Artificial rutile	Titanium dioxide	Carbon ferromanganese	Ilmenite	Feldspar	Mica	Marble	Sepiolite	Wood powder
a1	0	10	27	25	13	8	4	9	8	5	9	3
a2	2	8	27	25	13	8	4	9	8	5	9	3
a3	4	6	27	25	13	8	4	9	8	5	9	3
a4	6	4	27	25	13	8	4	9	8	5	9	3
a5	8	2	27	25	13	8	4	9	8	5	9	3
a6	10	0	27	25	13	8	4	9	8	5	9	3
a7	0	0	27	25	13	8	4	9	8	5	9	3

2.2. Microstructure and Properties Test

In the article, the test electrode efficiency was H, melting coefficient was R, deposition efficiency was f, and splash rate was ρ. The weld metal tissue was observed by Leica DM4000M digital microscope. The microstructure defects and weld phase of weld metal were analyzed by SEM (VEGA II XMH) and X-ray diffractometer (Empyrean) respectively. The MCE5582 universal material testing machine was used to test the welding joint. The hardness of welding joint was tested by digital micro-vickers hardness tester.

3. Test results and Analysis

3.1. Welding Efficiency Results and Analysis

The cladding efficiency, electrode efficiency and melting coefficient curves were showed in Fig.1and Fig.2, respectively.

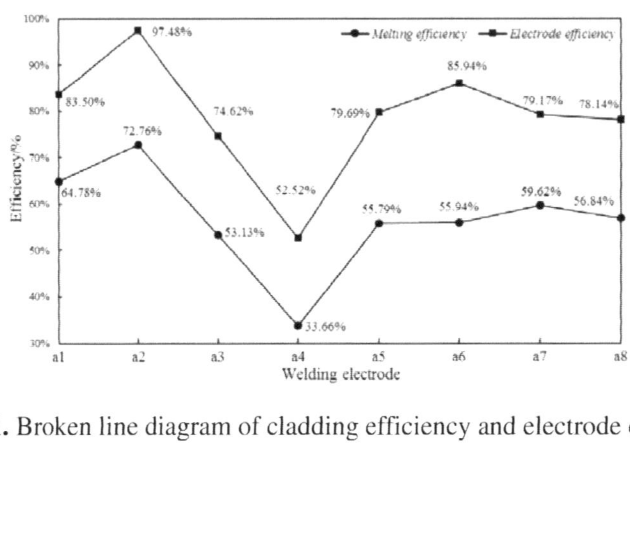

Figure 1. Broken line diagram of cladding efficiency and electrode efficiency

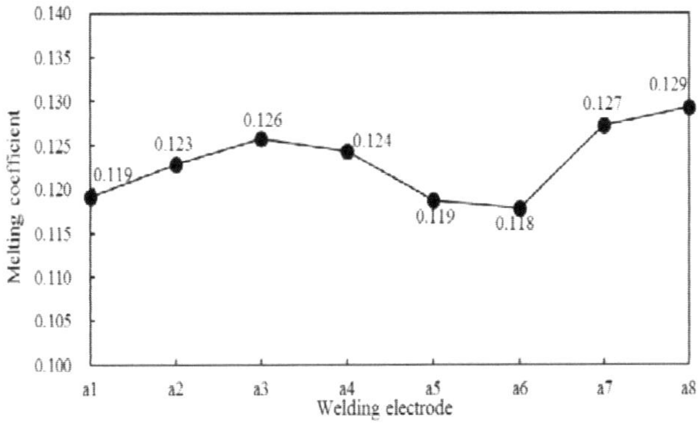

Figure 2. Broken line diagram of melting coefficient

When the total amount of cerium oxide and lanthanum oxide in the coating is 10%, with the increase of the amount of lanthanum oxide (the amount of cerium oxide decreased gradually), the deposition efficiency and electrode efficiency decreases first and then increases. The deposition efficiency and electrode efficiency of a2 is the highest, which is 97.48%.The melting coefficient of a3 is the highest, which is 0.126.The melting coefficient of a6 is the lowest, which is 0.118.As far as the melting coefficient is concerned, the E4303 electrode produced in this experiment is lower than a8. Therefore, the welding arc is extended and the melting coefficient is shortened by the rare earth elements in the weld metal, so the melting coefficient of the electrode added with rare earth oxide is lower than that of the common formula electrode.

3.2. Results and Analysis of Spatter Rate

The experimental results of welding spatter rate are shown in Table 2.When the amount of cerium oxide and lanthanum oxide is 10%, with the increase of lanthanum oxide (the amount of cerium oxide decreased), the spatter rate reflected a downward trend. The spatter rate of a1 is the largest, which is 5.11%, while that of a6 is the smallest, which is only 2.26%. When the content of lanthanum oxide increases to 10%, the content of cerium oxide is 0 and the spatter rate of the E4303 electrode is the lowest. The main reason for the increase of the spatter rate of the experimental sample is that the potential of the La element is lower than that of the Ce. However, rare earth elements are easy to adsorb on the surface of metal droplet, enter the molten pool, reduce the surface tension coefficient of the droplet, promote the formation of fine droplet, and make the spatter rate of the test electrode lower than that of a8. In addition, the surface tension of molten slag can not only be reduced by the surface activity of rare earth elements, but also decrease the interfacial tension between slag and liquid metal. At the same time, the slag coverage and spatter phenomenon can be improved. The spatter particles of most electrodes are smaller than that of a8, because in the welding process, the large amount of gas in the melting pool expands rapidly, the gas explodes, and the size of the spatter particles is smaller, so the particle size is small.

Table 2. Experimental results of welding spatter rate

number	a1	a2	a3	a4	a5	a6	a7	a8
Welding spatter rate (%)	5.11	4.56	3.76	3.89	3.66	2.26	5.44	4.48

3.3. Metallographic Analysis

The metallographic microstructure of Q235 carbon structural steel is showed in Figure 3. The microstructure is made up of mainly pearlite and ferrite. It can be seen from figure 3 that the base metal itself has defects such as voids and impurities.

IOP Conf. Series: Materials Science and Engineering **484** (2019) 012031 doi:10.1088/1757-899X/484/1/012031

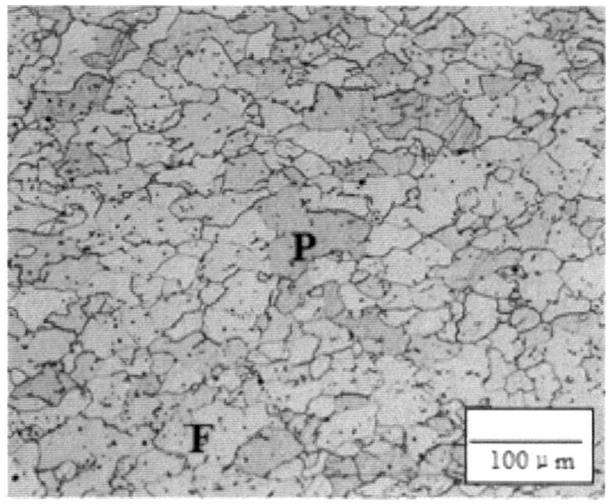

Figure 3. Metallographic microstructure of Q235 carbon structural steel

The metallographic structure of the weld is showed in Fig. 4. The metallographic structure of the weld is mainly composed of sheet ferrite, acicular ferrite, lateral plate stripe ferrite and some pearlite. There are finer acicular ferrite in a4 weld, coarser acicular ferrite in a6 weld, more lateral strip ferrite in a7 weld, and half acicular ferrite and side strip ferrite in a8 weld. It can be seen from figure 4 that the rare earth oxide is introduced into the weld through the coating can be reduced under the high temperature of welding, and react with sulfur, phosphorus and oxygen to desulphurize and deoxidize. At the same time, the austenite is dissolved into the austenite to form a crystalline core, which promotes the inhomogeneous nucleation. The content of ferrite in the side plate of a4 weld is less and the number of fine acicular ferrite

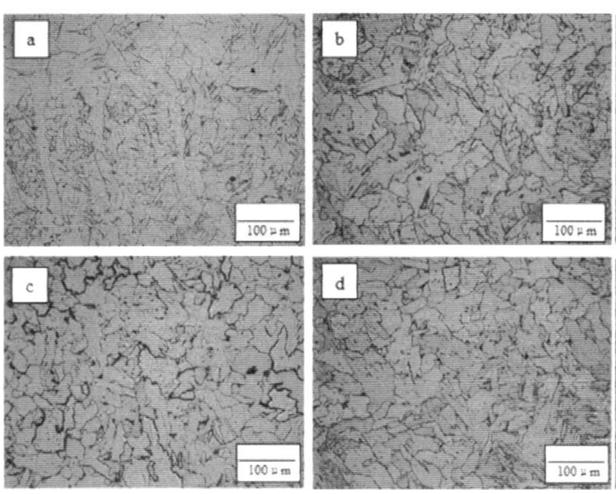

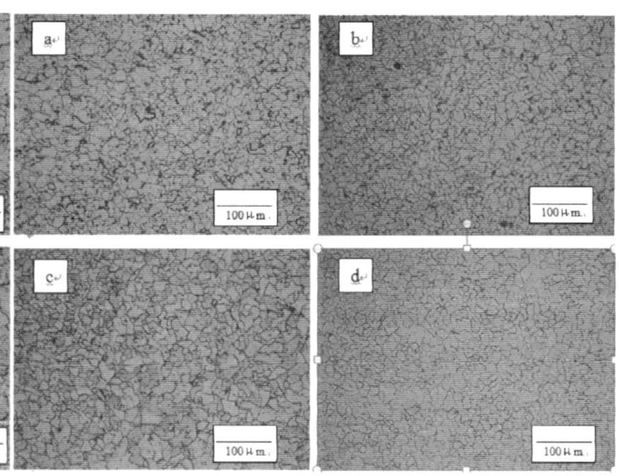

Figure 4. Metallographic structure of weld

a a4 weld bead b a6 weld bead
c a7 weld bead d a8 weld bead

Figure 5. metallographic structure of HAZ

a a4HAZ b a6HAZ
c a7HAZ d a8HAZ

It can be seen that the microstructure of the deposited metal in the weld can be refined by the addition of rare earth elements added. The fine and uniform acicular ferrite content increases. Compared with a4, the acicular ferrite grains of other electrodes are thicker and can't improve the properties of deposited metals.

The metallographic microstructure of the HAZ of the weld is showed in Fig. 5. It can be seen from Fig. 5 that the grains in the HAZ of a4, a6 and a8 are small and uniform, and that in the HAZ of a7 the grain distribution is not uniform. The pores in the heat-affected zone of the a7 joint are more impurity than that of a8, which is related to the compaction degree of the electrode and the amount of sodium silicate. It is difficult to control the amount of sodium silicate and the compaction degree in making electrode. Some inevitable defects are produced because of these uncontrollable factors.

IOP Conf. Series: Materials Science and Engineering **484** (2019) 012031 doi:10.1088/1757-899X/484/1/012031

3.4. Weld Phase Analysis

The diffraction pattern of the deposited metal is showed in figure 6 in the weld seam, from which three characteristic peaks appear in the weld of a1~a8 according to the analysis of the spectrum. The strongest diffraction peak is $Al_{0.5}Fe_3Si_{0.5}$ in a1, a2, a4, a7and a8. The strongest diffraction peak is $Al_{0.7}Fe_3Si_{0.3}$ in a3. While Fe_2SiTi, Al0.5Fe3Si0.5 and $Al_{0.3}Fe_3Si_{0.7}$ are found in a5. The strongest diffraction peak is $Al_{0.7}Fe_3Si_{0.3}$, secondly, $Al_{0.5}Fe_3Si_{0.5}$ in a6. Therefore, in the welding process, rare earth elements do not form a large number of compounds in the weld, but after embedded in the lattice, resulting in lattice distortion or promoting slag removal, desulphurization and other effects on the performance of weld joints.

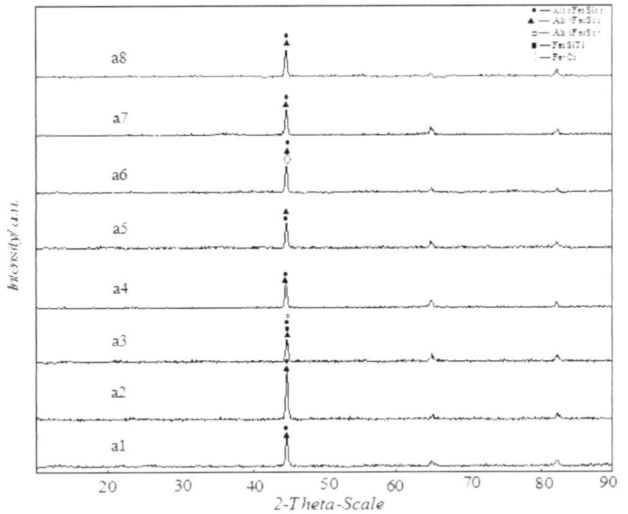

Figure 6. Spectrum of XRD

3.5. Tensile Results and Analysis

The tensile strength of the electrode is showed in figure 7. When the total amount of lanthanum oxide and cerium oxide in the electrode is 10%, the tensile strength of weld joint increases first and then decreases with the increase of lanthanum oxide content and the decrease of cerium oxide content. On the one hand, the rare earth elements are dissolved into the metal lattice and play a solid solution strengthening role. On the other hand, the size difference between lanthanum oxide and iron matrix is larger than that

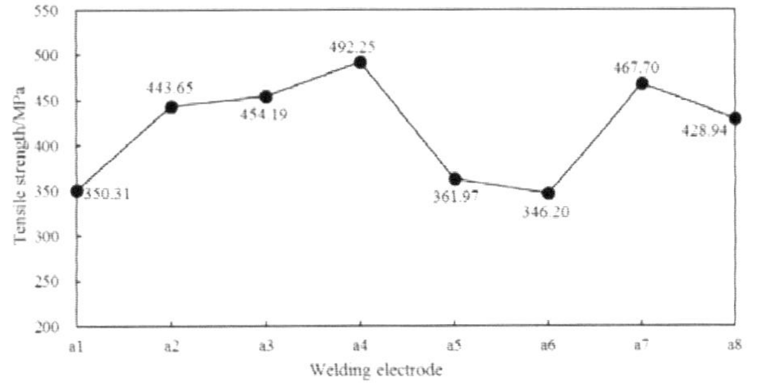

Figure 7. Distribution of tensile strength

Between cerium oxide and iron matrix, and the strengthening effect of solid lanthanum oxide is more obvious. Therefore, with the increasing of lanthanum oxide content (the content of cerium oxide gradually decreasing), the strengthening effect of rare earth element on weld deposited metal is enhanced. When lanthanum oxide content is 6%, the content of cerium oxide is 4%. The maximum tensile strength of the electrode is 492.25 MPA. With the increase of lanthanum oxide content, the content of lanthanum oxide and cerium oxide will exceed the solution degree, and the metallographic structure will distribute in the microstructure as impurity phase, which reduces the properties of the weld during drawing.

3.6. Microhardness and Analysis

The results of weld microhardness test are showed in figure 8. A certain amount of lanthanum oxide and cerium oxide contained in the coating of the welding rod can refine the grain to a certain extent during the welding process. With the increasing of lanthanum oxide content (the cerium oxide content gradually decreasing), the hardness of the weld increases. When the content of lanthanum oxide is 6%, the content of cerium oxide is 4%, the hardness of weld is the largest, the hardness of weld zone is HV145.40, and the hardness of heat affected zone is HV105.53. When the content of lanthanum oxide exceeds 6% and continues to increase, the hardness of weld decreases.

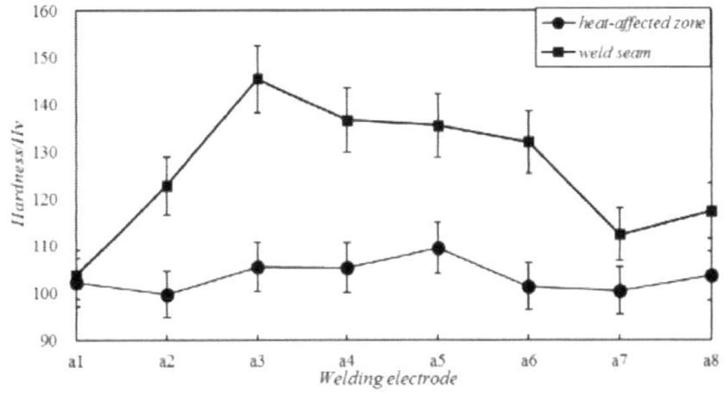

Figure 8. Distribution of microhardness

4. Conclusion

(1) By introducing a certain proportion of rare earth elements La and Ce into the weld metal, the welding arc can be raised and the melting coefficient can be reduced. When the content of lanthanum oxide is 10% and the content of cerium oxide is %0, Welding spatter rate is the lowest, which is 2.26%.

(2) When the content of lanthanum oxide and cerium oxide is 6% and 4%, the microstructure of weld deposited metal is the highest and the fine and uniform acicular ferrite content is the highest. In the welding process, rare earth elements do not form a large number of compounds in the weld, but are embedded in the lattice, and the properties of the weld joints are affected by lattice distortion or promotion of slag removal and desulphurization.

(3) With the increasing of lanthanum oxide content (cerium oxide content decreasing accordingly), the strength effect of rare earth element on weld deposited metal is enhanced. When lanthanum oxide content is 6%, cerium oxide content is 4%; the maximum tensile strength of the electrode is 492.25 MPA. The hardness of weld is the highest, the hardness of weld zone is HV145.40 and the hardness of heat-affected zone is HV105.53.

5. Acknowledgements

This work was financially supported by Provincial Innovation Training Program for College Students (201611360056).

6. References

[1] Y H Su, Z J Liu. Application and Development trend of Welding Materials at Home and abroad [J]. Modern welding, Vol.1 (2009):p1-p4.
[2] S X Wu. Guide for selection of electrode [M]. Chemical industry publishing house, 2010.
[3] L S Li, J Y Lan, X H Sun, etc. Brief Analysis on the Development of Welding Materials in China (I) [J]. Electrical industry, Vol.1 (2010):p10-p16.
[4] C F Li. Steel and Welding Materials in China [J]. Welding, Vol.6 (2006):p16-p26.
[5] J Wen, Y L Wen. Research and Development Technology of electrode [M]. University of Defense Science and Technology Press, 2001.

IOP Conf. Series: Materials Science and Engineering **484** (2019) 012031 doi:10.1088/1757-899X/484/1/012031

[6] J Wen. Research and application of new materials in J422 electrode [J] .Welding Technology, Vol.12 (2011): p46-p47.
[7] J Wen, G H Yang, G W Chen, etc. A J422 electrode [P]. China, 103240546, 2013.
[8] X Sun, W X Lu. Temperature rise cracking of stainless Steel electrode coating [J]. Journal of Welding, Vol.13 (1992): p205-p210.
[9] Z Zhang, W C Zhang, etc. Toughening effect of yttrium in the weld of alkaline welding rod [J]. Journal of rare Earth elements, Vol.13 (1995): p334-p337.
[10] S L Wang, W P Hu, B G Tang. Study on the improvement of Weld toughness by heavy rare Earth [J].Journal of Welding, Vol.7 (1986): p55-p62.
[11] D L Ji, Y L Li. Effect of rare earth on the technological properties of electrode and microstructure and properties of deposited metal [J] .rare Earth,Vol.19(1998):p19-p22.
[12] Y S Zuo, W Z Chen, W Liang. Modern analytical method of materials [M]. Beijing University of Technology Press, 2000.
[13] B Wang. Welding Arc Physics and electrode process Design [M]. Mechanical Industry Press, 1998.
[14] F Y Li. Metal Welding Foundation and material Welding [M]. Mechanical Industry Press, 2009.
[15] Norrish J, Richardson F. Metal transfer mechanisms [J].Welding and Metal Fabrication, Vol.56 (1988):p17-p22.
[16] J G Li, C L Yu, H T Xue, etc. Effect of rare earth on microstructure and properties of weld [J]. Science and Engineering of weapon Materials, Vol.24 (2001): p28-p30.

IOP Conf. Series: Materials Science and Engineering **484** (2019) 012032 doi:10.1088/1757-899X/484/1/012032

Delaunay-Based Aircraft Engine Casing Strain Reconstruction and Data Processing

Kun Wang[1] and Kejia Liu[1]
[1]School of Mechanical and Energy Engineering, Tongji University, shanghai, China, 201804
Email: kunwang@tongji.edu.cn

Abstract. In structural health monitoring, we focus on some critical parts to ensure safety. However, we are unable to install detection components in those parts, due to the structure or the working condition. In order to obtain the strain information of the entire detection area for the engine receiver, we propose a strain reconstruction method based on Delaunay. Besides, we visualize the data for manual processing. In this article, we take the engine fan case as an example. Firstly, we get data from pre-arranged sensors. Next, use EMD (empirically modal decomposition) to process data. Finally, with position data of the detection area, we accomplish strain reconstruction and data visualization. In the end, the validity of the result is analyzed.

1. Introduction

During the service of the aircraft, the harsh working environment is likely to cause damage to the structural components, which can result in the faults, mission failure, and even crashes. As an essential part of the aero engine, the engine casing is the base of the entire engine. It bears most of the external load. The engine casing mainly bears pneumatic load and mass inertial force under working conditions, and also, it can withstand the thermal load, acoustic load, and some assembly stress [1]. The state of the casing reflects the working state of the engine. Health monitoring of the engine casing can significantly improve flight safety.

The strain reconstruction refers to the discrete field information of the finite point obtained by the detection, and the strain field information of the detected portion is obtained by calculation. There are two common ways to achieve. One is to interpolate based on discrete information, and the other is to combine mechanical models for mechanical analysis. However, due to the complexity of the structure and the variability of the load in the working condition. We choose to use the interpolation method to perform strain reconstruction, using Delaunay triangulation interpolation.

2. Delaunay Triangulation

Delaunay triangulation plays a significant role in computational geometry and has considerable value in many fields. Delaunay triangulation algorithm is used to solve problems in many applications, such as the finite mesh generation, robust boundary simulation, image processing, and 3D solid geometry modeling [2].

In this paper, we choose the Triangulation growth method [3]. Using a spatial data point as a starting point, it gradually expands to form a triangulation covering all the point sets, forming a tetrahedral geometric network composed of only scattered points. After the entire area is tetrahedral, we can interpolate the entire scatter set (x_i, y_i, z_i, t_i) [4]. The data set (x_i, y_i, z_i, t_i) consists of two parts, one part of its position (x_i, y_i, z_i) and the other part of its measured value t_i.

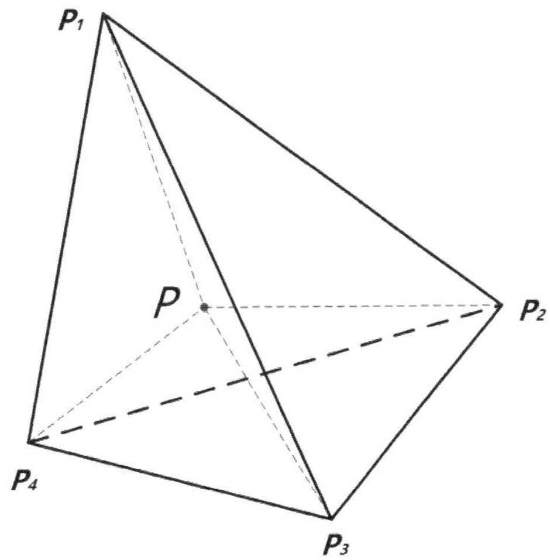

Figure 1. Tetrahedron $p_1p_2p_3p_4$ and inner point p

Suppose a tetrahedron $p_1p_2p_3p_4$, Given the four vertices p_1, p_2, p_3, p_4, namely: the locations (x_i, y_i, z_i) of four vertices and their values t_i. The value t at any arbitrary point $p(x, y, z)$ within this tetrahedron can be found by barycentric interpolation as a weighted average of the values t_i.

$$t = a_1t_1 + a_2t_2 + a_3t_3 + a_4t_4 \tag{1}$$

Where $a_1 + a_2 + a_3 + a_4 = 1$. The weights a_1, a_2, a_3, a_4 are the barycentric coordinates of the point p in the tetrahedron. It can be shown that a_1, a_2, a_3, a_4 are given by,

$$\begin{pmatrix} a_1 \\ a_2 \\ a_3 \\ a_4 \end{pmatrix} = \begin{pmatrix} x_1 & x_2 & x_3 & x_4 \\ y_1 & y_2 & y_3 & y_4 \\ z_1 & z_2 & z_3 & z_4 \\ 1 & 1 & 1 & 1 \end{pmatrix}^{-1} \begin{pmatrix} x \\ y \\ z \\ 1 \end{pmatrix} \tag{2}$$

3. Strain Reconstruction

Human brains process visual information much more easily than written information. When we analyze data, using graphs to summarize complex data ensures that the data is understood faster than disturbing reports or spreadsheets. Therefore, we transform the data into a graphical display in a visual form to enhance understanding.

The entire refactoring process is shown in Figure 2. First, we acquire the scatter coordinates from the monitoring area, and generate the grid coordinates and the surface of the detection area. Then, process the strain data acquired by the sensor. Finally, combined with the mesh surface, based on the tetrahedral interpolation, the strain reconstruction result is obtained.

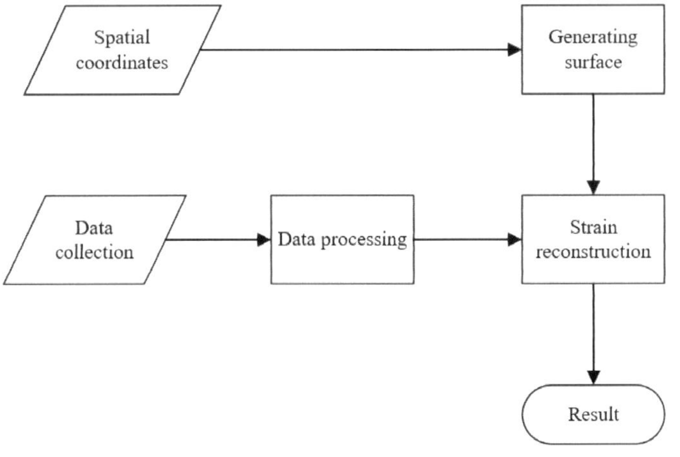

Figure 2. Refactoring process

We take the engine fan case as an example, as shown in Figure 3, for the selected surface in the figure to test.

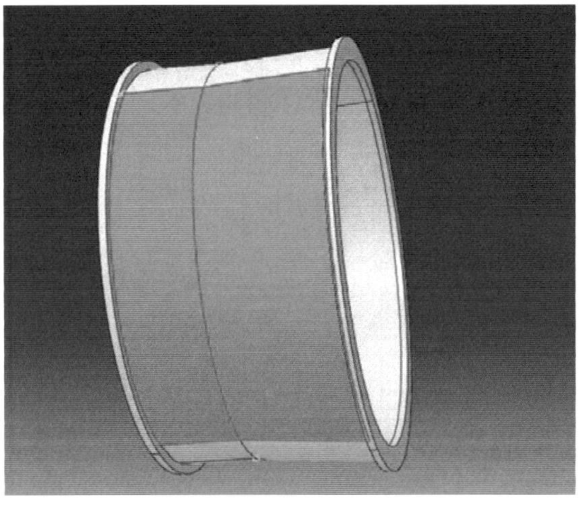

Figure 3. Engine casing and the selected surface

3.1. Entity Visualization

In order to realize the visualization of the detection area, the first step is to obtain the spatial coordinate information of the surface. Through the three-coordinate detection or the digital model, we can obtain the discrete coordinates of a large number of detection areas. According to the Delaunay triangulation, we can visualize the selected surface. In this paper, the spatial data of the monitoring surface is obtained by the digital model. The result is shown in Figure 4.

Figure 4. Delaunay-generated surface

3.2. Data Visualization

The data collected by the sensor contains the noise interference of the working environment. To this end, we also need to filter the collected signal. We use EMD filtering to process the data. The empirical mode decomposition (EMD) method is a signal analysis method proposed by Huang et al. in 1998. Breaking down signals into various components, EMD can be compared with other analysis methods such as Fourier transform and Wavelet transform. Using the EMD method, any complicated data set can be decomposed into a finite and often small number of components. These components form a complete and nearly orthogonal basis for the original signal. Without leaving the time domain, EMD is adaptive and highly efficient [5]. Based on these characteristics, we process the data at first.

Next, we combine the strain information obtained with the spatial mesh obtained in the previous. Finally, strain reconstruction is performed according to the Delaunay interpolation method. The result is shown in Figure 5.

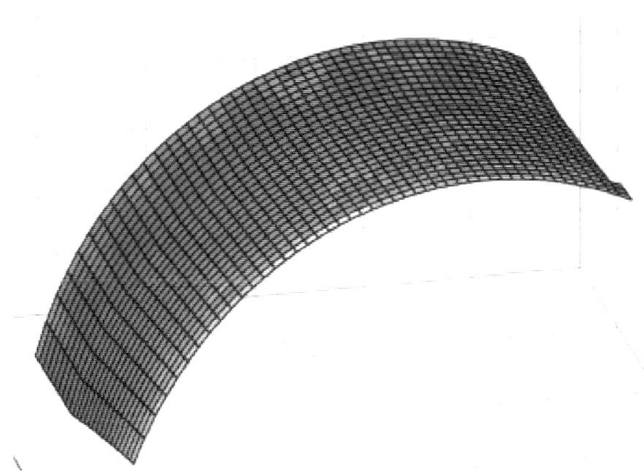

Figure 5. Visualization of strain reconstruction

We select 10 points from the result to compare with the preset control points. The results are shown in Figure 6. The cross is the actual detected value, and the circle is the calculated value. From the relative error table in Table 1, we can see that the measured values are similar to the calculated values, and the difference is small.

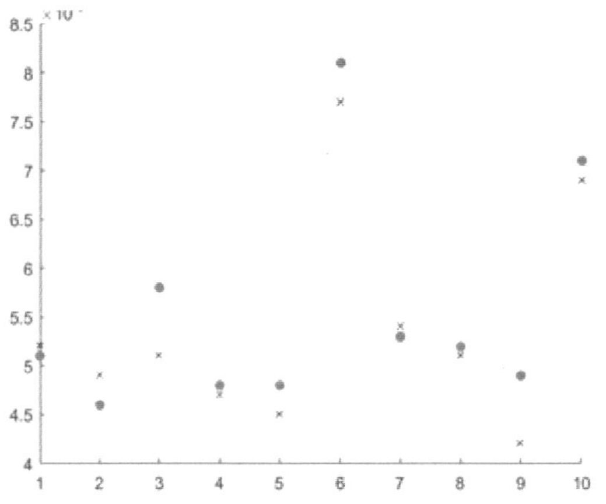

Figure 6. Data Comparison

Table 1. Relative Error

Number	Relative error	Number	Relative error
1	1.92%	**6**	5.19%
2	6.12%	**7**	1.85%
3	13.72%	**8**	1.96%
4	2.12%	**9**	16.67%
5	6.67%	**10**	2.90%

4. Conclusions

For the strain reconstruction of the engine casing, a strain field reconstruction method based on Delaunay triangulation is proposed, taking the surface of the casing as an example. The result shows that the method is in a great performance. However, at the end of the detection area, data accuracy gets worse, which is due to the characteristics of interpolation algorithms. Subsequent work is focused on the optimization of the measurement point layout and the processing of the endpoint data in order to obtain a more accurate spatial strain field.

5. References

[1] Wen D Z, Chen Y S. *Progress and Prospects of Aeroengine Dynamics Research* [J]. Journal of Dynamics and Control, 2013, 11(1):12-19.

[2] Berg M D, *Computational Geometry: Algorithms and Applications* [M]. Springer Publishing Company, Incorporated, 2000.

[3] Lili, *Study and application of three-dimensional Delaunay triangulation algorithm* [D], Dalian maritime university,2010.3.

[4] Amidror, Isaac. *Scattered data interpolation methods for electronic imaging systems: a survey* [J]. Journal of Electronic Imaging, 2002, 11(2):157.

[5] Chen Y. *Dip-separated structural filtering using seislet transform and adaptive empirical mode decomposition based dip filter* [J]. Geophysical Journal International, 2016, 206(1):ggw165.

IOP Conf. Series: Materials Science and Engineering **484** (2019) 012033 doi:10.1088/1757-899X/484/1/012033

Mechanical Property of Roof Plate Based on Engineered Cementitious Composites

Yuxin Gao, Rui Shen, Shuo Wang, Xinyi Yan, Baojun Cheng, Yaning Kong, Yao Bi and Jun Wang

Building Materials Science Academy of China West Construction Group Co. ,Ltd, Chengdu 610094, China.
Email:1039709168@qq.com

Abstract. In order to replace the steel roof in the Tibet of China, the roof plate based on engineered cementitious composites was studied. The results show that roof plate based on engineered cementitious composites is suitable for large deformation of structures. In order to reduce the weight of the roof plate, ribs are more suitable for upward.

1. Introduction

In Tibet of China, the climate is harsh and the temperature difference between day and night is large. In the aspect of building roof, if steel structure is chosen, the surface treatment is easy to wrinkle. Considering the multiple earthquakes in Tibetan area, engineered cementitious composites (ECC) is selected here to replace the steel roof. ECC, invented by Victor Li of the University of Michigan in the United States, have been designed and adjusted to produce high toughness cement-based composites based on the principles of fracture mechanics and micro-mechanics [1]. The ultimate tensile strain of the material can reach more than 2%, and the relationship between tensile stress and strain does not decrease with the increase of strain, which is commonly called strain hardening phenomenon. The formation of multiple micro-cracks during tension increases the macro-tensile strain by 100 to 300 times and the fracture energy by nearly 1000 times, which makes the cement-based materials change from traditional brittle materials to ductile materials similar to metals. In addition, ECC has a certain self-healing ability making it has waterproof ability [2]. Due to its excellent stress-strain properties, the application of this material in future civil engineering has attracted wide attention of scholars at home and abroad, and is considered as one of the main ways to improve the toughness of cement-based structural materials. The material has been successfully applied to repair the large-scale aging and cracking of Hiroshima Sanying Dam in Japan. At the same time, Japan, the United States, Europe and Korea have been widely used in many practical projects, including tunnel, railway, bridge, high-rise building, station reconstruction and reinforcement. Xu Shifeng's team also has applied this material to the construction of high roller compacted concrete cofferdam and conducted demonstration study on its application [3, 4]. Laboratory and field studies also show that ECC has excellent adhesion, impermeability and seismic performance [5-7].

In this paper, the effect of fiber content on the mechanical properties and bending properties of ECC was studied, and the bending properties of ECC under uniform load were studied by preparing 1:1 roof plate. Finally, the force distribution of ECC roof slab was simulated by FLAC.

2. Materials and Methods

2.1. Materials

P.O 42.5R cement and class I fly ash were used as binder. Quartz powder with particle size of 200 meshes was used as aggregate. Polyvinyl alcohol fiber (PVA) was used to increase the toughness. The density of PVA is 1300 kg/m^3, the ultimate tensile strength is 1540 MPa, the elastic modulus is 37.2 GPa, the diameter is 40 μm, the length of PVA is 12 mm. The dosages of PVA are 0%, 1%, 1.5% and 2% by volume. The self-developed polycarboxylate superplasticizer and deformer were used to adjust the working performance of ECC. The mix design of ECC is listed in Table 1. In this study, the mix with PVA content of 19.5 kg/m^3 was used to prepare the 1:1 roof plate.

Table 1. Mix design of ECC

No.	PVA	Cement	Fly ash	Quartz powder	Water	Deformer	Superplasticizer
1	0	550	660	498.2	302.5	0.182	26
2	13	550	660	498.2	302.5	0.182	26
3	19.5	550	660	498.2	302.5	0.182	26
4	26	550	660	498.2	302.5	0.182	26

2.2. Sample Preparation and Testing

Cement and fly ash were mixed in a forced mixer for 30 seconds, then water with superplasticizer and defoamer were added in the mixer stirring for 6 minutes, then PVA fiber was added and stirred for 3 minutes. Samples with sizes of 40×40×160 mm and 400 ×100 ×10 mm were prepared on a vibration table. After curing at 20 °C and relative humidity of >90% environment, the samples were used for the mechanical and bending testing. The size of actual proportional roof plate is 2 × 1 m. The specific shape is shown in Fig.1.

The compressive strength and flexural strength were tested on a hydraulic mortar machine. The four point bending was used for the bending property testing (Fig.2). The load is applied by weight to simulate the load distribution of the actual roof plate. The actual test scenario is shown in Fig.3. During the test, the mid-span deflection was recorded by micrometer. Finally, FLAC is used to simulate the stress and displacement of roof plate. Modulus of elasticity of ECC is selected as 15 GPa and Poisson's ratio is 0.27.

3. Results and Discussion

3.1. Mechanical Properties

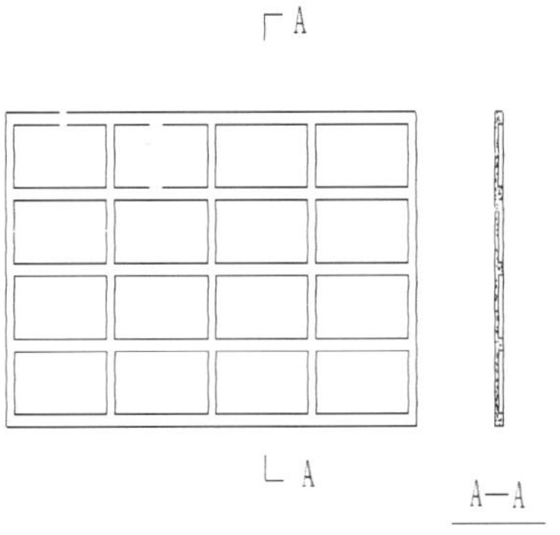

Figure 1. Size of actual roof plate

Figure 2. Four points bending testing

Figure 3. Load distribution of actual roof plate

The compressive strength and flexural strength of ECC at 14 day are listed in the Table 2. It can be seen from Table 2 that with the increase of PVA content the flexural strength increases significantly, but the compressive strength has a little effect. Therefore, the compressive-flexural ratio increases with the fiber increase, illustrating the toughness increase of ECC. In general, the compression flexural ratio of normal concrete is about 10. However, the ratio of ECC is much lower than normal concrete. This shows that the ECC has excellent crack resistance.

Table 2. Mechanical properties of ECC

No.	Flexural strength (MPa)	Compressive strength (MPa)	Compression-Flexural ratio
1	9.2	50.2	5.46
2	12.1	51.2	3.49
3	20.2	52.9	2.62
4	21.5	52.2	2.43

3.2. Four Points Bending Test

The results of four points bending test are listed in Table 3. It can be seen from tabe 3 that the ultimate failure load increases with the increase of fiber content. This is similar with the test of flexural strength. The mid-span deflection in failure increases significantly when the fiber content increases from 1% to 1.5%. The failure form also transfers from brittle fracure to high ductile failure. And when the fiber content lower than 1%, only one main crack was found when the penal failure. When the fiber content increases from 1.5% to 2%, the panals present multi-slit cracking effect and the crack spaceing decreases from 5 mm to 2.9 mm. The multi-slit cracing effect and fracture morphology of ECC with 1.5% PVA are shown in Figs.4 and 5. From Fig.5 it can be seen that Fiber pull-out failure happens during the failure process.

Table 3. The results of four points bending test

No.	Ultimate failure load (N)	mid-span deflection in failure (mm)	Failure form	Crack spacing (mm)
1	230	2	Brittle fracture	A main crack
2	271	5	Brittle fracture	A main crack
3	364	34	High ductile	5
4	415	40	High ductile	2.9

IOP Conf. Series: Materials Science and Engineering **484** (2019) 012033 doi:10.1088/1757-899X/484/1/012033

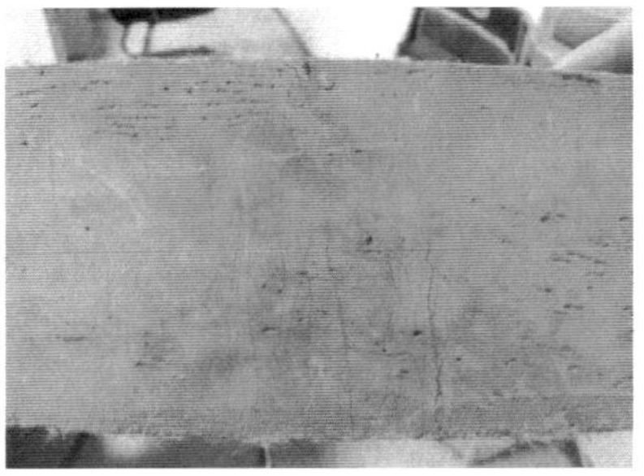

Figure 4. Multi-slit cracking effect diagram of ECC containing PVA of 1.5%

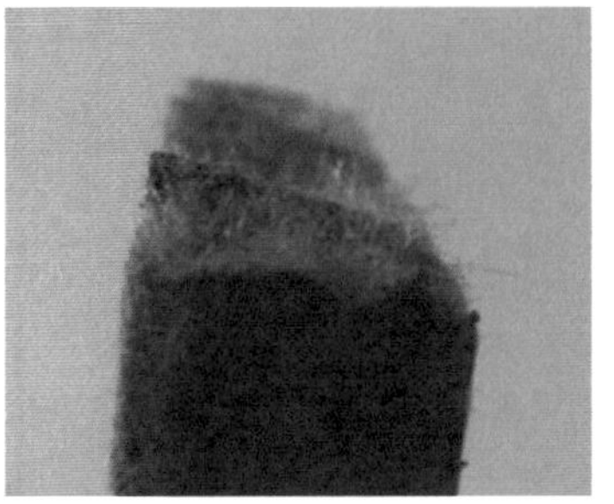

Figure 5. Fracture morphology of ECC containing PVA of 1.5%

3.3. Actural Roof Plate Bending Property

The load-deflection curves of the actural roof plate are shown in Figs.6 and 7. From Figs.6 and 7 it can be seen that the roof plate also present strain hardening during loading. When the plate ribs downloard, the initial crack load is about 250 kg, but when the ribs upward, the initial crack load is increased to about 550 kg. The cracks of plate with ribs downward and upward are shown in Figs.8 and 9. When the ribs downward, the main crack present at the ribs, but when the ribs upward, uniform cracks appear on the surface of the plate, and no main cracks occur.

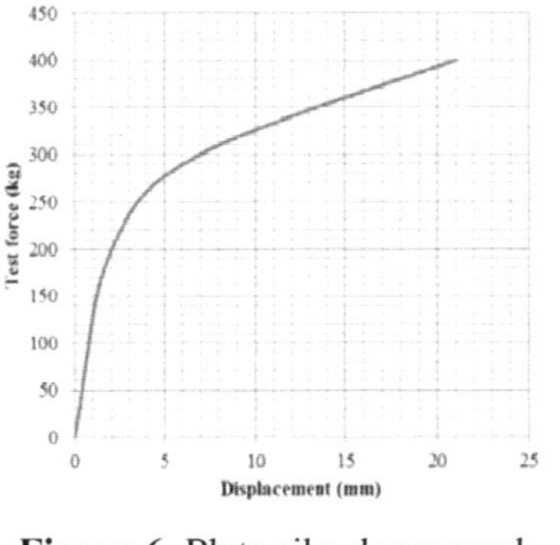

Figure 6. Plate ribs downward

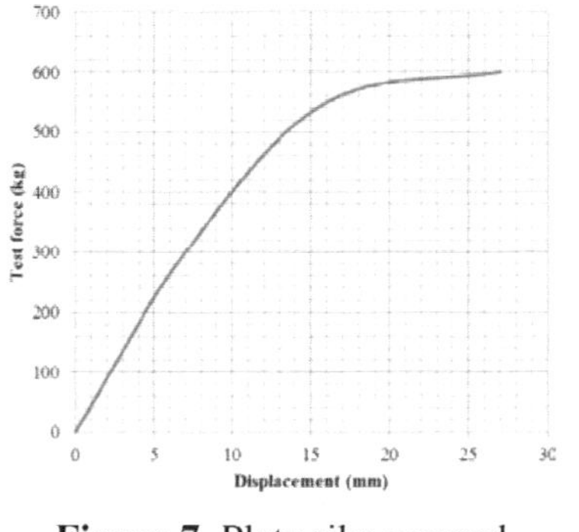

Figure 7. Plate ribs upward

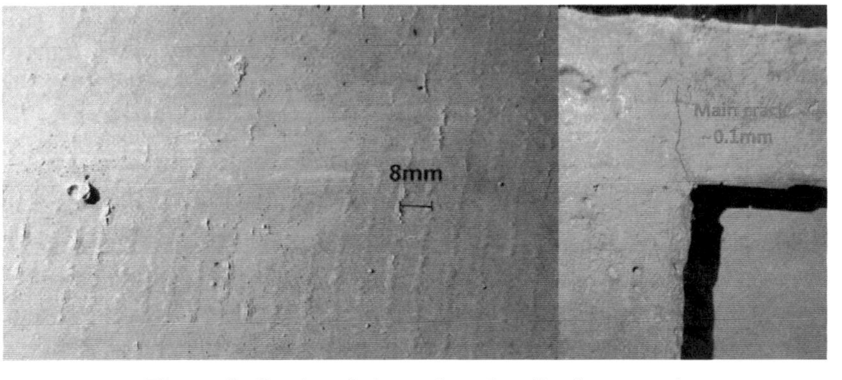

Figure 8. Cracks of plate when the ribs downward

Figure 9. Cracks of plate when the ribs upward

3.4. Numerical Simulation of ECC Roof Plate

The maximum principal stress of roof plate with simply supported ends boundary on the two short sides are shown in Figs.10 and 11. When the σ=12500 Pa was applied on the surface of roof plate with ribs downward, the maximum principal tensile stress occurs in the longitudinal rib near the midpoint. This is demonstrated by the Fig.8. When the σ=27500 Pa was applied on the surface of roof plate with ribs upward, the maximum principal compressive stress occurs in the longitudinal rib near the midpoint. Although the ECC has excellent tensile property, it is more resistant to pressure. Therefore, the mechanical property of roof plate with ribs upward is better.

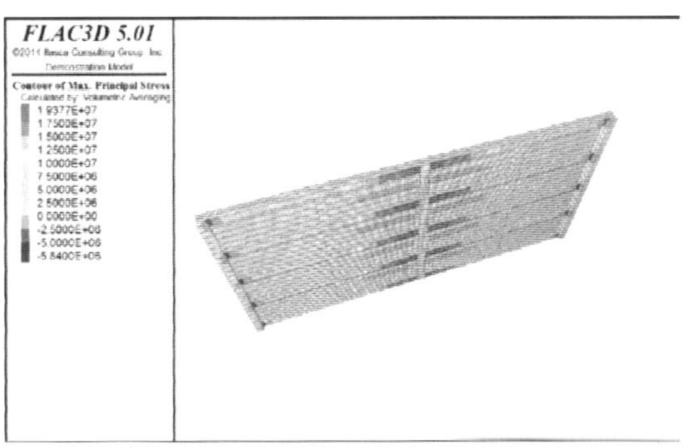

Figure 10. Maximum principal stress nephogram when the ribs downward

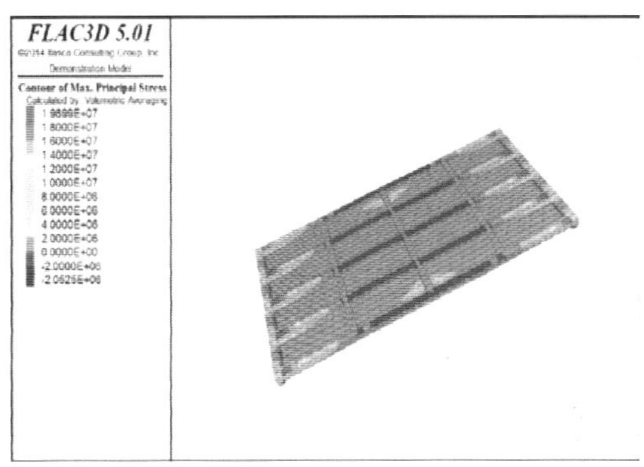

Figure 11. Maximum principal stress nephogram when the ribs upward

4. Conclusions

The compressive strength, flexural strength and bending property of ECC and the bending property of ECC roof plate were investigated. The stress of roof slab is also simulated numerically. The results show that ECC can be used for roof plate according to the results of compressive strength and flexural strength. It is suitable for large deformation of structures in earthquake area. In order to reduce the weight of the roof plate, ribs are more suitable for upward.

5. References

[1] V.C. Li, On Engineered Cementitious Composites (ECC) – A Review of the Material and its, China Association of Concrete and Cement Products Fiber Concrete Engineering Material 2012 Summit Forum, 2003.
[2] Y. Yang, M.D. Lepech, E.H. Yang, V.C. Li, Autogenous healing of engineered cementitious composites under wet–dry cycles, Cement & Concrete Research 39(5) (2009) 382-390.

[3] S. Xu;, Q. Li, Basic application of ultra-high toughness cementitious composites in advanced engineering structures, The Science Publishing Company, Beijing, 2010.
[4] M.L. Yun, V.C. Li, Durable repair of aged infrastructures using trapping mechanism of engineered cementitious composites, Cement & Concrete Composites 19(4) (1997) 373-385.
[5] M.D. Lepech, V.C. Li, Water permeability of engineered cementitious composites, Cement & Concrete Composites 31(10) (2009) 744-753.
[6] M. Maalej, T. Hashida, V.C. Li, Effect of Fiber Volume Fraction on the Off-Crack-Plane Fracture Energy in Strain -Hardening Engineered Cementitious Composites, Journal of the American Ceramic Society 78(12) (2010) 3369-3375.
[7] S. Yao, J.W. Zhang, Y.M. Tu, Application of Engineered Cementitious Composites for Earthquake Resistance and Disaster Mitigation of Structures, Industrial Construction 168-170 (2011) 1665-1669.

Effect of Pinellia Ageratum Decoction on Cisplatin-induced Vomiting and Its Mechanism

Xiaobing Li[1], Shu Wang[2], Wenjing Chen[2], Mengting Zhao[2], Xueqing Kong[2], Xiaoyu He[2], Yan Chen[3], Wenyu Chen[3], Tong Wu[2], Chunlei Xu[2], Jun Huang[2] and Shuanzhu Song [2]

1 College of Basic Medicine, Henan University of Chinese Medicine, room A473, 156 Jinshui East St., Zhengzhou, China.
2 The First school of clinical medicine of Henan University of Traditional Chinese Medicine, Zhengzhou, China
3 The Second school of clinical medicine of Henan University of Traditional Chinese Medicine, Zhengzhou, China
Email: baishaoyao@163.com

Abstract. The present study was to investigate the effect of Pinellia Ageratum Decoction on vomiting induced by cisplatin in rats and its mechanism. The experiment was divided into blank control group, cisplatin model group, ondansetron hydrochloride group, Major Pinellia Decoction group and Pinellia Ageratum Decoction group. The vomiting model was established by intraperitoneal injection of cisplatin in rats. At the end of the experiment, the amount of kaolin consumption, food intake, gastric residual rate, gastric binding mucus volume and small intestinal propulsion rate were observed and measured. Serum 5-hydroxytryptamine (5-HT) and dopamine levels were detected by ELISA. The results demonstrated that compared with the blank group, the amount of kaolin consumption in the model group increased significantly and the food intake decreased significantly. Compared with the model group, the amount of kaolin consumption in pinellia ageratum decoction group decreased significantly, the food intake increased significantly, the amount of gastric-binding mucus increased significantly, the rate of small intestinal propulsion increased significantly, and the differences were significant ($P < 0.05$), level of 5-HT decreased slightly, there was no significant difference, the level of dopamine decreased significantly, and the difference was significant ($P < 0.05$)pinellia ageratum decoction group has a therapeutic effect on cisplatin-induced vomiting in rats, and its mechanism may be related to its reduction of dopamine levels in rats.

1. Introduction

The cisplatin vomiting rate during tumor chemotherapy is as high as 90%. Strong nausea and vomiting not only affect the patient's eating and living, but also cause the body's electrolyte and acid-base balance disorder and exhaustion. General anti-vomiting drugs are difficult to work, easy to produce anxiety, depression and other emotions, is an important reason why many cancer patients can not tolerate chemotherapy [1]. Pinellia Ageratum Decoction comes from “The theory of the plague” and has a good anti-vomiting effect. It has been used in the treatment of vomiting for hundreds of years. This study explored the effect and mechanism of Pinellia Ageratum Decoction on chemotherapy-induced vomiting.

2. Materials and Methods

2.1. Animals
Healthy male SD, Weight 230-300 grams, were provided by the experimental animal center of Henan University of Chinese Medicine

2.2. Experimental Materials

2.2.1. Experimental Reagents. Pinellia ternate, ginseng, honey, pogostemon cablin, dried ginger, licorice root, white poria cocos, Guang tangerine, rhizoma atractylodis macrocephalae stir-fired with bran,(Development of Guangdong Efong Pharmaceutical Co., Ltd.), Cisplatin injection (Development of Jiangsu Haosen Pharmaceutical Group Co., Ltd., Batch Number: Guo Yao Zhun Zi H20040813), Ondansetron hydrochloride (Development of Qilu Pharmaceutical Co., Ltd., Batch Number: Guo Yao Zhun Zi H10970062), Alcian blue dyes (Development of Wuhan Saiwei Biotechnology Co., Ltd., Ltd., Item Number: G1027), Rats 5HT ELISA Kit, Rats DA ELISA Kit (96T,Wuhan Chundu Biotechnology Co., Ltd. ,Item Number: DRE20055).

2.2.2. Experimental Instruments. Laboratory animal clean breeding cabinet YX-1 type, electronic balance, enzyme mark instrument,722 type spectrophotometer, thermostat water bath.

2.3. Animal Treatment
The experiment was divided into blank control group, cisplatin model group, ondansetron hydrochloride group, Major Pinellia Decoction group and Pinellia Ageratum Decoction group.The vomiting animal model was established by intraperitoneal injection of 3mg/kg cisplatin 1hour prior to administration. The heterophilic behavior of mice was obvious, and the intake of kaolin increased compared with the blank control group, indicating successful modeling [2]. Mice with successful modeling were randomly assigned to the blank control group, the cisplatin model group, the ondansetron hydrochloride group, the Major Pinellia Decoction group and the Pinellia Ageratum Decoction group. The mice in ondansetron hydrochloride group were administered daily at a dose of 2.6mg/kg (gavage), the mice in Major Pinellia Decoction group were administered daily at a dose of 1.6g/kg (gavage) and the mice in Pinellia Ageratum Decoction group were administered daily at a dose of 1.6g/kg (gavage). The mice in blank control group and the mice in cisplatin model group were intragastrically administered with an equal volume of saline. Animals were sacrificed on the fifth day of the experiment.

2.4. Observations

2.4.1. The Amounts of Kaolin Eaten by Mice.

2.4.2. Gastric Residual Rate [3]. Dissolved 10g of sodium carboxymethyl cellulose in 250ml of distilled water, then added 16g of milk powder, 8g of sugar, 8g of starch and 2g of activated carbon, stirring evenly, finally configured into 300ml black semisolid Paste, about 300ml.Then put it in the refrigerator and kept it ready. Before the last administration, the mice in each group were fasted but water was given for 24 hours and after 30 minutes of the last administration, the mice in each group were given a nutritive semi-solid paste at a perfusion volume of 1ml/100g. After 30 minutes, anesthesia was performed, then the neck was sacrificed, the abdominal cavity was opened, and the stomach was separated. Cut the stomach from the pylorus and called the stomach full weight (M1), then cut the stomach along the big curve of the stomach, washing the stomach contents in physiological saline and wiping the water with the filter paper, finally weighed the stomach net weight (M2) and calculated the gastric residual rate.

Gastric residual rate = (M1-M2) / M1 × 100%

2.4.3. Detection of Gastric Binding Mucus Volume [4]. After collecting the gastric juice, the stomach was cut along the greater curvature of the stomach, rinsed, and the filter paper absorbs moisture. Then the stomach was immersed in the Alisin blue dye solution (10 ml) and incubated for 2 hours, and then the supernatant was collected by centrifuging the incubation solution at 3000 r/min for 15 minutes. The absorbance value (OD 650 nm) of each well was measured with the enzyme-labeling instrument, and the amount of gastric-bound mucus was calculated from the amount of the bound dye.

2.4.4. Detection of Small Intestine Propulsion Rate [3]. After taking the stomach, use the tweezers to gently extract the upper end to the pylorus, the lower end to the ileocecal intestine tube, spread on the experimental bench, without pulling, gently pull the small intestine into a straight line, and measure the small intestine propulsion index with a ruler. The small intestine propulsion rate was calculated by measuring the length of the small intestine (L1) and the length of the anterior end of the pylorus to charcoal (L2). Small intestine propulsion rate (%) = L2 / L1 × 100%

2.4.5. Detection of 5-hydroxytryptamine (5-HT) and dopamine levels in serum. Test according to the kit instructions

2.5. Statistical Analysis
Data are presented as the mean ± SD. Differences were evaluated using Statistical Package for Social Science 21.0. Statistical analysis was performed using One-way ANOVA followed by least-significant difference (LSD). $p < 0.05$ was considered to be statistically significant.

3. Results

3.1. Effect of Pinellia Ageratum Decoction on the Amount of Kaolin Consumption of Rats
Compared with the normal group, the amount of kaolin consumption in the model group was significantly increased in 24-28h and 48-72h, and the model group had significant statistical difference (P<0.05).Compared with the model group, the amount of kaolin consumption in the Major Pinellia Decoction and the Pinellia Ageratum Decoction was significantly decreased.

Table 1. Effect of Pinellia Ageratum Decoction on the amount of kaolin consumption of rats

group	before the experiment	0-24h	24-48h	48-72h
normal group	0.80±0.1	0.30±0.05	0.14±0.16	0.11±0.17
model group	0.71±0.01	2.93±2.10	3.03±0.74*	2.00±1.76*
western medicine group	0.73±0.10	2.67±1.53	2.50±1.64	1.27±0.59
Major Pinellia Decoction group	0.71±0.10	2.55±1.43	1.79±1.73	1.05±0.82
Pinellia Ageratum Decoction group	0.74±0.28	2.50±1.69	2.35±2.62	1.65±0.92

Note: Compared with the normal group, $^{**}P<0.01$, $^{*}P<0.05$; Compared with the model group, $^{\#\#}P<0.01$, $^{\#}P<0.05$.

3.2. Effect of Pinellia Ageratum Decoction on Rat Appetite
Compared with the normal group, the model group of rat appetite on 24h,24-48h and 48-72h were significantly reduced(*P<0.01*).Compared with the model group, the major pinellia decoction group of rat appetite was significantly increased(*P<0.01*), pinellia ageratum decoction group was significantly increased(*P<0.01*)

Table 2. Effect of pinellia ageratum decoction on rat appetite (g, $\bar{x}\pm s$)

group	before the experiment	0-24h	24-48h	48-72h
normal group	31.73±1.89	30.87±2.45	33.50±0.80	31.00±1.57
model group	32.73±1.37	20.00±2.86**	16.93±3.71**	13.40±2.19**
western medicine group	30.53±1.60	20.47±2.30	18.23±3.80	19.00±1.56$^{\#}$
Major pinellia decoction group	32.03±2.00	21.70±1.85	18.63±7.22	18.83±3.32$^{\#}$
pinellia ageratum decoction group	33.10±0.42	23.20±4.10	19.90±3.54	20.31±2.57$^{\#\#}$

Note: Compared with the normal group $^{**}P$<0.01, $^{*}P$<0.05, compared with the model group $^{\#\#}P$<0.01, $^{\#}P$<0.05.

3.3. Effect of Pinellia Ageratum Decoction on Gastric Residual Rate of Rats

Compared with the normal group, gastric residual rate in the model group was significantly increased, and the model group had significant statistical difference (*P*<0.01).Compared with the model group,the gastric residual rate in the Pinellia Ageratum Decoction group was decreased.

Table 3. Effect of Pinellia Ageratum Decoction on gastric residual rate of rats ($\bar{x}\pm s$)

group	*gastric residual rate*
normal group	42.20± 1.83
model group	74.50± 2.04**
western medicine group	70.97± 3.37
major pinellia decoction group	78.60± 2.62
pinellia ageratum decoction group	73.56± 18.63

Note: Compared with the normal group, $^{**}P$<0.01,$^{*}P$<0.05; Compared with the model group, $^{\#\#}P$<0.01, $^{\#}P$<0.05

3.4. Effect of Pinellia Ageratum Decoction on Gastric Binding Mucus Volume

Compared with the normal group, the model group of gastric binding mucus volume was significantly reduced (*P*<0.05). Compared with the model group, each drug administration group of gastric binding mucus volume was significantly increased (*P*<0.05), major pinellia decoction group was slightly lower than pinellia ageratum decoction group.

Table 4. Effect of pinellia ageratum decoction on gastric binding mucus volume (mg,$\bar{x}\pm s$)

group	*gastric binding mucus volume*
normal group	0.79±0.05
model group	0.52±0.13*
western medicine group	0.78±0.04$^{\#}$
major pinellia decoction group	0.77±0.03$^{\#}$
pinellia ageratum decoction group	0.79±0.05$^{\#}$

Note: Compared with the normal group $^{**}P$<0.01, $^{*}P$<0.05, compared with the model group $^{\#\#}P$<0.01, $^{\#}P$<0.05.

3.5. Effect of Pinellia Ageratum Decoction on Small Intestinal Propulsion Rate of Rats

Compared with the normal group, the small intestinal propulsion rate in the model group was significantly decreased, and the difference between groups had significant statistical difference (P<0.01). Compared with the model group, the small intestinal propulsion rate in the western medicine group and the Pinellia Ageratum Decoction group was significantly increased, and the difference between groups had significant statistical difference (P<0.05).

Table 5. Effect of Pinellia Ageratum Decoction on small intestinal propulsion rate of rats($\bar{x}\pm s$)

group	small intestinal propulsion(%)
normal group	100.00 ± 0.00
model group	40.70 ± 2.40**
western medicine group	67.40 ± 7.00$^{\#}$
major pinellia decoction group	55.43 ± 21.81
pinellia ageratum decoction group	70.70 ± 9.32$^{\#}$

Note: Compared with the normal group, $^{**}P$<0.01, $^{*}P$<0.05; Compared with the model group, $^{\#\#}P$<0.01, $^{\#}P$<0.05

3.6. Effect of Pinellia Ageratum Decoction on Serum 5-hydroxytryptamine and Dopamine Levels of Rat.

Compared with the normal group, the model group of Serum 5-hydroxytryptamine and dopamine levels were significantly increased (P<0.01). Serum 5-hydroxytryptamine level was significantly increased in Major pinellia decoction group and pinellia ageratum decoction group (P<0.05).

Table 6. Effect of pinellia ageratum decoction on Serum 5-hydroxytryptamine and dopamine levels of rat. ($\bar{x}\pm s$)

group	Serum DA(ng/L)	Serum 5-TH(ng/ml)
normal group	33.34±2.81	64.09±5.29
model group	52.31±2.40**	90.95±10.22**
western medicine group	42.53±4.57	82.73±12.62
major pinellia decoction group	37.48±6.57$^{\#}$	75.42±8.75
pinellia ageratum decoction group	38.66±1.63$^{\#}$	75.56±3.95

Note: Compared with the normal group $**P$<0.01, $*P$<0.05, compared with the model group $^{\#\#}P$<0.01, $^{\#}P$<0.05.

4. Discussion

Cisplatin can cause heterophile in a dose-dependent manner, i.e., ingestion of an inalimental substance such as kaolin. Studies have clearly pointed out that rats with heterophilic behavior can be equivalent to other nausea and vomiting reactions in animals with vomiting. Experimental chemotherapy vomiting and exercise vomiting often use the rat heterophilic model [5].

Studies have shown that the heterogeneity of kaolin in the Pinellia Ageratum Decoction group is lower than that in the model group, and its heterogeneity is close to half of the model group, indicating that Pinellia Ageratum Decoction can significantly improve the heterophilic behavior of rats and improve their vomiting symptoms. Studies have shown that Pinellia Ageratum Decoction can increase the food intake of vomiting rats, increase the amount of gastric-bound mucus, and increase the rate of small intestinal propulsion. It indicates that Pinellia Ageratum Decoction can enhance the gastrointestinal motility of rats with vomiting and alleviate the damage of gastrointestinal mucosa caused by cisplatin.

Dopamine is an important neurotransmitter in the body. The gastrointestinal mucosa is stimulated by chemotherapeutic drugs can release excess dopamine. After binding to the dopamine receptor, the signal is transmitted to CTZ via the vagus nerve. CTZ is impulsive and releases dopamine, which binds to the central dopamine receptor. Signaling to the vomiting center causes vomiting. 5-HT is another important vomiting signal and neurotransmitter that mediates chemotherapy-vomit. It is closely related to mental activity and participates in many physiological processes. The experimental results showed that compared with the model group, Pinellia Ageratum Decoction can lower serum 5-HT levels and significantly reduce serum dopamine content. It is suggested that the anti-vomiting effect of Pinellia Ageratum Decoction may be related to the regulation of dopamine neurotransmitter levels.

5. Acknowledgments

Supported by a project grant from The National Natural Science Foundation of China (Grand No.81603527), Science and technology project of Henan Province (Grand No.162102310466), Key scientific research projects of Henan Province Colleges and universities (Grand No.16A360010), Henan University of Traditional Chinese Medicine Scientific and technological innovation talent support program (Grand No.2015XCXRC05). Science and technology project of Zhengzhou City (Grand No.20150310), The Young Core Teacher of Henan Province (Grant No. 2016GGJS-080).

6. References

[1] WANG, X.H.; YUE, W.; WANG, L. Antiemetic effect and mechanism of mixed extracts from capsicum and gingers *Qingdao*, *Qingdao University*. **2008**, 1-17.

[2] WU, L.; WANG, X.L.; SONG, N. Preliminary study on the mechanism of Forsythia's intervention on chemotherapy-induced vomiting. *D. Zhengzhou*, *Henan University of Chinese Medicine*. **2010**, 11-13.

[3] CHEN, H.J.; WEI, L.X.; XIA, Z.J.; *et al.* Protective effects of anzhi jinghua san containing or not containing zuotai on gastric mncosal lesion in rats with functional dyspepsia. *J. Pharmacology and Clinics of Chinese Materia Medica*. **2018**, 34, 133-134.

[4] SUN, Q.W.; TENG, M.C.; HOU, Y. Preliminary study on the protective effect and mechanism of ginger on the gastric mucosa in rats. *J. Jiangxi Medical Journal*. **1997**, 27, 208.

[5] JIAO, M.N. Shenque Xue sticking ginger—processed rhizoma pinelliae experimental study of alleviate symptoms of nausea and vomiting. *D. Shanxi*, *Shaanxi University of Chinese Medicine*. **2015**, 14.

IOP Conf. Series: Materials Science and Engineering **484** (2019) 012035 doi:10.1088/1757-899X/484/1/012035

Distributed Optical Fiber Micro-vibration Sensing for Broken Fabric Bag Inspection in a Baghouse Filtration System

Bo Shi[1,2], Jun Li[1], Xu'an Liu[1,2], Yuquan Tang[1], Tao Pang[1] and Fengzhong Dong[1,2,*]

[1]Anhui Provincial Key Laboratory of Photonic Devices and Materials, Anhui Institute of Optics and Fine Mechanics, Chinese Academy of Sciences, Hefei 230031, China

[2]University of Science and Technology of China, Hefei 230029, China

Email: bshi@mail.ustc.edu.cn

Abstract. A method based on phase sensitive optical time domain reflectometer (φ-OTDR) distributed optical fiber micro-vibration sensor, is proposed to detect and locate the broken fabric bags in a baghouse filtration system. The mechanism is by detecting the laser intensity change in a standard optical fiber cable hang in the filter bag, and which is the result of the dynamic airflow's perturbation in the baghouse. While the location of broken bags is realized by optical time domain reflector method. Field trial in Hefei cement industry platform is conducted and test results show that the minimum detectable hole size is $1cm^2$. The method provides an alternative way for environmental dust emission monitoring and pre-warning in cement industry. Other application niches such as power plant, coalmine industry could also benefit from this new monitoring technology.

1. Introduction

The protection of environment is one of the most important objectives established by national and international organizations [1], and fabric filter baghouses, which may be one of the most important technologies, have been utilized for more than a century to control particulate emissions [2].

Because of its more and more powerful usage in environment protection, the leak detection of fabric filter baghouses has been more urgent and valuable than before. The methods of leak detection for baghouse filter mainly include direct visual inspection and indirect detection. Hayes J and Cabot C [3] used a label gas material to trace the leakage. Gernot P [4] developed a bag filter detection system based on light turbidity. Y T He, X D Wei and Q W Hu reported a technology of filter bag leakage based on capacitive sensing [5], and J Zhang analyzed the application of fluorescent powder detection technology in bag filter (in Chinese) [6].

In this paper, a method based on distributed optical micro-vibration sensor is proposed to monitor the health condition of the filter bags by detection of the dynamic vibration signals applied on the optic fiber cable placed inside the baghouse filtration system. The method provides a new route for smart positioning of the broken bags because of its unique advantages, such as electromagnetic interference resistant, capable of real-time and long distance monitoring.

2. Sensing Principle

The principle of a typical φ-OTDR distributed optical fiber micro-vibration sensor (DMVS) is shown in figure 1.

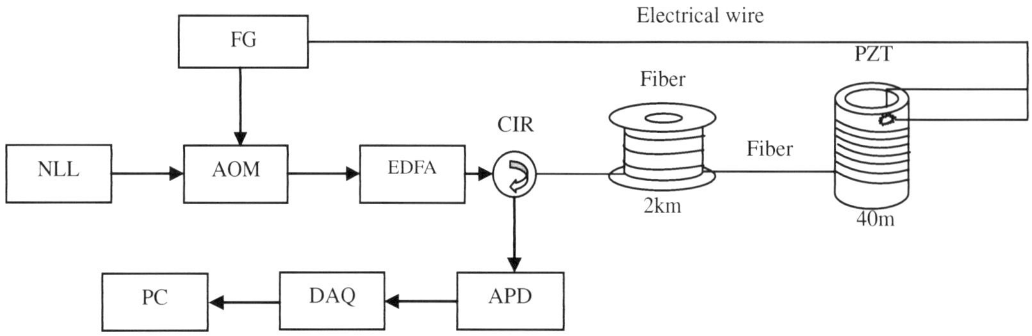

Figure 1. Laboratory set-up of a φ-OTDR distributed optical fiber micro-vibration sensing structure.

A continuous narrow linewidth fiber laser with a center wavelength set at 1550.12 nm is converted into pulsed light through an Acousto-Optic Modulator (AOM), and then amplified by Erbium-doped Fiber Amplifier (EDFA). Ultra narrow linewidth and low frequency drift can effectively improve the signal-to-noise ratio and location accuracy. The amplified pulsed laser is coupled into a 2km single mode optical fiber via an optical circulator and the Rayleigh backscattering signals are converted to electronic signal by an Avalanched Photodiode. A digital function generator with a 30 kHz repetition rate triggers a Data Acquisition card (DAQ) with a sampling rate of 100M to acquire the Rayleigh backscattering signal from the laser pulse. In order to accurately locate the micro-vibration, the spatial resolution of sensor is designed to be ±5m.

The experiment to test the frequency response of DMVS was carried out under 1kHz triggering rate when the PZT tube's driving frequency is set at 10Hz, 50Hz, and 100Hz respectively. The result is shown in figure 2

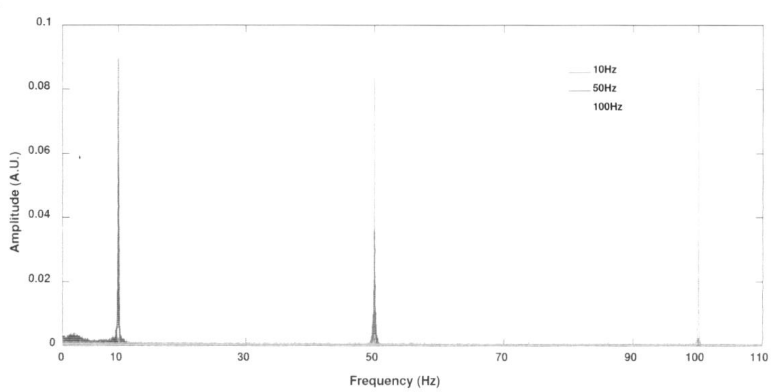

Figure 2. FFT spectrum with perturbation frequencies set at 10Hz, 50Hz, and 100Hz respectively under 1 kHz triggering rate.

3. Field Setup and Experiment

In this section, field test was carried out in a pulsejet baghouse filter system at Hefei Zhongya Environmental and Protection Co.Ltd., Anhui, China. The optical fiber cable with a total length of 1.2km was folded twice every 20m to produce a 3m optical fiber segment and tied up in place. Then the cable was placed into each fabric bag with a length of 4m as shown in figure 3(b). The 3m optical fiber cable segment hangs freely in the fabric bags and was labeled from 1 to 5 as shown in figure 3(b). There is a 20m optical fiber between each bag. It is noted that the spatial resolution of the DMVS is ±5m for our field trial, therefore, 20m will be long enough to distinguish micro-vibration signals from two adjacent bags. The cable is then connected to the DMVS with fusion splicing process to reduce reflection at optical fiber connection points. To simulate the broken bag in the baghouse filtration system, holes with a size of 1×1, 2×1, 3×1 cm^2 are manually created respectively. If there is a hole filter bag was broken, the dust air will flow inside through the leak point, and this airflow will apply a perturbation onto the sensing fiber. The small perturbation will change the optical phase difference at the leak point. Therefore, by using the demodulation algorithm, the broken bags can be detected.

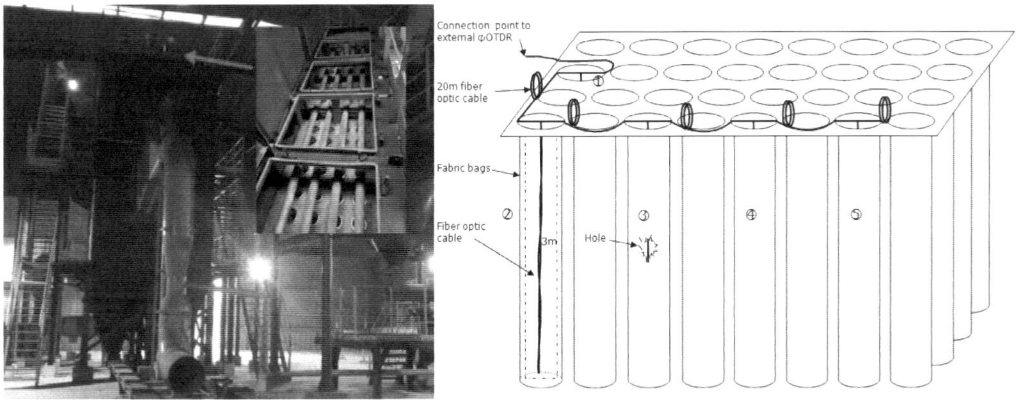

(a) (b)

Figure 3. (a) Fabric baghouse filter experimental platform at Hefei Zhongya Environment&Protection Co. Ltd; (b)Schematic diagram of distributed optical fiber micro-vibration sensor deployment structure at cement plant platform.

4. Result and Discussion

The temporal stability correlation factor *sta_cor* and temporal difference correlation factor *dif_cor* is defined as follow equations (1) and (2), to describe the stability of the vibration signal in a fixed filter bag, and the difference of vibration signal between normal filter bag and broken filter bag in different time slice, respectively.

$$sta\_cor(i) = \frac{1}{n-1}\left(\sum_{j=1}^{n} corrcoef(f\_sig(i), f\_sig(j)) - 1\right) \tag{1}$$

$$dif\_cor(i) = \frac{1}{n}\sum_{j=1}^{n} corrcoef(f\_ns(i), f\_ds(j)) \tag{2}$$

Where i is the serial number of the time slices, n is the total number of the time slices, $f\_sig(i)$ is the frequency of the vibration signal in i^{th} time slice, $f\ ns(i)$ is the frequency of the normal bags' vibration signal in i^{th} time slice, $f\_ds(i)$ is the frequency of the damage bags' vibration signal in i^{th} time slice and $corrcoef$ is the correlation coefficients function.

Regarding real-time monitoring, 51200 ms is chosen as the length of time slice for its temporal stability, which has been experimentally proved, and for the convenience of data processing, 3rd filter bags is selected to be the presentation of other 4 filter bags.

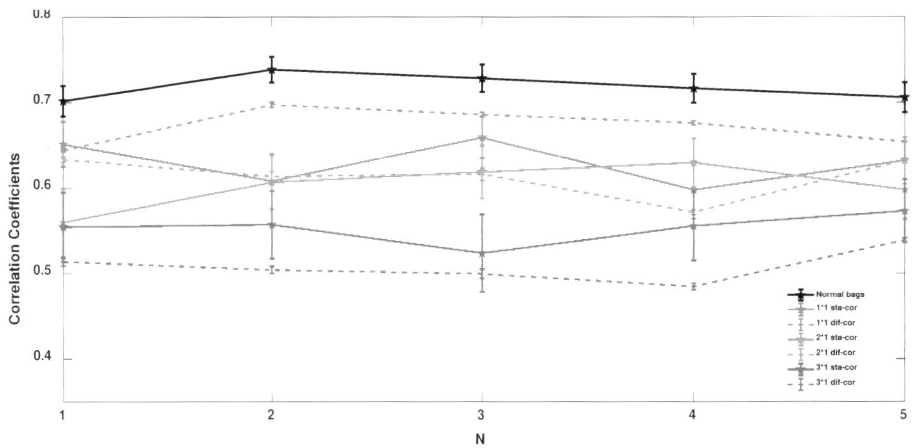

Figure 4. The error bar of temporal stability correlation factors and temporal difference correlation factors of 3rd filter bag under 4 situations mentioned below.

In figure 4, the temporal stability correlation factors and temporal difference correlation factors of 3rd filter bag under 4 situations: normal, broken with $1\times 1cm^2$ hole, $2\times 1cm^2$ hole $3\times 1cm^2$ hole respectively, is plotted. The temporal stability correlation factors between the normal bag and broken filter bags demonstrate that the signal's stability of the former is stronger in general. Comparing the temporal difference correlation factors, the broken filter bags are smaller than the normal bag overall, while the variances are almost the same, which means the differences between of normal bag and the broken bags can be easily identified.

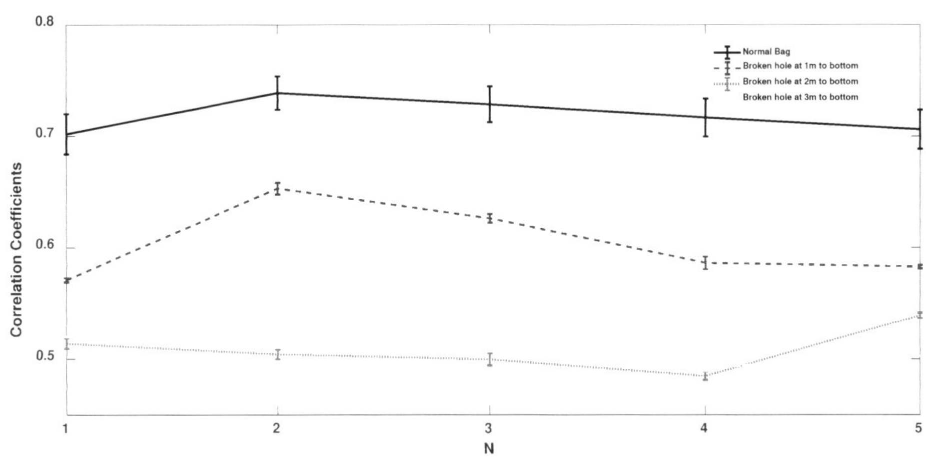

Figure 5. The error bar of temporal difference correlation factors of the 3rd filter bag under 4 situations.

The results of the temporal difference correlation factors of the 3rd filter bag under 4 different situations are shown in figure 5: normal filter bag, broken filter bag with 1×3 cm^2 hole at 1 meter, 2 meters, 3 meters to the bottom of the filter bag respectively. The difference between the normal filter bag and the broken one is obvious; furthermore, the location of the hole has complex effects on the signal, which may be caused by the irregular distribution of the airflow inside the filter bag due to the leak.

5. Conclusion and Future Work

In this paper, we propose a method to monitor the states of fabric bags in a baghouse filtration system. The method uses a distributed micro-vibration sensor to detect airflow perturbation signals through the

IOP Conf. Series: Materials Science and Engineering **484** (2019) 012035 doi:10.1088/1757-899X/484/1/012035

hole on the broken bags. Cross correlation algorithm is employed to identify the characteristics of the broken bags. On-site measurements at Hefei Environmental and Protection Co. Ltd experiment test platform is carried out for long term monitoring and results are found that broken bags with hole diameter as small as 1cm can be identified with correlation algorithm. The technique has provided a new way for dust emission control in cement industry as well as other industry environment.

6. Acknowledgement
This work is supported by Chinese Academy of Sciences under the STS scheme (Grant No.KFJ-STS-SCYD).

7. References
[1] M Pham, C Clark and J McKenna, The evolution and impact of testing baghouse filter performance, *Journal of the Air & Waste Management Association*, 62:8, pages 916-923, DOI: 10.1080/10962247.2012.687360
[2] A Lay-Ekuakille,S Ikezawa, M Mugnain, iR Morello and C De Capua, Detection of specific macro and micropollutants in air monitoring: Review of methods and techniques, *Measurement*, vol 98, pages 49-59, 2017.
[3] Hayes J R, Cabot C, Bag filter leak detection [P]. US. 1976.
[4] M A E Plinke, G Pranghoferg and U Virnich, Remote leak detection sensing method and device [P]. US. 1996.
[5] Y T He, X D Wei and Q W Hu, Technology of bag -house filter leakage detection by capacitive sensor, *Industrial Safety and Environmental Protection*, vol 39, No.1, pages 44-46, January 2013.
[6] J Zhang, Application of Fluorescent Powder Leak Detection Technology in Bag Filter, *Journal of electric power*, vol 28, No.3, June 2013.

IOP Conf. Series: Materials Science and Engineering **484** (2019) 012036 doi:10.1088/1757-899X/484/1/012036

The Development and Improvement of the Coal Chemical Circular Economy System in China: a Case Study

Shuo Yang [a,*], Yun Fu [a], Wanying Zhao [a], Lili Wu [b] and Jianfei Wang [c]

[a] Resource and Environmental Branch, China National Institute of Standardization, Beijing 100191, China
[b] Planning & Design Research Institute Co., LTD
[c] Institute of Scientific and Technical Information, Shandong Academy of Agriculture Sciences, Jinan 250001, China
Email: yangshuo_1984@126.com

Abstract. Circular economy plays an essential and useful role in China in considering economic and ecological sustainability. Individual firms, as the primary implements of a circular economy, are facing many obstacles particularly regarding the resource efficiency. The study presents an analytical method to help decision makers better solve these problems and to guide resource-based chemical industries in sustainable development. First, this study describes the SFA of a selected coal chemistry firm and identifies the roles and correlations among relevant modules. Furthermore, by applying economic benefit assessment, key potential opportunities for improvement were identified. The results demonstrate that the economic benefit of coal chemical products presents a cascade as methanol > wash oil and asphalt blending component > crude benzene > thermal power product > coke. For the goal of improving resource efficiency of the whole system, materials shall flow to the product categories with high resource benefits. Thus, expanding the production capacity of methanol to make full use of the coke oven gas and establishing a utilization module of coal ash in consistence with the market needs can lead to a more sustainable system for the target company. The method proposed in this paper can be applied, duplicated, and spread not only within the coal chemical industry but also in other resource-based chemical industries.

1. Introduction

China's extraction of resources is increasing dramatically. In response to resource constraints, population pressure and widespread environmental damage from the current development pattern, China implemented the strictest environment and resource strategies, which pose great pressure upon traditional industrial sectors. The coal chemical industry has been regarded as one of the biggest energy and resource consumer, as well as major pollution emitter. Hence, it is more important than ever for relevant firms and companies to seek a more sustainable development path.

BTL Company is a large-scale coal chemical enterprise including coal mining, raw coal washing and dressing, coking, coke-oven gas-to-methanol, coal tar hydrogenation (trial production stage), coke dry quenching, and heating. Since its establishment in 2003, BTL has endeavored to implement circular economy. The raw coal mined can be used to produce metallurgical coke, coal chemical products and building materials after deep processing and can also be used to generate power and heat; and an industrial chain of coal - coke - chemical - power - heat - building materials, etc. has been established, which is presented by Figure 1

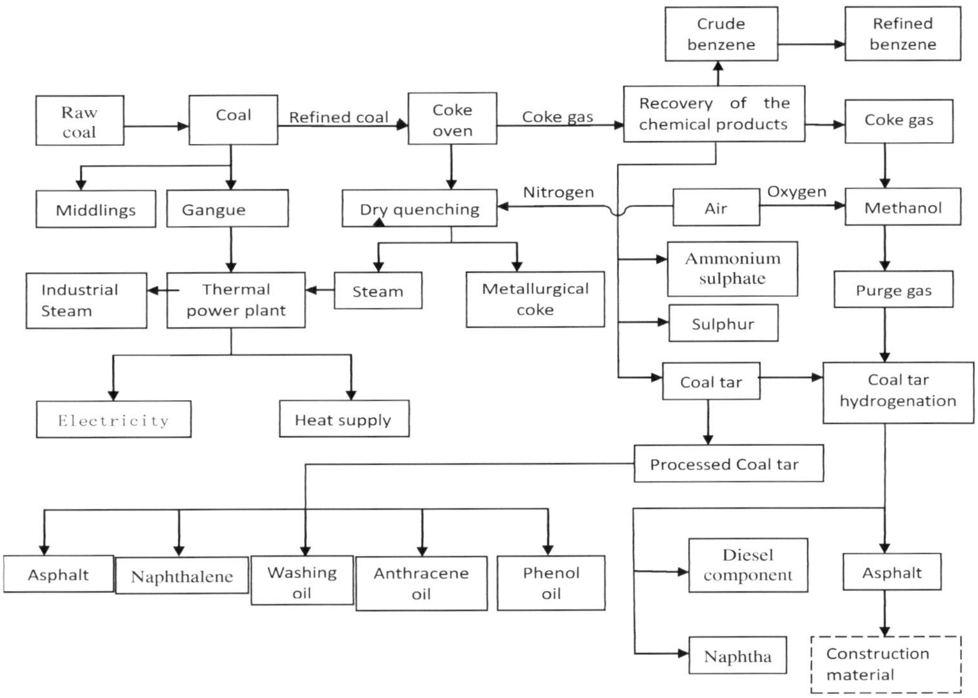

Figure 1. The coal chemical circular economy system illustration

In order to extend the coal-chemical industrial chain to pursue the highest profit, BTL not only increases the added values of the products but also tries to realize the efficient utilization of resources and energy as well as the reduction of emission [1]. In BTL, the raw coal produced from the mine is transported to the plant. Refined coal separated by 4 mt/a dressing plant is fed in 1.58 mt/a coke oven to produce high-quality metallurgical coke which would be supplied to Fushun Steel Company and other plants. By-products of coking including crude benzene, ammonium sulphate, coal tar, coke oven gas are either used onsite or sold to other users. The nitrogen, a by-product produced by air separation of methanol-oxygen, is recovered and used as the nitrogen source for coke dry quenching (CDQ). The waste heat of CDQ and by-products of coal dressing, namely the coal gangues, are used for power generation (CHP). Part of the power generated was self-used, while the rest was connected to the grid. The industrial steam produced by thermal power plant is used for production of the plant and the waste heat is provided for the nearby residence. Through the coal tar hydrogenation process, gasoline, diesel components and asphalt are produced. As for the 300,000 t/a coal tar deep processing project, main products include phenol oil, industrial naphthalene, wash oil, anthracene oil with the by-product of asphalt which is further used as the raw material for the production of needle coke. The wastewater, flue gas and solid wastes generated from the above-mentioned processes are collected together. The wastewater is treated in a comprehensive wastewater treatment system with a processing capacity of $100m^3/h$ and reused as the supplemental water for quenching and washing processes, so as to achieve the “zero emission" state.

In 2012 and 2013, BTL realized the energy conservation of 10,532 tce and 11,923 tce, respectively. In comparison with other coal chemical plants both domestic and international in terms of energy and resource efficiency, the comprehensive energy consumption and the fresh water usage for per unit product, the utilization rate of coke oven gas, and the water recycling rate in BTL all have reached state-of-the-art level.

In such a complex system as coal chemistry, the material flow, energy flow, and value flow are deeply correlated and mutual relied. The system can only operate smoothly and efficiently when a well-functioned symbiotic relationship is formed. The present research studied the circular economy system of BTL by adopting the industrial metabolism methodology. The coal flow throughout the system was analyzed with the association of the economic profits. Some advices were given on basis of the study to improve the current system. The authors believe both the analytical methodology and the conclusion can provide some insights for the establishment and optimization of similar circular economy systems.

2. Research Method

The traditional ecology is the science that studies the relationship between biology and environment. With more and more frequent industrial activities of human beings (mainly the industry) and the ever-increasing impacts on natural systems, the original thought of industrial ecology was formed at the end of 80s. The biological metabolism processes as well as the structure and function units in ecosystem were started to compare and use as analogies to industrial activities. The concept of "industrial metabolism" was first proposed by Robert A Frosch to refer to the use of metabolic process of organisms to simulate the modern industrial production which converts raw materials, energy and labor into products and wastes. N.E.Gallopoulos, et al. then put forward the concepts of "Industrial Ecosystem" and "Industrial Ecology" from the point of view of ecosystem. In 1991, the National Academy of Sciences of the USA and the Bell Labs jointly organized the world's first "Industrial Ecology" forum, which comprehensively and systematically summarized the concept, content, method and application prospects of industrial ecology. The conceptual framework of industrial ecology was basically formed. According to the research carried out by Bell Labs, the industrial ecology is "an interdisciplinary study that identified the relationships among various industrial activities and products with the circumstantial environments" [9].

Industrial metabolism focuses on the simulation and analysis of the material flow and energy flow in the industrial system. Analogic to the natural ecosystem, the industrial ecosystem shall also contain 3 basic components, namely producer, consumer, and decomposer (as shown in Table 1). The metabolic mechanism and cybernetics of industrial ecosystem are studied by analyzing the changes of system structure, functional simulation and analysis of substance flow. The commonly used methods are "supply chain" network analysis (similar to the "food chain network") and material balance accounting [10].

Table 1. Comparison of Components of Industrial and Natural Ecosystems

Component	Natural Ecosystem	Industrial Ecosystem
Producer	Utilizing solar energy or chemical energy to turn inorganic matters into organic matters or converting solar energy into chemical energy to meet their own growth and development needs and provide food and energy for other species (including human beings) at the same time, such as green plants, algae, and chemoautotrophs.	Primary: utilizing basic environment factors (air, water, soil, rocks, minerals and other natural resources) to produce primary products, such as mining plants, and smelters. Senior: deep processing of primary products and production of senior products, such as chemical, fertilizer manufacturing, clothing, food processing, machinery and electronics industries.

Table 2. Comparison of Components of Industrial and Natural Ecosystems (Cont.)

Consumer	Utilizing organic matters and energy supplied by the producers for growth and development as well as secondary production of organic matters and production of metabolites for the use of decomposers, such as animals (herbivorous animals, carnivorous animals, etc.) and human beings.	Utilizing raw materials provided by producers for their own operation and development, while producing productivity, products and services and other functions, such as administration, commerce, finance, entertainment and service industries.
Decomposer	Decomposing excrement and residual bodies of animals and plants into simple compounds for the use of producers, such as decomposer microorganisms, bacteria, fungi and micro animals.	Disposal, conversion, reuse and so on of by-products and "wastes" produced by industrial enterprises, such as waste recycling companies and renewable resources companies.

3. Substance Flow Analysis

The original business of BTL was coke production. However, with the continuous decline of the coke prices in recent years, the industrial chain kept extending and more products were thus included. At present, besides coke, other coal chemical products such as crude benzene, methanol, wash oil and asphalt blending components, and electricity and heat were also produced. Table 2 present the major inputs and outputs of BTL in the last five years (2011-2015) [2,3,4,5,6].

Table 3. Inputs and outputs of BTL from 2011 to 2015

Raw Material (t)	2011		2012		2013		2014		2015	
Raw coal	2.93E+06		2.16E+06		1.92E+06		2.41E+06		2.73E+06	
Refined coal	1.34E+06		1.01E+06		9.03E+05		1.31E+06		1.37E+06	
Product (t)	Production	Sales	Production	Sales	Production	Sales	Production	Sales	Production	Sales
Coke	1.22E+06	1.23E+06	1.14E+06	1.10E+06	1.09E+06	1.05E+06	1.13E+06	1.11E+06	1.07E+06	1.09E+06
Crude benzene	1.71E+04	1.71E+04	1.64E+04	1.64E+04	1.61E+04	1.65E+04	1.63E+04	1.48E+04	1.49E+04	1.63E+04
Methanol	9.83E+04	9.89E+04	8.57E+04	8.73E+04	6.74E+04	6.95E+04	9.36E+04	9.11E+04	8.65E+04	8.59E+04
Wash oil and asphalt blending component	5.95E+04	5.81E+04	6.56E+04	6.76E+04	6.83E+04	6.48E+04	8.86E+04	8.54E+04	7.25E+04	7.69E+04
Power (kWh)	3.57E+08	1.57E+08	3.46E+08	1.55E+08	3.22E+08	1.45E+08	3.45E+08	1.39E+08	3.61E+08	1.58E+08
Heating (GJ)	4.92E+05	4.92E+05	9.74E+05	9.74E+05	1.60E+06	1.60E+06	2.13E+06	2.13E+06	2.13E+06	2.13E+06

Based on the investigation on technical process of BTL, substance flow analysis was carried out

IOP Conf. Series: Materials Science and Engineering **484** (2019) 012036 doi:10.1088/1757-899X/484/1/012036

with the focus placed on coal. The layout of the coal flow throughout the BTL circular system is shown in Figure 2 [7].

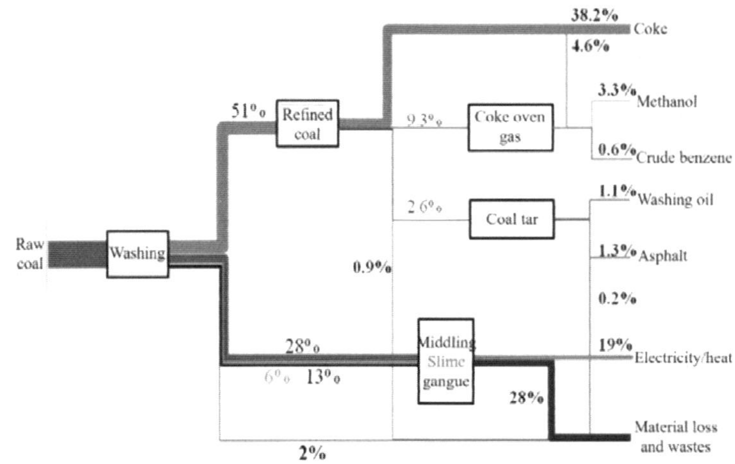

Figure 2. Substance flow analysis of BTL's coal chemical process

From the viewpoint of material flow, "raw coal → refined coal → coke" is the main process of the coal chemical production system of BTL which accounts for 42.8wt% of the coal input. The second major flow of raw coal is the material loss and waste including the emission of gaseous matters (CO_2) as well as fly ash, dust and so on. After coal washing, middlings (28wt%), coal slime (6wt%) and gangue (13wt%) enter the thermal power plant. If the ash contents of the three substance are excluded (20wt% for middlings, 50wt% for slime, and 80wt% for gangue), the part that really contribute to power and heat generation accounts for 19wt% of raw coal mass. Methanol, crude benzene, wash oil and asphalt are produced by using the by-products of coke oven gas (9.3wt%) and coal tar (2.6wt%) during the coking as the raw materials. After exported, half of the coke oven gas returned to the coke oven as fuel and the other half is used for the production of methanol (3.3wt%) and crude benzene (0.6wt%). Coal tar is mainly used to produce wash oil (1.1wt%) and asphalt (1.3wt%).

4 Assessment of Coal Chemical Circle Economy System

4.1. Layout and Roles

On the basis of substance flow analysis, the key links in the existing system can be identified by applying relevant theories of industrial ecology and ecological design tools. The links which are not strongly related with the main material flow are ignored, whereas the ones of greater correlation are merged as illustrated by Figure 3. Furthermore, all modules are classified as producer, consumer, and decomposer according to their roles in the industrial ecosystem. By extending the industrial chain, upgrading the key processes or reusing/recycling the wastes, the overall efficiency of the resources and ecological benefits of the entire system can be improved [8].

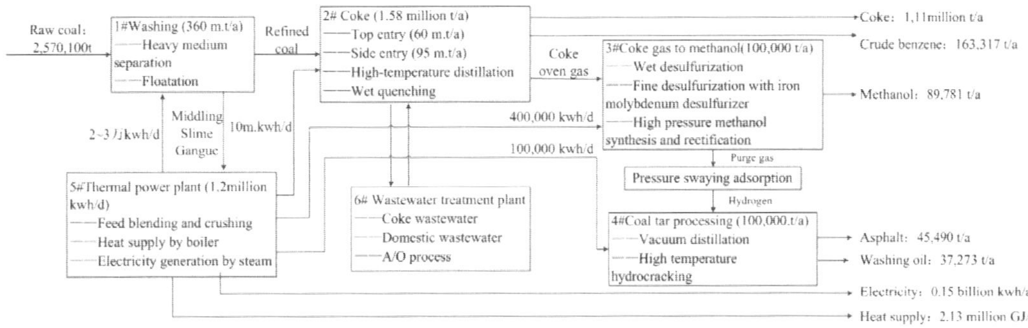

Figure 3. The coal chemical economy system

1# coal washing module (producer): BTL currently has two coal washing plants, namely Plant 1 and Plant 2 with capacities of 1.2 mt/a and 2.4 mt/a respectively. The core technology of coal washing process of BTL is heavy medium separation and flotation. When the raw coal enters the coal washing session, iron, large gangues and woods is manually removed firstly and then the coal is crushed and separated (particle size of less than 50mm). the broken product enters the heavy medium cyclone to separate the raw material into refined coal (51wt%), middlings (28wt%), gangue (13wt%), and coal slime (6wt%). The refined coal enters the coking module, while the middlings, coal slime and gangue enter 5# thermal power plant module.

2# coking module (consumer / producer): Two coke ovens are currently operated in BTL. Oven 1 has a capacity of 600,000 t/a with the entry from top. Oven 2 has a scale of 980,000 t/a with a tamping side entry. The refined coal firstly enters the coaling tower and then is pushed into the coke oven by a coal cart. The refined coal turned into coke in the carbonization chamber for 30h of anaerobic decomposition under 1,280~1,320°C with the by-products of coke oven gas and coal tar. Among them, the conversion rate of refined coal to coke is approximately 75%. The generation of coke oven gas is 400~430m^3/(t· refined coal). The coke is quenched by the reclaimed water from the sewage treatment plant is used for quenching. Beside water quenching, BTL also set up a CDQ system which is not running for now due to the relatively high cost.

3# methanol synthesis module (consumer): the production of coke oven gas to methanol of BTL is 100,000 t/a. The conversion rate is approximate 2,000m^3 coke oven gas for 1t methanol. The purge gas from the methanol synthesis process is of high content of hydrogen (70~80%). After separation and purification, the hydrogen content can raise up to 99.99%, which was transported to 4# coal tar deep-processing module for the hydrogenation of coal tar.

4# coal tar deep-processing module (consumer): the production scale of BTL's deep processing of coal tar is 100,000 t/a. After the vacuum distillation, the coal tar is separated as refined oil (43%) and heavy oil (i.e. asphalt, 55%). After the refined oil is obtained, hydrogenation (purge gas from the methanol plant) is performed and atmospheric pressure separation is used to obtain oil absorbing products (70%); unconverted oil (30%) is used as a starting material for the cracking reactor for circulating use. Asphalt can be used as the raw material for needle coke production in the plant.

5# thermal power plant module (decomposer): the coal slime, middlings and gangue from the coal washing module are supplied to the thermal power plant module. The feed is firstly blended to make its calorific value higher than 2,900kal and then enters the drying shed. After crushing, the product will be delivered to the coal bunker, and finally to the boiler for combustion. The boiler water is provided by the water plant. The superheated steam produced by the boiler is mostly used to propel the turbine to generate power. The generation efficiency of steam turbine is 372kwh/ standard coal (converted to heat value) and the daily power generation is 1.2 million kWh/d totally.

6# Sewage treatment module (decomposer): the coking wastewater and some domestic wastewater are treated in this module. The different influent water was firstly mixed to adjust the COD to less than 6,000mg/L, NH_3 less than 350mg/L, and sulfur content less than 20mg/L. The core process of the sewage treatment module is A/O process. After treatment, the effluent water meets national level II discharge standard and is used for coke quenching.

IOP Conf. Series: Materials Science and Engineering **484** (2019) 012036 doi:10.1088/1757-899X/484/1/012036

4.2. Assessment on the System

The present system in BTL is assessed according to the national standard *Specification for Performance Evaluation of Circular Economy of Industrial Parks,* which mainly measures two aspects, namely, resource output rate and resource recycling rate. The resource output rate includes 3 specific indexes as energy output rate, land output rate and water resource output rate. The resource recycling rate includes 2 required indexes, i.e. comprehensive utilization rate of industrial solid waste and reuse rate of industrial water, and 3 optional indexes, i.e. reuse rate of reclaimed water, recycling rate of waste heat resources and recycling rate of waste gas resources.

Compared with industrial parks which normally involve various processes and inputs, the circular economy system in BTL is based on the coal chemistry with coal as the major raw material. Secondly, the circular economy industrial chain of BTL is the extension of original coking. Therefore, all modules are closely correlated. Thirdly, coal is not only the raw material but also the energy source for the system. In order to assess the system more accurately so as to direct the optimization, the economic benefit according to the material distribution are calculated.

1) Evaluation of economic benefit

In this study, the economic benefit is calculated as presented below:

$$EB = \sum_{i=1}^{N} \sum_{j=1}^{N_i} (S_{ij} M_{ij} - P_{ij} C_{ij}) \tag{1}$$

Where, EB (economic benefit) represents the economic benefit of the entire system; i refers the number of module; j represents the product category; N is the number of module; N_i means the number of product categories in module i; S_{ij} and M_{ij} represent the sales and product price of product j in module i respectively; P_{ij} and C_{ij} represent the production and production costs of product j in module i respectively. The relevant data are collected and compiled in Table 3.

Table 4. The average production and sales of the main products

Product *i*	Annual Production *P* (t)	Annual Sales *S* (t)	Production Costs *C* (RMB/unit product)	Product Price *M* (RMB/unit product)
Coke	1,110,041.935	1,101,744.92	764.3023253	750.420365
Crude benzene	15,639.08	15,535.65	420.6551576	3,056.524802
Methanol	90,062.13	88,487.54	563.2234056	1,687.982058
Wash oil and asphalt blending component	80,558.665	81,146.24	1,159.445015	2,039.272673
Power (kWh)	353,043,459.5	148,755,309.5	0.091204743	0.321269827
Heating (GJ)	2,129,747	2,129,747	33.23740214	37.76828272

The economic profits of each products are shown in Figure 4.

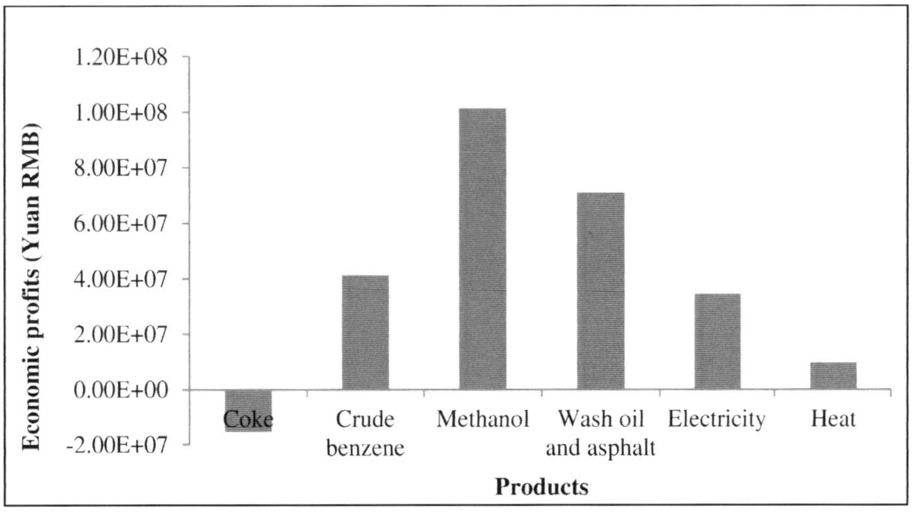

Figure 4. Economic Benefits of Main Products of BTL

As the price of coke has dropped continuously in recent years, production cost per unit coke is higher than the sale price. While the other coal chemical products, as well as power and heating, present better economic benefits, and among which methanol is the most profitable product.

2) Calculation of material distribution

The distribution of raw coal in different materials (including waste and loss) is calculated as below:

$$M_0 = \sum_{i} \sum_{j=1}^{N_{ip}} M'_{ij\,ij}\delta_{ij} \quad (2)$$

Where M_0 is the amount of raw coal entering the target system; M'_{ij} is the amount of raw coal required for product j of module i; δ_{ij} is the conversion coefficient between product j of module i and raw coal.

According to the material flow analysis results, the material distribution of the raw coal entering the coal chemical system of BTL in different products is shown in Table 4. Among them, material loss and waste, coke and heat / power are the top consumers in terms of the raw coal.

Table 5. Material distribution of raw coal in different products

Raw coal (t)	Module i	Product j	Conversion Coefficient $1/\delta_{ij}$	Consumption of Raw Coal by Product M'_{ij}(t· raw coal)
2,570,100	Coking module	Coke	0.428	1,100,002.8
	Methanol synthesis module	Methanol	0.033	84,813.3
		Crude benzene	0.006	15,420.6
	Coal tar deep -processing module	Wash oil and asphalt blending product	0.024	61,682.4
	Thermal power plant module	Heat/power	0.19	488,319
	Others	Material loss and waste	0.319	819,861.9
			1	2,570,100

3) Evaluation of resource benefit

The evaluation result of resource benefit is calculated by applying the formula as below:

$$R_{ij} = \frac{EB_{ij}}{M'_{ij}} \tag{3}$$

Where, R_i is the resource benefit of product j of module i and EB_{ij} is the economic benefit of product j in module i.

The resource benefits of the products of the coal chemistry circular economy system in BTL can be seen in Table 5.

Table 6. The resource benefits of the different products

	Economic Benefit (RMB)	Consumption of Raw Coal by Product $M'_{i\ j}$ (t· raw coal)	Product Resource Benefit R_i (RMB/ t· raw coal)
Coke	-15,409,558.16	1,100,002.8	-14.00865358
Crude benzene	4,122,2563.06	84,813.3	486.0389002
Methanol	101,298,160	15,420.6	6,569.015472
Wash oil and asphalt blending component	70,877,741.57	61,682.4	1,149.075613
Thermal power product	43,873,032.22	488,319	89.84502389
Total	241,861,938.7	2,570,100	94.10604205

5. Conclusion

From the point of view of resource benefits, a cascade among the different products exist as methanol > wash oil and asphalt blending component > crude benzene > thermal power product > coke. For the goal of improving the resource output efficiency of the whole system, materials shall flow to the product categories with high resource benefits as far as possible when designing and optimizing the industrial chain of circular economy. For example, in the current industrial chain system of BTL, only half of the coke oven gas produced by the coking module enters the methanol synthesis module and the rest coke oven gas is used as fuel, resulting in a decrease of the overall the resource benefit. Therefore, expanding the production capacity of methanol to make full use of all the coke oven gas can be profitable. With the expansion of methanol production scale, the amount of by-product purge gas will also increase. The extra hydrogen can also be used to produce more wash oil, so the resource benefit of the whole system can be further enhanced.

Although the resource benefits of coke and thermal power product are relatively low, the coking module and the thermal power plant module are special in the circular economy system since they play the roles of producer and decomposer respectively. Among them, the production scale of coke is directly related to the production of coke oven gas and coal tar. Therefore, this part constrains the scales of methanol synthesis and coal tar deep-processing amount. The inputs of thermal power product are middlings, coal slimes and other impurities, whereas the power and steam generated by the thermal power plant module contribute to other modules. Therefore, its actual resource benefit is far higher than the evaluation results may indicate.

In addition, the proportion of material loss and waste to the total input of coal is more than 30% and most of them are in the form of fly ash (28%). Although BTL has established a comprehensive utilization facility of fly ash, the product is not as expected due to depressing market of construction material. Hence, a pressing task at this stage is to add a utilization module of coal ash which is in consistence with the market needs.

IOP Conf. Series: Materials Science and Engineering **484** (2019) 012036 doi:10.1088/1757-899X/484/1/012036

6. Acknowledgments

The authors disclosed receipt of the following financial support for the research, authorship, and/or publication of this article: This work was financially supported by Chinese National Institute of Standardization Funding Program (grant number 542017Z-5453).

7. Reference

[1] Su, B., Heshmati, A., Geng, Y., Yu, X., 2012. A review of the circular economy in China: moving from rhetoric to implementation. J. Clean. Prod. 42, 215-227.

[2] BTL Annual Report 2011, http://app.finance.ifeng.com/data/stock/ggzw/601011/14009751

[3] BTL Annual Report 2012, http://guba.eastmoney.com/news,601011,74525835.html

[4] BTL Annual Report 2013, http://stock.stockstar.com/notice/JC2014032400004194.shtml

[5] BTL Annual Report 2014, http://app.finance.ifeng.com/data/stock/ggzw.php?id=14948425&symbol=601011

[6] BTL Annual Report 2015, http://app.finance.ifeng.com/data/stock/ggzw.php?id=15591507&symbol=601011

[7] Brunner, P.H., Rechberger, H., 2004. Practical handbook of material flow analysis. Int. J. Life Cycle Assess. 9 (5), 337-338.

[8] Rob_ert, K.H., Schmidt-Bleek, B., Aloisi de Larderel, J., et al., 2002. Strategic sustainable development selection, design and synergies of applied tools. J. Clean. Prod. 10 (3), 197-214.

[9] Yuan Zenghua, Bi Jun, et al., Industrial Ecology. Science Press. 1990.

[10] Singh, R.K., Murty, H.R., Gupta, S.K., Dikshit, A.K., 2009. An overview of sustainability assessment methodologies. Ecol. Indic. 9 (2), 189-212.

[11] Binder, C.R., Graedel, T.E., Reck, B., 2006. Explanatory variables for per capita stocks and flows of copper and zinc. J. Industrial Ecol. 10 (1e2), 111-132.

Operation Performance Evaluation of Municipal Wastewater Treatment Plant by Analytic Hierarchy Process

Xiaoxin Zhang [1, 2], Jin Huang [1], Ling Lin [1], Yueping Wang [3] and Xinheng Zhang [3]

[1]China National Institute of Standardization, 100191 Beijing, China
[2]School of Environment, Tsinghua University, 10084 Beijing, China
[3]MCC HUATIAN Engineering & Technology Corporation, 243011 Ma'anshan, China
Email: zhangxx@cnis.gov.cn

Abstract. The operation performance evaluation of the municipal wastewater treatment plants in China is a very significant and complex issue. Based on China's existing municipal wastewater treatment plant operation performance evaluation indicator system as well as the principles and features of analytic hierarchy process (AHP), this paper utilizes the excel tool to conduct the programming calculation, calculating the index weight at each level of municipal wastewater treatment plant operation performance evaluation index system in China.

1. Introduction

The municipal wastewater treatment and recycling facilities are the indispensible infrastructure for the urban development and the significant guarantee of economic development as well as the residents' safe and healthy life. During the "Thirteenth Five-Year Plan" period, China has further made the overall planning, increased the input, and tried to realize the transition of the construction of municipal wastewater treatment facilities from "scale growth" to "higher quality and enhanced efficiency", from "Sewage treatment primary, sludge treatment secondary" to "equal treatment of sewage and sludge", and from "sewage treatment" to "recycling and reuse", so as to fully enhance the support capacity and service level of China's municipal wastewater treatment facilities.

At present, some cities have just completed the concept change of sewage treatment from "pursuit of sewage treatment volume" to "pursuit of sewage treatment quality". As the centralized treatment unit of pollutants, the municipal wastewater treatment plant plays a significant role in China's pollutant emission reduction strategy. In order to fully realize the function of existing facilities of sewage treatment plant, and enhance the operation and management level of municipal wastewater treatment plant, it is urgent to strengthen the operation performance assessment of sewage treatment plant.

2. Municipal wastewater Treatment Plant Operation Performance Evaluation Index System

According to G1obersOil's assessment index system, eight basic principlesshall be followed [1]. Combined with the operation features and realities of China's municipal wastewater treatment plant, we have proposed the municipal wastewater treatment plant operation performance evaluation index system [2], which consisting of three levels, five categories and 23 indexes, as shown in Table 1.

IOP Conf. Series: Materials Science and Engineering **484** (2019) 012037 doi:10.1088/1757-899X/484/1/012037

Table 1. Municipal wastewater Treatment Plant Operation Performance Evaluation Index System

Target Level	Primary Index Level	Secondary Index Level
Municipal wastewater Treatment System Operation Performance Evaluation A	Environmental Performance B1	Annual average water quality compliance rate C1 Annual average ambient air quality compliance rate C2 Annual average sludge quality compliance rate C3 Annual average pollutant integrated reduction rate C4 ...
	Resources & Energy Consumption B2	Power consumption per unit water treatment C5 Chemical consumption per unit water treatment C6 Fresh water consumption per unit water treatment C7 ... Technical performance requirements C8*
	Technical & Economic Performance B3	Operation cost per unit water volume C9 Annual average unit oxygen consumption pollutant reduction power consumption C10 Annual maintenance fees C11 Annual labor fee C12 ...
	Production Management B4	Rules & regulations C13* Personnel training C14* Emergency response plan C15* Operation & maintenance ledger & record C16* Surveillance & analysis record C17* ...
	Major Equipment Status B5	Annual operation rate C18 Operation hydraulic load C19 Operation COD load C20 Structure integrity rate C21 Critical equipment integrity rate C22 Critical equipment advancement C23

Note: * means that this index is the determinative assessment index.

3. Municipal wastewater Treatment Plant Operation Performance Assessment with AHP Method

AHP method was a simple and practical multi-target decision-making analytic method proposed by USA operational research expert T·L Saaty in 1970s [3]. It utilizes the order of excellence and degree of all the alternative plans judged by the experienced person (experts and etc.) to quantify the qualitative problems and simplify the complex problems, and is extensively applied to the production decision-making, education, medicine, agriculture, economy, safety and other fields. For AHP, it must firstly define the problem to be solved, and then simplify it to the orderly and multi-level hierarchical structure, and finally correspond to many alternative plans [4]. After carefully analyzing and sorting out the essence of complex problem, impact factor and the internal relationship of all factors, AHP shall construct the hierarchical structure model, and then utilizes the less quantitative information to mathematize the decision-making process, so as to solve the multi-target, multi-criteria or structureless complex problem decision-making. According to the basic steps of AHP and the study results of JI Nan et al., this study utilize the excel programming to calculate the weight of municipal wastewater treatment plant operation performance with the AHP method [5]. Taken the calculation and judgment of characteristic value and characteristic vector of matrix with the square root method as an example, the specific steps of calculation with the excel programming is as shown below [6].

3.1. Excel Workbook Establishment

Assume the judgment matrix is 3×3 N-matrix B, and input its matrix elements into the position as shown in the Table below. The calculation results taken according to the corresponding equation are as shown in Table 2.

Table 2. Characteristic Value & Characteristic Vector Calculated with Excel

	A	B	C	D	E	F
1	b11	b12	b13	D1	E1	F1
2	B21	B22	B23	D2	E2	F2
3	B31	B32	B33	D3	E3	F3
4	A4	B4	—	—	—	—
5	A5	B5	—	—	—	—
6	A6	B6	—	—	—	—
7		B7	C7	—	—	—

Note: A1~C3 are judgment matrixes, while F1~F3 are judgment vectors and C7 are characteristic values.

3.2. Corresponding Equations

The calculation equations that are used to calculate the relevant elements of matrix with excel are as shown below:

(1) D1=A1×B1×C1, D2=A2×B2×C2, D3=A3×B3×C3;

(2) E1=POWER (D1, 1/3) E2=POWER (D2, 1/3) E3=POWER (D3, 1/3) E4=SUM (E1:E3);

(3) F1= E1/ E4, F2= E2/ E4, F3= E3/ E4;

(4)A4=MMULT (A1:C1, F1:F3), A5=MMULT (A2:C2, F1:F3), A6=MMULT (A3:C3, F1:F3);

(5) B4=A4/F1, B5=A5/F2, B6=A6/F3, B7=SUM (B4:B6);

(6) C7=B7/3

Among these, function POWER (a, 1/n) intends to return the 1/n root of numeric value a, while function MMULT ( ) intends to return the product of two array matrixes, and function SUM ( ) intends to return the sum 0 of all numeric values within the cell range.

4. Weight Calculation of Municipal wastewater Treatment Plant Operation Performance Assessment Index

In order to ensure the scientificity and practicability of this assessment, when the index weight is determined, the experts and scholars engaged in the ecological protection and environmental contamination study, the relevant directors of the municipal wastewater treatment plants and others are invited to provide the mutually independent judgment. Meanwhile, the relative importance assessment is conducted for the relevant factors of sewage treatment plant in combination with the survey results of China's municipal wastewater treatment plants. Thus, the corresponding weight values are taken. The primary index three-scale matrix of municipal wastewater treatment plant operation performance evaluation is as shown in Table 3.

Table 3. Primary Index Three-Scale Matrix of Municipal wastewater Treatment Plant Operation Performance Evaluation

Integrated Assessment A	B1	B2	B3	B4	B5
Environmental Performance B1	1	2	2	2	2
Resources and Energy Consumption B2	0	1	0	1	0
Technical and Economic Performance B3	0	2	1	2	1
Operation Management B4	0	1	0	1	0
Major facilities and Equipment Status B5	0	2	1	2	1

IOP Conf. Series: Materials Science and Engineering **484** (2019) 012037 doi:10.1088/1757-899X/484/1/012037

According to equation (1)-(3), the three-scale matrix is converted into nine-scale matrix, and the maximum characteristic root and corresponding characteristic vector of judgment matrix are solved with the excel, as shown in Table 4.

Table 4. Primary Index Nine-Scale Matrix of Municipal wastewater Treatment Plant Operation Performance Assessment

Integrated Assessment A	B1	B2	B3	B4	B5	W	Consistency Check
Environmental Performance B1	1	8	4	8	4	0.479	λmax=5.135
Resources and Energy Consumption B2	1/8	1	1/5	1	1/5	0.041	C.R=C.I./R.I
Technical and Economic Performance B3	1/4	5	1	5	1	0.173	=0.339/1.12 =0.030
Operation Management B4	1/8	1	1/5	1	1/5	0.041	
Major facilities and Equipment Status B5	1/4	5	1	5	1	0.173	

CR is less than 0.1, which conforms to the consistency.

4.1. *Determination of Environmental Performance Index Weight*

The secondary index (environmental performance index) three-scale matrix of municipal wastewater treatment plant operation performance evaluation is shown in Table 5.

Table 5. Secondary Index (Environmental Performance Index) Three-Scale Matrix of Municipal wastewater Treatment Plant Operation Performance Evaluation

Environmental Performance B1	C1	C2	C3	C4
Annual average water quality compliance rate C1	1	2	2	1
Annual average ambient air quality compliance rate C2	1	1	1	1
Annual average Sludge quality compliance rate C3	1	1	1	1
Annual average pollutant integrated reduction rate C4	1	2	2	1

According to equation (1)- (3), the three-scale matrix is converted into nine-scale matrix, and maximum characteristic root and corresponding characteristic vector of judgment matrix are solved with the excel, as shown in Table 6.

Table 6. Secondary Index (Environmental Performance Index) Nine-Scale Matrix of Municipal wastewater Treatment Plant Operation Performance Evaluation

Environmental Performance B1	C1	C2	C3	C4	W	Consistency Check
Annual average water quality compliance rate C1	1	3	3	1	0.375	λmax=4.9
Annual average ambient air quality compliance rate C2	1/3	1	1	1/3	0.125	C.R=C.I./R.I =0/0.89 =0
Annual average Sludge quality compliance rate C3	1/3	1	1	1/3	0.125	
Annual average pollutant integrated reduction rate C4	1	3	3	1	0.375	

CR is less than 0.1, which conforms to the consistency.

4.2. Determination of Resources & Energy Consumption Assessment Index Weight

The secondary index (resources & energy consumption index) three-scale matrix of municipal wastewater treatment plant operation performance evaluation is as shown in Table 7.

Table 7. Secondary Index (Resources & Energy Consumption Index) Three-Scale Matrix of Municipal wastewater Treatment Plant Operation Performance Evalutaion

Resources & Energy Consumption B2	C5	C6	C7
Power consumption per unit water treatment C5	1	1	1
Chemical consumption per unit water treatment C6	1	1	1
Fresh water consumption per unit water treatment C7	1	1	1

According to equation (1)-(3), the three-scale matrix is converted into nine-scale matrix, and the maximum characteristic root and corresponding characteristic vector of judgment matrix are solved with the excel, as shown in Table 8.

Table 8. Secondary Index (Resources & Energy Consumption Index) of Municipal wastewater Treatment Plant Operation Performance Evaluation

Resources & Energy Consumption B2	C5	C6	C7	W	Consistency Check
Power consumption per unit water treatment C5	1	1	1	0.333	λ_{max}=3.0
Chemical consumption per unit water treatment C6	1	1	1	0.333	C.R=C.I./R.I =0/0.58 =0
Fresh water consumption per unit water treatment C7	1	1	1	0.333	

CR is less than 0.1, which conforms to the consistency.

4.3. Determination of Technical & Economic Performance Index Weight

The secondary index (technical & economic performance index) three-scale matrix of municipal wastewater treatment plant operation performance evaluation is as shown in Table 9.

Table 9. Secondary Index (Technical & Economic Performance Index) Three-Scale Matrix of Municipal wastewater Treatment Plant Operation Performance Evaluation

Technical & Economic Performance B3	C8	C9	C10	C11	C12
Technical performance requirements C8	1	1	1	1	1
Operation cost per unit water volume C9	1	1	1	1	1
Unit COD Reduction Cost C10	1	1	1	1	1
Annual maintenance fees C11	1	1	1	1	1
Annual labor fee C12	1	1	1	1	1

According to equation (1)-(3), the three-scale matrix is converted into nine-scale matrix, and the maximum characteristic root and corresponding characteristic vector of judgment matrix are solved with the excel, as shown in Table 10.

Table 10. Secondary Index (Technical & Economic Performance Index) Nine-Scale Matrix of Municipal wastewater Treatment Plant Operation Performance Evaluation

Technical & Economic Performance B3	C8	C9	C10	C11	C12	W	Consistency Check
Technical performance requirements C8	1	1	1	1	1	0.200	λ_{max}=5.0
Operation cost per unit water volume C9	1	1	1	1	1	0.200	
Unit COD Reduction Cost C10	1	1	1	1	1	0.200	C.R=C.I./R.I =0/1.12 =0
Annual maintenance fees C11	1	1	1	1	1	0.200	
Annual labor fee C12	1	1	1	1	1	0.200	

CR is less than 0.1, which conforms to the consistency.

4.4. Determination of Operation Management Index Weight

The secondary index (operation management index) three-scale matrix of municipal wastewater treatment plant operation performance evaluation is as shown in Table 11.

IOP Conf. Series: Materials Science and Engineering **484** (2019) 012037 doi:10.1088/1757-899X/484/1/012037

Table 11. Secondary Index (Operation Management Index) Three-Scale Matrix of Municipal wastewater Treatment Plant Operation Performance Evaluation

Operation Management B4	C13	C14	C15	C16	C17
Rules & regulations C13	1	1	1	1	1
Personnel training C14	1	1	1	1	1
Emergency response plan C15	1	1	1	1	1
Operation & maintenance ledger & record C16	1	1	1	1	1
Surveillance & analysis record C17	1	1	1	1	1

According to equation (1)-(3), the three-scale matrix is converted into nine-scale matrix, and the maximum characteristic root and corresponding characteristic vector of judgment matrix are solved with the excel, as shown in Table 12.

Table 12. Secondary Index (Operation Management Index) Nine-Scale Matrix of Municipal wastewater Treatment Plant Operation Performance Evaluation

Operation Management B4	C13	C14	C15	C16	C17	W	Consistency Check
Rules & regulations C13	1	1	1	1	1	0.200	λ_{max}=5.0
Personnel training C14	1	1	1	1	1	0.200	
Emergency response plan C15	1	1	1	1	1	0.200	
Operation & maintenance ledger & record C16	1	1	1	1	1	0.200	C.R=C.I./R.I =0/1.12 =0
Surveillance & analysis record C17	1	1	1	1	1	0.200	

CR is less than 0.1, which conforms to the consistency.

4.5. Determination of Major Facilities & Equipment Status Index Weight

The secondary index (major facilities & equipment status index) three-scale matrix of municipal wastewater treatment plant operation performance evaluation is as shown in Table 13.

IOP Conf. Series: Materials Science and Engineering **484** (2019) 012037 doi:10.1088/1757-899X/484/1/012037

Table 13. Secondary Index (Major Facilities & Equipment Status Index) Three-Scale Matrix of Municipal wastewater Treatment Plant Operation Performance Evaluation

Major Facilities & Equipment Status B5	C18	C19	C20	C21	C22	C23
Annual operation rate C18	1	1	1	1	1	0
Operation hydraulic load C19	1	1	1	1	1	0
Operation COD load C20	1	1	1	1	1	0
Structure integrity rate C21	1	1	1	1	1	0
Critical equipment integrity rate C22	1	1	1	1	1	0
Critical equipment advancement C23	2	2	2	2	2	1

According to equation (1)-(3), the three-scale matrix is converted into nine-scale matrix, and the maximum characteristic root and corresponding characteristic vector of judgment matrix are solved with the excel, as shown in Table 14.

Table 14. Secondary Index (Major Facilities & Equipment Status Index) Nine-Scale Matrix of Municipal wastewater Treatment Plant Operation Performance Evaluation

Major Facilities & Equipment Status B5	C18	C19	C20	C21	C22	C23	W	Consistency Check
Annual operation rate C18	1	1	1	1	1	5/11	0.139	λ_{max}=5.305
Operation hydraulic load C19	1	1	1	1	1	5/11	0.139	
Operation COD load C20	1	1	1	1	1	5/11	0.139	C.R=C.I./R.I =0.076/1.26 =0.060
Structure integrity rate C21	1	1	1	1	1	5/11	0.139	
Critical equipment integrity rate C22	1	1	1	1	1	5/11	0.139	
Critical equipment advancement C23	11/5	11/5	11/5	11/5	11/5	1	0.306	

CR is less than 0.1, which conforms to the consistency.

4.6. Total Hierarchy Ordering

As calculated above, the single hierarchy ordering is accomplished. As a result, the total hierarchy ordering should be conducted, while the order of excellence between each level and its upper level will be concluded, as shown in Table 15.

Table 15. Total Hierarchy Ordering Results

Level	B1	B2	B3	B4	B5	Total Ordering Weight at Level A
	0.479	0.041	0.173	0.041	0.173	
C1	0.375					0.178
C2	0.125					0.060
C3	0.125					0.060
C4	0.375					0.178
C5		0.333				0.014
C6		0.333				0.014
C7		0.333				0.014
C8			0.200			0.035
C9			0.200			0.035
C10			0.200			0.035
C11			0.200			0.035
C12			0.200			0.035
C13				0.200		0.008
C14				0.200		0.008
C15				0.200		0.008
C16				0.200		0.008
C17				0.200		0.008
C18					0.139	0.024
C19					0.139	0.024
C20					0.139	0.024
C21					0.139	0.024
C22					0.139	0.024
C23					0.306	0.053

The consistency check results are as shown below:

$CT = \sum_{i=1}^{5} B_i \times CI_i = 0.479\times0 + 0.041\times0 + 0.173\times0 + 0.041\times0 + 0.173\times0.076 = 0.013$

$RT = \sum_{i=1}^{5} B_i \times RI_i = 0.479\times0.089 + 0.041\times0.058 + 0.173\times1.12 + 0.041\times1.12 + 0.173\times1.26 = 0.721$

CR'=CT/RT=0.013/0.721=0.018<0.1. Therefore, it conforms to the consistency. As shown from the weight results of each index, the impact of environmental performance is the largest (0.479), followed by technical & economic performance and major facilities & equipment status, which is 0.173 separately. Meanwhile, the impact of resources and energy consumption and operation management is the smallest, which is 0.041 separately.

5. Conclusions

The study utilized the excel programming to conduct the integrated hierarchy analysis and evaluation of the municipal wastewater treatment plant operation performance, determined the relative importance degree of each index in this integrated evaluation index system. The conclusion is as follows: the impact of environmental performance is the largest (0.479), followed by technical & economic performance and major facilities & equipment status, which is 0.173 separately. Meanwhile, the impact of resources and energy consumption and operation management is the smallest, which 0.041 separately.

6. Acknowledgement

This work was supported by the National Key R&D Program of China (2018YFF0213200) and Foundation of President of China National Institute of Standardization (542018Y-5926).

7. References

[1] Zhang S C and Zhou W J 2008 *Value Engineering*, 27 (7): 70-73. (In chinese)

[2] Zhang X X, Huang J, Lin L, Wang Y P and Zhang X X. Research of AHP/DEA evaluation model for operation performance of municipal wastewater treatment plants. *International Conference on Advances in Energy and Environment Research (ICAEER),2018.*

[3] Xu S B 1987 *Analytic hierarchy process* (Tianjin University Press, Tianjin). (in chinese)

[4] Chang J and Jiang T L 2007 *Journal of Wuhan University of Technology* (1): 153-156. (in chinese)

[5] Ji N 2011 *Research on Comprehensive Evaluation Index System and Method in Wastewater Treatment Plant* (Harbin Institue of Technology).

[6] Chen T 2008 Establishing and Application of Gray Multi-level Comprehensive Evaluation Model, *Journal of Daqing Normal University*, 28(5): 79-81.

IOP Conf. Series: Materials Science and Engineering **484** (2019) 012038 doi:10.1088/1757-899X/484/1/012038

Technical Assessment of Industrial Solid Waste Recycling Process Based on the Willingness-to-Pay LCIA Model

Huiting Guo[a], Shuo Yang[a,*], Xiuteng Wang[a] and Jing Liu[a]

[a] Resource and Environmental Branch, China National Institute of Standardization, Beijing 100191, China

Email: yangsh@cnis.gov.cn

Abstract. The recycle and reuse of solid waste must consider both the economic and environmental respect. For most of the present environmental technologies, the biggest obstacle for their practical implementation is the economic barrier. For some cases, even the technical process demonstrates the good performance in energy-saving, emission-reduction etc., but it is still hard to be applied due to the high operating cost. For most of the technology assessment and comparison, they are carried out only from the environmental or the economic perspective, and rarely combine the two respects. For assessing the recycle and reuse of solid waste in a more comprehensive manner, the present paper adopted the monetized LCIA method based on the Willingness-to-Pay theory. Based on the proposed method, a LCIA case study on the production process of sintered brick with coal fly ash was carried out. Through monetizing the environmental impact during its production process, it was found that the environmental cost resulted from the production of sintered brick is 8.76E-3Yuan/kg, namely, 8.76 Yuan/ton. The major environmental impact category is global warming, followed by the acidification and suspended particulate matter. In contrast, due to the reuse of industrial wastes as raw materials, the consumption of natural resources only accounts for a minor part of the entire environmental impacts. Therefore, the control of GHG emissions shall be prioritized for the technical optimization.

1. Introduction

Recycle and reuse of the solid waste are effective ways to tackle down the current resources and environmental crisis in China. The solid wastes recycled as the raw materials for production can replace natural resource, but it may also pose extra pollutions to the environments and threats the human health. Therefore, it is necessary to access and optimize the solid waste recycling process from the comprehensive perspective, so as to make full use of the resource properties and avoid the occurrence of secondary pollution.

At present, life cycle assessment (hereinafter referred to as "LCA") is the well-accepted method for the measurement of the environmental burden. According to the definition of United Nations Environment Programme, "LCA is a tool to assess environmental impacts associated with all the stages of a product system's life cycle from raw material extraction through materials processing, manufacture, packaging, marketing, use, reuse and maintenance, and disposal or recycling." Although different institutes and organizations may have slightly different understandings about LCA, the general LCA procedure follows similar framework. According to the ISO 14040 standards, a Life Cycle Assessment is carried out in four inter-correlated phases, namely, goal and scope definition, inventory analysis, impact assessment and interpretation, as illustrated in Figure 1 [1]:

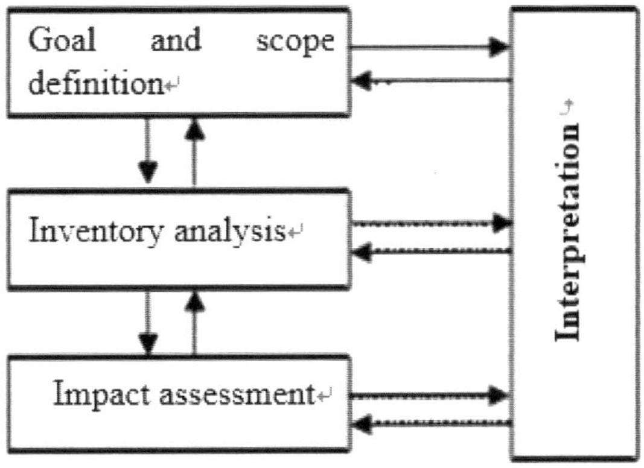

Figure 1. Technical Framework of LCA

Theoretically, the assessment of industrial solid waste recycle process shall follow the LCA framework as shown in Figure 2. However, due to the substitution of conventional materials with solid wastes, the waste recycle process must be taken into consideration. The extra transportation and energy required due to the usage of the solid wastes also need to be included as shown in Figure 2. It is also important to note that these processes usually do not occur in sequence or go on simultaneously. Instead, they are interrelated and exist in the very complex manner, while the boundaries of all links are not clear. Therefore, sometimes, appropriate hypothesis and simplification shall be made for the relevant system during the assessment process.

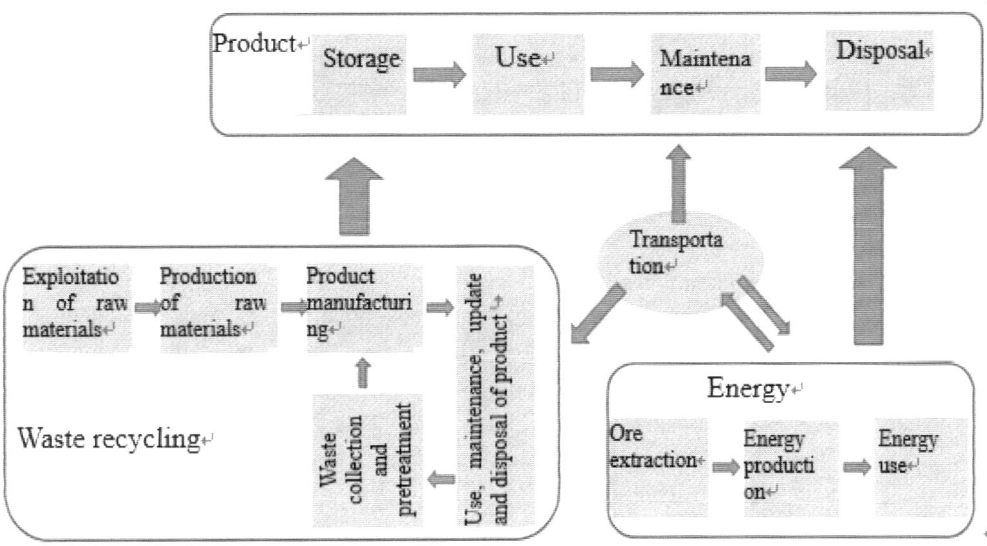

Figure 2. Life Cycle of Industrial Solid Waste Recycling Product

LCA study needs to define the system boundary in accordance with the purpose. To ensure the completeness of assessment, the indispensable processes must be incorporated. However, the scope mustn't be overextended. The collection of solid waste, for instance, can be excluded, whereas the pretreatment of the wastes specifically for the production must be taken into consideration. The consistency matter and energy flow also need to be ensured throughout the entire system.

For the case study, the present research analyzed the production process of sintered bricks with coal fly ash. China has always been the major producer of bricks, with the annual output of more than 800 Billion bricks. With the emerging resources problem in recent years, the production of sintered solid bricks is subject to certain restrictions, while the percentage of hollow bricks is gradually rising every year. Waste slag, coal fly ash and other solid wastes are also commonly used to replace the natural resources. This study intends to calculate the environmental impacts posed by the typical sintered brick production, so as to provide the suggestions for the technical improvement of sintered bricks.

2. Study Method

The relevant study and application of LCA in China has been lag behind compared to the developed countries and regions. On the basis of the ISO 14040 series standards, China enacted the GB/T 24040 *Environmental Management - Life Cycle Assessment - Principles and Framework* [2], GB/T 24042 *Environmental Management--Life Cycle Assessment - Life Cycle Impact Assessment* [3], and GB/T 24043 *Environmental Management - Life Cycle Assessment- Life Cycle Interpretation* in 1999 [4]. At present, China also has a few LCA databases [5,6], which incorporate the background LCA data in terms of energy [7,8,9], iron & steel [10,11], non-ferrous metals & building materials [12,13,14], high polymer materials [15,16] and other basic products. In recent years, China has also gradually carried out the LCA study on cars, TVs, Air Conditioners, and other complex products. The inventory allocation method, data quality, uncertainty analysis and other methodology research have also been conducted [17, 18, 19].

For the life cycle impact assessment, there are three models have been proposed in China, namely, target distance model, willingness-to-pay model and expert choice model. The major differences among them are the weighting basis.

2.1. Target Distance Model

Based on the concept framework at the LCA impact assessment of SETAC & ISO, Yang et al established the assessment model as shown in Figure 3. This method firstly converts the inventory data into the environmental impact categories via the characterization, and then standardizes the environmental impacts with the national per capita index (or unit space indicator) of all environmental impacts as the reference value, and obtains the relative value. The weight of different impact categories are determined by the target distance method. The severity of certain environmental effects shall be demonstrated by the distance between the current level and target level (standard or capacity) of such effect [20], which is shown in Table 1.

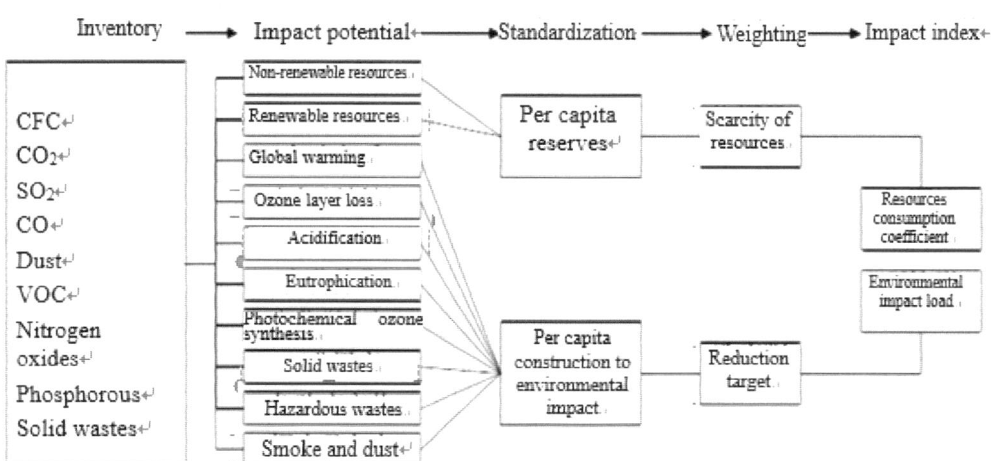

Figure 3. Target Distance Model of Environmental Impact Assessment

Table 1. Environmental Impact Weight Determined by the Target Distance Method

Impact Category	Global Warming	Ozone Layer Loss	Acidification	Eutrophication	Photochemical Ozone Synthesis	Solid Wastes	Hazardous Wastes	Smoke & Dust
Weight	0.83	2.7	0.73	0.73	0.53	0.62	0.45	0.61

Table 2. Resources Weight Determined by the Target Distance Method

Resources	Service Life (a)	Weight	Resources	Service Life (a)	Weight
Oil	43	0.023	Copper	36	0.027
Coal	172	0.006	Nickel	52	0.019
Lignite	387	0.003	Manganese	86	0.012
Natural Gas	61	0.016	Lead	21	0.049
Iron	119	0.008	Antimony	30	0.034
Aluminum	195	0.005	Timber Production Forest	16	0.063
Zinc	20	0.051	Water	1	1.000

2.2. Willingness-to-Pay Model

Zhang et al construct their assessment model as shown in Figure 4. In the weighting process, the significance of environmental impacts is represented with the economic cost that shall be paid to mitigate the ecological damage resulted from such impact. Such method is based on the willingness-to-pay theory in the environmental economics, where the environmental tax and resources tax are used as the evidence to determine the weight [21].

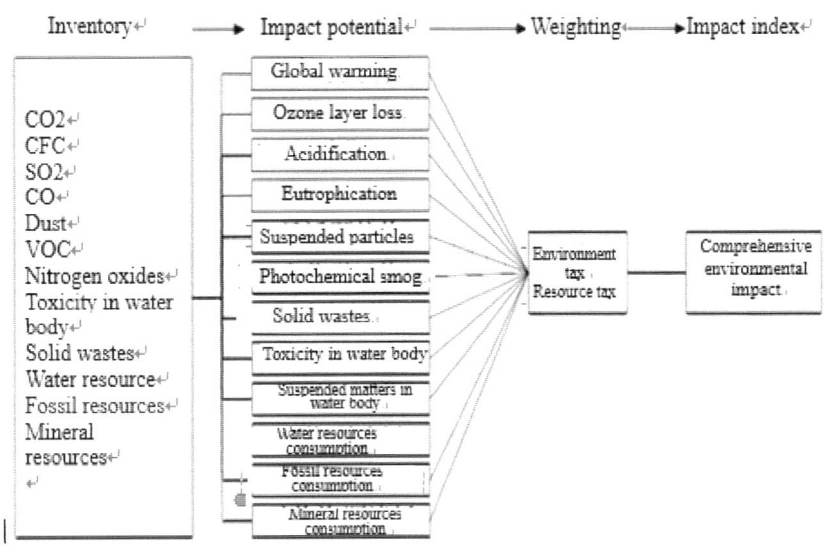

Figure 4. Willingness-to-Pay Model of Environmental Impact Assessment

For the project that does not determine the environmental tax and resources tax, the calculation process of its weighting indicator is as shown in the equation below:

$$w_i = \sum_j (e_{ij} \cdot c_{ij}) \quad (i,j = 1,2,3\ldots\ldots) \tag{1}$$

Where, w_i refers to the weight factor, while c_{ij} refers to currency factor, namely, the discharge fee of such pollutant, but its unit of measurement shall be measured with the impact factor (for instance, the unit of measurement for the currency factor of SO_2 is RMB/ kgeq.SO_2); Besides, e_{ij} refers to the overall impact potential coefficient of pollutant number j under the environmental impact number i. Obviously, it is not only related to the contamination capacity of pollutant, but also related to the total discharge amount of such pollutant. An environmental impact may be resulted from variable pollutants. Therefore, during the weight calculation process, only several typical pollutants will be selected for an impact category. For instance, only SO_2, NO_x & NH_3, these three substances are selected for the acidification calculation. The calculation method of impact potential coefficient is as shown in the equation below:

$$e_{ij} = \frac{f_j \cdot a_j}{\sum_j (f_j \cdot a_j)} \quad (i,j = 1,2,3......) \tag{2}$$

Where, f_j refers to the unit impact potential coefficient of pollutant number j, while a_j refers to the annual discharge amount of pollutant number j. Partial environmental impact and resources consumption category weight taken from the aforesaid method is as shown in Table 3. Such weight has the dimension, while its demonstration is the price of unit environmental impact. The unit of integrated environmental impact calculated and taken from such method is Yuan or RMB.

Table 3. Environmental and Resources Weight (Partial) Determined by Willingness-to-Pay Method

Impact Factor	Weight (RMB/kg)	Impact Factor	Weight (RMB/kg)
Global Warming (CO_2 Equivalent)	0.01	Water Resources	Regional Variety
Ozone Layer Loss (ODP Equivalent)	15.36	Iron Ore	0.0171
Acidification (SO_2 Equivalent)	0.74	Limestone	0.002
Eutrophication (NO_3^- Equivalent)	0.58	Bauxite	0.02
Suspended Particles	0.19	Zinc	0.003
Solid Wastes	0.06	Copper	0.0014
Photochemical Pollution (C_2H_4 Equivalent)	3.41	Manganese	0.002
Toxicity of Water Body (Pb Equivalent)	6.04	Kaolin	0.003
Suspended Matter in Water Body	0.175	Fluorite	0.02
Fossil Energy (Standard Coal Equivalent)	Regional Variety	Saline Rock	0.025

2.3. Expert Choice Model

Gu adopts the endpoint damage model to quantify the environmental impact, as shown in Figure 5 [22]. This method selects four endpoint environmental impact categories, namely, resource depletion, energy depletion, human health damage and ecological damage. When the mineral resources consumption is calculated, this method defines the scarcity factor of iron ore as 1, while the scarcity factor of other depletable resources is the relative value comparing to the iron ore. For the farmland, forest and water resources which are theoretically renewable but extremely scarce at present, their scarcity factors are independently analyzed. The resource scarcity factors in this method are as shown in Table 4. The resource depletion amount is the sum product of weight of all resources and their scarcity factors. This method makes a difference between the fossil energy and normal mineral resources. When the consumption of fossil energy is calculated, it defines the energy quality coefficient, namely, the specific value of energy's ability to do work and its total energy, as shown in Table 5. The depletion amounts of all energies are jointly calculated and taken as per their caloricity and energy quality coefficient, which not only considers the caloricity variation of different energies, but also considers the quality difference of different energies. Then, after conducting the questionnaire survey to 130 plus experts from the relevant fields, it determines the weight of four categories of endpoint environmental impact, as shown in Table 6, while the integrated assessment indicator, namely, environmental load is further taken via weighting, and its unit of measurement is defined as point (hereinafter referred to as "pt").

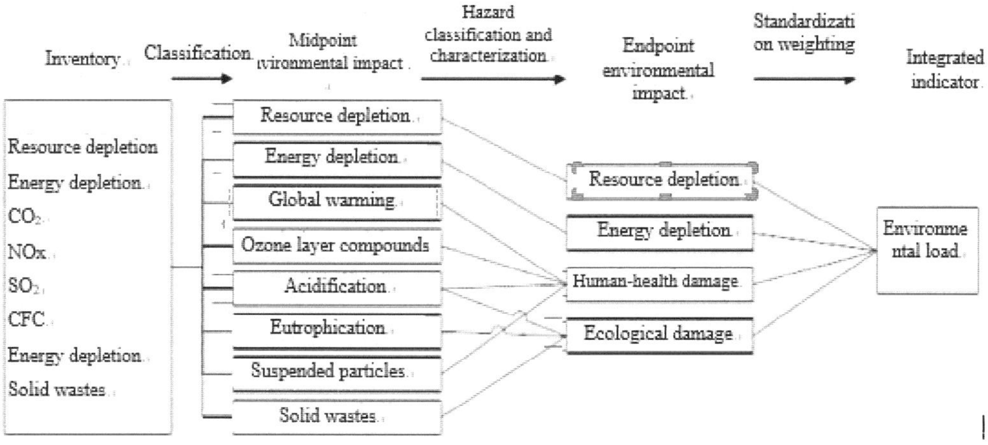

Figure 5. Expert Marking Model of Environmental Impact Assessment

Table 4. Environmental and Resources Weight Determined by Willingness-to-Pay Method

Resource	Service Life (a)	Weight	Resource	Service Life (a)	Weight
Iron Ore	190	1.00	Cement Limestone	500	0.38
Manganese Ore	64	2.98	Glass Silica Gel	434	0.44
Zinc Ore	94	2.02	Fluorite	500	0.38
Tin Ore	84	2.25	Gypsum	500	0.38
Copper Ore	76	2.51	Marble	500	0.38
Bauxite	345	0.55	Farmland	43	4.37
Saline Rock	500	0.38	Forest	-	3.84
Kaolin	500	0.38	Water Resources	-	0.0075

Table 5. Energy Quality Coefficient of Three Fossil Resources

Resource Category	Coal Ore	Oil	Natural Gas
Energy Quality Coefficient	0.35	0.46	0.52

Table 6. Weight of Four Endpoint Environmental Impacts

Endpoint Impact	Resource Depletion	Energy Depletion	Human Health Damage	Ecological Damage
Weight	0.27	0.28	0.22	0.23

All abovementioned three methods are developed based on China's context. The Target-Distance and willingness-to-pay models have more environmental impact categories and it is hard to weight, but the quantitative mechanism is much simpler. Whereas, the third method makes the secondary category. Therefore, there is less endpoint environmental impact number, and it is easy to determine the weight, but the quantitative mechanism is very complex. For weight decision, the target distance method may be related with the national policy and highlight the key social problems at present. However, the environmental policies generally have much stronger phased and regional characteristics and are prone to be affected by the non-environmental factors. Once any change occurs to the target, great changes may occur to the results. As mentioned above, for the expert marking method, its biggest problem is that it has much higher requirements upon the expert's personal quality and

IOP Conf. Series: Materials Science and Engineering **484** (2019) 012038 doi:10.1088/1757-899X/484/1/012038

requires the experts to have the much stronger ability of macroscopically handling the interdisciplinary or multidisciplinary problems. Therefore, this method also has a certain limitation or controversy for its application. Combined with the reality of China's industrial solid waste recycling, this paper finally selects the willingness-to-pay method as the LCIA method, which has the advantages and disadvantages as shown in Table 7.

Table 7. Advantages & Disadvantages of Conducting the LCIA of Industrial Solid Waste Recycling Product with the Willingness-to-Pay Model

Pro	Con
1. It directly introduces the economic indicator in the form of weight, while it is the only one with the dimension among all three methods. Besides, it is simple and intuitive to assess the environmental impact in currency unit, which can be directly correlated with the cost, revenue and other data of the product; 2. It has many data sources, such as environmental tax, energy tax, resource tax, pollution discharge fee. Meanwhile, it has multi-layers within the temporal and spatial scope, such as the worldwide/ nationwide/ within the province, city or region and even the industry; 3. It has the timeliness property, which can demonstrate the current hot-spot issues of concern, especially regarding resource consumption. It can directly reflect the market impact. Therefore, the obtained assessment results is much easier to be accepted by the market; 4. It is featured in simple structure and low technical threshold. Relatively speaking, it is the easiest manner to realize the modularization and software-based operation among all three analysis methods, easy to be promoted and convenient to be applied.	1. China's environmental tax and resource tax system is still not perfect. Especially for environmental tax, many environmental impacts are still not levied with taxes. Meanwhile, the existing environmental tax amount is generally much lower, while the environmental tax is often subject to the macro-control policy and cannot completely equal to "economic cost required for environmental recovery". 2. It is difficult to reflect the priority relationship of all environmental factors in a long-term and objective manner. Instead, it is much suitable to solve the current short-term prominent problems. However, when the long-term planning problem is encountered, it will be subject to the much larger temporal and spatial restrictions; 3. Compared with the other two methods, willingness-to-pay method lacks the professional theoretical support (experts do not participate during the whole process.)

3. LCA Analysis of Sintered Brick with Coal Fly Ash

3.1. Scope of Study

The determination of study scope is to define the system boundaries of LCA assessment. Theoretically, the life cycle assessment shall include the whole process of the assessment object, from its cradle to its grave, namely from extraction of raw materials to final disposal of wastes. However, due to time and data source limitations, generally, a certain definition shall be made for the system boundaries. This study defines the scope of study as the production process of sintered brick. Therefore, the system boundary starts from extraction of raw materials and ends with the termination of production process, as shown in Figure 6. Compared with other materials, the production process of sintered brick is much simpler, while the process flow is much shorter. The raw materials involved in the production mainly include the shale, coal gangue, coal fly ash, slag, etc. Meanwhile, the consumed fuels mainly include the coal and electricity, while a certain amount of CO_2, SO_2 and other environmental pollution gases are excluded.

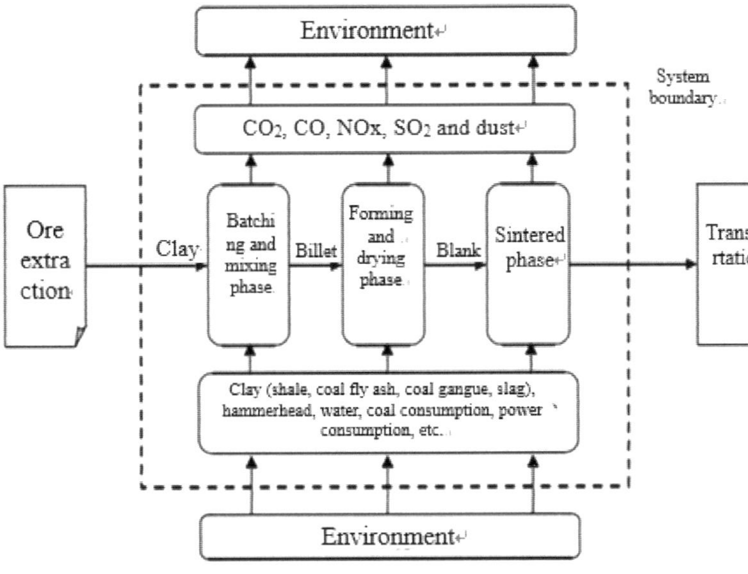

Figure 6. System Boundary of LCA Assessment in the Sintered Brick Manufacturing with Coal Fly Ash

3.2. Inventory Analysis

During the manufacturing process of sintered bricks, the energy and resources are consumed, while all sorts of environmental discharge will also be followed. These variable environmental impacts shall be accounted for. The relevant data in the case study mainly comes from 30 plus varying-scale sintered brick manufacturers located in different regions, such as Inner Mongolia, Sichuan, Guizhou, Guangxi, Hunan, Shandong, Henan, Lanzhou, Tianjin, Jilin and etc. After the data supplement and correction, it can almost reflect the present production level and energy consumption level of China's sintered wall materials. The study period, namely, the data acquisition period was in 2007.

3.2.1. Energy Consumption Analysis. The sintered wall brick industry not only consumes a great amount of natural resources, but also consumes a great amount of fossil energies during its production process, which may have impacts upon China's coal, power and other energy industry. In recent years, due to the introduction of "prohibition of solid bricks and restrictions of clay use" and other acts, China's sintered wall materials enterprises have paid more attention to the recycling of coal fly ash, coal gangue and other calorific wastes, which leads to the reduction of coal and power consumption in 2007. The integrated energy consumption was reduced from the 988kgce/ 1×10^4 standard bricks in 2000 to 900kgce/ 1×10^4 standard bricks in 2005, and then to 812kgce/ 1×10^4 standard bricks in 2007, decreased by 8.9% & 9.7% respectively. Anyhow, due to the great output of China's sintered brick, the absolute amount of energy consumption is still very frightening. As per the statistics, the energy consumption of wall materials industry accounts for about 5% of the national process energy consumption, while the sintered wall materials industry is also the major energy consumer of wall materials. As per the typical survey and integrated measurement, and with reference to the calculation methods in the national resource consumption estimates, the major energy consumption details of sintered brick & tile industry in 2007 is taken as shown in Table 8.

Table 8. Major Energy Consumption Conditions of Sintered Brick Manufacturing with Coal Fly Ash

Energy Consumption Items	Coal (kg/kg Brick)	Electricity (KWh/kg Brick)	Fuel (kg/kg)
Amount	7.61E-03	4.36E-02	9.31E-05

3.2.2. Resource Consumption Analysis. During the entire manufacturing process of sintered bricks, totally 7 major raw materials are consumed, as shown in Table 9.

Table 9. Major Raw Materials Consumption Conditions of Sintered Brick Manufacturing with Coal Fly Ash

Resource Item	Shale	Coal Gangue	Coal Fly Ash	Slag	Hammerhead
Amount (kg/kg)	1.30E+00	2.92E-01	1.49E-01	1.25E-01	2.71E-04

Among which, the hammerhead is one of the steel crushing tool used during the raw materials crushing process, while its damage is the consumption of iron resources in fact. For the hammerhead manufacturing process inventory, please refer to the iron & steel manufacturing process mentioned in the master thesis of ZHOU Hemin, while its resource consumption and discharge inventory is as shown in Table 10.

Table 10. Functional Resource Inventory of Hammerhead Manufacturing

Impact Category	Resource Item	Amount (kg/kg Hammerhead)
Resource Consumption	Dolomite	8.52×10^{-2}
	Fluorite	1.80×10^{-2}
	Steel Scrap	1.72×10^{-1}
	Iron Ore	2.05
	Limestone	1.78×10^{-1}
	Manganese Ore	4.64×10^{-2}
	Silicon Ore	9.90×10^{-4}
Energy Consumption	Coal	2.59
	Fuel Oil	1.25×10^{-3}
Pollutant Emission	CO_2	2.42
	SO_2	3.69×10^{-3}
	NOx	8.63×10^{-4}
	CO	2.05×10^{-4}
	Industrial Dust	5.67×10^{-3}

3.2.3 Pollutant Emission Analysis. The emissions resulted from the enterprise production process mainly *include* the SO_2, HF, NOx, CO_2 and others contained in solid waste dust, smoke dust and flue gas. Due to the variable raw materials & coal quality and equipment level used by enterprises, and the different emission amount, as well as the imperfect surveillance mechanism, there are some sintered brick manufacturers without providing the emission data and the emission amount of their manufacturing wastes is unknown. The emission amount of the relevant wastes is herein estimated with the empirical formula, as shown in Table 11.

Table 11. Pollutant Emission Conditions of Sintered Brick Manufacturing with Coal Fly Ash

Pollutant Emission	CO_2	SO_2	NOx	CO	CH_4	Industrial Dust
Amount (kg/kg)	9.61E-02	3.41E-04	5.43E-05	3.20E-05	1.84E-05	8.38E-04

After combining the resource consumption, energy consumption and pollutant emission of all the aforesaid processes, the life cycle inventory of sintered brick manufacturing process is taken, as shown in Table 12. Among while, the relevant data of shale extraction, power generation process and coal production come from Gu's *Studies on the Environmental Impact of the Building Industry in China*

Based on the Life Cycle Assessment [23] and Di's *Several Fundamental Researches in the Life Cycle Assessment for Mineral Resources and Materials* [24].

Table 12. Input & Output Inventory during the process of Sintered Brick Manufacturing with Coal Fly Ash (kg/kg Brick)

	Sintered Brick Manufacturing	Shale Extraction	Hammerhead Production	Coal Production	Power Generation	In Total
Shale	1.30E+00					1.30E+00
Coal Gangue	2.92E-01					2.92E-01
Coal Fly Ash	1.49E-01					1.49E-01
Slag	1.25E-01					1.25E-01
Dolomite			2.31E-05			2.31E-05
Fluorite			4.88E-06			4.88E-06
Steel Scrap			4.66E-05			4.66E-05
Iron Ore			5.55E-04			5.55E-04
Limestone			4.82E-05			4.82E-05
Manganese Ore			1.26E-05			1.26E-05
Silicon Ore			2.68E-07			2.68E-07
Coal		3.97E-04	7.01E-04	5.63E-03	3.79E-03	1.05E-02
Oil		9.17E-07	3.39E-07	4.47E-06	8.82E-05	9.39E-05
Natural Gas		3.33E-05		2.67E-10	6.59E-05	9.92E-05
CO_2	9.61E-02	3.56E-03	6.55E-04	3.22E-05	7.28E-03	1.08E-01
SO_2	3.41E-04	7.86E-06	9.99E-07	3.88E-08	4.14E-05	3.91E-04
NOx	5.43E-05	1.60E-05	2.34E-07	2.23E-07	4.34E-05	1.14E-04
CO	3.20E-05	5.85E-06	5.55E-08	2.69E-08	1.10E-05	4.89E-05
Dust	8.38E-04	2.56E-06	1.54E-06	4.43E-08	1.81E-05	8.60E-04
Industrial Wastes			3.58E-06	2.56E-04		2.60E-04
CH_4	1.84E-05	4.91E-06		4.54E-05	2.20E-05	9.07E-05
COD			2.59E-09		4.99E-07	5.02E-07

As shown from the life cycle inventory results, the major resources consumption during the manufacturing process of sintered brick with coal fly ash is shale consumption, with the amount of 1.30kg/kg brick, while the major energy consumption is raw coal, crude oil and natural gas, and the major gas emission is CO_2, with the amount of 0.129kg/k brick. If analyzed as per each production phase, the sintered brick manufacturing and power generation process are the major source of pollutants during the whole life cycle. It is noteworthy that the water consumption is not incorporated into this inventory, due to the great variation of relevant data and the extensive application of recycling water or intermediate water in all processes. Therefore, this paper temporarily does not consider the water resources consumption during the life cycle of sintered bricks.

IOP Conf. Series: Materials Science and Engineering **484** (2019) 012038 doi:10.1088/1757-899X/484/1/012038

4. Monetized Life Cycle Impact Assessment (LCIA)

The inventory analysis results only reflect the environmental consumption amount and pollutant emission amount involved in the life cycle of sintered brick. In order to more clearly demonstrate the environmental problems related to the energy & resources consumption and pollutant emission of sintered brick manufacturing, it is required to allocate the inventory results into the variable environmental impact categories and characterize the selected environmental problems, characterize the use category parameters, and also interpret the life cycle inventory results.

4.1. Classification

As the technical framework of the LCA specifies, it is required to incorporate the resources consumption and pollutant emission inventory data into the variable environmental impact categories. This paper adopts the internationally almost generally accepted impact category, category parameters and environmental load category. Meanwhile, it also adopts the equivalent assessment model extensively applied in the LCA at present, and classifies the resources and pollutants involved in the inventory of sintered bricks into the twelve environmental impact categories determined in the aforesaid Willingness-to-Pay model, while the results are as shown in the Table below.

Table 13. Environmental Impact Category at the Life Cycle Phases of Sintered Brick with Coal Fly Ash

Environmental Impact Category	Environmental Load Item	Environmental Impact Category	Environmental Load Item
Global Warming	CO_2,CH_4	Solid Waste	Industrial Solid Waste
Ozone Layer Loss	N/A	Toxicity in Water Body	N/A
Acidification	SO_2,NOx	Suspended Matters in Water Body	N/A
Eutrophication	COD	Water Resources Consumption	Water (not considered temporarily)
Suspended Particles	Industrial Dust	Fossil Resources Consumption	Coal, Oil & Natural Gas
Photochemical Smog	CO	Mineral Resources Consumption	Dolomite, Iron, and etc.

4.2. Characterization

As per the properly classified environmental impact category and selected computational model, the relevant data of life cycle inventory are converted into the corresponding environmental impact indicators, which are known as characterization. Herein, the characterization method adopts the effect equivalent factor mentioned in Yang's *Methodology and Application of Life Cycle Assessment*. Take acidification as an example. The pollutants with the acidification effect in the life cycle of sintered brick are SO_2 & NO_X, while the characterization indicator of acidification is SO_2 equivalent. As shown in Yang's *Methodology and Application of Life Cycle Assessment*, the equivalent indicator of NOx is 0.7kgSO_4eq./kg. Therefore, the overall acidification and characterization results in the life cycle of coal fly ash can be calculated, as shown in Table 14.

Table 14. Acidification Characterization Results at the Life Cycle Phases of Sintered Brick with Coal Fly Ash

Emission	Emission Amount a (kg/kg Brick)	Characterization Factor b ($kgSO_2eq./kg$)	Characterization Results a*b ($kgSO_2eq./kg$ Brick)
SO_2	3.91E-04	1.00E+00	3.91E-04
NOx	1.14E-04	7.00E-01	7.99E-05
In Total			4.71E-04

As per the similar process, the characterization calculation may be made for environmental impact at all phases of life cycle, while the results are as shown in Table 15. Among which, the environmental impact of shale is demonstrated in the form of energy consumption and pollutant emission during its extraction process. The steel scrap, coal gangue, coal fly ash and slag belong to waste recycling. Therefore, their consumption is not incorporated into the scope of environmental impact assessment. Due to the adoption of willingness-to-pay method, the weight of dolomite, fluorite and other minerals may be directly taken from environmental tax. Therefore, during the characterization process, they are not converted into the form of single equivalent indicator.

Table 15. Characterization Results at the Life Cycle Phases of Sintered Brick with Coal Fly Ash (/kg Brick)

Environmental Impact Category	Unit	Sintered Brick Manufacturing	Shale Extraction	Hammer head Production	Coal Production	Power Generation	In Total
Acidification	$kgSO_2eq.$	3.79E-04	1.91E-05	1.16E-06	1.95E-07	7.18E-05	4.71E-04
Eutrophication	$kgNO_3eq.$			5.96E-10		1.15E-07	1.15E-07
Solid Waste	Kg			3.58E-06	2.56E-04		2.60E-04
Photochemical Smog	$kgC_2H_4eq.$	9.60E-07	1.76E-07	1.67E-09	8.07E-10	3.30E-07	1.47E-06
Global Warming	$kgCO_2eq.$	1.02E 01	5.13E-03	6.55E-04	1.46E-02	1.43E-02	1.37E-01
Suspended Particles	Kg	8.38E-04	2.56E-06	1.54E-06	4.43E-08	1.81E-05	8.60E-04
Fossil Resources	kgSCEeq.		3.29E-04	5.01E-04	4.03E-03	2.92E-03	7.78E-03
Mineral Resources	Dolomite (kg)			2.31E-05			2.31E-05
	Fluorite (kg)			4.88E-06			4.88E-06
	Iron Ore (kg)			5.55E-04			5.55E-04
	Limestone (kg)			4.82E-05			4.82E-05
	Manganese Ore (kg)			1.26E-05			1.26E-05
	Silicon Ore (kg)			2.68E-07			2.68E-07

IOP Conf. Series: Materials Science and Engineering **484** (2019) 012038 doi:10.1088/1757-899X/484/1/012038

5. Conclusions

During the characterization process, the environmental load value of all environmental impact categories are taken, but they only represent the absolute total amount. In order to further demonstrate the comparability of variable impact category data, it is required to integrate the characterization results. When the determined weighting coefficient is adopted for normalization calculation, all environmental impact results during the life cycle process of sintered bricks may be taken. When calculated as per China's environmental tax items, the weight of dolomite is RMB 0.003 /kg. When the quartz sand is counted, the weight of silicon ore is RMB 0.003 /kg. The calculation results are as shown in Table 16.

Table 16. Normalization Results at the Life Cycle Phases of Sintered Brick with Coal Fly Ash (RMB/kg Brick)

Environmental Impact Category	Sintered Brick Manufacturing	Shale Extraction	Hammerhead Production	Coal Production	Power Generation	In Total
Acidification	2.69E-04	1.35E-05	8.26E-07	1.38E-07	5.10E-05	3.35E-04
Eutrophication			3.34E-10		6.43E-08	6.46E-08
Solid Waste			2.15E-07	1.54E-05		1.56E-05
Photochemical Smog	3.27E-06	5.98E-07	5.68E-09	2.75E-09	1.13E-06	5.01E-06
Global Warming	6.12E-03	3.08E-04	3.93E-05	8.74E-04	8.59E-04	8.20E-03
Suspended Particles	1.59E-04	4.86E-07	2.93E-07	8.42E-09	3.44E-06	1.63E-04
Fossil Resources		1.25E-06	1.90E-06	1.53E-05	1.11E-05	2.95E-05
Mineral Resources			9.83E-06			9.83E-06
					In Total	8.76E-03

As shown from the normalization results, the environmental impact resulted from the sintered brick manufacturing process is RMB 8.76E-3 /kg, namely, RMB 8.76/ton. Among which, the environmental impact dominated by global warming, namely, CO_2 emissions, followed by acidification and suspended particles. In contrast, due to the adoption of a great amount of industrial wastes as the raw materials, the consumption of mineral resources only accounts for a minor part of the entire environmental impact. This shows that it is the key task of next-step ecological design to control the GHG emissions, especially, reducing the generation of carbon dioxide during the sintered brick manufacturing process.

6. Acknowledgements

The authors disclosed receipt of the following financial support for the research, authorship, and/or publication of this article: This work was financially supported by Chinese National Institute of Standardization Funding Program (grant number 542017Z-5453).

7. References

[1] ISO 14040. Life Cycle Assessment - Principles and Framework, Geneva: International Organization for Standardization, 1997.

[2] General Administration of Quality Supervision, Inspection and Quarantine of the People's Republic

of China, GB/T 24040-1999 Environmental Management - Life Cycle Assessment - Principles and Framework, Beijing: Standards Press of China, 1999.
[3] Standardization Administration of the People's Republic of China, GB/T 24042-2002 Environmental Management--Life Cycle Assessment - Life Cycle Impact Assessment, Beijing: Standards Press of China, 2002.
[4] Standardization Administration of the People's Republic of China, GB/T 24043-2002 Environmental Management - Life Cycle Assessment- Life Cycle Interpretation, Beijing: Standards Press of China, 2002.
[5] GONG Xianzheng, NIE Zuoren, WANG Zhihong, Study on the Database Framework for Typical Materials Life Cycle Assessment, Journal of Wuhan University of Technology, 2004, 26(3): 12-14.
[6] Liu XL, Wang HT, Chen J, et al. (2010) Method and basic model for development of Chinese reference life cycle database. Acta Scientiae Circumstantiae 30: 2136–2144.
[7] DONG Zhiqiang, MA Xiaoqian, ZHANG Ling, et al., Life Cycle Assessment for Influence of Using Natural Gas on Environment, Natural Gas Industry, 2003, 23(6):126-130.
[8] ZOU Zhiping, MA Xiaoqian, ZHAO Zengli, et al., Life Cycle Assessment on the Hydropower Project, Water Power, 2004, 30(4): 53-55.
[9] JIANG Jinliang, MA Xiaoqian, Comparison on Different Power Source Effect on Environment Based on LCA, Power System Engineering, 2004, 20 (3): 26-28.
[10] ZHOU Hemin, Life Cycle Assessment on Iron and Steel Processes [D], Beijing: Beijing University of Technology, 2001.
[11] LIU Jianglong, Environmental Impact Assessment of Materials, Beijing: Science Press, 2002.
[12] LIU Shunni, LIN Zongshou, ZHANG Xiaowei, Studies on the Life Cycle Assessment of Portland Cement, China Environmental Science, 1998,18 (4):328-332.
[13] CHEN Wenjuan, NIE Zuoren, WANG Zhihong, Life Cycle Inventory and Characterization of Flat Glass in China, China Building Materials Science & Technology, 2006,25(3): 54-58.
[14] CHEN Qingwen, MA Xiaoqian, Life Cycle Assessment on the Building Ceramic, China Ceramics, 2008, 44(7): 35-39.
[15] CHEN Hong , HAO Weichang , SHI Feng , et al., Life Cycle Assessment of Several Typical Macromolecular Materials, ACTA SCIENTIAE CIRCUMSTANTIAE, 24(3): 545-549.
[16] GONG Xianzheng, NIE Zuoren, WANG Zhihong, Study on the Database Framework for Typical Materials Life Cycle Assessment, Journal of Wuhan University of Technology, 2004, 26(3): 12-14.
[17] YANG Jianxin, WANG Shoubing, XU Cheng, Allocation Rule in Life-cycle Inventory, China Environmental Science, 1999,19(3):285-288.
[18] MO Hua, ZHANG Tianzhu, Data Quality Assessment of Life Cycle Inventory Analysis, Research of Environmental Sciences, 2003, 16(5): 55-58.
[19] LIU Tao, HUANG Zhijia, Selection for Key Data in Uncertainty Analysis of Life Cycle Inventory, Journal of Anhui University of Technology, 2006, 23(1): 91-95.
[20] YANG Jianxin, XU Cheng, WANG Rusong, Methodology and Application of Life Cycle Assessment, Beijing, China Meteorological Press, 2000.
[21] LI Xiaodong, WU Xing, ZHANG Zhihui, Study on Social WTP for Environmental Impacts based on the LCA Theory, Journal of Harbin Institute of Technology, 2005, 37(11):1507-1510.
[22] GU Daojin, Life Cycle Assessment for China Building Environment Impacts [D], Beijing: Tsinghua University, 2006.
[23] GU Lijing, Studies on the Environmental Impact of the Building Industry in China Based on the Life Cycle Assessment, [D], Beijing: Tsinghua University, 2009.
[24] DI Xianghua, Several Fundamental Researches in the Life Cycle Assessment for Mineral Resources and Materials [D], Beijing: Beijing University of Technology, 2005

Response of the Plant and Soil Features to Degradation Grades in Semi-arid Grassland of the Inner Mongolia, China

Xia Xu *, Honglei Jiang, Xiaoyu Tian, Mengxi Guan and Lingfei Wang

State Key Laboratory of Earth Surface Progress and Resources Ecology, Beijing Normal University, 100875, Beijing, China
School of Natural Resources, Faculty of Geographical Science Beijing Normal University, 100875, Beijing, China
Email: xuxia@bnu.edu.cn

Abstract. As a result of climate change and human activities over the past century, grassland degradation has a profound impact on the carbon storage process. We aimed to research the effects of grassland degradation on plant and soil features and soil carbon emission rate. The carbon content decreased in the leaves but increased in the roots as the degradation gradient increased. The net ecosystem exchange was -8.29 $\mu mol.m^{-2}.s^{-1}$, -3.14 $\mu mol.m^{-2}.s^{-1}$ and -1.57 $\mu mol.m^{-2}.s^{-1}$ in LD (light degradation), MD (moderate degradation) and SD (severe degradation) grassland, respectively. The carbon emission rate (7.22, 6.56 and 4.22 $\mu mol.m^{-2}.s^{-1}$) decreased significantly as the degradation grade increased.

1. Introduction

Grassland vegetation is an important part of the terrestrial ecosystem. Carbon stocks account for approximately 15.2% of the total terrestrial ecosystem (Ajtay, 1979). The grassland ecosystem is widely distributed in China and covers 41% of the land area, with Inner Mongolia occupying 25% of the grassland (Ni, 2002). Grassland ecosystems have profound impacts on soil sinks and sources (Piao et al., 2009), and enhanced ability for carbon and nitrogen sequestration, thereby influencing global environmental change (Wright et al., 2004). However, 90% of grasslands (3.99 billion ha) have been degraded by climate change and human activities over the past several decades (Ren et al, 2007a). Grassland degradation is defined as the process of environmental deterioration and leads to desertification caused by overexploitation, poor management, climatic warming and drought (Zhou et al., 2005; Li et al., 2011). Grassland degradation causes altered composition of the ecosystem and biodiversity, decreased primary productivity, and renewed soil carbon cycling (Schippers &Joenje, 2002; Harris, 2010; Wang et al., 2011). It also affected the dynamics of ecosystem services (Zhang et al., 2015; Zhang et al., 2016). Moreover, degradation can indicate plant species diversity and productivity by changing soil carbon cycling (Wang et al., 2014). Most previous studies on grassland degradation have focused on the vegetation status under enclosure in Inner Mongolia; Grassland degradation also has a crucial influence on soil respiration, which further affects carbon stocks and soil nutrient availability. Previous research has focused on macroscopic monitoring and evaluation to identify problems associated with degradation areas and grades (Tong et al., 2004; Wu et al, 2011). Degenerated grassland is extremely sensitive to the atmosphere and soil drought. Degradation is the main way to damage the ecosystem and alter its structure-process-functions.

We assume that there are sharp variations in carbon content of organs, soil water content in different levels of degradation, and that the carbon emission rate-soil feature (soil temperature and soil water content) relationship may be dramatically altered by changes associated with grassland degradation.

Based on these assumptions, we conducted a study at Xilingol League, Inner Mongolia, China,to research the dynamics change of plant and soil carbon stocks along gradients of grassland degradation and examine the effects of grassland degradation on carbon content of organs, soil water content and carbon emission rate. Grassland degradation is a complex process that integrates various aspects. Therefore our research focused on the essential traits of plants and soils to provide comprehensive assessment of grassland degradation and potential restoration methods.

2. Material and Methods

2.1. Study Sites and Experimental Design

According to plant coverage conditions(Wang et al, 1996a; Wang et al, 1996b),The grassland are classified into three groups: light degradation (LD) grassland with plant coverage >50%; moderate degradation (MD) grassland with plant coverage between 30%-50%; and severe degradation (SD) grassland with plant coverage <30%.

The field experiment on soil and vegetation were conducted from July to August 2015. In the research area, 74 sampling plots (100 m × 100 m) were randomly chosen every 20 km along a major road (Fig 1). Three typical sampling quadrats (1 m × 1 m) were placed to sample the vegetation according to plant coverage in each plot. The quantitative features included plant coverage and height, above-ground biomass, species composition, numbers and density, soil bulk density and soil water content. According to the division of the degree of degradation, 13 sampling plots were divided into LD grassland, 37 into MD grassland, and 24 into SD grassland of the 74 observed points.

2.2. Respiration Rate Measurements

The respiration rate was also measured in the sample plot with photosynthetic rate determined. The sample plots were classified into three types: bare land (Soil respiration rate), grass-cutting (Soil respiration rate plus root respiration rate) land and grass-retained land (Net carbon emission rate). An automated soil CO_2 efflux measurement system (model LI-8150 multiplexer, LI-COR, Nebraska, USA) was used to measure the respiration rate. Two permanent PVC collars were installed in each type of land between July and August 2015. The respiration rate was continuously measured during 8:00-12:30 each day. Soil temperature and soil water content were also measured at a 10 cm depth outside of each chamber using the 8150-203 soil temperature sensor and soil moisture sensor, respectively.

3. Results

3.1. Response of Vegetation Features to Different Degeneration Grade

3.1.1. The Organ Carbon Content of Different Vegetative Organs. The organ carbon content in the root increased and the leaf organ carbon content and the stem organ carbon content decreased with grassland degradation gradient. The root carbon content was 375.7 mg/g, 395.6 mg/g and 400.6 mg/g in LD, MD and SD grassland, respectively. The leaf carbon content was 454 mg/g, 448 mg/g and 436mg/g in LD, MD and SD grassland, respectively. The stem carbon content was 442.4 mg/g, 435.5 mg/g and 430.8 mg/g in LD, MD and SD grassland, respectively. The root organ carbon content was significantly lower than that in the stem and leaf, and the root organ carbon content have the maximum range of change in the degradation gradient.

3.1.2. The Carbon Sequestration Ability in the Degeneration Grade. Gross ecosystem productivity decreased from 15.51 $\mu mol.m^{-2}.s^{-1}$ in LD grassland to 9.7 $\mu mol.m^{-2}.s^{-1}$ in MD grassland and to 5.8 $\mu mol.m^{-2}.s^{-1}$ in SD grassland. The differences among the three levels of degeneration grasslands were significant ($p<0.05$). The marked decrease in ecosystem productivity in SD grassland showed that ecosystem productivity is seriously influenced by severe degradation. The net ecosystem exchange was -8.29 $\mu mol.m^{-2}.s^{-1}$, -3.14 $\mu mol.m^{-2}.s^{-1}$ and -1.57$\mu mol.m^{-2}.s^{-1}$ in LD, MD and SD grassland, respectively, It suggesting that the carbon sequestration ability of the ecosystem decreases with the degree of degradation. The plant respiration rate decrease also decreased significantly with a higher level of degradation. The respiration rate was 7.22μmol.m^{-2}.s^{-1}, 6.56 μmol.m^{-2}.s^{-1}and 4.22 μmol.m^{-2}.s^{-1} in LD, MD and SD grassland, respectively. The difference in respiration rate between LD and SD grasslands was significant (Fig.1).

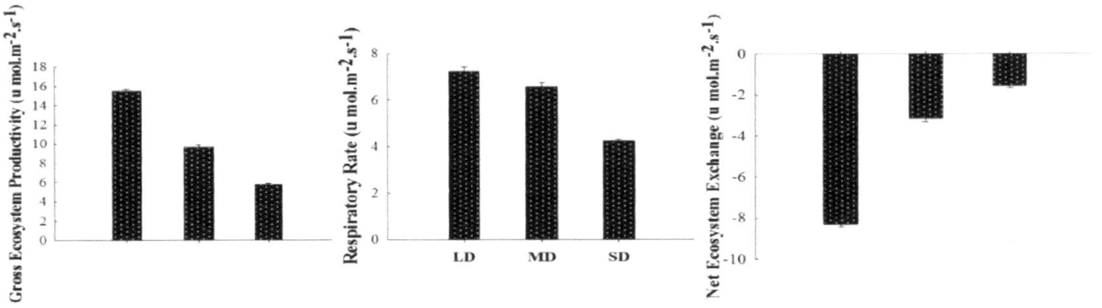

Figure 1. CO_2 exchange rate for different degeneration grade steppes

3.2. Response of Soil Features to Degeneration Grade

3.2.1. Soil Water Content. Overall, the soil water content in SD grassland was dramatically lower than in the LD or MD grassland of each soil layer (Fig. 2). At the 0-10 cm depth, the soil water content decreased from 13.1% in LD grassland to 9.1% in MD grassland and to 6.3% in SD grassland, with the difference between that in LD grassland being significantly higher than that in SD grassland. These results indicate that the surface soil water content decreased significantly as a result of severe degradation. In the second to fifth soil layers, the soil water content did not differ significantly between the different levels of degradation, suggesting that degradation does not affect the subsoil water content.

The soil water content decreased with soil depth in LD and MD grasslands. In LD grassland, the soil water content in the first layer was significantly higher than in the second to fourth layers. The soil water content in the first layer was the highest. There were no significant differences between the five layers in MD grassland. In SD grassland, the soil water content did not exhibit a decreasing trend. Overall, the surface soil had more water content than the subsoil, and have the maximum range of change in the degradation gradient.

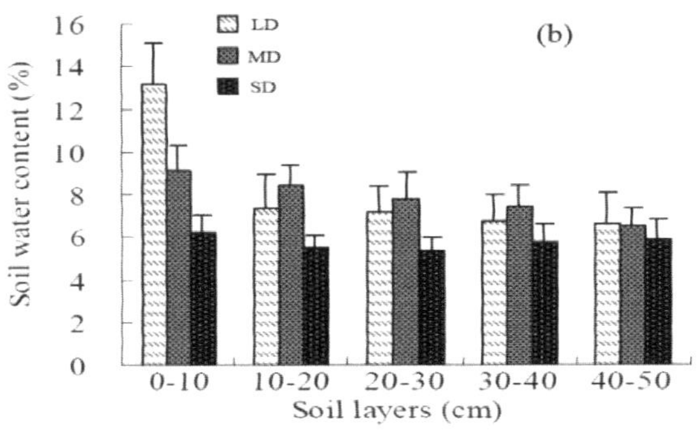

Figure 2. Soil water content in five layers of different degradation grades

*3.2.2. Soil Organic Carbon Content.*In the five soil layers, the soil organic carbon content decreased dramatically with degradation gradient, especially in LD grassland, which was significantly higher than in MD and SD grasslands (Fig. 3). the soil organic carbon content in MD grassland differed significantly with SD grassland at the 0-10 cm depth. These results suggest that the surface soil organic content differs significantly in different degradation levels.

At the same degradation level, the soil organic carbon decreased with soil depth. In LD grassland, the soil organic carbon exhibited a fast decrease, (i.e., 2.08 times from 18.04 mg/g at the 0-10 cm depth to 8.65 mg/g at the 40-50 cm depth). The soil organic carbon content at the 0-10 cm depth wad significantly higher than at the 20-30 cm, 30-40 cm, and 40-50 cm depths ($p<0.05$). In MD grassland, the soil organic carbon was 2.95 times higher in the first layer than in the fifth layer, and was significantly higher than the other four depths ($p<0.05$). In SD grassland, the soil organic carbon content decreased 1.95 times (i.e., from 11.53 mg/g at the 0-10 cm depth to 5.85 mg/g at the 40-50 cm depth).

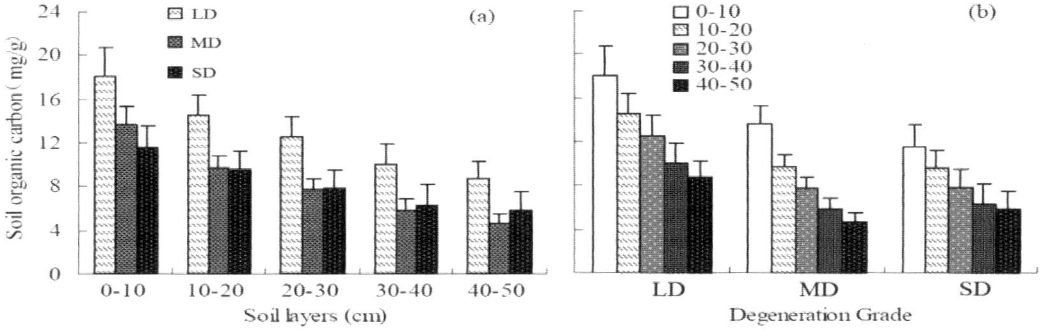

Figure 3. Soil organic carbon content in different degradation grades

3.2.3. Soil Total Carbon Content. The soil total carbon content decreased significantly with degradation gradient at the 0-10 cm depth, which decreased from 26.03 mg/g in LD grassland to 17.66 mg/g in MD grassland and to 13.08 mg/g in SD grassland (Fig. 4). At the 10-20 cm depth, the soil total carbon content was highest in MD grassland and was significantly higher than in SD grassland ($p<0.05$). At the 20-30 cm depth, the soil total carbon content was 13.42 mg/g, 16.01 mg/g, 13.94 mg/g in LD, MD and SD grassland, respectively, although the differences were not significant. In the fourth to fifth layers, the soil respiration rates and soil total carbon content were significantly higher in SD grassland than in LD and MD grasslands ($p<0.05$). In LD grassland, the soil total carbon content decreased significantly with soil depth which in the first layer was 2.5 times higher than in the fifth layer and was significantly higher than the other four depths in LD grassland ($p<0.05$). In MD grassland, the soil carbon content showed a decreasing trend with soil depth, which decreased from 17.66 mg/g at the 0-10 cm depth to 10.71 mg/g at the 40-50 cm depth. In SD grassland, the soil total carbon content exhibited a clear increasing trend with soil depth, which was significantly higher in the fourth layer than in the second layer ($p<0.05$).

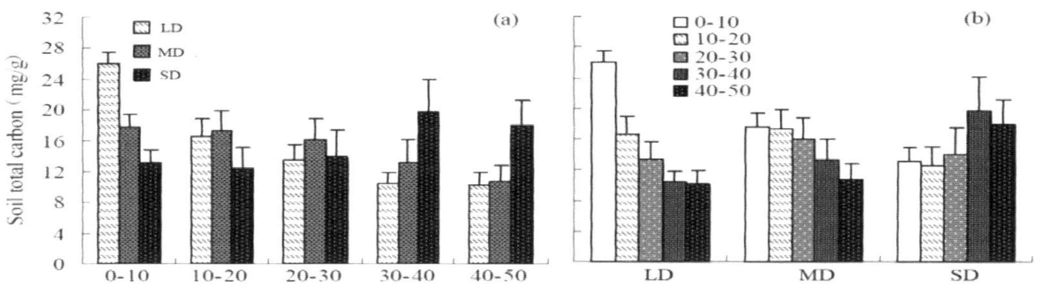

Figure 4. Soil total carbon content in different degeneration grades

3.3. The Carbon Emissions from Different Components of the Ecosystem

In LD grassland, the soil respiration rates showed clear increasing trend from 8: 00 -11:00, with the maximum soil respiration rate (11.88 $\mu mol.m^{-2}.s^{-1}$) occurring at 11:00 in grass-retained land. In MD and SD grasslands, the maximum soil respiration rate reached 7.79 $\mu mol.m^{-2}.s^{-1}$ and 5.91 $\mu mol.m^{-2}.s^{-1}$ at

12:00, respectively, In MD grassland, the maximum soil respiration rate reached 8.32 μmol.m^{-2}.s^{-1} in grass-retained land at half past eleven, and reached 5.28 μmol.m^{-2}.s^{-1} and 2.78 μmol.m^{-2}.s^{-1} in grass-cutting and bare land at twelve o'clock, respectively. In SD grassland, the soil respiration rate exhibited the same trend as LD and MD grasslands, which reached 1.11 μmol.m^{-2}.s^{-1}, 2.95 μmol.m^{-2}.s^{-1} and 3.42μmol.m^{-2}.s^{-1} in bare land, grass-cutting land and grass-retained land, respectively. The soil respiration rate in the grass-retained land was similar to that in the grass-cutting land, it suggesting that the plant respiration rate rarely affects grass-retained land of SD grassland. Overall, the soil respiration rate in grass-retained land was significantly higher than in grass-cutting and bare land.

In LD grassland, the soil respiration rate decreased significantly from 9.72 μmol.m^{-2}.s^{-1} in grass-retained land to 6.27 μmol.m^{-2}.s^{-1} in grass-cutting land and to 4.89 μmol.m^{-2}.s^{-1} in bare land (p<0.05). The same trend was observed in MD and SD grasslands. The soil respiration rate decreased significantly with degradation grade in the different land cover types. In bare land, the soil respiration rate in LD grassland was 5 times higher than in MD grassland. In the grass-cutting and grass-retained land, the soil respiration rate was 3 times higher in LD grassland than in MD grassland, suggesting that vegetation roots play an important role in the soil respiration rate in LD grassland.

4. Discussion

4.1. Changes in Carbon Content

It is well known that grassland degradation can easily cause decreased soil fertility (Wang et al., 2014). In this study, the soil organic carbon content was decreased significantly with increasing grassland degradation and the total soil carbon content also decreased at the 0-20 cm depth with degradation grade. The soil organic carbon content was extremely constrained by the vegetation conditions; the decrease in biomass with degradation grade immediately caused decrease in soil organic carbon content. Overall, the soil bulk density increased significantly with degradation gradient and the soil water content was significantly lower in SD grassland than in LD and MD grasslands. Moreover, the abrupt changes in these values between LD grassland and SD grasslands show that severe degradation has a profound effects on soil fertility and soil bulk density. Many studies have researched the effects of soil fertility on plant productivity (Clark&Tilman, 2008; De Deyn et al., 2009). The nutrient supply is a major limiting factor for community productivity (Reynolds&Haubensak, 2009). In the present study, community productivity was higher in LD grassland, which can be attributed to the sufficient fertility supply. The fraction of root biomass increased significantly with the grassland degradation gradient, suggesting that the allocation of resources changed from aboveground to underground. Community plant respiration decreased with degradation grade, which may be explained by the following two reasons: first, the community plant respiration may actually include the soil respiration, and grassland degradation caused the decline of root distribution in the soil, resulting in the decline of soil respiration; second, grassland degradation caused decreased plant coverage, which directly decreased plant respiration. The carbon sequestration ability of the *Stipakrylovii* community in SD grassland is higher than the *Stipa grandis* community in LD grassland (Wang et al, 2012).Soil respiration decreased significantly with degradation grade, indicating decreased carbon emission ability. These results correlate with the vegetation carbon absorption ability. Moreover, the difference in vegetation carbon absorption ability caused variations in root, stem and leaf carbon contents.

4.2. Relations between Carbon Emissions And Soil Water Content And Soil Temperature

Quadratic models were used to quantify the relationship between soil respiration rates and soil water content in each surficial vegetation from the three levels of degradation. The R^2 of the models were higher than 0.6 and were significant at the 0.01 level. All second-order item coefficients of the quadratic curve equations were negative, indicating a soil water content threshold; the soil respiration rate increased with soil water content before the threshold value and decreased thereafter. The threshold differed slightly under various conditions. Each degradation grade displayed a linear relationship between soil respiration rate and soil temperature at 0-10 cm. The soil respiration rate exhibited a significant positive correlation with soil temperature at 0-10 cm, with R^2 values ranging from 0.75 to

0.95. All regressions were significant at the 0.01 level. In LD and MD grasslands, the largest slope of regression was observed in grass-retained land. In SD grassland, the slope of regression was significantly lower in bare land than in grass-cutting and grass-retained land.

Soil respiration was controlled mainly by soil temperature in temporal variations and soil water content and soil organic carbon in spatial variations (Fan et al., 2015). Many studies have shown that significant exponential relationship between soil respiration and temperature (Zhang et al., 2010; Yu et al., 2011; Zhou et al., 2013). The effects of soil moisture on CO_2 efflux from soils have been described by numerous equations, including: linear, logarithmic, quadratic and parabolic functions (Davidson et al., 2000). The soil respiration quadratic is related to soil water content and exhibits a significant positive correlation with soil temperature in different degradation grades. Soil water content also affects soil respiration, mostly by altering the soil structure to change soil permeability, water circulation, soluble organic matter and effects the soil microbial activity (Moyano et al., 2013; Zhou et al., 2014). The soil respiration rate was mostly related to surface soil moisture in the growing season (Qi et al., 2010).Temperature also influenced soil respiration, and soil water content enhanced the response of respiration to temperature (Zhang et al., 2010). Wan et al (2005) found that an increase in temperature enhanced the soil respiration rate by promoting above-ground biomass, demonstrating a significant positive correlation between soil respiration rate and above-ground biomass. Moreover, an increase in temperature accelerated the decomposition rate of soil and litter fall, which increased CO_2 efflux (Luo et al., 2010). Degradation reduced plant-available water and decreased grassland productivity, which are associated with changes in soil structure and soil moisture regime (Zhao et al., 2011). Vegetation type can also affects soil respiration by controlling the root biomass and distribution of photo assimilates (Han et al., 2007; Bahn et al., 2010). The soil bulk density was the highest and soil water content was the lowest in SD grassland; therefore, soil respiration was the lowest in SD grassland.

5. Conclusions

Grassland degradation directly caused significant changes in the productivity, organ carbon content of different vegetative organs, soil bulk density and soil water content. The grassland community structure and plant characteristics changed significantly with degradation grade, which also affected the plant carbon content. Plant biomass dramatically affected the soil organic content. Ecosystem productivity was seriously affected by severe degradation. The relationship between carbon emission and soil features was influenced by degradation grade. The carbon sequestration ability of the ecosystem decreased with degradation grade.

6. Acknowledgement

This research was supported by the National Key Research and Development Program of China(2017YFA0604902,2017ZX07301-001-03)and The Foundation for Innovative Research Groups of the National Natural Science Foundation of China(41621061), State Key Laboratory of Earth Surface Processes and Resource Ecology.

7. References

[1] Ni J (2002). Carbon storage in grasslands of china. Journal of Arid Environments 50 (2): 205-218.
[2] Piao S, Fang J, Ciais P, Peylin P, Huang Y, Sitch S ,Wang T. The carbon balance of terrestrial ecosystems in china. Nature (2009).458 (7241): 1009-1013.
[3] Wright AL, Hons FM, Rouquette FM. Long-term management impacts on soil carbon and nitrogen dynamics of grazed bermudagrass pastures. Soil Biology and Biochemistry (2004). 36 (11): 1809-1816.
[4] Ren H, Du WB, Wang J, Yu ZY, Guo QF Natural restoration of degraded rangeland ecosystem in Heshan hilly land. Acta Ecol Sin (2007).27:3593–3600
[5] Li, G. Y., Liu, Y. Z., Frelich, L. E., & Sun, S. C. .Experimental warming induces degradation of a Tibetan alpine meadow through trophic interactions. Journal of Applied Ecology, (2011).48, 659–667.

[6] Wang, C. T., Wang, G. X., Liu, W., &Wu, P. F.. The effects of plant–soil–enzyme interactions on plant composition biomass and diversity of alpine meadows in the Qinghai–Tibetan Plateau. International Journal of Ecology, (2011a).41, 1–10.
[7] Zhang X, Niu J, Buyantuev A, Zhang Q, Dong J, Kang S, Zhang J. Undesoil respiration ratestanding grassland degradation and restoration from the pesoil respiration ratespective of ecosystem services: a case study of the xilin river basin in Inner Mongolia, china. Sustainability (2016). 8 (7): 594.
[8] Zhang Y, Huang D, Badgery WB, Kemp DR, Chen W, Wang X, Liu N. Reduced grazing pressure delivesoil respiration rates production and environmental benefits for the typical steppe of north china. Scientific Reports (2015). 5: 16434.
[9] Wang X, Dong S, Yang B, Li Y, Su X. The effects of grassland degradation on plant divesoil respiration ratesity, primary productivity, and soil fertility in the alpine region of asia's headwatesoil respiration rates. Environmental Monitoring and Assessment (2014). 186 (10): 6903-6917.

Study on Extreme Microbial Degradation of Petroleum Hydrocarbons

Hong Li, Xiaokang Li, Tao Yu, Feifei Wang and Chengtun Qu*

Shaanxi Oil and Gas Pollution Control and Reservoir Protection Key Laboratory, Xi'an Shiyou University, 710065, Xi'an
E-mail: xianquct@163.com

Abstract. Some oil-contaminated areas are often accompanied by extreme environments including high salinity, low temperature, and high temperature. The petroleum-degrading microorganisms isolated from these extreme environments can quickly adapt to extreme environments, grow and metabolize in extreme environments, and show great potential in the field of oil pollution control. This article reviews species, adaptation mechanism and research progress of oil-degrading bacteria in three extreme environments of hypersaline, low temperature and high temperature, so as to provide theoretical basis for further research and application.

1. Introduction

Oil as one of the most important energy in the industry, the product has been extensively used in life. The huge demand for oil has led to a growing range of energy exploitation and oil spills occur frequently simultaneously. Large quantities of petroleum pollutants enter the environment causing serious degradation of land and water ecosystems. Especially in recent years, the Polar Regions, the salt lakes and the marine areas are often contaminated with high levels of petroleum hydrocarbons. These systems have considerable economic, ecological and scientific value.

Bioremediation technology utilizes microorganisms to degrade toxic pollutants to harmless products such as CO_2, H2O and other inorganic compounds [1]. Microbial degradation of petroleum hydrocarbon pollutants has long been considered as an environmentally friendly, cost-effective and efficient biological treatment. There are different kinds of microorganisms to degrade petroleum hydrocarbon, and their physiological characteristics such as salt resistance, temperature resistance and saline-alkali resistance are also different. The conventional microorganisms have significantly less metabolic capacity and less degradation, and are unable to function in extreme environments. Extremophiles are adapted to grow and thrive under these adverse conditions. Hydrocarbon degrading extremophiles are thus ideal candidates for the biological treatment of polluted extreme habitats and scientific value. It plays an important role in the control of oil pollution in extreme environments.

2. Microbial Degradation of Petroleum Hydrocarbons in Hypersaline Environments

High salinity concentration has a great impact on the physiological characteristics and life activities of microorganisms, including disruption cell membrane, denaturation of enzyme, and reduction of solubility of oxygen and hydrocarbons. Therefore, bioremediation of saline environments without costly dilution of salt-laden soil and water requires halophilic or halotolerant organisms that tolerate high salt concentrations.

2.1. Species of Halophilic Microorganism

Microorganisms that need salt to grow are called halophiles. Halophiles are classified into three groups according to their optimal salt concentration for growth: slightly halophilic, moderately halophilic, and extremely halophilic. The slightly halophiles are mostly marine microorganisms, and the optimum salt concentration is 1% to 3%(w/v); The moderately halophiles are mainly fungi, and the optimum salt concentration is 3% to 15%(w/v); The extremely halophiles are basically archaea and the optimum salt concentration is 15% to 32%(w/v) [2].

2.2. Adaptations of Halophilic Microorganism to Their Environments [3, 4]

2.2.1. Accumulation of inorganic salt ions. Halophile accumulates $K^+$and Cl^- in its cells to maintain its osmotic balance with the environment. Extreme halophiles of the archaeal Halobacteriaceae family such as H. salinarum, Haloarcula marismortui, Halococcus morrhuae maintain their osmotic balance by concentrating K^+inside cells.This is achieved by the concerted action of the membrane-bound proton-pump bacteriorhodopsin, ATP synthase, and the Na^+/H^+ antiporter and results in an electrical potential that drives the uptake of K^+ into cells via a K^+-uniport mechanism.

2.2.2. Accumulation of compatible organic solutes. The water potential inside the cell is higher than outside in hypersaline environment. In order to avoid cell dehydration due to water outflow from the cell, low compatible molecular organics are synthesized by the halophile, e.g., glutamine, glycerol, betaines, proline, and glutamate. In addition, the sugars sucrose and trehalose are osmolytes vital to stabilizing microbial membranes.They protect microbial proteins from denaturation in saline environments and maintain cell structure stability.

2.2.3. Changes in protein structure. The halophile protein tissue has a unique adaptability to the high salt environment and promotes folding of the riparophilic peptide chain to form a functional protein, ensuring proper functioning of the enzyme activity.

2.3. Microbial Degradation of Petroleum Hydrocarbons in Hypersaline Environments

In the last two decades there has been impressive progress in the area of hydrocarbon degradation in hypersaline environments. Li et al.[5] have isolated a *Planococcus* capable of degrading BTEX in the presence of 0.5%~25% NaCl from oil-contaminated soil. Wang et al.[6] confirmed that the halophile group had no significant effect on the removal of phenanthrene in the presence of 0.1% NaCl, while the phenanthrene could be completely degraded within 5 days in the presence of salt concentrations 5%. Khemili-Talbi et al.[7] have isolated an extremely halophilic archaea from highly mineralized wastewater. It was found that the halophilic archaea had the potential of degrading phenol, pyrene and naphthalene at a concentration of 25% of NaCl and produced biosurfactant.

3. Microbial Degradation of Petroleum Hydrocarbons in Low Temperature Environments

In the past, most studies on oil-degrading bacteria have focused on normal temperature environment. In recent years, it has been reported that petroleum hydrocarbon components can also be degraded by indigenous microorganisms in various land and marine ecosystems, including arctic, alpine, Antarctic soil and sediment. Such microorganisms, which are normally living in low temperature environments, are known as cold-adapted microorganisms. They are able to rapidly adapt to the pollution environment and therefore, cold-adapted microorganisms are becoming increasingly concerned.

3.1. Species of Cold-adapted Microorganism

Cold-adapted microorganism is classified into two groups according to the growth temperature: psychrophilies and psychrotrophs. Both of these microorganisms are able to grow at temperatures around 0°C. Psychrophiles have an optimum growth temperature below 15°C and do not grow above 20°C, whereas psychrotrophs (cold-tolerant) have optimum and maximum growth temperatures above 15°Cand 20°C. At present, there are known cold-adapted microorganisms including bacteria, cyanobacteria, fungi and psychrophilic archaea[8].

3.2. Adaptations of Cold-adapted Microorganisms to Their Environments

At extreme conditions, low-temperature microorganisms develop a series of unique vital signs adapted to the environment. The unsaturated fatty acids and short-chain fatty acids increased and the cell membrane fluidity was enhanced under low temperature. The cold-adapted microorganism is capable of producing degradation enzymes with specific environmental adaptation mechanisms at low temperatures. These enzyme molecules contain enzyme genes capable of being expressed at low temperatures, so that the cryogenic enzyme molecule is more flexible, reduces the activation energy of the petroleum hydrocarbon degradation reaction, and improves the catalytic activity of the enzyme.

Some cold-adapted microorganisms have antifreeze proteins and cold protective substances, among which antifreeze proteins creat a thermal lag by binding with ice crystals on the surface of the proteins, reducing the freezing point of body fluids or cellular water, so that the cold-adapted microorganisms can survive at low temperatures; Cold protective substances can prevent protein denaturation and polymerization, improve the physicochemical reaction environment in the cells of microorganisms, make nutrients stay and concentrate in the cells, and prevent extracellular enzymes from cold denaturation and inactivation.

The cold adaptation capacity of psychrophilic microorganisms is also related to their genetic material. On the one hand, substances such as ribosome extract and RNA polymerase in the psychrophilic microorganisms have good activity at low temperature due to their unique physical and chemical properties, which can maintain protein folding rate.Various enzymes can be normally expressed at low temperature. On the other hand, the cold-adapted microorganism will produce a large amount of cryoprotein when it is impacted by low temperature. In cell responses such as transcription, translation and protein folding of genetic material, low temperature makes cryoprotein express rapidly. These unique physical and chemical properties enable the microorganism to decompose normally at low temperature [9, 10]

3.3. Microbial Degradation of Petroleum Hydrocarbons in Low Temperature Environments

The results have shown that the oil degradation rate can be improved by optimizing the environmental conditions, such as temperature and PH. In addition, biostimulation and bioaugmentation of contaminated cold sites has been tested as a bioremediation strategy. Biostimulation means to stimulate the activity of enzymes in microorganisms by adding different levels of nutrient elements and electronic receptors, such as O_2 and H_2O_2, and the degradation rate of microorganisms to petroleum hydrocarbon pollutants is improved. Romina et al.[11] have studied chronically contaminated Antarctic soil at Carlini Station. Total hydrocarbon removal (71 %) was detected after 30 days when adding fish meal. Paudyn et al.[12] increased oxygen content through soil tillage in the Arctic diesel contaminated soil , and it was found the removal rate of petroleum hydrocarbon pollutants reached above 80% after three years. Bioaugmentation refers to the enhancement of microbial degradation of petroleum contaminants by inoculating low temperature petroleum hydrocarbon degrading bacteria and changing the original quantity and composition of the bacteria. Mohn et al.[13] confirmed over 80 percent of petroleum hydrocarbons were degraded by the addition of cold-adapted microorganisms in soil of polar cold regions remediation tests.

4. Microbial Degradation of Petroleum Hydrocarbons in High Temperature Environments

In 1965, Thomas D. Brock has isolated an extremely thermophilic bacterium in hot spring water at 80°Cfor the first time. Since then, it has been found that some microorganisms can often survive in hot environments, such as deserts and volcanoes. Such microorganisms are called thermophilic microorganisms.

4.1. Species of Thermophilic Microorganism

The optimal growth temperature of thermophilic microorganisms can reach above 40°C, and the maximum growth temperature of most thermophiles ranges between 50°Cand 70°C. Optimal growth of extreme thermophiles and hyperthermophiles is 70–80°C and above 80°C, respectively. At present, the known thermophiles are basically eubacteria and archaea [14].

4.2. Adaptations of Thermophilic Microorganism to Their Environments

The adaptation mechanism of thermophilic bacteria in high-temperature environment is related to various factors, including special cell membrane structure, thermophilic protein, thermophilic enzyme and its genetic material. Firstly, there's a lot of structurally specific complex fats in the membrane, which can make the cell membrane more stable at high temperature. Secondly, thermophilic microorganisms can produce some thermophilic proteins with stable structure in high temperature environment. Thirdly, the chemical modification of certain metal ions in the thermophiles may also enhance the thermal stability of the protein. In addition, thermophilic bacteria have unique thermophilic enzyme, which contains special amino acid, hydrogen bond, ion pair or disulfide bond, which improves thermophilic enzyme thermal stability. Finally, most thermophilic microorganisms have significantly higher G and C bases contents in DNA double helix structure or tRNA than conventional microorganisms. More hydrogen bonds are produced to maintain the stability of the structure. It is the combination of these factors that makes the thermophilic microorganisms have high thermal stability [15, 16].

4.3. Microbial Degradation of Petroleum Hydrocarbons in High Temperature Environments

With the increase in demand for petroleum energy, the problem of oil pollution comes as the scope of oil development expands to high-temperature areas. The most known oil-degrading bacteria are medium temperature microorganisms, which have poor degradation effect on petroleum hydrocarbon under the high temperature environment. Therefore, thermophilic microorganisms are of great importance to the treatment of oil pollution in high temperature environments. Jiang et al.[17] have isolated six thermophilic oil-degrading strains from the oil-contaminated soil at Daqing Oilfield, China, including the Bacillus licheniformis WY2. The study found that the strain WY2 optimal growth in the presence of 0.5% v/v crude oil and pH 7 was observed at 52-80°Cand the content of total petroleum hydrocarbons was reduced by 52%. Wang et al.[18] have reported the isolation thermophilic oil-degrading bacteria group capable of degrading saturated hydrocarbon, aromatic hydrocarbon components, asphaltene and asphaltene as the source of carbon at temperature 45°Cfrom the oil-contaminated soil at Daqing Oilfield . Audrius et al.[19] have isolated a number of thermophilic bacteria capable of utilizing naphthalene as a sole source of carbon from a high-temperature oilfield in Lithuania. These isolates were able to utilize several other aromatic compounds, such as anthracene, benzene, phenol, benzene-1, 3-diol, protocatechuic acid as well.

5. Conclusions

So far, extremophiles capable of degrading petroleum hydrocarbons have been found in oil-contaminated areas under high pressure, high temperature, low temperature or high salt. The use of extremophiles to treat petroleum hydrocarbon pollution has been taken seriously by researchers in the world. However, the degradation of petroleum hydrocarbon by these extremophiles is still limited to its degradation characteristics and related degradation enzymes. The mechanism of these microorganisms to adapt to the extreme environment and the specific ways to degrade petroleum hydrocarbons still need to be further researched. The great potential to exploit microbes in the field of oil pollution and how to apply them efficiently to the site is still a key issue that needs to be studied further in the future.

6. Reference

[1] Philp J, Bamforth S, Singleton I, et al. Environmental Pollution and Restoration: A role for Bioremediation. Bioremediation: Applied Microbial Solutions for Real-World Environmental Cleanup. 2005.

[2] Fathepure B Z. Recent studies in microbial degradation of petroleum hydrocarbons in hypersaline environments. Frontiers in Microbiology, 2014, 5(2):173.

[3] Mao X, Guo J F, Functional protein research progression in extremophiles. Chinese Bulletin of Life Sciences. 2018, 30(01), 107-112

[4] Kong F J, Wang X J, Halophilic microoganisms in salt lakes on Tibet Plateau and their potential application. Science & Technology Review. 2017, 35(12), 27-3.
[5] Li H, Liu Y H, Luo N, et al. Biodegradation of benzene and its derivatives by a psychrotolerant and moderately haloalkaliphilic Planococcus sp. strain ZD22. Research in Microbiology, 2006, 157(7): 629-636.
[6] Wang C, Huang Y, Zhang Z, et al. Salinity effect on the metabolic pathway and microbial function in phenanthrene degradation by a halophilic consortium. AMB Express, 2018, 8: 1-13.
[7] Khemili-Talbi S, Kebbouche-Gana S, Akmoussi-Toumi S, et al. Isolation of an extremely halophilic arhaeon Natrialba sp. C21 able to degrade aromatic compounds and to produce stable biosurfactant at high salinity. Extremophiles Life Under Extreme Conditions, 2015, 19(6):1109-20.
[8] R Margesin, F Schinner. Biodegradation and bioremediation of hydrocarbons in extreme environments. Appl Microbiol Biotechnol , 2001 , 56 (5-6) :650-663.
[9] Phadtare S. Recent developments in bacterial cold-shock response. Current Issues in Molecular Biology, 2004, 6(2): 125-136.
[10] Xin M, Zhou P. Advance of research for microbial life in low temperature environments. Acta Microbiologica Sinica, 1998, 38(5): 400-403.
[11] Dias R L, Ruberto L, Calabró A, et al. Hydrocarbon removal and bacterial community structure in on-site biostimulated biopile systems designed for bioremediation of diesel-contaminated Antarctic soil. Polar Biology, 2015, 38(5): 677-687.
[12] Paudyn K, Rowe K, et al. Remediation of hydrocarbon contaminated soils in the Canadian Arctic by landfarming. Cold Regions Science and Technology, 2008, 53:102-114.
[13] Mohn WW, Radziminski CZ, Fortin MC, et al. On site bioremediation of hydrocarbon-contaminated Arctic tundra soil in inoculated biopiles. Applied Microbiology and Biotechnology,2001,57:242-247.
[14] Stetter KO, Hyperthermophiles: isolation, classification, and properties. In: Horikoshi K, Grant WD (eds) Extremophiles: microbial life in extreme environments. Wiley-Liss,New York, 1998: 1-24.
[15] Shen J S, Shen B, Thermostability mechanism of thermophilic enzyme.Journal of Microbiology. 2010, 2(30): 80-85.
[16] Li H Y, Shao J Y, The thermophilic mechanism of thermophilic baeteria. Chinese Bulletin of Life Sciences. 2000, 12(1):30-33.
[17] Jiang Y, Ma H X, et al. The screening and capability of heat resistant oil-degrading bacteria. Chinese Journal of Microecology. 2016, 28(6):674-677.
[18] Wang Y, Study on Composting Remediation of Oil Contaminated Soil in Daqing Oil Field and Structure of the Microbial Community. Northeast Agricultural University.
[19] Audrius B , Gražina G, et al. Degradation of naphthalene by thermophilic bacteria via a pathway, through protocatechuic acid. Central European Journal of Biology. 2008, 3(1):61-68.

IOP Conf. Series: Materials Science and Engineering **484** (2019) 012041 doi:10.1088/1757-899X/484/1/012041

Synthesis and Degradation Properties of Heat-resistant Polymer based on Cyclotriphosphazene

Yani Guo[1*], Liyun Liang[2] and Chengmei Liu[2]

[1] School of Material Science and Engineering, Wuhan Institute of Technology, Wuhan 430073, P R China
[2] School of Chemistry and Chemical Engineering, Huazhong University of Science and Technology, Wuhan 430074, P R China
E-mail: guoyani@wit.edu.cn

Abstracts. The thermal performance and the decomposition mechanism of the cyclotriphosphazene based homopolymer were investigated by means of TGA, DTG and integrated TG/FTIR analysis. The thermal decomposition of the homopolymer between 100 and 800 °C includes two stages; the initial degradation temperature of the first stage is about 377 °C with the 292 kJ/mol of average decomposition activation energy. Accordingly, the possible thermal decomposition mechanism of the homopolymer was proposed. In addition, the LOI (34.3 %) calculated from the amount of char remaining indicates the homopolymer is self-extinguishing in air.

1. Introduction
The phosphazene based polymers have received a great interest thanks to their superior properties, such as superior thermal resistances, excellent biodegradability, chemical stability and high refractive index. Polyphosphazene materials have broad application fields and development space due to their chemical structure designability and material function diversity [1-3].

Several optical resins were prepared by copolymerizing cyclotriphosphazenes and olefine monomers in our previous research [4-6]. In this paper, the thermal and degradation performance of the homopolymer based on cyclotriphosphazene were investigated and its thermal decomposition mechanism was discussed. The experiment data show clear evidence for the heat-resistant and halogen-free flame retardant of cyclotriphosphazene units.

2. Experimental

2.1. Polymer Synthesis
The homopolymer based on cyclotriphosphazenes was synthesized in this laboratory according to the literature [4], as illustrated in Scheme 1.

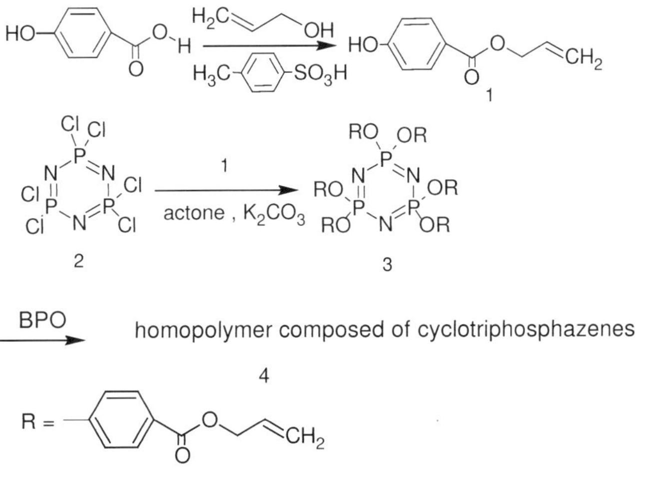

Scheme 1. Synthetic route of homopolymer composed of cyclotriphosphazenes

2.2. Analytical Methods

The thermogravimetry analysis was performed on a Netzsch thermal analyzer at the heating rates of 5, 8, 10, 15, 20, 30 and 40 °C/min from ambient temperature to 900 °C under the nitrogen flow. A combination of Simultaneous DSC-TGA Instrument and Nicolet IR-380 infrared spectrometer was used for decomposition mechanism analysis. The homopolymer sample was heated from 40 °C to 900 °C at 10 °C/min heating rate under nitrogen flow, while the decomposed gas product was characterized online by infrared spectrometer.

3. Results

3.1 Thermal Decomposition Behavior of the Homopolymer

Fig. 1a and 1b show the TGA and DTG curves of the thermal decomposition of the homopolymer composed of cyclotriphosphazenes at different heating rate in N_2, and the corresponding experimental data are listed in Table 1. The initial degradation temperature of the homopolymer (t_0) is all above 326 °C in different heating rate, which is much higher than that of ordinary thermoplastic optical resins such as polystyrene and polymethylmethacrylate [5-6].

Two inflection points can be found from the TGA curves, and the corresponding DTG curves have two minima, which indicates that the thermal decomposition of the homopolymer between 100 °C and 800 °C includes a two-stage reaction. For example, when the heating rate is 10 °C/min, the homopolymer undergoes the first stage thermal degradation between 300 °C and 400 °C, reaching the maximum degradation rate at 341.6 °C, the weight loss at this stage is 40.9 %. The second stage of thermal degradation is between 400 °C and 800 °C, reaching the maximum degradation rate at 451 °C, the weight loss at this stage is 9.83 %. The average activation energy of decomposition for the first stage is 292 kJ/mol calculated by the Flynn-Wall-Ozawa method.

According to the semi-empirical formula [4], the limiting oxygen index (LOI) of the homopolymer could be calculated from the char yield at 850 °C. The char yield of the homopolymer at 850 °C is 40.1 % (Table 1), accordingly the calculated LOI value is 34.3%, which means the homopolymer may be self-extinguishing in air atmosphere.

IOP Conf. Series: Materials Science and Engineering **484** (2019) 012041 doi:10.1088/1757-899X/484/1/012041

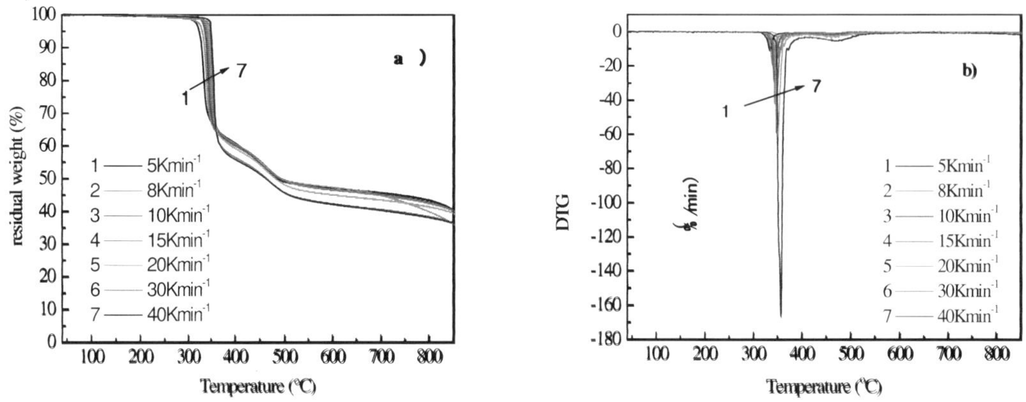

Figure 1. a) Dynamic TGA and b) DTG curves for the homopolymer at different heating rates in N_2

Table 1. Parameters of TG and DTG curves of cyclotriphosphazene homopolymer in N_2

β/ °C ·min^{-1}	t_0 / °C	t_{max} / °C	w/ %
5	326.1	333.0	40.4
8	333.0	339.5	36.1
10	336.6	341.6	40.1
15	340.2	345.3	40.2
20	344.4	348.8	39.4
30	348.7	352.1	36.5
40	349.9	356.3	36.4

β:Heating rate; t_0:Initial degradation temperature; w:Char yield at 850 °C

3.2. Thermal Decomposition Products of the Homopolymer

Two infrared spectra shown in Fig. 2 indicate the volatile products of the two-step thermal decomposition of the homopolymer in N_2, respectively. In the infrared spectrum of the volatile products of step I (29.59-30.40 min), strong absorption peaks at 669 and 2358 cm^{-1} come from carbon dioxide (CO_2), medium strong absorption peaks at 3646, 1608, 1508, 1267 cm^{-1} come from phenol, weak absorption peaks at 3580 and 1758cm^{-1} come from benzoic acid, which indicates the main volatile products are CO_2, phenol and a little of benzoic acid. In the infrared spectrum of the step II (41.53 min), there are only a few weak absorption peaks, indicating little phenol and little water are released.

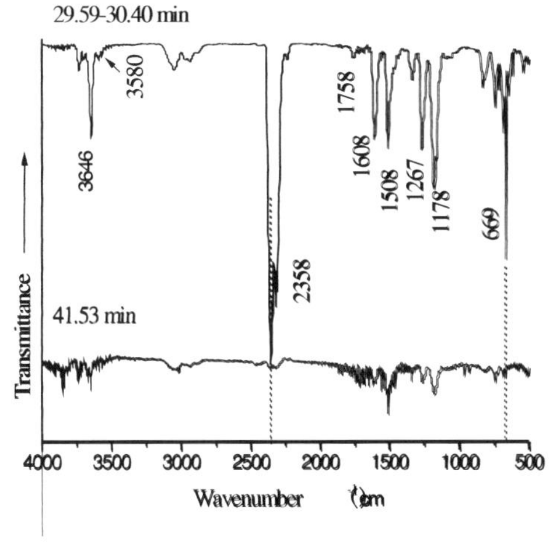

Figure 2. Infrared spectra of the volatile products of decomposition in N_2

3.3. Thermal Decomposition Mechanism of the Homopolymer

Based on the results from TGA, DTG and infrared spectra analysis of the volatile gas, the decomposition model of the homopolymer in N_2 is proposed, as shown in scheme 2. In the first stage of thermal degradation of the homopolymer, the C-O and P-O bonds of the cyclictriphosphazene unit's break, releasing a large amount of CO_2, phenol and little benzoic acid, and simultaneously forming phosphate. In the second stage, the C-C bonds are broken, a small amount of phenol and water are released, carbonaceous char and some phosphorus compounds are left.

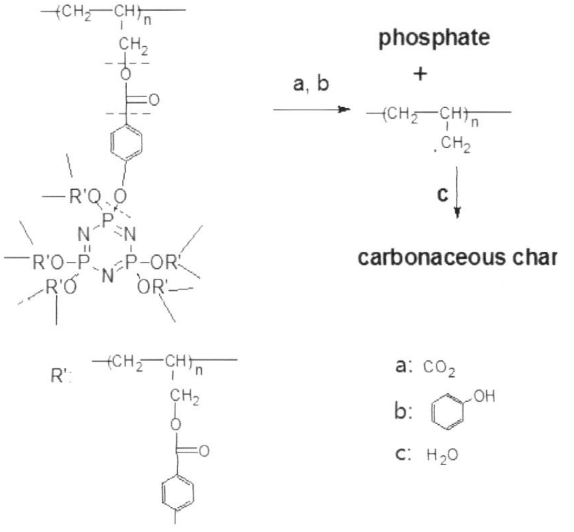

Scheme 2. Possible mechanism of the two-stage thermal decomposition of the homopolymer in N_2

4. Conclusions

A heat-resistant homopolymer based on cyclotriphosphazene was prepared by free radical polymerization. As compared with ordinary thermoplastic optical resins, the homopolymer exhibits much higher thermal stability due to the crosslinked structures. The LOI value (34.3 %) calculated by char yield according to the semi-empirical method indicates the homopolymer is self-extinguishing in air.

The decomposition behavior of the homopolymer includes two stages between 100 °C and 800 °C in N_2. In the first stage, the breaking of the C-O and P-O bonds of the cyclic triphosphazene units releases CO_2 and phenol, in the second stage the breaking of C-C bonds results in the mass loss.

5. Acknowledgements.

This work was supported by the Scientific Research Fund (Wuhan Institute of Technology, 16QD27) and Opening fund of Hubei Key Laboratory of Material Chemistry and Service Failure (No. 2017MCF02).

6. References

[1] Allcock H R. *Curr Org Chem*, 2006, 10: 231-240
[2] De Jaeger R and Gleria M. *Prog. Polym. Sci.* 1998, 23: 179-276
[3] Mathew D, Nair CPR and Ninan KN. *Polym. Int.* 2000, 49: 48-56
[4] Guo YN, Zhao C, Liu SZ, Li D, Wang S J, Qiu JJ and Liu CM. *Polym. Bull.* 2009, 62: 421-31
[5] Guo YN, Qiu JJ, Tang HQ and Liu CM. *J. Appl. Polym. Sci.* 2011, 121: 727-34
[6] Guo YN, Ming JY and Li CY. *J. Appl. Polym. Sci.* 2011, 121: 3137-44

Development of Small Intestinal Submucosa as Biomaterial in Tissue Engineering

Yu Wang[1], Peng Chen[1] and Jianbo Qin[1]*

[1]Chongqing Center for Biomedicines and Medical Equipments, Chongqing Academy of Science and Technology, Chongqing, China.

Email: wanyu200311@163.com

Abstract. Small intestinal submucosa as a kind of natural ECM was widely studied as biological material from now on. Attachment on SIS, the proliferation, migration and differentiation of numerous cells were found be influence. Due to its excellent biocompatibility, biomechanical properties and biological activity, SIS had been extensive used as biomaterial in regenerative medicine. This article reviews the recent progress in the characterization and medical application of SIS respectively.

1. Introduction

Small intestinal submucosa (SIS) is a kind of biomaterial common used in clinical. The construction, biological activity and immune response of SIS were suit for implanting in body. Nowadays, it often used as scaffolds in tissue engineering and regenerative medicine to organs such as the artery [1], bladder [2], intestinal tract [3], valve [4], esophagus [5], and tendon [6]. The SIS has approved by FDA which could used in humans for urogenital procedures. cystoplasties, ureteral reconstructions and so on [7]. The scholars discussed to process SIS to numerous types such as injection, patch, scaffold and so on, in order to enlarge its utility.

2. The Characteristics of the SIS

2.1. Biochemical Properties of SIS

Lei shi and Vincent Ronfard concluded the structural properties and biochemical composition of SIS as documented in Table 1[8]. The basic component of SIS is collagen. And the collagen in SIS, were mainly collagen type I and a few of other types [9]. Small amount of elastin, fibronectin, laminin glycosaminoglycans and proteoglycans also were found in SIS [10-13]. The collagen contracted wound and promoted healing. The multidomain glycoproteins regulated cell attachment on ECM. And the proteoglycans supplied cell adherence binding site, inhibited substrate degrading enzyme [14-16]. Additionally, the growth factors such as fibroblast growth factor-2 were released by SIS too [17, 18]. The compositions of SIS above-mentioned were influence tissue healing that is why SIS becomes popular biomaterial on tissue engineering.

IOP Conf. Series: Materials Science and Engineering **484** (2019) 012042 doi:10.1088/1757-899X/484/1/012042

Table 1. Characteristic and Component of Small Intestinal Submucosa [8]

characteristic	Description
Thickness	50 to 220 μm
Ultrastructure	0.020 to 0.030 mm
Mechanical properties	1 layer: 20.6 MPa (dry) and 7.2 MPa (wet) 4 layer: 36.7 MPa (dry) and 22.4 MPa (wet)
component	Collagen, Elastin ,Fibronectin ,Laminin Glycosaminoglycans and proteoglycans Fibroblast growth factor-2 (FGF-2), Transforming growth factor-beta1 (TGF-ß1), Vascular endothelial growth factor (VEGF)

2.2. SIS Effects on Cell Proliferation and Differentiation

Cells and ECM related and interacted with each other, with reaching a kinetic balance [19].These interactions happened while cellular microenvironment transformed, during tissue repair, embryogenesis, malignant transformation and angiogenesis [20].

The interaction between cells and ECM was thought to be mediated through integrin. And the mechanism, how SIS influences cells, is related to integrin binding. The peptide sequences of fibronectin were researched. And researchers thought fibronectin played a significance role in adherence of cells to Small intestinal submucosa [21]. The study displayed that RCG and REDV were important for adherence on SIS. Furthermore, the integrin which influence RCG and REDV may mediate cell adherence. Some researchers suggested that the adherence of endothelial cells to Small intestinal submucosa hinged on its structure and formation [22]. Additionally, the β1, β4, and α6 integrins changed the attachment of cancer cells on SIS [23]. This study suggested neutralization of integrin entire effect on adherence of cells.

The growth factors, from Small intestinal submucosa, may mediate cell differentiation and angiogenesis which were carried on [21]. Hodde found that the differentiation of pc12 affected on small intestinal submucosa was controlled by neutralizing antibody against FGF-2. This study showed a new regulatory approach that FGF-2 perhaps adjust the biological response of Small intestinal submucosa [24].

2.3. Mechanical Properties between Different Matrix Layers of Sis

The structure of 1-4 layer small intestinal submucosa were researched. The scholars contrast the mechanical properties with different layers of SIS [25]. The thickness of one layer is about 50μm. In wet condition, the ultimate tensile strength and elastic modulus were measured. The same detection indexes also were measured in dry condition. Determined in dry conditions, peak strength enhanced while the number of SIS layer rise. But the material elongated when the number of layers increased. The elastic modulus of different layers of SIS was similar which shown they had resemblance mechanical property as biomaterial.

Testing in wet conditions, the results of peak strength were lower than dry conditions [25]. And the peak strengths of the multi-layer small intestinal submucosa specimens, in wet conditions, were obviously higher than the 1 layer small intestinal submucosa specimens. But compared with multi-layers each other, there were not obviously difference of peak strength. In wet conditions, when the number of layers and peak strength increased, the elongation was declined. The one layer small intestinal submucosa elongated the most, about 130%. However, all of multi-layers small intestinal submucosa elongated 30%-40% [25].

IOP Conf. Series: Materials Science and Engineering **484** (2019) 012042 doi:10.1088/1757-899X/484/1/012042

3. Medical Application of SIS

Abdominal wall defects were common clinical symptoms which usually caused serious complication by unfit repair method. The previous studies for the material used in abdominal wall defects were limitation. Recently, the researchers used SIS patches for restoring abdominal wall defects in dogs. And contrast the biological and mechanical properties between SIS and SIS + polypropylene mesh. The dogs were undergone surgery for constructing a model of abdominal wall defects and use the two kinds of material to repair defects. To evaluate the mechanical property, biocompatibility, toxicology of SIS and SIS + polypropylene mesh, the main purpose was shown the effectiveness and safety of two kinds of biomaterials. The research results display that SIS had better biocompatibility and histocompatibility than SIS + polypropylene mesh [26].

SIS also used in bone tissue engineering generally, in order to improve bone healing of animal experiment. Small intestinal submucosa scaffold had excellent biocompatibility which usually applied in tissue engineering. Bone cell could be repaired by SIS, because of its biological characteristics which could influence migration, proliferation, differentiation and adherence of cells. In animal models, mice were subjected to surgical resection to bring about calvarial defect. The growth of defective bone was significantly improved with the mice implanted in small intestinal submucosa scaffold. The researcher found SIS could upregulate the expression of BMP-2 and CD31; simultaneously increase the downstream target gene ID-1. The results of this research point out that, because of the biological and mechanical property, regulating in regeneration of cells which participate in bone regeneration, SIS maybe a kind of excellent biomaterial used in bone tissue [27].

Preliminary researchers found that small intestinal submucosa scaffold can be used as superior biomaterial in vascular transplantation. The artificial vessel that manufactured by small intestinal submucosa transplanted into the canine carotid artery of dogs [28]. Then, the artificial vessel performance vascular mechanical properties and gradually evolve into carotid artery. Other studies found SIS also can be excellence material used as a basement membrane because of its biological properties to promote epidermal cells attaché fibroblasts [11]. The research shown small intestinal submucosa encourages proliferation of endothelial progenitor, rather than prevent. The endothelial progenitor growth on SIS which has excellent cell morphology and the rate of proliferation was increased. Furthermore, seeded on SIS surface, the endothelial progenitor could grow, migrate and differentiate. All the results shown that SIS was excellence biomaterial without any cytotoxicity, even promote proliferation of epidermal cells.

The restore of full thickness corneal wound in animals, which used SIS as corneal replacement material, were suggested for many years. Nowadays, the preliminary evaluation about biocompatibility of SIS as rabbit corneal replacement biomaterial was completed. The results display that corneal base transparent while used SIS with producing only a small amount of scar tissue [29,30]. The small intestinal submucosa was an excellent material could be recommend used in veterinary ophthalmology.

It is a complex process of repairing tracheal defects by surgery when the gold standard treatment with primary end-toend anastomosis is not possible. SIS has been used as graft material for bioengineering applications and to promote tissue regeneration. There are sixteen rabbits used as animal models that randomly divided into two groups. To cut the trachea by surgery, one group of rabbit use SIS material as transplanting alternatives cover the incision and another group use noting as control. The area of perimeter of tracheal gap narrower by 22% and 15%, respectively, in the SIS group than in the control group. Histological analysis revealed immature cartilage, pseudostratified ciliated epithelium, and connective tissue about 55% of the SIS group, while no cartilaginous regeneration was observed in the control group.Although tracheal SIS engraftment could not prevent stenosis in a rabbit model of tracheal injury; it produced some remarkable changes, efficiently facilitating neovascularization, reepithelialization, and neoformation of immature cartilage [31].

4. Conclusion

Small intestinal submucosa as a kind of natural ECM was widely studied as biological material from now on. With the biological property, outstanding improving the cure effect, SIS common suggested which applied in tissue repair. Attachment on SIS, the proliferation, migration and differentiation of numerous cells were found be influence. Simultaneously, growth factor, encouraged secreting by SIS, could induce angiogenesis to affect new blood vessel. And the biodegradability of SIS also showed illustrious histocompatibility without cytotoxicity in vivo. At present, SIS products have been applied in different forms, including tablet, granular, spongy and gel, but it is still difficult to meet different clinical needs. In order to satisfy the needs of various tissue repairs, the combination of SIS with other materials, cells and drug factors to improve the properties of SIS, is an important direction for future development.

5. References

[1] Pavcnik D, Obermiller J, Uchida BT et al. Angiographic evaluation of carotid artery grafting with prefabricated small-diameter, small-intestinal submucosa grafts in sheep, Cardiovasc. Intervent. Radiol. Vol.32 (2009) No.1, p. 106–113.

[2] Ayyildiz A, Akgül KT, Huri E et al. Use of porcine small intestinal submucosa in bladder in bladder augmentation in rabbit: long-term histological outcome. ANZ J. Surg. Vol.78 (2008) No.1-2, p.82–86.

[3] Lee M, Chang PC, Dunn JC. Evaluation of small intestinal submucosa as scaffolds for intestinal tissue engineering. J. Surg. Res. Vol.147 (2008) No.2, p.168–171.

[4] Metzner A, Stock UA, Iino K et al. Percutaneous pulmonary valve replacement: autologous tissue-engineered valved stents, Cardiovasc. Res. Vol.88 (2010) No.3, p. 453–461.

[5] Tan B, Wei RQ, Tan MY et al. Tissue engineered esophagus by mesenchymal stem cell seeding for esophageal repair in a canine model, J. Surg. Res. Vol.182 (2013) No.3, p. 40–48.

[6] Liang R1, Woo SL, Nguyen TD et al. Almarza, Effects of a bioscaffold on collagen fibrillogenesis in healing medial collateral ligament in rabbits. J. Orthop. Res. Vol.26 (2008) No.8, p. 1098–1104.

[7] Franklin M, Russek K. Use of porcine small intestinal submucosa as a prosthetic material for laparoscopic hernia repair in infected and potentially contaminated fields: long-term follow-up assessment. Surg. Endosc. Vol.25 (2011) No.5, p. 1693–1694.

[8] Shi 1, Ronfard V. Biochemical and biomechanical characterization of porcine small intestinal submucosa (SIS): a mini review. Int J Burn Trauma. Vol. 3(2013) No.4, p.173-179

[9] US Pharmacopoeia, Inc. USP Monographs: Small Intestinal Submucosa Wound Matrix. Available at: http://www.pharmacopoeia.com.cn/v29240/usp29nf24s0_m541.html.

[10] McPherson TB, Badylak SF. Characterization of fibronectin derived from porcine small intestinal submucosa. Tissue Eng Part A.Vol. 4 (1998) p. 75-83.

[11] Lindberg K, Badylak SF. Porcine small intestinal submucosa (SIS): a bioscaffold supporting in vitro primary human epidermal cell differentiation and synthesis of basement membrane proteins. Burns. Vol. 27 (2001) p. 254-266.

[12] Hurst RE, Bonner RB. Mapping of the distribution of significant proteins and proteoglycans in small intestinal submucosa by fluorescence microscopy. J Biomater Sci Polym Ed. Vol.12 (2001) p. 1267-1279.

[13] Hodde JP, Badylak SF, Brightman AO et al. Glycosaminoglycan content of small intestinal submucosa: a bioscaffold for tissue replacement. Tissue Eng. Vol.2 (1996) p. 209-217.

[14] Mulloy B, Rider CC. Cytokines and proteoglycans: an introductory overview. Biochem Soc Trans. Vol. 34 (2006) p. 409-413.

[15] Baum CL, Arpey CJ. Normal cutaneous wound healing: clinical correlation with cellular and molecular events. Dermatol Surg. Vol.31 (2005) p. 674-686.

[16] Gallo RL. Proteoglycans and cutaneous vasculardefense and repair. J Investig Dermatol Symp Proc. Vol.5 (2000) p. 55-60.

[17] McDevitt CA, Wildey GM, Cutrone RM. Transforming growth factor-beta1 in a sterilized tissue derived from the pig small intestine submucosa. J Biomed Mater Res A.Vol.67 (2003) p. 637-640.

[18] Voytik-Harbin SL, Brightman AO, Kraine MR et al. Identification of extractable growth factors from small intestinal submucosa. J Cell Biochem. Vol.67(1997) p. 478-491.

[19] Bissell MJ, Aggeler J. Dynamic reciprocity: how do extracellular matrix and hormones direct gene expression? Prog Clin Biol Res. Vol.249(1987) p.251-262.

[20] Nelson CM, Bissell MJ. Of extracellular matrix, scaffolds, and signaling: tissue architecture regulates development, homeostasis, and cancer. Annu Rev Cell Dev Biol. Vol.22(2006) p. 287-309.

[21] Hodde J, Record R, Tullius R et al. Fibronectin peptides mediate HMEC adhesion to porcine-derived extracellular matrix. Biomaterials. Vol. 23(2002) p. 1841-1848.

[22] Badylak S, Liang A, Record R et al. Endothelial cell adherence to small intestinal submucosa: an acellular bioscaffold. Biomaterials. Vol.20(1999) p. 2257-2263.

[23] Hurst RE, Kyker KD, Bonner RB et al. Matrix-dependent plasticity of the malignant phenotype of bladder cancer cells. Anticancer Res. Vol.23(2003) p. 3119-3128.

[24] Hodde JP, Record RD, Liang HA et al. Vascular endothelial growth factor in porcine-derived extracellular matrix. Endothelium. Vol.8(2001) p. 11-24.

[25] Ermis R, Shi L, Carson D. Comparison of mechanical properties between small intestinal submucosa (SIS) with varying layers. Poster presented as part of the Wound Healing Society Annual Meeting, Dallas, TX. Vol.8(2011) p.14-17.

[26] Wang ZL, Wu SZ, Li ZF et al. Comparison of small intestinal submucosa and polypropylene mesh for abdominal wall defect repair. J Biomater Sci Polym Ed. Vol.29(2018)No.6, p. 663-682.

[27] Li M, Zhang C, Cheng M et al. Small intestinal submucosa: A potential osteoconductive and osteoinductive biomaterial for bone tissue engineering. Mater Sci Eng C Mater Biol Appl. Vol.75(2017) No, p.149-156

[28] Roeder R, Wolfe J, Lianakis N et al. Compliance, elastic modulus, and burst pressure of small-intestine submucosa (SIS), small-diameter vascular grafts. J Biomed Mater Res. Vol.47(1999) p. 65-70.

[29] Bussieres, M., Krohne, S., Stiles, J et al. The use of porcine small intestinal submucosa for the repair of fullthickness corneal defects in dogs, cats and horses. Veterinary Ophthalmology. Vol.7 (2004) p. 352–359.

[30]Grigue F, Raymond I, Regnier A. Preliminary evaluation of the biocompatibility of the small intestinal submucosa (SIS) biomaterial with the rabbit cornea. Revue de Médecine Vétérinaire. Vol.152(2001)No.8, p. 597–604.

[31] Bergonse Neto N, Jorge LF, Francisco JC et al. Regeneration of Tracheal Tissue in Partial Defects Using Porcine Small Intestinal Submucosa. Stem Cells Int. Vol.26(2018) P.50-61.

IOP Conf. Series: Materials Science and Engineering **484** (2019) 012043 doi:10.1088/1757-899X/484/1/012043

Polyurea Materials and Their Environmental Applications

Xiaojing Dong, Xubao Jiang, Shusheng Li and Xiang Zheng Kong

College of Chemistry and Chemical Engineering, University of Jinan, Jinan, China
Email: xzkong@ujn.edu.cn

Abstract. With the rapid economy development, environmental pollution becomes an urgent issue. Porous polymer materials have been a main player to this issue because their structures are easy to be designed to address these issues. Here we present different protocols for preparations of a variety of polyureas (PU), all through polymerization of a diisocyanate with a diamine in aqueous phase, added in water or in-situ formed from the reaction of water with the diisocyanate involved. Two PU materials of different types are discussed in this study: (1), PU under form of microspheres was prepared through precipitation polymerization using isophorone diisocyanate as the monomer in a mixed solvent of water-acetonitrile. Aldehyde groups were attached to the surface of the PU microspheres. Immobilized of an enzyme, laccase, was carried out. The immobilized enzyme was used as biocatalyst in the process of degradation of organic dyes, and demonstrated high performance with good reusability. (2), porous PU (PPU) was also prepared through precipitation polymerization of toluene diisocyanate with a diamine or via the reaction of water. PPU was used directly to adsorb organic dyes and metal ions from aqueous systems with high performance and high selectivity.

1. Introduction

With the fast growth of chemical, textile and manufacture industries, environmental pollution has become an urgent challenge [1]. Among a vast variety of pollutants, heavy metal ions and organic dyes are the most common ones [2, 3, 4], and great efforts have been never ceased for their removal from polluted environment. Up to date, the main strategies to deal with these issues include adsorption separation, membrane separation, catalytic degradation, ion exchange and chemical precipitation etc [5,6]. In all the related studies or industrialized processes, polymers materials have been playing important roles because of their high specific surface area and low density, and in particular the easiness of control for their pore size and functionalization by a great variety of groups [7]. Separations by adsorption on an adsorbent have been the most widely used for metal ions and dye removals. For this purpose, diverse polymer porous absorbents have been prepared. The preparation involves generally two steps, namely, the synthesis of the base polymer and its chemical modification, which are usually comprised of several steps. The whole process is in general complicated and also time and energy consuming.

Uniform microspheres of polyurea (PU) have been fabricated with a high yield, through a process of precipitation polymerization of isophorone diisocyanate (IPDI) as the only monomer in a binary mixture of water-acetone [8,9]. The study demonstrated that the microspheres thus prepared were highly thermally stable with good resistance against common organic solvents. In this study, the preparation of PU microspheres is optimized in a binary mixture of water-acetonitrile (H_2O-AN). In order to make a biocatalyst for degradation of dyes molecules, immobilization of enzyme laccase was conducted on the surface of these microspheres, and the immobilized enzyme was used as biocatalyst in the degradation of dye molecules. Porous polyurea (PPU) was successfully prepared also through a one-step process, either by precipitation polymerization of toluene diisocyanate (TDI) with its diamine counterpart derived from its reaction with water when only TDI was present as the monomer, or by its copolymerization with ethylene diamine (EDA) in a mixed solvent of H_2O-AN (acetonitrile). PPU as-prepared was employed as an absorbent for dyes molecules; high adsorption performance and good reusability were observed. This PPU was also used as adsorbent for a selected group of metal ions. The adsorption selectivity and the reusability were also investigated.

2. Experimental

2.1. Synthesis of Polyurea Microspheres and Immobilization of Laccase

A typical protocol for the preparation of PU microspheres is as following: To a reaction bottle, a binary mixture of H_2O/AN (20/80 by mass) was charged, followed by introduction of isophorone diisocyanate (IPDI). The bottle was sealed, shaken by hand to make sure that the mixture was homogenized and installed in a water bath set at 30 °C for 2 h, without any stirring or shaking. The microspheres were separated by centrifugation, rinsed twice with H_2O/AN mixture of 20/80 mass ratio, and dried up for 12 h in a vacuum oven at 80 °C. Microspheres were also prepared at different polymerization temperature or with changed IPDI concentration and in solvent of different H_2O/AN ratio. For laccase immobilization, the microspheres suspension was first mixed with an aqueous solution of glutaraldehyde (GA) to attach aldehyde functional groups onto the surface of the microspheres, through the reaction between amine on the microspheres and aldehyde group of GA, followed by enzyme immobilization by the reaction between the newly attached aldehyde groups from GA with the amine groups of laccase, a common practice. Catalytic activity of the immobilized laccase was tested in the degradation of organic dyes, and the performance was compared with the free laccase.

2.2. Preparation of Porous PU, Dyes and Metal Ions Adsorption

For the preparation of PPU adsorbent, a H_2O-acetone mixture (H_2O/acetone of 30/70 by mass) was first located into a round bottom flask (250 mL), which was then installed in a water bath at 30 °C. The polymerization was started by addition of TDI at a constant rate under constant stirring. Upon completion of TDI addition, the polymerization was allowed to continue for another 2 h, followed by separation of the microspheres by centrifugation and drying the polymer at 70 °C under vacuum. For the test of dye removal from aqueous solution by adsorption, an accurately known amount of PPU was first added into a glass bottle, followed by addition of an aqueous solution of the dye for test with known concentration at a pre-adjusted pH. The adsorption was done with the bottle located in a water bath shaker at 30 °C and shaken at 300 osc/min for 30 min. The bottle content was then centrifuged to separate PPU out from the dye solution. The amount of the no-adsorbed dyes in the solution was determined by analyzing the supernatant using a spectrophotometry at a given wavelength for a given dye. The adsorption was done for two dye molecules, anionic acid fuchsine and Remazol brilliant blue R, i.e. RBBR. For a typical process for adsorption of metal ions on PPU, a Cu^{2+} aqueous solution (using $CuSO_4 \cdot 5H_2O$), of known concentration with pH pre-adjusted at 5.5, was first added to a glass flask of 25 mL capacity, and a known amount of PPU was added in a second step. The adsorption was allowed to proceed for 3 h at 30 ºC. The content was centrifuged and the no-adsorbed Cu^{2+} in the supernatant was determined. Different adsorption processes were carried out under different experimental conditions.

3. Results and Discussions

3.1. Optimization of PU Microspheres Preparation

The preparation was optimized with regard to different experimental conditions, such as the monomer concentration, composition of the solvent (H_2O/organic solvent ratio), polymerization temperature etc. It is reported that the formation and the productivity of the uniform microspheres were closely related to the solubility of IPDI in the binary solvent. Uniform microspheres were obtained only when initial IPDI concentration was equal or below its solubility in the solvent. In this case IPDI molecules were homogeneously dissolved in, and neither IPDI droplets nor phase separation were present. The uniformity of the outcome microspheres was deteriorated whenever the initial IPDI concentration used was higher than its solubility at that given temperature of polymerization and in the solvent concerned.

In a previous report [8], with H_2O-acetone as the binary solvent, 11 wt% of IPDI was the upper limit achieved to have uniform PU microspheres. To further increase the productivity of the uniform microspheres, polymerization was carried out by substituting acetone by different solvent in the binary mixture. Solvent selection was done with objective to find a solvent system, in which the solubility of IPDI is increased and PU is insoluble. As to the insolubility, experiments showed that IPDI-based PU was insoluble only in AN, THF and water. Water was straightly excluded because IPDI is not soluble in; and aggregation of the primary PU particles was always observed in early stage of the polymerization conducted in THF. Nevertheless, experiments showed that the solubility of IPDI was higher in AN than in acetone, and IPDI-based PU was insoluble in AN. Therefore, a set of polymerization was done at different temperature in H_2O/AN of different composition. Corresponding IPDI solubility, microsphere yield, their size (D_n) and size distribution (D_w/D_n) were determined and listed in Table 1. Preparation of PU microspheres at high yields was achieved, under experimental conditions optimized with regard to solvent composition, temperature of polymerization and IPDI concentration while keeping a high uniformity ($D_w/D_n \leq 1.10$) for the microspheres. It was concluded that the highest productivity of the uniform microspheres was reached by polymerization at 50 °C with monomer concentration of 23 wt%, which gave 95.12% for the yield of the microspheres.

Table 1. Comparison of the Preparation of PU microspheres in H_2O-acetonitrile to that in H_2O-acetone at different polymerization temperature and IPDI concentration

Solvent	H_2O (wt%)	Temperature (°C)	IPDI (wt%) solubility	IPDI (wt%) concentration	Sphere yield (wt%)	Sphere size D_n (μm)	D_w/D_n
H_2O-Acetone	30	50	8.6	8.6	88.75	6.91	1.006
H_2O-Acetone	30	70	11.0	11.0	88.46	10.5	1.006
H_2O-AN	20	30	17.0	17.0	93.68	7.15	1.008
H_2O-AN	20	50	23.0	21.0	93.79	10.95	1.008
H_2O-AN	20	50	23.0	23.0	95.12	11.22	1.010
H_2O-AN	20	70	29.0	21.0	93.13	8.11	1.008

3.2. Immobilization of Laccase on PU Microspheres and Catalytic Activity in Dyes Degradation

Following their preparation, PU microspheres in this work were directly used as the support for immobilization of laccase. The microspheres were firstly activated using GA to attach aldehyde groups on their surface, as described in previous section. As a typical result, immobilization of 20.63 mg of laccase was achieved for each gram of the microspheres. The activity of the immobilized enzyme was determined at different pH and temperature, and compared with that of the free laccase. A highest activity was observed at 50 °C for the immobilized laccase, whereas a highest activity was detected at 40 °C for the free laccase. This shift toward higher temperature for the activity of the immobilized enzyme was attributed to its increased activation energy to adopt an optimal conformation for the enzyme when it was immobilized on the support. The optimum pH for the activity was observed at 3.0 for both the free and the immobilized laccase. It was found that free enzyme was more sensitive to pH than the immobilized counterpart, in particular in alkaline medium. When pH was below 5.5, a lower activity was observed for the immobilized laccase than for the free enzyme; this activity was inversed once pH was increased to above 5.5. For instances, the activity of the free laccase was 32.0 U/mg at pH 3.0, and it was 8.3 U/mg for the immobilized laccase, i.e. only 25.9% of activity was retained after immobilization; whereas at pH 7.2, the activity of immobilized enzyme was 3.09 U/mg, which was 3.8 times higher than that of the free laccase (0.64 U/mg) at this pH. In addition, a much better alkali resistance was also observed for the immobilized laccase.

Thermostability at 55 °C was also measured for both the free and the immobilized enzyme (Fig. 1). After 120 min of incubation, 89% of the initial activity was retained for the immobilized one, whereas only 27% of the initial activity was retained for the free laccase, a thermal stability significantly lower than the immobilized enzyme. This was attributed to a more restricted conformational mobility for the enzyme in their immobilized form. The immobilized enzyme showed also higher storage stability than the free one. For the free enzyme, 63% of its initial activity was lost after storage of 10 day and 80% after storage of 30 days, always at room temperature. In contrast, for the immobilized enzyme, the corresponding values were found to be 40% and 60%, respectively.

The immobilized enzyme was employed as catalyst for RBBR degradation (decolorization). The test results are shown in Fig. 2 for both the free and the immobilized enzyme. To assure that this decolorization was due to the catalytic effect of the enzymes, a reference test was also done, in which exactly the same procedure for RBBR degradation was followed, using the virgin PU microspheres as-prepared prior to their activation by GA and enzyme immobilization. It is clearly seen that no any sign of RBBR decolorization was detected in the blank test, excluding therefore possible catalytic activity of PU microspheres. The results in Fig. 2 demonstrate that the immobilized enzyme was performing better in the whole degradation process of RBBR. With the immobilized enzyme, 64.1% of RBBR was degraded at 300 min, while only about 54.7% of RBBR was decolorized at the same time with the free enzyme.

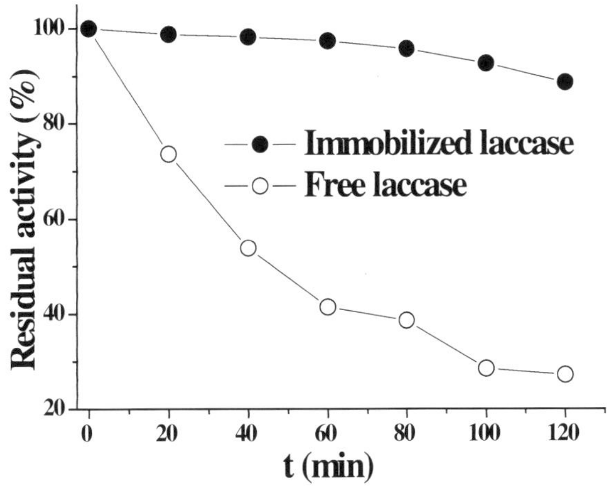

Figure 1. Thermostability of the immobilized and the free laccase at 55 °C.

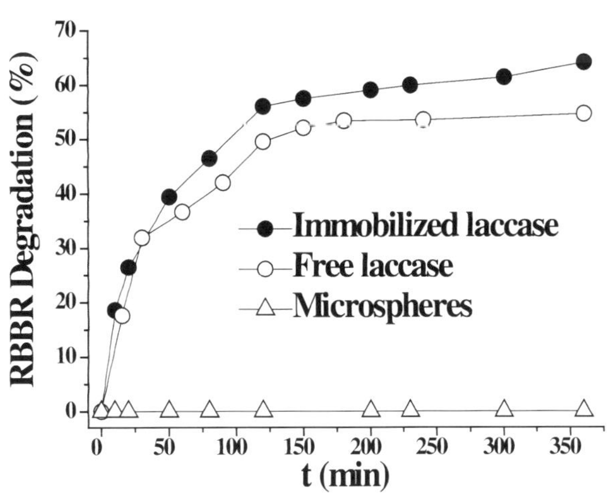

Figure 2. Comparison of degradation of RBBR by free laccase to that by immobilized laccase.

The importance and interest of enzyme immobilization are the reusability because this will reduce significantly the cost of the enzyme. The reusability of the immobilized laccase was tested through the same procedure, which demonstrated that, in comparison with the amount of RBBR degraded in a first run with freshly prepared immobilized enzyme, 92% of the initial degradation was obtained in the third consecutive recycled use of the immobilized enzyme, and about 65% of the initial degradation was detected in the seventh recycled use. From these results, it was concluded that the immobilized laccase was of fair reusability as a catalyst in RBBR degradation.

3.3. Direct Adsorption of Organic Dyes and Copper Ions on Porous PU

Porous PU (PPU) was prepared using the process described above in Experimental section. The chemical structure and morphology were characterization to show that this material is a typical porous material. As described above, the surface of PPU is rich in amine groups. It is therefore believed that this PPU may be used as an adsorbent for organic dyes and metal ions. Adsorption on PPU of two anionic dyes (RBBR and AF) in their aqueous solutions was done at 30°C with 0.25 wt% of PPU and at initial concentration of 100 mg/L for the dyes in test. To assure an equilibrated adsorption for the dye molecules, an adsorption time of 4 h was allowed for all the tests. The results showed that the pH of the aqueous dye solution imposed important effect on the adsorptions for the two dyes involved. With pH controlled between 2 and 4, an adsorption of about 38 mg/g was detected for AF; whereas this adsorption was slightly decreased at pH=1, and once pH shifted to above 4, this adsorption was sharply declined. A similar adsorption profile was seen for RBBR. At pH=2, a highest adsorption of 39.39 mg/g was obtained, and a steep decline in the adsorption was clearly seen once pH was increased to a value above 4.

For the adsorption of metal ions on PPU, the test was done for five ions, i.e. Cu^{2+}, Ni^{2+}, Pb^{2+}, Cd^{2+} and Cr^{3+}. The highest selectivity combined with the high adsorption for Cu^{2+} ions was observed, which was interpreted by the formation of multifunctional complexation of Cu^{2+} ions, based on XPS analysis on the PPU absorbent before and after the adsorption of the metal ions. The multifunctional complexation assumes that a coordination complex was formed not only between Cu^{2+} ions and the terminal amine of PPU, it occurred also between Cu^{2+} ions and the oxygen and nitrogen atoms of the carbamido units of PPU. Simulations of the adsorption kinetics showed that this adsorption followed the pseudo-second-order model; the simulation of the adsorption process demonstrated that a better fitting was found for the experimental data with Freundlich isothermal adsorption in comparison to Langmuir isothermal adsorption.

4. Conclusions

Optimization of the preparation of PU uniform microspheres was done with regard to different experimental conditions. In comparison with what achieved in the mixed solvent of H_2O-acetone, higher yield of the uniform PU microspheres was indeed obtained in the binary solvent of H_2O-AN, because of higher IPDI solubility in H_2O-AN than in binary solvent of H_2O-acetone. A highest productivity of 95.12% was obtained for the uniform microspheres with monomer concentration of 23 wt% and polymerization at 50 °C. Using these PU microspheres as support, enzyme laccase was immobilized on their surface using GA as the coupling agent in order to chemically attach aldehyde groups. Laccase was then immobilized through the reaction of the attached aldehyde groups with the amine groups of laccase. The highest activity was observed at 50 °C for the immobilized laccase, instead of 40 °C for the free enzyme. At pH 5.5 or below, the immobilized laccase manifested a lower activity than the free enzyme; and a higher activity than the free laccase was detected for the immobilized enzyme once pH increased to above 5.5. A distinctly higher thermal stability was also observed for the immobilized enzyme, owing to most likely a more restricted conformational mobility when enzyme was immobilized. Employed as the biocatalyst in RBBR degradation, 64.1% of RBBR was degraded within 300 min using the immobilized laccase, a better performance than the free laccase. Porous PU (PPU) was also prepared by precipitation polymerization using TDI as the monomer, and used as adsorbent in adsorption separation for two anionic dyes, RBBR and AF. For AF, an adsorption of up to 38 mg/g was achieved with pH controlled between 2 and 4. For RBBR, a high adsorption of 39.39 mg/g was achieved at pH=3, and a slight decrease was detected with pH shifted to either direction (pH<3 or pH>3). PPU was also directly used as adsorbent for metal ions. Among 5 ions tested, PPU was found to have a higher adsorption with better selectivity for Cu^{2+} ions. Based on XPS analysis, the higher performance for Cu^{2+} adsorption was explained by the formation of

IOP Conf. Series: Materials Science and Engineering **484** (2019) 012043 doi:10.1088/1757-899X/484/1/012043

multifunctional complexations of the ions with the nitrogen atoms of its carbamido units and the terminal amine of PPU, as well as with the oxygen atoms of the carbamido units.

5. References

[1] Taskin O S, Ersoy N, Aksu A, Kiskan B, Balkis N and Yagci Y 2016 Polym. Int. 65 439
[2] Karim Z, Claudpierre S, Grahn M, Oksman K and Mathew A P 2016 J. Membrane Sci. 514 418
[3] Hase H, Nishiuchi T, Sato T, Otake T, Yaita T, Kobayashi T andYoneda T2017 J. Hazard. Mater. 329 49
[4] Li S, Kong X Z, Jiang X and Zhu X 2013 Chinese Chem. Lett. 24 287
[5] Azarova Y A, Pestov A V andBratskaya S Y2016 Cellulose 23 2273
[6] Guo N, Su S J, Liao B, Ding S L and Sun W Y 2017 Carbohydr. Polym. 165 376
[7] Hasanzadeh R, Moghadam P N, Bahri-Laleh N and Sillanpää M2017 J. Colloid Interface Sci. 490 727
[8] Jiang X, Zhu X and Kong X Z 2012 Chem. Eng. J. 213 214
[9] Jiang X, Kong X Z and Zhu X 2011 J. Polym. Sci. Part A: Polym. Chem. 49, 4492

IOP Conf. Series: Materials Science and Engineering **484** (2019) 012044 doi:10.1088/1757-899X/484/1/012044

$Co_{1-x}Zn_xIn_2S_4$ Photocatalysts for Photocatalytic H_2 Production Activity

Renzhang Wang, Zhiyun Liang, Fuying Li, Qiyou Wu, Peijing Xu and Yu Niu*

Fujian Provincial Collaborative Innovation Center for Clean Coal Gasification, Sanming University, Sanming 365004, China.
E-mail: niuyu200704@163.com;

Abstract. In this work, $Co_{1-x}Zn_xIn_2S_4$ nanocomposites photocatalysts were synthesized by hydrothermal method, and characterized by UV-visible DRS. The prepared $Co_{0.5}Zn_{0.5}In_2S_4$ nanocomposites, which exhibits the highest H_2 production rate, the value is 36.08 umol·h^{-1}·g^{-1}, exceeding more than 8 times compared with the pure $ZnIn_2S_4$ and $CoIn_2S_4$. $Co_{0.5}Zn_{0.5}In_2S_4$ nanocomposites also has good reuse and stability. Two reasons lead to higher photocatalytic H_2 production activity are enough visible-light absorption capacity and suitable conduction band potential of the $Co_{0.5}Zn_{0.5}In_2S_4$. In this research, it is proven that $Co_{0.5}Zn_{0.5}In_2S_4$nanocompositesphotocatalystscan enhance the photocatalytic H_2 production activity. The significant improvement in the performance of $Co_{0.5}Zn_{0.5}In_2S_4$ nanocomposites is mainly due to effective charge separation.

1. Introduction

Using semiconductor as photocatalyst to produce hydrogen by water cracking under sunlight is an ideal way to convert solar energy into chemical energy efficiently [1-3]. However, TiO_2 has a band gap width of 3.2 eV and cannot respond to visible light [4, 5]. Since visible light accounts for a larger proportion of solar energy than ultraviolet light, in order to make more efficient use of solar energy resources, many studies have focused on the development of visible light responsive photocatalysts[6, 7]. On the basis of previous studies, metallic sulfide compounds [8-10] with suitable band gap width and edge position can make use of visible light, which has attracted more attention in the production of photocatalytic H_2.

In this study, we reported for the first time the preparation of $Co_{1-x}Zn_xIn_2S_4$ composite nanoparticles by hydrothermal method, and tested its performance of photocatalytic water decomposition in visible light irradiation to produce hydrogen. This study provides a new perspective for the understanding of photocatalyst response to visible light and the mechanism of promoting catalytic activity and provides a new idea for the design of new ternary metal sulfide photocatalytic materials.

2. Experimental

2.1. Catalyst Preparation

All the chemicals were of reagent grade and used as received without any further purification.

$Co_{1-x}Zn_xIn_2S_4$ solid solutions were prepared by the hydrothermal method. In a typical procedure, given amounts of $In(NO_3)_3$, $Co(NO_3)_3 \cdot 6H_2O$ and $Zn(CH_3COO)_2$ in different Co/Zn molar ratios were dissolved in 5 mL deionized water and 20 mL ethyl alcohol were stirred at room temperature for 0.5 h. The total amount of Co $(NO_3)_3 \cdot 6H_2O$ and Zn $(CH_3COO)_2$ was 0.5 mmol, $In(NO_3)_3$ was 1 mmol. At the same time, 2.5 mmol of thiourea ($(NH_4)_2S$) was added to the above solution as the S precursors, and in order to prepare a well-mixed solution, stirring was continued for 0.5 h and sonicated for 0.5 h. Subsequently, the mixture was transferred to a Teflon lining of an autoclave and heated at 180 ° C for 12 h. Finally, the obtained precipitate was collected by centrifugation and washed three times with distilled water and ethanol to remove unbound impurities, and then dried in air at 60 ° C for 12 h, and then ground for 1 h. The x of the prepared $Co_{1-x}Zn_xIn_2S_4$ solid solution is 0, 0.3, 0.5, 0.7 and 1.0, respectively.

2.2. Characterization of Catalysts

The UV-vis diffuse reflectance spectroscopy (DRS) of the samples were recorded using a Varian Cary 500 Scan UV-vis-NIR spectrometer with $BaSO_4$ as the reference. The reflectance spectra were transformed to absorption intensity by using KubelkaeMunk method.

2.3. Catalytic Performance

The photocatalytic reactions were carried out in a sealed 25 mL quartz tube reactor using a 300 W Xe lamp with a UV-cutoff filter (≥ 400 nm) as the light source. Under ultrasonic conditions, 25 mg of the solid catalyst powder was dispersed in 5.0 mL of a mixed solution containing 76 wt% CH_3OH and 24 wt% H_2O. The air in the reactor was evacuated and filled with high purity (99.999%)nitrogen, and then subjected to a photocatalytic reaction at room temperature for 12 h.The H_2 produced by the reaction were analyzed using an Agilent Micro GC3000 with a 5A molecular sieve column and a thermal conductivity detector.

3. Results and Discussion

3.1. UV–visDiffuse Reflection Spectra (DRS)

A comparison of UV−visible diffuse absorption spectra of the $Co_{1-x}Zn_xIn_2S_4$ (x = 0, 0.3, 0.5, 0.7 and 1.0) samples is displayed in Fig. 1. In the spectra of, with the increase of cobalt content in $Co_{1-x}Zn_xIn_2S_4$ nanoparticles, there is a significant and continuous red shift of absorption edge of the samples. By controlling the molar ratio of cobalt and zinc in the composite nanoparticles, the band gapof $Co_{1-x}Zn_xIn_2S_4$photocatalysts absorption edge redshift can be achieved. The spectra have been converted from corresponding diffuse reflectance spectra by means of the Kubelka-Munk function. Fig. 2 shows the curves of $(ahv)^{1/2}$vs (hn) for the samples, by extrapolating the linear portion of the curves to $(ahv)^{1/2} = 0$, the E_g values of $Co_{1-x}Zn_xIn_2S_4$ were evaluated to be less than 3.0 eV. As a result, the narrow band gap enables the photocatalysts to have visible light hydrogen production activity. This result is consistent with the increasing wavelength range absorption edge in semiconductor is related to the optical absorption edge energy decreases.

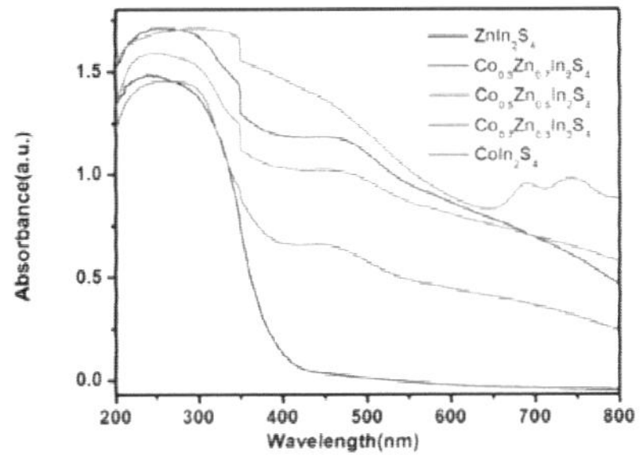

Figure 1. UV–vis DRS of the samples.

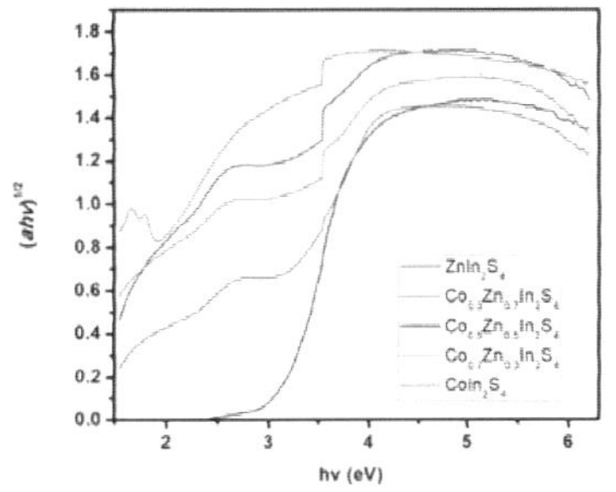

Figure 2. Plots of $(ahv)^{1/2}$vs (hv) for estimating optical band gaps of the samples.

3.2. PhotocatalyticPerformance of Samples

The photocatalytic activities of $Co_{1-x}Zn_xIn_2S_4$samples were evaluated by using the photocatalytichydrogen generation under methanol aqueous solution. The results are shown in Fig. 3.

As it can be seen from the figure, the comparison of the H_2 production rate ofphotocatalysts.

Compared with the samples of $Co_{1-x}Zn_xIn_2S_4$ (x = 0, 0.3, 0.5, 0.7, and 1.0), the visible-light photocatalytic H_2 activity of pure $ZnIn_2S_4$ and $CoIn_2S_4$ are almost negligible, and the H_2 production rate are 4.23 $umol{\cdot}h^{-1}{\cdot}g^{-1}$ and 4.91 $umol{\cdot}h^{-1}{\cdot}g^{-1}$ respectively. When the value of x is less than 0.5, the H_2 production rate of $Co_{1-x}Zn_xIn_2S_4$ increases with the increase of cobalt content; When the value of x is equal to 0.5, it reaches the highest 36.08 $umol{\cdot}h^{-1}{\cdot}g^{-1}$; However, when the value of x continues to increase, as the cobalt content is further increased, the photocatalytic H_2 production performance is rapidly deteriorated.

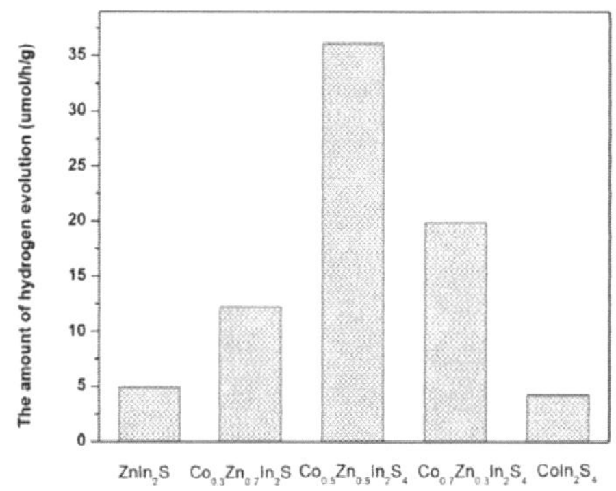

Figure 3. Photocatalytic H_2 production activity of the samples

4. Conclusion

In summary, $Co_{1-x}Zn_xIn_2S_4$catalysts were synthesized using the hydrothermal methods. The results show that $Co_{1-x}Zn_xIn_2S_4$ had the photocatalytic activity in the H_2 production reactions under 300 W Xe lamp irradiation. When the cobalt and zinc molar ratio is 1:1, the $Co_{0.5}Zn_{0.5}In_2S_4$ sample exhibits theoptimal H_2 production rate of 36.08 $umol{\cdot}h^{-1}{\cdot}g^{-1}$. This study not only used a simple and novel method to prepare $Co_{1-x}Zn_xIn_2S_4$ composite nanoparticles with high photocatalytic water decomposition activity for hydrogen production, but also provided a new perspective for understanding visible light absorption and photocatalytic enhancement mechanism.

5. Conflict of Interests

All authors of this article declare that there is no conflict of interests in this publication.

6. Acknowledgments

This work was financially supported by the Fujian Provincial Collaborative Innovation Center for Clean Coal Gasification Technology, the Fujian Provincial University Student Innovation Project (201611311044), the Key Project Young Natural Science Foundation of Fujian Provincial University (JZ160478), Outstanding Youth Scientific Research Talent Incubation plan in Universities of Fujian Province ([2017]52).

7. References

[1] KUDO A. Catalysis Surveys from Asia, 2003, 7(1): 31-8.
[2] CARLSSON J M. Nature Materials, 2009, 8(1): 76-80.
[3] MAEDA K, TERAMURA K, LU D, et al. Nature, 2006, 440(7082): 295.
[4] OHNO T. Applied Catalysis A General, 2004, 265(1): 115-21.
[5] S I, A O, R B, et al. Journal of the American Chemical Society, 2007, 129(45): 13790-1.
[6] ASAHI R, MORIKAWA T, OHWAKI T, et al. Science, 2001, 293(5528): 269-71.
[7] ANPO M, TAKEUCHI M. Journal of Catalysis, 2003, 216(1): 505-16.
[8] CHENG F, XIANG Q. Rsc Advances, 2016, 6(80)

[9] JING D, LIU M, CHEN Q, et al. International Journal of Hydrogen Energy, 2010, 35(16): 8521-7.
[10] ZONG X, YAN H, WU G, et al. Journal of the American Chemical Society, 2008, 130(23): 7176-7.

Electrodeposition of PPy/HNO_3 Film and its Capacitive Behaviour

Yong Chen[a], Xia Zhang[b], Juan Juan Li[c] and Hui Xu[d]

College of Perochemical Engineering, Lanzhou University of Technology, Lanzhou 730050, China

E-mail: [a]yongchen2003@126.com, [b]1316468736@qq.com, [c]1436678441@qq.com, [d]xuhui@lut.cn

Abstract. Through cyclic voltammetry polypyrrole/nitric acid (PPy/HNO_3) film was electrodeposited on stainless steel mesh. Its special structure like corals was demonstrated by scanning electron microscopy (SEM). Various approaches such as cyclic voltammetry (CV), galvanostatic charge-discharge (GCD) and electrochemical impedance spectrum (EIS) were imployed to investigate the electrochemical performance of PPy/HNO_3 film. The results showed that the specific capacitance of PPy/0.5M HNO_3 can reach to 596 $F\cdot g^{-1}$ at a current density of 5 $mA\cdot cm^{-2}$; even 1000 cycles the specific capacitance still maintained 96% at a scan rate of $50mV\cdot s^{-1}$. All data illustrate that PPy/HNO_3 film may become a proming candidate for supercapacitor applications.

1. Introduction

Supercapacitors, featuring excellent power density, long cycling life and fast rate of charge–discharge, become more important category of energy storage devices than rechargeable batteries and conventional capacitors [1, 2]. As we all known that one of crucial factors determining the supercapacitor performance is electrode materials where transition metal oxides, carbon materials and conducting polymers are commonly used. Among these materials, PPy has attracted more attentions in view of its high conductivity, high specific capacitance (SC) and low cost [3].

Usually, two methods including electrochemical and chemical polymerization can be used to synthesize PPy[4]. In terms of polymer films, electrodeposition is a better preparation way in solutions containing pyrrole monomer. However, most investigations prove that PPy exibits poor SC without dopants because the high SC of PPy largly depends on the doping/dedoping process of dopants that allows the charge storage in the material[5].Therefore, many inorganic and organic acids are employed as dopants during polymerization process. For instance, Zhang et. al.[6] prepared PPy film on the stainless steel electrode via pulse galvanostatic method, and the specific capacitance was 403 $F\cdot g^{-1}$ when p–toluene sulphonic acid was used as dopants. Furthermore, high active PPy owned SC of 480 $F\cdot g^{-1}$ was electrodeposited on Ti foil by using oxalic acid as dopants through cyclic voltammetry at a scan rate of 200 mV/s[7]. R. Ramya et. al. [8] employed p-toluene sulphonic acid as dopants to synthesize PPy on stainless steel electrode via the galvanostatic polymerization, and SC of 200 $F\cdot g^{-1}$ was achieved at 0.8 $mA\cdot cm^{-2}$. X. Li[9] electrodeposited PPy films with 5-sulfosalicylic acid (SSA) as dopants on stainless steel substrates by galvanostatic and pulse method, and 545 $F\cdot g^{-1}$ was the highest SC obtained at a scan rate of 2 $mV\cdot s^{-1}$. Unfortunately, up to now, the SC of PPy reported by literatures is lower than the theroiotical value(620$F\cdot g^{-1}$) [10]. Thus, more efforts are on the way to enhance the elecetrochemical property of PPy.

In this work, HNO_3, one of inorganic acids, was utilized as not only electrolyte but also dopants to fabricate PPy films on low cost stainless steel mesh substrates by cyclic voltammetry. As expected, a SC as high as 596 F·g^{-1} can be provided by the PPy films at a current density of 5 mA·cm^{-2}, which is higher than above reported literatures.

2. Experimental

Pyrrole was purified before use. Electropolymerization was carried out on CHI660B electrochemical work station by using HNO_3 with different concentration of 0.1, 0.3, 0.5, 0.7 and 0.9 M, containing 0.1M pyrrole monomer. During electrodeposition, a standard calomel electrode(SCE) was used as reference electrode, and counter electrode and working electrode were supported by two stainless steel(316L,1×1cm), respectively. At a scanning rate of 5 mV·s^{-1}, 31 segments CV curves were performed to obtain the PPy films under the scanning potential from -0.2 to 0.9 V. After that, 0.5 mol·L^{-1} HNO_3 and distilled water were used to rinse the deposited PPy films to remove the soluble species. Then, the obtained PPy films were dried for 12h at 60 °C

The samples used in all tests were scraped from stainless steel mesh substrates. Scanning electron microscopy (SEM: J EOL JSM-6701F) was utilized to characterize the morphology of PPy films. The electrochemistry properties of samples were conducted in 0.5M H_2SO_4 electrolyte with a SCE eletrode as the reference electrode and a platinum sheet as the counter electrode. The potential window was chosen between -0.3 and 0.7V when cyclic voltammograms(CV) and galvanostatic charge–discharge (GCD) were tested. Electrochemical impedance spectra (EIS) were recorded with the frequency ranging from 10^5 to 10^{-2} Hz under the amplitude of 0.5mV.

3. Results and Discussion

Figure 1 shows the relationship of deposition yield of PPy films and the concentration of HNO_3. It can be seen that the depositon yield increases with an increase of H^+ concentration at the beginning, because the higher acidity the solution, the higher protonation degree of pyrrole. It suggested that H^+ played an important role during the eletrochemical polymerization of PPy[12]. However, with the further increasing of H^+ concentration, the depositon yield is decreased, which is contributed to the decline of doping amount and the hydrolyzation of PPy. The morphology of PPy films is displayed in Figure 2. Obvilusly, PPy films exhibit coralline structure, which looks like an aggregation of PPy pariticles. The structure makes material loose and porous, benefitting for the improvement of its SC.

CV curves of PPy with different concentration of HNO_3 are depicted in Figure 3(A) at a scan rate of 5mV·s^{-1} in 0.5M H_2SO_4 electrolyte between -0.3 and 0.7V. It is evident that the PPy films show good capacitive behavior, as indicated by distorted box shape CVs without obvious redox peaks. Furthermore, it is obviously observed that the area of CV curve of PPy/0.5M HNO_3 is much larger than others, implying the existence of a higher SC. At the same time, the conclusion can be proved by GCD test at current density of 5mA·cm^{-2}, which is presented in Figure 3(B). It is easily found that the shape of all lines are bended seriously rather than ideal straight lines, which derives from the faradic reaction between the surface of electrode and the electrolyte. The following equation is used to calculate the specific capacitance (C_m) of PPy films according to the discharge time in the GCD curves[13].

$$C_m = I\Delta t/m\Delta V \qquad (1)$$

Where C_m(F·g^{-1}), I(A), Δt(s) and ΔV(V) represtent specific capacitance, charge/discharge current, the discharge time and the potential drop in the discharge process, respectively. m(g) obtained by subtraction method is the mass of active material. Therefore, the correspondingly calculative results of SC of PPy films are attributed to 93, 355, 596, 242 and 135 F·g^{-1} with the HNO_3 concentration of 0.1, 0.3, 0.5, 0.7 and 0.9M, respectively. Obviously, the highest SC of 596 F·g^{-1}of PPy film is obtained in 0.5 M HNO_3, which is in accordance with the result of CV test.

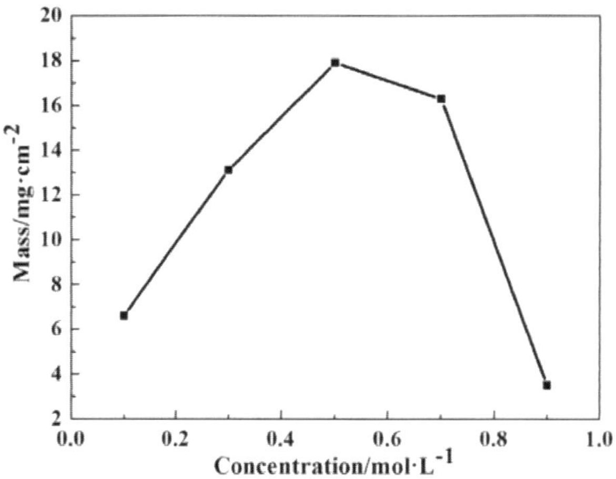

Figure 1. Deposited mass of PPy on stainless steel.

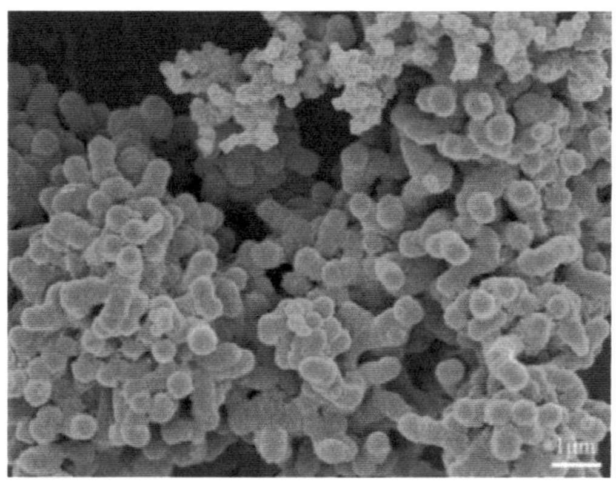

Figure 2. SEM image of PPy film prepared in 0.5 M HNO_3.

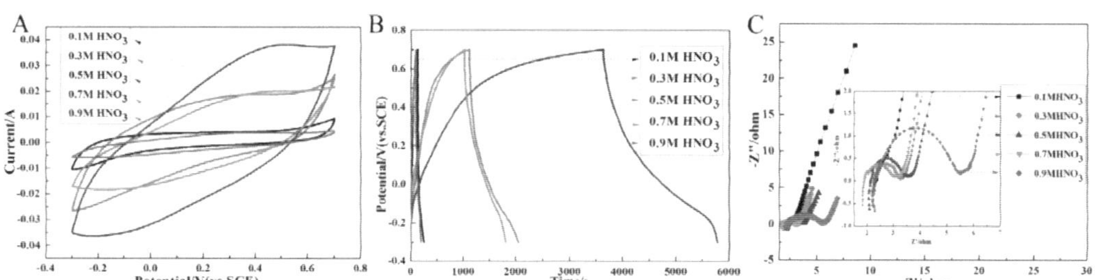

Figure 3. (A) CV curves of PPy/HNO_3 films at a scan rate of $5mV\cdot s^{-1}$. (B) Charge–discharge curves of PPy/HNO_3 films at current density of $5mA\cdot cm^{-2}$. (C) EIS of PPy/HNO_3 films.

Only having the analyses of CV and GCD is not enough to estimate the electrochemical properties of PPy films. Another indispensable approach is EIS test, which can offer more information about charge-discharge kinetics. As is well-known, the curves are usually composed two parts in different frequency region, one is a semicircle, the other is a titled straight line. Through the intercept of the semicircle on the real axis in high frequency, we can easily tell the solution resistance. Accordingly, the diameter of the semicircle directly points out the charge transfer resistance (Rct) that exists on the electrodes/electrolyte interface. Moreover, in low frequency region the tilted straight lines are used to verify a limiting diffusion process that happens in electrolyte to explain the behavior of ideal capacitive[14]. The Nyquist plots of all samples are shown in Figure 3(C). We can clearly observe that the film prepared in 0.5 M HNO_3 have a smaller charge transfer resistance and nearly vertical line, which is beneficial for capacitive behaviour.

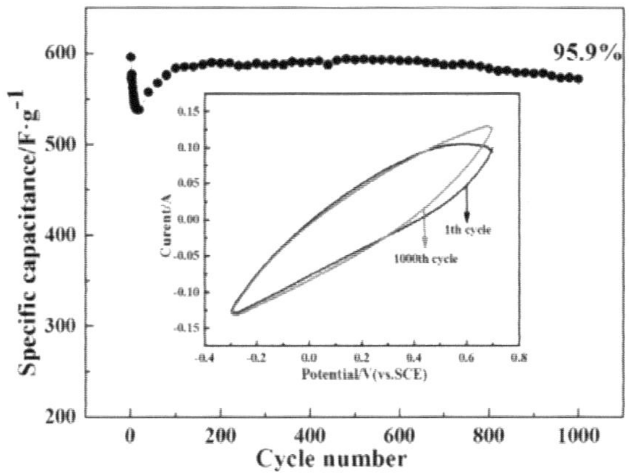

Figure 4. Sepefic capacitance versus cycle numbers at a scan rate of 50 mV·s^{-1}(insert depicts the difference of CV curves of first cycle and 1000th cycle).

Another important indicator to use as electrodes in supercapacitor is the cycle stability of materials. Employed voltammetric cycles, we evaluated the cycle stability of PPy films in 0.5M HNO_3 solution. It can be seen in Figure 4 that an obvious decrease of specific capacitance that tied with the degradation of incompletely formed polymer or some un–doped PPy, occurs at the beginning, after that increase slightly because of the changes of electrode microstructure[15]. After 1000 cycles, a decrease of only 4% of the SC is observed, in other words, the SC still remains 96% of the initial value, indicating a good cycling stability. Moreover, the shapes of the 1st and the 1000th CV curves are similar each other, which can be observed in the inset in Figure 4.

4. Conclusions

Polypyrrole films with a high-performance were successfully synthesized by the cyclic voltammetry technique in nitric acid solution. It was found that, during the elctrodeposition process, the nitric acid not only provides a acidic condition, but also employs as dopant ions, which is favourable to improving electrochemical properties of PPy. When the concentration of HNO_3 is 0.5 M, the highest specific capacitance was reached to 596 F·g^{-1} at a current density of 5 mA·cm^{-1}. Furthermore, a decrease only 4% of the specific capacitance was observed after 1000 cycles. Therefore, a conculsion can be drawn that the PPy films deposited on low cost stainless steel substrates are excellent candidates for the electrode of supercapacitors.

5. Acknowledgements

This work was supported by the National Natural Science Foundations of China (51503092) and the Natural Science Foundation of Gansu Province (2017GS10857).

6. References

[1] Gogotsi Y, What nano can do for energy storage, 2014 Acs Nano 8 5369-71.
[2] Liew S Y, Thielemans W and Walsh D A, Electrochemical Capacitance of Nanocomposite Polypyrrole/Cellulose Films, 2010 Journal of Physical Chemistry C 114 17926-33.
[3] Lou X W, Deng D, Lee J Y, Feng J and Archer L A, ChemInform Abstract: Self-Supported Formation of Needlelike Co3O4 Nanotubes and Their Application as Lithium-Ion Battery Electrodes, 2008 Advanced Materials 39 258-262.
[4] Chen S and Zhitomirsky I, Capacitive behaviour of polypyrrole, prepared by electrochemical and chemical methods, 2014 Materials Letters 125 92-95.

[5] Khomenko V, Frackowiak E and Béguin F, Determination of the specific capacitance of conducting polymer/nanotubes composite electrodes using different cell configurations, 2005 Electrochimica Acta 50 2499-2506.
[6] Zhang J, Kong L B, Li H, Luo Y C and Kang L, Synthesis of polypyrrole film by pulse galvanostatic method and its application as supercapacitor electrode materials, 2010 Journal of Materials Science 45 1947-54.
[7] Fan L Z and Maier J, High-performance polypyrrole electrode materials for redox supercapacitors, 2006 Electrochemistry Communications 8 937-40.
[8] Ramya R and Sangaranarayanan M V, Analysis of polypyrrole-coated stainless steel electrodes-Estimation of specific capacitances and construction of equivalent circuits, 2008 Journal of Chemical Sciences 120 25-31.
[9] Li X and Zhitomirsky I, Capacitive behaviour of polypyrrole films prepared on stainless steel substrates by electropolymerization, 2012 Materials Letters 76 15-17.
[10] Liu A S and Oliveira M A S, Corrosion control of aluminum surfaces by polypyrrole films: influence of electrolyte, 2007 Materials Research 10 205-209.
[11] Roberts M E, Wheeler D R, Mckenzie B B and Bunker B C, High specific capacitance conducting polymer supercapacitor electrodes based on poly(tris(thiophenylphenyl)amine), 2009 Journal of Materials Chemistry 19 6977-79.
[12] Mandić Z, Duić L and Kovačiček F, The influence of counter-ions on nucleation and growth of electrochemically synthesized polyaniline film,1997 Electrochimica Acta 42 1389-1402.
[13] Zhang J, Kong L B, Wang B, Luo Y C and Kang L, Electrochemical polymerization of polyaniline doped with Zn2+ as the electrode material for electrochemical supercapacitors, 2014 Journal of Solid State Electrochemistry 18 813-19.
[14] Wei H, Zhu J, Wu S, Wei S and Guo Z, Electrochromic polyaniline/graphite oxide nanocomposites with endured electrochemical energy storage, 2013 Polymer 54 1820-31.
[15] Shi K and Zhitomirsky I, Influence of current collector on capacitive behavior and cycling stability of Tiron doped polypyrrole electrodes, 2013 Journal of Power Sources 240 42-49.

IOP Conf. Series: Materials Science and Engineering **484** (2019) 012046 doi:10.1088/1757-899X/484/1/012046

Market-Oriented Operation Pattern of Regional Power Network Integrated With High Penetration Level of Distributed Energy Resources

Qing Duan[1], Jian Le[2*], Xi Ye[2], Zhipeng Lv[1] and Qian Zhou[2]

[1]China Electric Science Research Institute Co., Ltd., Beijing, 100192, China
[2]School of Electrical Engineering, Wuhan University, Wuhan 430072, Hubei Province, China
* lej01@mails.tsinghua.edu.cn

Abstract. This paper presents a novel market-oriented operation pattern for regional power grid to meet the demand of market integration at the electricity retail side, especially under the background of high penetration of distributed energy resources accessed. Firstly, the market organizational structure is designed for the proposed market-oriented operation pattern, and characteristics of this operation pattern such as the full marketization and the openness to distributed energy resources are analyzed through the introduction of operation mechanism. Then, an energy management system, which is suitable for the proposed market-oriented operation pattern, is realized by the design of regional network energy management, regional source/load energy management and the effective interaction of these two parts. The energy management processes based on electricity trading plan are proposed. Finally, the main technical problems needs to be addressed are presented. This paper provides a reference for promoting the market integration process of electricity retail side with high penetration of distributed energy resources accessed.

1. Introduction

Awareness to the demands for reducing environmental pollution, achieving sustainable and diverse energy systems and improving energy efficiency, the penetration of new-energy power generation has been growing rapidly in power systems across the world [1, 2]. By the end of 2014, the growth rate of the global new-energy generation capacity had reached 19%, accounting for 6.2% of the total generation capacity. From 2010 to 2011, the EU launched the ambitious"20-20-20" energy strategy for member countries, aiming to increase the share of new energy and renewable energy in energy consumption to 20% by 2020[3]. Meanwhile, the new energy development strategy and action plan of China has proposed that non-fossil fuels and natural gas should account for more than 15% and 10%, respectively, of primary energy consumption in 2020.This paper will cover modular, Distributed Generation (DG) with small capacity, such as renewable energy source-based wind power, solar power and hydropower, and will discuss Distributed Energy Storage (DES)represented by electric cars that operate in the Vehicle-To-Grid(V2G)model. It will also consider Demand Response (DR) program. These topics are collectively known as Distributed Energy Resources (DERs) [4].

At the same time, the liberalization process of the electricity market that is characterized by full competition and complete deregulation also attracts more and more attention. Many countries have initiated market-oriented reforms of power generation and transmission. Currently, marketizing the sales side of electricity has become a trend and a popular research topic in power system reformation and development across the world [5].Central power reformation document No.9 of China intends to

IOP Conf. Series: Materials Science and Engineering **484** (2019) 012046 doi:10.1088/1757-899X/484/1/012046

"liberate the sales side of electricity and cultivate major market players through multiple channels". Currently, more than 30 electricity sales companies have been founded and registered nationwide, and a number of state-owned enterprises are actively preparing for electricity sales. Thus, electricity sales market reformation will be a trend in the future. Considering that most DERs will be accessing low-voltage distribution networks, we selected regional power network (RPN), such as a large city power grid, to study and design the market operation mechanism of its electricity sales.

A comprehensive analysis of these two aspects suggests that a market-oriented operation with high DER penetration is the inevitable new form and requirement faced by future RPN systems. Such DER penetration will bring new challenges and opportunities to its market-oriented operation, as discussed below.

•Novel market-oriented operation pattern of RPN is required to facilitate the integration of DER. Currently, there are two situations standing in the way of marketizing DER access. On the one hand, DER access is facing low status in the market; for instance, setting capacity constraints for market participation [6] prevents DER from participating in the wholesale electricity market competition [7], while excessively high economic penalties for power imbalances [8] discourage DERs from participating in the market. Moreover, technologies that place cutting out DERs as the primary consideration during operation scheduling affect the operational economy of DERs. These challenges are mainly due to power system operation safety considerations, the small capacity of DERs, the random and fluctuating power output of DERs, and the lagging development of DER scheduling and control technology. On the other hand, DER access is also facing high status in the market, for instance, because of excessively high subsidy programs and the requirement for renewable energy to be entirely integrated and consumed, thereby resulting in unilateral, unregulated growth of DER penetration that is separated from the actual electricity utilization scale and from the operation capability of the supply network.

•DER access must be considered in RPN marketization. For example, there are large numbers of DERs with different interest demands and highly distributed ownership/management/control; these represent new conditions that were neither present nor foreseen during the marketization of conventional power grids. Thus, DER-adapted market-oriented operation mechanisms and implementation methods must be designed for RPNs. In contrast, since marketization began, regional grid operators have long been caught in situations in which they can only purchase electricity from external grids; the sales of electricity is liberated while the power sources for purchase are very limited. Therefore, DER access provides a good base and creates conditions that will help to solve this dilemma.

This paper focuses on research into business operation models and the associated implementation programs for marketizing RPNs to meet the needs of RPN market-oriented operation in the context of high DER penetration. First, we analyze the current research on electricity sales marketization with DER access, and then, we propose an operation pattern for RPN marketization based on the above analysis, discussing its organizational structure and operational mechanism. We design an energy management system (EMS) that fits the business model and is not only based on but also achieves organic integration of the regional sources/loads and the regional network energy management; we further propose an energy management procedure based on the electricity trading plan. Phased steps to implement the proposed market-oriented RPN business model are given, and the key technical issues that must be addressed are highlighted.

2. Research Status of Der Access And Electricity Sales Marketization

To eliminate the impacts on the power grid and on users caused by simply integrating into the grid a large number of DERs that are scattered in different locations and have various forms and different characteristics and to improve the market participation of DER and its capability to provide ancillary services, the primary idea is to aggregate multiple DERs into an independent integrated unit in external markets and electricity networks. Currently, the Micro Grid (MG) [9] and Virtual Power Plant (VPP) [10] are recognized to be the most effective way in this domain and have been widely studied.

MGs can work in either the grid-paralleling mode or the island mode through control and coordination [11-13]. Reference [14] used a centralized controller to maximize the benefits of MGs by considering the price of electricity exchange with external grids and the internal energy demand of MGs under capacity constraints for power exchange with external grids and the active power contract between micro-sources and the main controller. Additional related details can be found in [15].

In contrast to MGs, which are more concerned with their own coordination/control and internal economic operation, the concept of VPP has been proposed as a way to improve the economy of DER operation and to help achieve the safe and economical operation of power systems [16]. Based on the functions that a VPP can fulfil, there are two main types of studies on this subject [17]. The first is a Commercial VPP (CVPP), whose basic function is to develop optimal generation schemes based on user-demanded load forecasting and power potential prediction and to participate in market bidding. A Technical VPP (TVPP) ensures that the entire system operates in a safe and optimized way; accounting for physical limitations and the potential services a VPP can provide.

CVPP market operation has bi-level meaning; on the one hand CVPP is regarded as an ordinary participant and competes with other participants in a variety of external markets. Reference [18] analyzed the advantages and disadvantages of the three markets that are accessible to CVPPs, namely, the long-term futures market, the day-ahead market (DAM) and the continuous intraday market (CIM), and noted that uncontrollable DERs fit better in a DAM. Currently, DAM and CIM are most often considered in CVPP studies. The overall time frame and the corresponding operation mechanism of these markets are introduced in [19]. In situation when the total capacity of VPP is considered to be small enough, the VPP is considered to be a price taker and its bid is thus always accepted, and the market price is defined by the other participants [20].The more popular way is that all market participants including VPPs submit bids and offers with price and MW pairs based on the estimation of its cost, potential income and risk appetite. [21] is concentrated on the participation of a VPP in a joint market of energy and spinning reserve service. A non-equilibrium model is proposed to design the optimization bidding strategy of VPP simultaneously for both markets. An optimal bidding strategy for a CVPP to compete with traditional thermal power plant in the wholesale market was designed based on Nash equilibrium game theory in [22]. A more detailed study and analysis on the bidding strategies of participants in the electricity market can be found in [23].

Currently, VPP scheduling optimization strategies mainly aim to maximize the revenue and to minimize the operation cost of VPP, and they use optimization algorithms to solve for the given output power of each DER by considering equality constraints, such as the active/reactive/power/heating power balance, and inequality constraints, such as the DER output power limit, the limit of the output power change rate, the network line flow limit and the node voltage limit [24]. The benefit/cost calculation for electricity purchase/sale is also typically considered.

To solve the practical scheduling problem that VPP output power may deviate from a previously promised level and thus result in economic penalties, controllable DG, DR and DES are often aggregated into VPPs to reduce the fluctuation and randomness in the output power of VPPs [25]. Another important method to reduce the output power imbalance is phased scheduling. Phased scheduling methods are generally divided into two steps, namely, advance offline planning, which is based on prediction data, and online scheduling, which uses real-time data during operation [26]. Reference [27] employed a two-phase scheduling method to establish an optimized stochastic planning and scheduling scheme that includes new-energy generation and a DR; the scheduling strategy does not require knowledge about the stochastic process probability distribution of the new-energy generation and DR, thus minimizing the mathematical expectation of economic penalties for imbalances.

Studies on the market-oriented operation of electricity sales mainly focus on basic theories, such as the natural monopoly theory, control theory and electricity market theory, on market structure and operational mechanism design [28], and on market participants and power acquisition optimization decisions [29]. Increasing numbers of studies have begun to pay attention to electricity market operations with new-energy generation access [30, 31].

Reference [32] studied the general evaluation and planning framework required by electricity distributors, focusing on DG access. Reference [33] proposed a new system structure and scheduling/control algorithm that can self-recover when there are multiple MGs accessing the system. The top-level goal is to connect the EMSs of multiple MGs to achieve global optimization and to support the fault area when there is a system error by coordinating multiple MGs. Reference [34] considered both the increased DER penetration in distribution networks and the demands of electricity marketization and designed a double-auction electricity market for a residential area. Reference [35] proposed a two-level decision-making model for distribution companies to participate in the DAM. The model involves two additional energy resources that distribution companies can take advantage of, namely, interruptible loads and distributed generation.

On the whole, current studies on electricity sales marketization have begun to focus on the impact and utilization of DERs; however, there are no effective ways to integrate a large number of scattered small-capacity DERs into the market operation framework. Nonetheless, when using MGs and VPPs to aggregate DERs, most studies have focused on the control and scheduling strategies of VPPs and MGs themselves, and therefore, the methods that are employed are suitable only for specific scenarios and fail to consider VPPs in the context of market-oriented electricity sales.

3. Organizational Structures and Mechanisms of Market-Oriented Operation Pattern

This paper describes the design of a new market-oriented regional power grid (RPN) operation pattern to meet the demands of electricity sales-side marketization with high DER penetration. The organizational structure of the market operation with this model is shown in Fig.1.

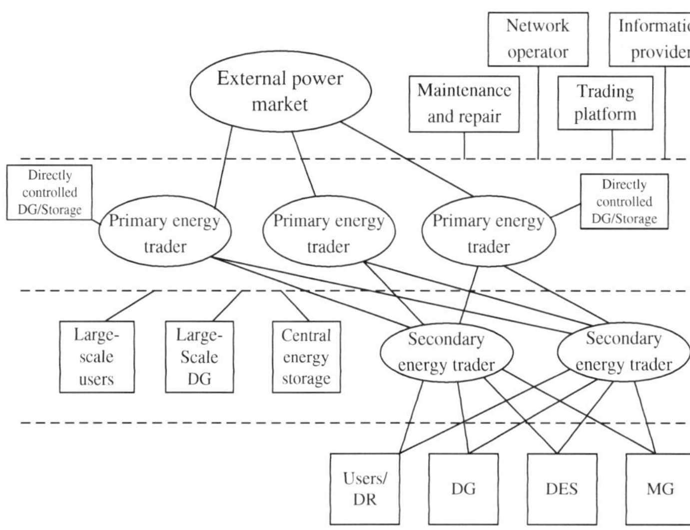

Figure 1. Organizational structure of the market-oriented operation pattern of RPN

Primary energy traders that are directly involved in trading with external power markets and are qualified to supply power to the RPN users are located at the first level. In contrast to traditional energy traders, the primary energy traders in this structure are pure interest entities and, under ideal conditions, need not have any ownership in or control of any network, power source, energy storage system, users or information sources. The operating mechanism is to use motivational incentives to dynamically aggregate the large-capacity DGs, centralized energy storage systems, secondary energy traders, and large-capacity users/DRs that are shown in the second level of Fig.1 to form an energy portfolio that participates in trade with external markets and supplies power to users. The number of primary energy traders can be set according to the size of the RPN; access mechanisms similar to the "license" can also be developed, and such licenses may be publicly auctioned to attract the social capital for participation.

The second level primarily consists of a large number of DGs, centralized energy storage systems, secondary electricity traders and large-capacity users/DRs. Each can decide for themselves whether to join the electricity portfolios of one or more primary electricity traders through long-term bilateral contracts. They may choose to do so base on their own situations, their particular objectives or the incentives (mainly price) of programs in the energy portfolio, thereby cutting their costs or increasing their revenues for electricity generation. Specifically, secondary electricity traders can also use business models resembling those of primary traders to aggregate lower-level users/DRs, DESes, DGs or even MGs with motivational incentives to form their own portfolios. Similarly, various entities at the lower level can also dynamically choose for themselves to join one or more of the electricity portfolios of the upper-level traders until all of the DERs and users in the RPN are covered.

In addition, as shown in Fig.1, we also propose the establishment of multiple third-party professional agencies that are adapted to the RPN marketization business model and that focus on handling certain global business interests of the RPN. These third-party agencies are also entities with clear interests.

It is clear that the market-oriented RPN operation pattern shown in Fig.1 will be able to solve the major problems future power grids will face. In terms of marketization, all of the market participants are clearly interest entities that will enhance their revenues and reduce costs and risk of loss through competition. For example, electricity traders at all levels must compete against other traders at the same level in the upper-level electricity market; given that the electricity demands in their own portfolios are met, they can expand their gains by optimizing their participation in the upper-level electricity generation portfolio. Meanwhile, they also compete with other traders at the same level in the lower-level market and motivate interest entities at the lower level to participate in their electricity portfolios by optimizing their price quotes in the lower-level market. In terms of handling access to DERs with high penetration, a market-layering mechanism can be used to expand the size of the energy portfolio created by DER aggregation, thereby improving the opportunities of DERs to participate in market trading, reducing the risk of individual power imbalances, and promoting coordination among the various DERs. Meanwhile, various DERs will be able to independently choose whether to participate in the upper-level energy portfolio plan and to adjust and optimize their operation schemes based on the market, thus motivating capital investment in DERs. Meanwhile, DER investment strategies can also be optimized through market-based mechanisms to assist in making DER investment decisions that are appropriate for the current electricity demand and the grid size; such strategies will also automatically suppress DER penetration beyond its actual level.

Due to space limitations, the following discussion will include only the key components involved in the operation pattern, and only the implementation steps and major technical problems to be solved in this model will be covered.

4. Market-Oriented Rpn Energy Management

4.1. EMS Design

The market-oriented RPN EMS is designed to meet the demand for the safe and economical operation of the regional grid and to implement energy management for DERs and users in the context of market-oriented operation. To meet these requirements, the regional grid EMS is divided into two parts in this paper, namely, energy management for the regional network and energy management for the sources/loads in the region. The regional network EMS (NEMS) is responsible for the safe and economical operation of the regional network, and the regional source/load EMS (SLEMS) is designed to achieve market-oriented energy scheduling and control. Meanwhile, the two EMSs can interact and coordinate while controlling the network and its sources and loads, thereby allowing energy management for the entire RPN.

Fig.2 shows the schematic structure of the proposed EMS that is adapted to the market-oriented RPN operation pattern.

A network EMS (NEMS) is based on a dedicated monitoring communication network and can be realized using a network-based control model design [36]. Communication protocols, such as IEC61850 standards [37], can be used to enhance the scalability and compatibility of the system. All of the measuring units used for regional network scheduling and control send the monitoring information to a network information database through the communication network. The network information database integrates processes and stores the information. The regional network operator can then make energy management-related decisions based on the network information database and can send appropriate control commands through the equipment in the monitoring communications network, thus enacting scheduling and control tasks.

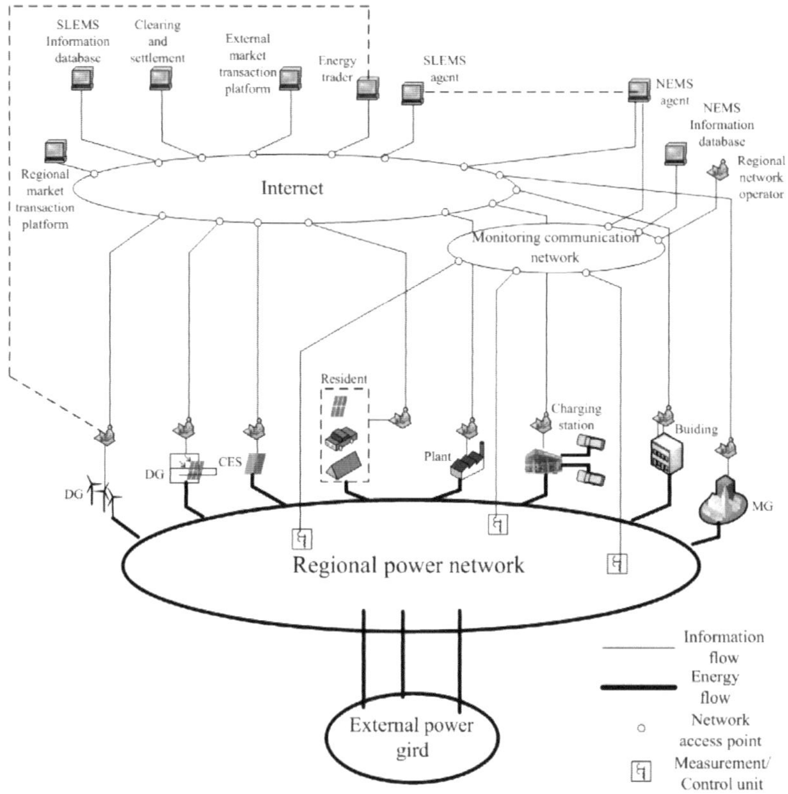

Figure 2. Schematic diagram of the energy management system architecture

A regional SLEMS uses the Internet as the communications network and is realized by using a fully distributed control model design. Energy traders at all levels and DERs/users all connect to the Internet as clients, where they can conduct a variety of electricity-trading activities on the regional electricity market trading platform and can make their own control decisions, based on market transactions they have completed, to fulfill their scheduling and control tasks. In addition, electricity traders may have direct lines of communication with some of the DERs, shown as a dotted line in Fig.2, that allow them to directly control and schedule these DERs through the ownership protocol for their own needs. Information produced by the SLEMS during energy management is organized and maintained by the SLEMS information database.

The communications between a NEMS and a SLEMS are managed by a NEMS agent that connects to both the Internet and the monitoring information network. This agent is owned, managed and controlled by the NEMS. This arrangement is designed mainly to facilitate compatibility with existing regional EMSand to meet the rigorous security requirements of NEMS systems. A dedicated SLEMS agent can also be set up inside the SLEMS, and the two agents can communicate through a direct communication line (shown as the dotted line connecting the two agents in Fig.2).

Given the above connection, the two EMSes can interact in the following ways. From the viewpoint of the SLEMS, the NEMS can initiate market trading according to its own considerations, such as balancing the market, providing auxiliary services to the market, and restoring market trading for the electricity network; a NEMS can accomplish these tasks using its own market-trading mechanisms. Thus, in this way, the NEMS can be treated as an ordinary electricity trader in the SLEMS. A NEMS can provide the historical/real-time network status information that is required for

IOP Conf. Series: Materials Science and Engineering **484** (2019) 012046 doi:10.1088/1757-899X/484/1/012046

making control decisions to various parties trading in the market inside the NEMS; thus, in this way, the NEMS can be treated as a data source in parallel to the SLEMS information database. A NEMS can use its monitoring capabilities over power network equipment to participate in market trades that other electricity traders initiate within the SLEMS, for example, in the highly reliable power supply market; in this way, the NEMS can be treated as a special DER. A more important possible function of a NEMS is to approve the trading plans of the market and to monitor/schedule the implementation of energy management in the SLEMS in the context of the safe and economical operation of the RPN; in this way, the NEMS can be treated as a special supervisor of the SLEMS.

Within a NEMS, a SLEMS can likewise be considered as a special monitoring information source and a controllable scheduling resource. Users and DERs within the SLEMS all have local monitoring capabilities, and historical/real-time monitoring information can be transmitted to the NEMS and serve as a source of supplemental information or of backup/verification data. A market-oriented business model typically does not implement programs that use a NEMS to directly control loads/sources within the SLEMS; instead, indirect source/load scheduling and control are achieved by initiating the corresponding market trading within the SLEMS.

4.2. Energy Management Implementation

The regional grid energy management process in the market-oriented operation pattern can be initiated by different objects at different management phases and with different management objectives. In this paper, the regional grid energy management process is designed to use the electricity trading plan as the carrier. The energy management process that is based on the electricity trading plan is shown in Fig.3.

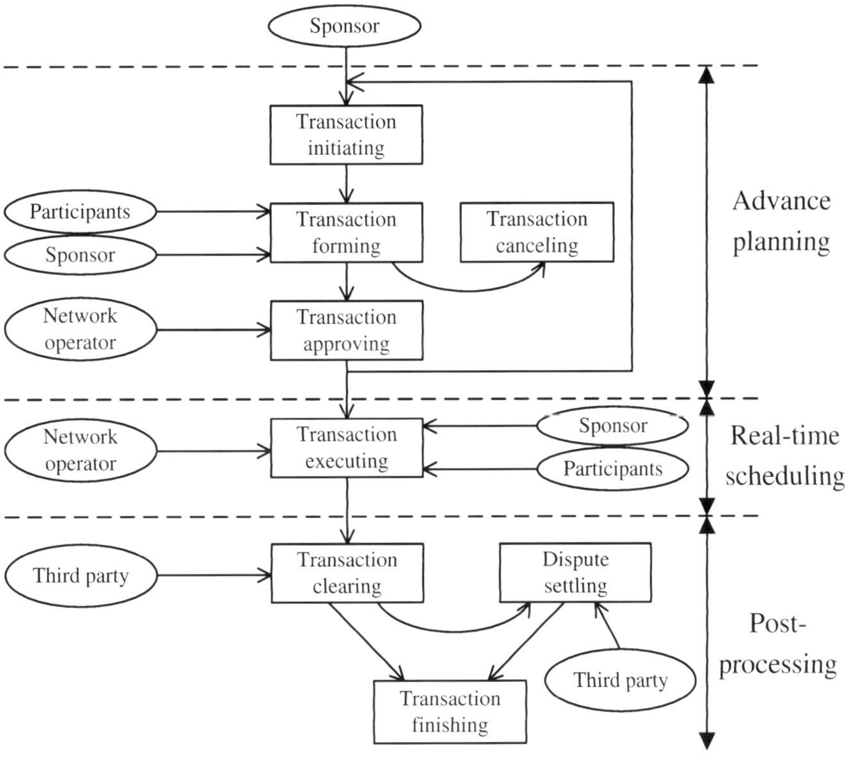

Figure 3. Energy management process based on transaction plan

The energy management process is divided into 3 phases, namely, advance planning, real-time scheduling and post-processing.

The advance-planning phase of energy management can be further divided into phases such as transaction initiating, transaction forming and transaction approving, according to the status of the electricity trading plan. Typically, the energy management process is triggered by a trader at any level sending out an energy trading plan. By using data such as historical trading information in their local database and the SLEMS database, in addition to historical/real-time network operation information in the NEMS database, traders at all levels can autonomously make control decisions and release electricity portfolio purchase plans on the regional market trading platform. The plan contents can vary depending on the market involved. For example, for long-term contract markets, a plan may include the contract period, the electricity, and the price of the contract and default clauses; for a DAM, the plan may include pricing for phased electricity and the imbalance penalty. Capacity limits may be set to prevent entities that do not meet the conditions from participating, and different prices can be set according to the capacity levels and lengths of participation to attract high-quality entities to participate in the released electricity purchase portfolio plan as soon as possible. Traders at all levels will also release electricity sale portfolio plans, which may include electricity sales plans in the form of long-term contracts for large-capacity users and short-term electricity sales plans for small-capacity users. In addition, different prices can be set for electricity sales plans with different reliability levels to achieve the objective of high price for high quality.

In the formation phase of a transaction, given that the limiting conditions set by the plan are met, users and DERs, or even energy traders at all levels, may participate in the trading plan and choose to join a certain plan in the regional market trading platform based on historical information in their local database, the electricity purchase portfolio, or the electricity sales portfolio plan. The program selection decisions typically aim to maximize gains, and they are based on the available data and capability of a participant. For large-capacity users and DERs, their local EMSes can implement the decision-making process.

The plan sponsors can conduct real-time tracking and assessment of the implementation of the trading plan that they initiated. Without prejudice to the interests of the plan participants, they can modify the trading plan to maximize the expected benefits and to achieve early implementation of the trading plan. Participants in the plan can also perform real-time tracking and evaluation on other existing trading plans or opt-out of their current plan and join another trading plan. Through the process described above, various trading plans will eventually be implemented, and those that cannot be implemented will be withdrawn or re-launched.

At the trading plan approval phase, the power grid operator can comprehend the various electricity purchase and sales portfolio plans that have been implemented, together with the power supply status of the RPN, to evaluate and predict the operation of the network. They can also approve trading plans that meet the network operation requirements and develop appropriate plans that adjust the network to ensure that the approved trading plans can be successfully implemented; for trading plans that do not meet the requirements for network operation, they can ask the trading plan sponsor to modify the plan, noting for the plan sponsor's reference conflicts such as time, location, and the amount of power. For plans that meet the operational requirements but do not permit the economical operation of the network, they can induce the traders to modify their plans by means such as adjusting the fees for using the regional network. Once approved, a trading plan can enter the actual implementation phase.

The real-time scheduling phase of energy management mainly refers to the actual implementation of the trading plan. The EMSes of DERs and users can make independent decisions, based on the trading plans in which they participate, about how to control DERs and electrical equipment to optimize their own energy management while ensuring that their electricity in the network is compliant with the requirements of the trading plans so as to reduce the financial penalties caused by discrepancies with the trading plan. In the actual scheduling process, real-time measurements of local power will also be sent to the trading plan initiators to serve as evidence of transaction completion. After receiving real-time measurements from various plan participants, the plan initiator will also make real-time scheduling and control decisions using their own data, thereby reducing the discrepancy between the actual scheduled power and the power promised in the trading plan and avoiding the corresponding financial penalties.

The roles and tasks of a RPN operator at the real-time scheduling phase of energy management are to control and schedule the grid equipment to ensure the safe and economical operation of the RPN. The coordination of the sources and loads in the RPN needed to reach the scheduling and control targets can be achieved by initiating within the market appropriate trading plans, such as active power imbalances, reactive power and voltage regulation, and power quality regulation. In an emergency situation, when the safe and stable operation of the RPN is threatened, any measure or approach may be taken to ensure the safe operation of the power grid.

The post-processing phase of energy management mainly involves third-party fund settlements and profit distribution based on the actual implementation of the trading plan and the advance agreements. Disputes that arise during this process are handled by a third party, together with the various parties involved. The trading is completed after all of the related matters are processed, and the energy management process also ends at that point. The various trading parties will process all of the information involved in the trade to generate historical data that will support subsequent trades they will initiate or in which they will participate.

The above-described EMS and energy management process will have the features and advantages discussed below.

It will be adapted to the requirements of a market-oriented RPN business model: the parties involved in the energy management process are entities with clear responsibilities and rights, and the entire energy management process is realized in a market-oriented manner. It will meet the requirements for high DER penetration: through market-oriented operation, the DER penetration level will eventually adapt to the consumption level and to the network operation capability, and it will also induce the growth of electricity consumption and network operation capabilities, thereby resulting in the dynamic and coordinated development of the three. The operations and interactions of NEMS and SLEMS can ensure both the economy and safety of the RPN operation, as they are highly compatible and scalable: by accessing the unified regional market trading platform through the Internet, both DERs and users can immediately participate in the energy management process. The EMS is characterized by completely distributed and intelligent control: the control decisions of electricity traders, DERs, and users are made based on their own intelligence, without any centralized control. The EMS will achieve simultaneous competition and collaboration among individuals: electricity traders, DERs and users aim to maximize their own interests when participating in the market competition, while also collaborating to implement the global trading plans initiated by the regional network operator.

5. Conclusions

In this paper, a market-oriented RPN operation pattern and its EMS are designed and studied in the context of high DER penetration. The operation pattern can accommodate the requirements for both RPN marketization and DER access, and it has the advantage and characteristics of a high degree of compatibility and scalability, fully distributed intelligent control, and organic integration of individual competition and coordination. We intend for this study to be a useful reference for promoting electricity sales-side marketization when DERs are integrated into the operation.

Due to space limitations, the macroscopic problems (e.g., national policy and the establishment of regulations and laws) and technical problems (e.g., trading platform construction, interest entities' pricing mechanisms, and communication and information platform construction) involved in the implementation of the market-oriented RPN business model are not discussed in this paper. These issues must be solved and improved in subsequent studies. Meanwhile, in the proposed business model framework, it is clear that by organically integrating innovative ideas from other industries and effectively combining the latest technologies, such as Internet connectivity and big-data analysis, additional new service and business models can be created. For instance, funds or equipment may be provided to users to build distributed power generation units, and prepaid electricity fees may be pooled to allow participation in trading in the financial markets. All of these topics warrant further in-depth study.

6. Acknowledgments

The research was supported by the Basic and Prospective program PD71-14-003, funded by the State Grid Corporation of China.

IOP Conf. Series: Materials Science and Engineering **484** (2019) 012046 doi:10.1088/1757-899X/484/1/012046

7. References

[1] P.Barker. Determining the impact of distributed generation on power systems: part 1-radial distribution systems. Proc. 2000 IEEE power engineering society summer meeting, Seattle. 2000, 3:1645-1656.

[2] The operating situation of photovoltaic industry in China in the first half of 2015. Ministry of Industry and Information Technology of China. Online Available at: http://www.miit.gov.cn/n1146285/n1146352/n3054355/n3057643/n3057648/c3625316/content.html.

[3] An energy policyfor Europe: communicationfromthe commission to the European council and the European parliament. Online Available at: ec.europa.en/energy/ energy_policy/doc/01_energy_policy_for_europe_en.pdf. 2007.

[4] A.Molderink, V.Bakker, M.G.C.Bosman, et.al. A three-step methodology to improve domestic energy efficiency. Innovative Smart Grid Technologies (ISGT), 2010: 1-8.

[5] LIU Junyong, DUAN Dengwei, WU Jiguang. Models and operation of distribution market in several countries and suggestions for distribution side deregulation in China. Electric Power Automation Equipment. 2004, 24(7): 7-14.

[6]S.Ropenus, K.Skytte. Regulatory review and barriers for the electricity supply system for distributed generation in EU-15. Future Power Systems, 2005 International Conference on. 2005: 1-6.

[7] L.Toma, M.Eremia, D.Bica. Economic operation of distributed energy resources in a microgrid. PowerTech (POWERTECH), 2013 IEEE Grenoble. 2013: 1-6.

[8] M.Milligan, B.Kirby, J.King, et.al. Operating reserveimplication of alternative implementations of an energy imbalanceservice on wind integration in the Western Interconnection. Proc.of IEEE PES General Meeting, 2011:1-8.

[9] YANG Zhichun, LE Jian, LIU Kaipei, et.al. Study on the standard of the grid-connected microgrids. Power System Protection and Control. 2012, 40(2): 66-71.

[10] LIU Jizhen, LI Mingyang, FANG Fang, et.al. Review on Virtual Power Plants. Proceedings of the CSEE. 2014, 34(29): 5103-5108.

[11] N.Hatziargyriou. Microgrids-The future of small grids. Key note speech, Future Power Systems, FPS2005, Amsterdam, 2015: 16-18.

[12] G.Celli,F.Pilo, G.Pisano, et.al. Optimal participation of a microgrid to the energy market with an intelligent EMS. IEEE-7th International Power Engineering Conference, Singapore, 2005: 663-668.

[13] N.D.Hatziargyriou, A.Dimeas, A.G.Tsikalakis, et.al. Management of micro-gridsin market environment. International Conference on Future Power Systems. 2005: 1-7.

[14] Z.Jiang,R.A.Dougal. Hierarchical microgrid paradigm for integration of distributed energy resources. IEEE Power and Energy Society General Meeting-Conversion and Delivery of Electrical Energy in the 21st Century. 2008: 1-8.

[15] J.Giraldez, D.Heap. Overview of microgrids in a market environment. Green Technologies Conference (GreenTech), 2015 Seventh Annual IEEE. 2015: 36-43.

[16] D.Pudjianto, C.Ramsay, G.Strbac, et.al. The virtual powerplant: enabling integration of distributed generation and demand. FENIX Bulletin 2, 2008.

[17] S.Lukovic, I.Kaitovic, MarcelloMura, et.al. Virtual power plant asa bridge between distributed energy resourcesand smart grid. System Sciences (HICSS), 2010 43rd Hawaii International Conference on, 2010:1-8.

[18] B.Zwaenepoel,T.LVandoorn,J.I.Laveyne, et.al. Solar commercial virtual power plant day ahead trading. PES General Meeting, IEEE, 2014: 1-5.

[19] Shi You, C.Traeholt, B.Poulsen. Amarket-basedVirtualPowerPlant. Clean ElectricalPower, 2009 International Conference on, 2009: 460-465.

[20] F.Bourry,L.M.Costa,G.Kariniotakis. Risk-Based strategies for wind pumped-hydro coordination under electricity markets. PowerTech, 2009 IEEE Bucharest. 2009: 1-8.

[21] E.Mashhour,S.M.Moghaddas-Tafreshi. Bidding strategy of Virtual Power Plant for participating in energy and spinning reserve markets Part I: problem formulation. IEEE Trans. On Power Systems. 2011, 26(2): 949-956
[22] H.Nezamabadi, P.Nezamabadi, M.Setayeshnazar, et.al. Participation of virtual power plants in energy market with optimal bidding based on Nash-SFE equilibrium strategy and considering interruptible load. ThermalPowerPlants(CTPP), 2011 Proceedings of the 3rd Conference on, 2011: 1-6.
[23] A.David, F.Wen. Strategic bidding in competitive electricitymarket: A literature survey. IEEE Power Eng. Society. Summer Meeting, Seattle, WA, 2000, 4: 2168-2173.
[24] M.A.Salmani, A.Anzalchi, S.Salmani. Virtual power plant: new solution for managing distributed generations in decentralized power system. Management and Service Science (MASS), 2010 International Conference on, 2010: 1-6.
[25] J.Mohammadi, Ashkan Rahimi-Kian, M.S.Ghazizadeh. Joint operation of wind power and flexible load as virtual power plant. Environment and Electrical Engineering (EEEIC), 2011 10th International Conference on, 2011: 1-4.
[26] A.Molderink, V.Bakker, M.G.C.Bosman, et.al. On the effects of MPC on a domestic energy efficiency optimization methodology. Energy Conference and Exhibition (EnergyCon), 2010 IEEE International, 2010: 120-125.
[27] Guo Yuanxiong, Gong Yanmin, Fang Yuguang, et.al. Stochastic minimization of imbalance cost for a virtual power plant in electricity markets. Innovative Smart Grid Technologies Conference (ISGT), 2014 IEEE PES, 2014: 1-5.
[28] ZHAO Yu, YU Erkeng. Study on retail electricity market part two: Participants of the retail electricity market. Automation of Electric Power Systems[J]. 2003, 27(10): 35-38.
[29] E.F.Bompard, Bei Han. Market-based control in emerging distribution system operation. IEEE Trans. on Power Delivery. 2013, 28(4): 2373-2382.
[30] ZHAO Yu, YU Erkeng. Study on retail electricity market part six: Influence on the electric power system by distributed generation[J]. Automation of Electric Power Systems. 2003, 27(15): 25-28.
[31] R.Palma-Behnke,J.L.C.A, L.S.Vargas, et.al. A distribution company energy acquisition market model with integration of distributed generation and load curtailment options. IEEE Trans. onPower Systems. 2005, 20(4): 1718-1727.
[32] G.W.Ault,Mcdonald, R.James. Planning for distributed generation within distribution networks in restructured electricity markets. Power Engineering Review, IEEE. 2000, 20(2): 52-54.
[33] Z.Wang, B.Chen, J.Wang, et.al. Networked microgrids for self-healing power systems. IEEE Trans. on Smart Grid. 2016, 7(1): 310-319.
[34] M.Ampatzis, P.H.Nguyen, W.L.Kling. Implementation and evaluation of an electricitymarket operated at district level. PowerTech, 2015 IEEE Eindhoven. 2015: 1-6.
[35] H.Khazaei,B.Vahidi, S.H.Hosseinian, et.al. Two-level decision-making model for adistribution company in day-ahead market. Generation, Transmission & Distribution, IET. 2015, 9(12): 1308-1315.
[36] T.C.Yang. Networked control system a brief survey. IEE Proc.-Control Theory Appl., 2006, 153(4): 403-412.
[37] N.Etherden, V.Vyatkin, M.Bollen. Virtual Power Plant for Grid Services using IEC 61850. IEEE Trans. onIndustrial Informatics. 2015, PP(99): 1-1.

IOP Conf. Series: Materials Science and Engineering **484** (2019) 012047 doi:10.1088/1757-899X/484/1/012047

Investigation Aging Behavior of High Energy Propellant Materials

Yan Gu[1*], Wenliang Liu[1], Qiong Wang[1], Jiaojiao Du[1], Fangfang Wang[1] and Silong Yu[1]

[1]Xi'an Modern Chemistry Research Institute, Xi'an 710068
Email: e-mail: guyan0506@126.com

Abstract. NEPE propellants were exposed to aging at 55°C, 65°C, 70°C, and 75°C, respectively. The relative contents of stabilizer (MNA), tensile strength and are measured to analysis the aging characteristic of NEPE propellants. Based on aging mechanism of the stabilization depletion and the physical aging properties of NEPE propellants, the whole aging process can be separated into two processes: stabilizer gradual degradation at first stage (chemical aging stage) and degradation reaction of polymer at seconded stage. The stabilizer degradation occurred in the first process, therefore the storage life prediction at normal temperature for NEPE propellant ought to be based on the first process during high temperature aging stage. The storage life of the NEPE propellant at room temperature has been evaluated via Arrhenius and Berthelot approach.

1. Introduction

NEPE (Nitrate Ester Plasticized Polyether) propellants are cross-linked solid propellants, which use polyether polyurethane as binder and nitrate esters as plasticizer, filled with a large amount of high-energy oxidizer, such as HMX and AP [1]. It is well known that nitrocellulose-based propellants decompose gradually during storage, which may lead the decreasing of the chemical stability. Researches about aging process are important in evaluating the reliability and safe employability of NEPE propellants. Sometimes, hydrolysis reaction will occur when nitrate esters absorb the moisture in the air [2]. In the process of hydrolyzation, the decomposition of nitrate esters will be accelerated by NO_x free radicals, which can catalyze the further decomposition. Hence stabilizers are usually employed into NEPE propellant to neutralize acid and nitric oxide. Meanwhile, the self-catalytic decomposition of nitrate esters is restrained, and the degradation of plasticizer framework is reduced [3]. Accordingly, the depravation of polymer network and stabilizer are the criterions to evaluate the safe storage life and chemical stability of propellant [4].

In this paper, the safe storage life of NEPE propellant was predicted by the Arrhenius and Berthelot equation via thermal aging storage. The aging process has been separated into three stages based on the chemical stability caused by the decomposition of nitrate esters and the mechanical aging caused by the degradation of cross-linked network structure.

2. Experimental Sections

2.1. Accelerated Aging Test

The typical composition of NEPE propellant contents of polyether, NG- BTTN and MNA. The fuel density was $0.846 g/cm^{-3}$. The propellant samples were cut into cubic shape (120mm×25mm×10mm). The samples were sealed and aged at 55°C, 65°C, 70°C, and 75°C for different times.

IOP Conf. Series: Materials Science and Engineering **484** (2019) 012047 doi:10.1088/1757-899X/484/1/012047

2.2. Content of Stabilizer Analysis

The relative migration ratio of stabilizer (MNA) is characterized by HPLC. NEPE propellant lamella of different aging time weighting 0.2 g were put into 20 mL Erlenmeyer flask and soaked in chromatographic acetone/H_2O mixture for 8 h. Afterwards the soaking liquid was detected by HPLC. The relative content of MNA is obtained by calculating the ratio of peak area.

2.3. Determination of Cross–linking Density

The variety of cross-linked densities during NEPE propellant aging process was characterized via magnetic resonance cross-linking instrument. The cross-linking density can be calculated via rubber elasticity statistics theory.

2.4. Determination of Tensile Strength

The tensile strength of NEPE propellant has been characterized via uniaxial tension test according to GJB 770B-2005. Figure. 1 shows the general sample for the uniaxial tension test.

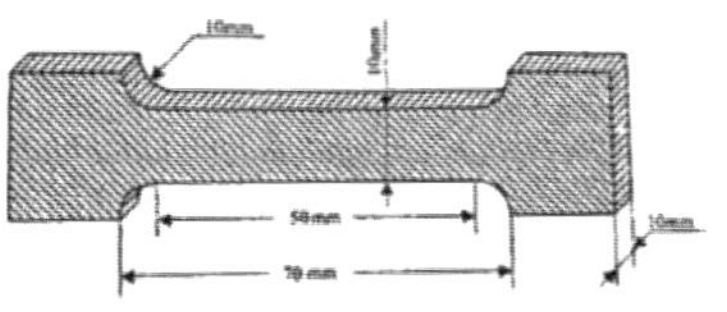

Figure 1. Tensile specimens.

2.5. Storage Life Prediction Methods

Prediction of safe storage life based on accelerated aging dates commonly needs scaling up of the storage life from high temperature to that at normal temperature. The Berthelot and Arrhenius approaches are extensively used for such extrapolation.

2.5.1. Arrhenius Approach. Arrhenius approach has been usually used to analyse accelerated aging dates of propellants at elevated temperature, and the results can be calculated for storage predictions at any storage temperature [5]. The Arrhenius equation is given by Eq. (1):

$$k = A\exp(-E_\alpha/RT) \quad (1)$$

Which can be rearranged in this form:

$$\ln k = \ln A - \exp(-E_\alpha/RT) \quad (2)$$

Where k represents the reaction rate constant, a represents pre-exponential factor, Ea represents the activation energy (kJ/mol), R represents the universal gas constant, and T represents the temperature (K). t_1 or t_2 is the storage life at temperature T_1 or T_2. Correspondingly, t is proportional inversely to k.

2.5.2. Berthelot Approach. Berthelot equation is another common approach for storage life estimation. This approach depicts curves reflecting the correlation between variation in reaction rate and temperature [6]. Thus, logarithm of time based 10 is in directed proportion to temperature in Eq. (3):

$$\log k = aT + b \quad (3)$$

where k represents the reaction rate constant, a and b are constants, and T represents the temperature in Kelvin (K). For zero-order reactions, the rate constant k can be expressed as:

$$k = c/t \quad (4)$$

where c is defined as the concentration decrease of the substrate, and t represents the required time for achieving a certain change (50% of stabilizer according to GJB 770B-2005) in the concentration of substrate. By superseding expression of k in Eq. (4), Eq. (3) can be written as:

$$\log c/t_{0.5} = aT + b \quad (5)$$

Therefore, t_1 or t_2 can be defined as the storage life under temperature T_1 or T_2. Correspondingly, the following equation can be deduced by subtracting T_1 from T_2 in Equation (5):

$$\log(t_1/t_2)=a(T_2\text{-}T_1) \quad (6)$$

3. Results and Discussions

3.1. Aging Behaviour of Chemical Stability

Figure 2 shows the relationship of MNA stabilizer depletion and aging time under 55°C, 65°C, 70°C, and 75°C. The values of MNA depletion obtained from different accelerated aging temperatures can then be used to define the effect of temperature on the degradation rate or to predict the storage life of propellant.

According to the data in Figure 2, the trend of MNA stabilizer consumption exhibited almost linear and the concentration of MNA have independence on MNA stabilizer degradation, so the degradation of stabilizer accords with the zero-order reaction [7]. Concentration of MNA stabilizer has liner functional relationship with aging time and the slope of stabilizer content-time line denotes reaction rate. The degradation rate of MNA under different accelerated aging temperatures can be utilized to predict the safe storage life of NEPE propellant. Accordingly, the prediction of safe storage life at ambient temperature will be carried out via both Berthelot and Arrhenius approach.

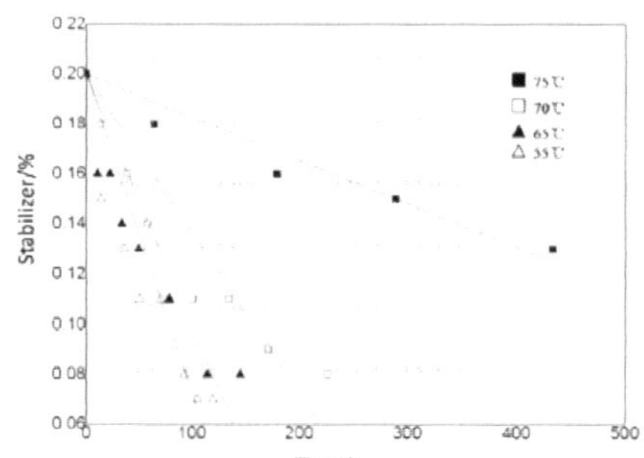

Figure 2. Prediction lines of degradation stabilizer (MNA) concentration based on experimental dates collected in the range of 55°C-70°C.

The slope of prediction lines in Figure 2 represents the reaction rate constants (k) under 55°C-75°C (328-348k). According to Eq. (2), the plot of lnk as a function of 1/T gives a straight line. Thus, the values of reaction rate constants at 55°C-75°Ccan be calculated the slope of MNA stabilizer depletion trend line under different accelerated aging temperatures in Figure 2. Meanwhile, Eq. (2) can be rearranged in this form:

$$\frac{t_1}{t_2}=\frac{k_2}{k_1}=\exp((E_\alpha/R)(1/T_1-1/T_2)) \quad (7)$$

T represents the temperature (K). t_1 or t_2 is the storage life at temperature T_1 or T_2. Correspondingly, t is proportional inversely to k.Therefore, lnk under various aging temperatures can be compute and the Arrhenius plot of the degradation reaction based on Eq. (2) is obtained, which is shown in Figure 3.

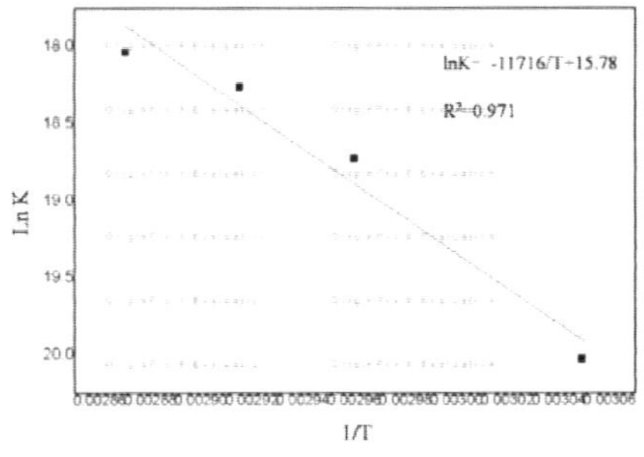

Figure 3. Arrhenius plot for the zero-order reaction

At the same time, the aging times ($t_{0.5}$) of MNA concentration decreasing to 50% also can be obtained from Figure 2. According to Berthelot's mathematical expression in Eq. (2), T and lg(t) present straight line functional relationship. Then the Berthelot plot of the MNA stabilizer degradation based on Eq. (2) is shown in Figure 4.

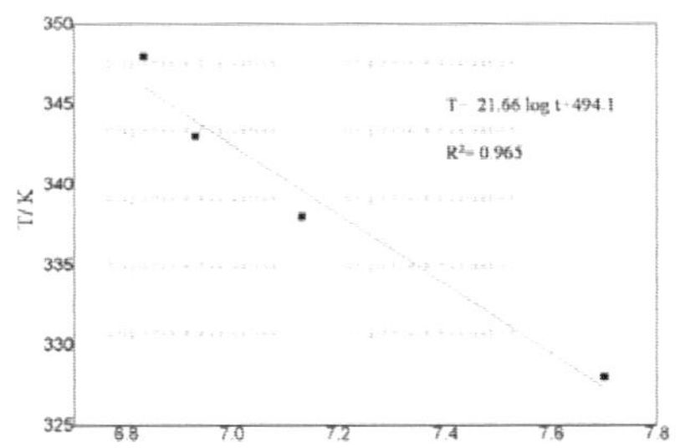

Figure 4. Berthelot plot of MNA.

The results of safe storage life of NEPE propellants at normal temperature calculated via Arrhenius and Berthelot approaches were compared in Table 1, which shows the safe storage life estimated by means of Arrhenius and Berthelot plots in the normal temperature range and high temperature.

Table 1. Comparison of safe storage life of NEPE propellants prediction via Arrhenius and Berthelot plots.

T/K	T/°C	storage life by Arrhenius plot/y (lnk=-11716/T+15.78)	storage life by Berthelot plot/y (T=21.66logt=494.1)
303	30	137.9483277	21.08
298	25	263.9207194	35.87
293	20	516.2342493	61.03

As can be seen from Table 1, the numerical value of safe storage life at normal storage temperatures estimated by Berthelot approach is obviously lower than that calculated from Arrhenius plot; nevertheless, at high temperature, the values calculated by two approaches have proximity in number, which is accord with several former investigations [8]. According to the kinetic researches, the depletion of MNA stabilizer belongs to zero-order reaction. Since Arrhenius approach follow the exponential variation while Berthelot approach follow the power law variation, the reaction rate constants (K) in Berthelot and Arrhenius plot are not relevant to temperature. Besides, the Arrhenius approach is governed by the activation energy (Ea) while the Berthelot approach is depending on the values of 10% reaction rate factor. The storage life prediction strongly rests with the depletion of MNA stabilizer concentration. Considering the life prediction is more significant in relatively lower temperature condition, the lower storage life (predicted by Berthelot) is recommended as the storage life owing to safety aspect.

3.2. Aging Regularity of Binder Network Structure

Figure 5 show the change behaviour of MNA degradation, tensile strength and cross-linking density with aging time. The declination of tensile strength and cross-linking density are similar with the extension of storage time. Theoretically, tensile strength is given in Eq. (8) according to thermodynamics [9]:

$$\boldsymbol{\sigma_m = 3\alpha kT\varepsilon} \tag{8}$$

Where $\boldsymbol{\sigma_m}$ represents the tensile strength, T represents temperature in Kelvin (K), ε represents the elongation, k represents the Boltzmann constant, and α represents cross-linked density.

Equation 8 indicates the tensile strength is proportional inversely to cross-linked density, which can reflect the change of inner network system in NEPE propellant. The cross-linking network system will be reinforced on the condition of the number of hard molecular chains increase, which leads to the concentration of hard polyether molecular chain and tensile strength of NEPE propellant increase congruously.

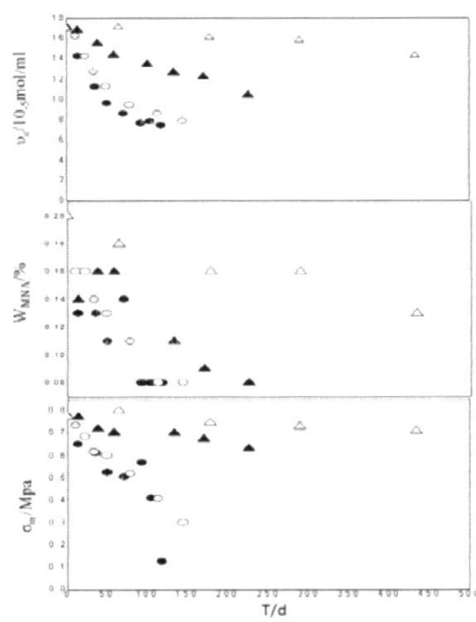

Figure 5. The change behaviour of cross-linking density, MNA degradation and tensile strength with aging time (●-75°**C**,○-70°**C**,▲-65°**C**,△-55°**C**).

Figure 5 also show that tensile strength and cross-linking density changed gradually with the consuming of MNA. After the stabilizer gradual decreased while the chemical aging accelerated, the network structure parameters decline logarithmically with the decreasing of the MNA content. The closer of MNA content to lowest value, the faster tensile strength and cross-linking density decrease. The network structure performance becomes worse, while the decomposition of nitrate esters speeds up and the contents of BTTN/NG have cubic function relationships with cross-linking density.

3.3. The Relationship of Chemical Stability and Network System

The aging behaviour of chemical stability can be depicted in Figure 6: In the initial stage of storage, decreasing of chemical stability is dominant. Although products of nitric esters (NG and BTTN) decomposed, stabilizers can neutralize NOx so that the tensile strength and cross-linked density of propellant decreased indistinctively. The function of the stabilizer is to neutralize free radical NOx released from the initial nitrate esters decomposition, then to prevent the auto-catalytic reaction [10]. Therefore, decomposition of molecular chain in binder network structure can be restrained in this stage. Simultaneously, the stabilizer (MNA) is consuming during the process. With MNA stabilizer consumed, the decomposition of NG and BTTN accelerated gradually.

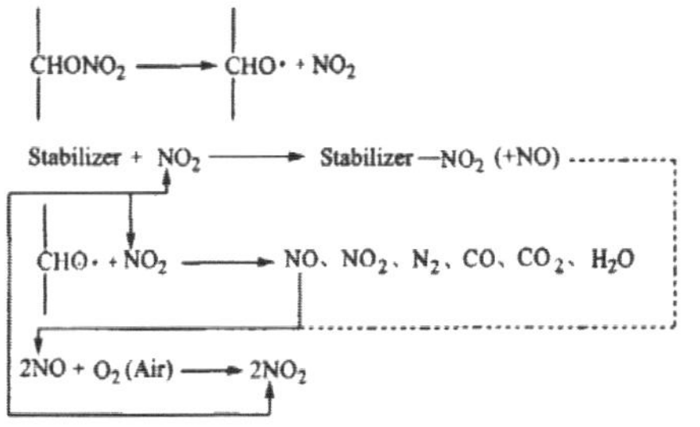

Figure 6. The aging behaviour of chemical stability of NEPE propellant

In the second stage, the NO_x concentration in propellants increases for lacking stabilizer. This progress will lead to the polymer chain decomposed in the thermal and oxidative way. The increasing of NO_x accelerates the decomposition of polyether polyurethane binder structure, which leads to the rapid declining of tensile strength and cross-linking density [10]. The aging of structural performance lagged the aging of chemical stability, the existence of MNA, effectively reduces the self-catalytic decomposition of nitrate esters, then slows down the breaking and decomposition of binder polymer chains, thereby inhibits the decreasing of cross-linking density. So MNA plays a positive role in maintaining good network structure during the process of propellant storage. The network structure parameters decrease logarithmically with the decline of the MNA content; the contents of nitrate esters have cubic function relationships with cross-linking density.

The existence of MNA effectively reduces the self-catalytic decomposition of NG and BTTN, and then restricts the degradation of binder polyether polyurethane system. Therefore, the initial stage (decreasing of chemical stability) is dominant in evaluation the storage life of the NEPE propellant sample. Thereby, the safe storage life prediction should be established in variety of MNA content.

4. Conclusions

(1) Based on change behaviour of cross-linking density, MNA degradation and tensile strength with aging time, the whole aging process of NEPE propellant can be separated into two processes: stabilizer gradual degradation at first stage (chemical aging stage) and degradation reaction of polymer at seconded stage.

(2) Degradation of polymer structure lagged decreasing of chemical stability. The function of the stabilizer is to neutralize free radical NOx, and then restrains the decomposition of binder network system.

(3) The stabilizer degradation occurred in the first stage, therefore predicting the safe storage life of NEPE propellant at normal temperature rests with the chemical aging stage durıng hıgh temperature aging. Thus, the storage life of the NEPE propellant at 25°C is 35.87 years.

5. References

[1] X. P. Fan, Z. R. Liu, L. X. Sun Chin. 2003 J. Explos. Propellants, 26, 43–46.
[2] M. Liu, L. Du, J. Su Chem. 2009 Propellants Polymeric Mater, 4, 31–33.
[3] W. Xu, 2010 Hunan University, Master Thesis.
[4] H. Zhang, A. M. Pang, S. Peng 2005 J. Solid Rocket Technol, 28, 271–275.
[5] M. D. Judge 2003 Propell. Explos. Pyrotech, 28, 114.
[6] K. C. Raha, S. S. Adhav, N. M. Bhide, A. D. Yewale, G. K. Gupta, J. S. Karir 2002 Def. Sci. J, 52, 165.
[7] O. Levenspiel, 1999 Chemical Reaction Engineering, vol 3(John Wiley & Sons, New York) p 517.
[8] H. Shekhar, 2011 Propellants Explos. Pyrotech. 36, 356.
[9] N. Wang, Y. Xiong. 2014 Propellants Explos. Pyrotech, 39, 102.
[10] M. Bohn 2001 J. Therm. Anal. Calorim. 65, 103.

Development of Environment-Friendly Plastering Gypsum

Xingyu Wang, Shuling Sun, Xiaohua Tian, Dongji Liu and Fengqing Zhao*

College of Chemical and Pharmaceutical Engineering, Hebei University of Science & Technology, Shijiazhuang, 050018, China
Email:zhaofq3366@126.com

Abstract. A new type of plastering gypsum which has good operability was prepared from calcined desulphurization gypsum, ground steel slag and appropriate additives. The proper material ratio is: calcined desulphurization gypsum 95%, ground steel slag 4.5%, and water-retaining additive 0.5%. The product meets the requirements of GB/T 28627-2012. The effects of energy saving and emission reduction of this product were analysed. Compared with the common cement-based or lime-based materials, it has the advantages of better volume stability, lower application cost and less carbon emission.

1. Introduction

Plastering gypsum has a lot of advantages as fire prevention, sound insulation, light insulation and humidity conditioning, etc [1]. It belongs to a sort of non-toxic and pollution-free green construction materials. However, the application of semi-hydrated gypsum is restricted because of the short setting time and poor water resistance [2]. Some retarders are used to adjust the setting time currently [3-5].

The commonly used setting retarders include organic acids and their soluble salts, alkaline phosphates and proteins, etc. Peng J-H [6] thought that the crystal obviously coarsening, and the strength of harden body was decreased evidently after adding citric acid. Ding Y [7] found that both sodium hexametaphosphate and sodium citrate could have retarding function on gypsum, but the influence on mechanical strength was relatively large. Zhang M-M [8] discovered that protein retarders from waste mycelium had the merit of low intensity, but the preparation was complex comparatively. Singh M [9] studied the effects of several kinds of setting retarders on the setting time and hardening characteristics of gypsum under various pH, and discovered that the system' pH had significant influence on compressive strength and microstructure of gypsum paste. These explain that great majority setting retarders not only can change the setting time of gypsum, but also reduce the mechanical strength drastically at the same time. Besides, the cost of retarders used in gypsum is high. Consequently, it is imperative to seek a cheap material instead of conventional retarders.

2. Experimental

2.1. Material

β-semi-hydrated gypsum was provided by Xibaipo Power Plant, which is composed of di-hydrate gypsum 9.83%, semi-hydrated gypsum 86.93% and anhydride 3.24% ; ground steel slag and standard quartz sand were from Pingshan Bai Hua ao Building Materials Limited Co. , the specific surface area of steel slag is 450 m^2/kg. Table 1 and Table 2 give the composition of the main raw materials and the physical and mechanical properties of β-semi-hydrated gypsum respectively.

Table 1. Composition of some raw materials.

Component	SiO2 (%)	Fe2O3 (%)	Al2O3 (%)	CaO (%)	MgO (%)	MnO (%)	SO3 (%)	Other
β-semi-hydrated gypsum	7.45	0.42	1.13	38.16	0.19	—	48.41	4.24
Steel slag	13.96	27.18	5.28	33.19	8.18	5.82	0.33	6.06

Table 2. Physical and mechanical properties of semi-hydrated gypsum.

Standard consistency water demand (%)	Setting time		Dry intensity	
	Initial time (min)	Final time (min)	Flexural strength (MPa)	Compressive strength (MPa)
0.62	5	8	5	13

2.2. Method

Determined the setting time and mechanical strength of β-semi-hydrated gypsum according to GB/T 28627-2012. A certain proportion of ground steel slag was put into gypsum pastes to determine the desired setting time. The gypsum slurry containing ground steel slag and additive was poured into molds of 40 × 40 × 40mm and de-mold after 2 h. The samples were then cured at 55°C for 3 days. Put the samples into oven and dried at (40±2°C) to constant weight, tested its flexural and compressive strength.

3. Results and Discussion

3.1. Effects of Grinding Steel Slag on Setting Time of Gypsum

Grinding steel slag was used to adjust the setting time of β-semi-hydrated gypsum. The results are shown in Figure 1 and Figure 2.

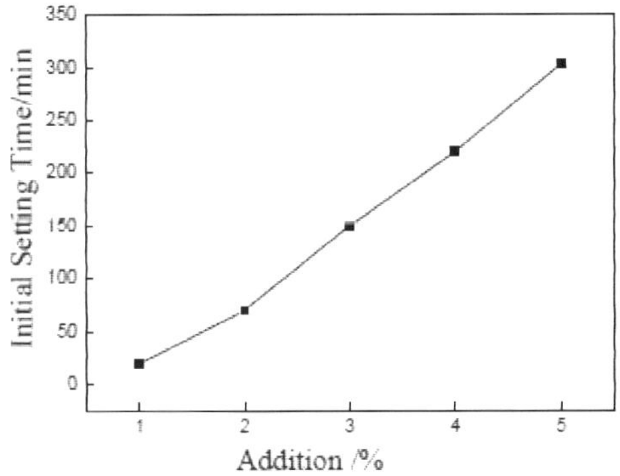

Figure 1. Effects of steel slag on initial setting time.

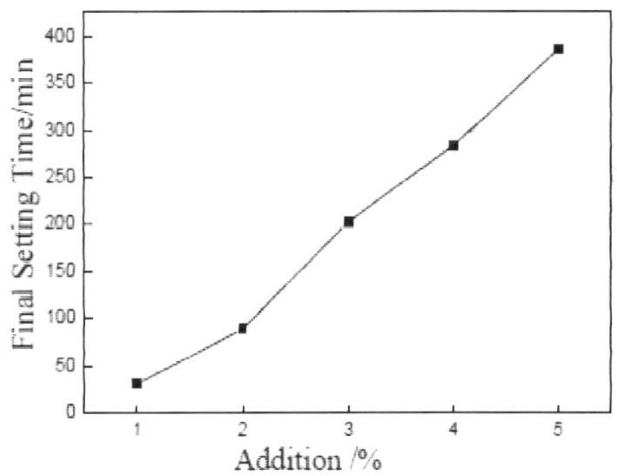

Figure 2. Effects of steel slag on final setting time.

As can be seen from Figure 1 and Figure 2, with the increase in the addition of ground steel slag, the setting time of β-semi-hydrated gypsum was prolonged. When the addition of steel slag was 3%, the initial setting time was 150 minutes and the final setting time arrived to 205 minutes. This explains that ground steel slag has a certain retarding effect on gypsum.

IOP Conf. Series: Materials Science and Engineering **484** (2019) 012048 doi:10.1088/1757-899X/484/1/012048

3.2. Effects of Retarders on Mechanical Strength of Gypsum

Flexural strength and compressive strength were investigated against the addition of ground steel slag in β-semi-hydrated gypsum. The results are shown in Figure 3 and Figure 4.

From Figure 3 and Figure 4, we discovered that with the increase in steel slag addition, the mechanical strength of gypsum also reduced. When the addition of steel slag was 3%, the flexural strength was 3.5MPa and the compressive strength was 6.8MPa. But the loss of mechanical strength is mild and the cost is much lower compared to frequently-used citrates and phosphates.

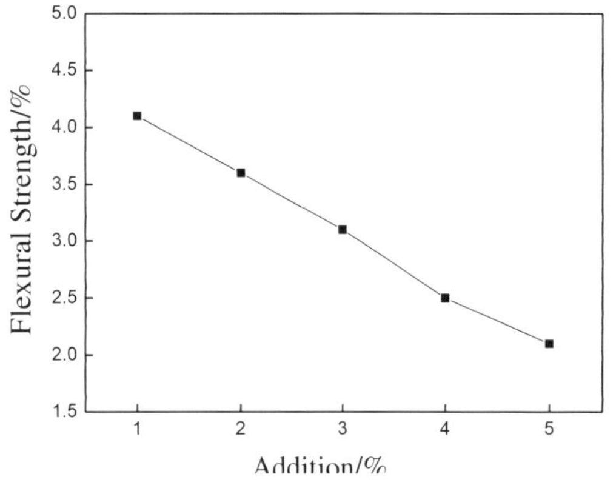

Figure 3. Effects of steel slag on flexural strength

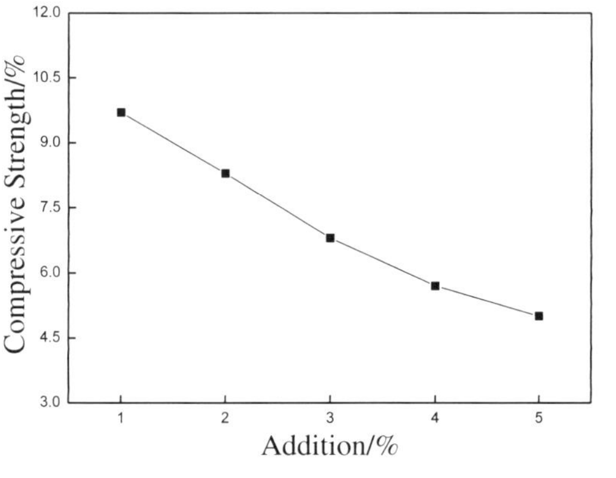

Figure 4. Effects of steel slag on compressive strength

3.3. Preparation of Plastering Gypsum

The optimized material proportion (mass ratio) of plastering gypsum was 95% calcined desulphurization gypsum, was 4.5% ground steel slag, and water-retaining additive was 0.5%. The flexural strength of the product was 2MPa, and the compressive strength was 4MPa. The other performances of this product conform to GB/T 28627-2012. Figure 5 and Figure 6 give the photos of raw materials and product.

Figure 5. Plastering gypsum materials.

Figure 6. Plastering gypsum.

4. Analysis of Energy Saving and Emission Reduction

This product can reduce the carbon dioxide emissions in terms of energy conservation compared to cement- or lime based plastering. The Life Cycle Assessment (LCA) was carried out and the results are shown in Figure 7.

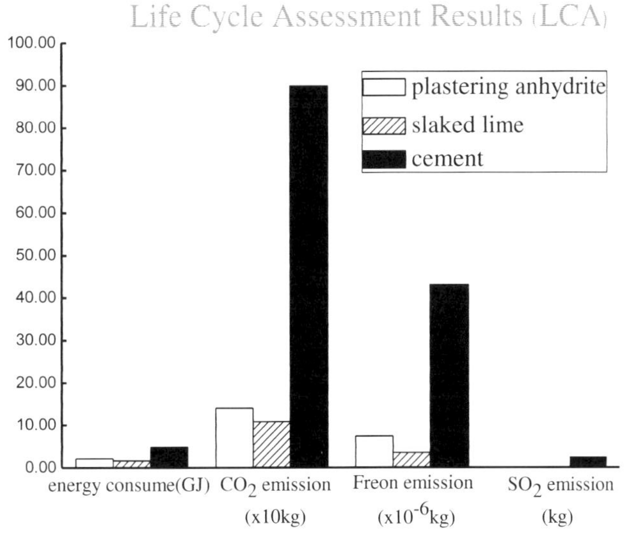

Figure 7. Energy saving comparison chart.

As can be found from Figure 7, plastering gypsum had obvious advantages over cement and lime based product; the energy consumption, carbon dioxide and sulphur dioxide emissions were lower than the later; thus the economic and environmental benefits were higher than the former as well. Besides, gypsum has the advantages of stable performance, less shrinkage and short construction period. Therefore, plastering gypsum is worthy of promotion as wall painting materials.

5. Conclusion

(1) With calcined desulphurization gypsum, ground steel slag as major materials to prepare plastering gypsum conforms to principles of the circular economy laws. This product meets GB/T 28627-2012, and has the characteristics of mild mechanical loss, low cost and high economic benefits.

(2)When the addition of steel slag was 3%, the initial setting time was 150 minutes and the final setting time reached 205 minutes. The addition of ground steel slag may change the hydration kinetics process of gypsum, reduce the formation of dihydrate gypsum and extend the setting time accordingly.

(3) LCA results indicated that compared with cement and lime-based product, plastering gypsum has prominent advantages. Its energy consumption, carbon dioxide and sulphur dioxide emissions are much lower.

6. Acknowledgements

We are grateful to Yan-cong Jiang, Chen-yuan Hu, Hong-ying Ji, Tian-ben Fan and Xiong-mu Chen for their help in the experiment. Thanks for the support of Hebei Engineering Research Center of Solid Waste Utilization.

7. References

[1] Li H 2012 *Technology & Market* vol 3(19), pp 3-4
[2] Feng Q-B, Chen F-M and Tong D 2006
[3] Cao X-M, Mou X-A and Qian Z-Q et al, 2013 *Research Progress in Building Gypsum Retarder* vol 27, pp 298-301
[4] Chen M-M, Zhang M-M and Zhao F-Q 2017 *Chinese of Envirnmental Engineering* vol 11(6), pp 3747-3752
[5] Feng C-H, Chen M-M and Li D-X 2014 *Bulletin of the Chinese Ceramic Society* vol 05, pp 1231-1235
[6] Peng J-H, Chen M-F and Zhang J-X et al, 2005 *Journal of Building Materials* vol 8(1), pp 94-99

[7] Ding Y, Fang Y-C and Ren Q-F et al, 2016 *Materials Review* vol 30(3), pp 121-125
[8] Zhang M-M and Zhao F-Q 2017 *Chemical Industry and Engineering Progress* vol 36(6), pp 2275-2281
[9] Singh M and Garg M 1997 *Cement and Concrete Research* vol 27(6), pp 947-950
[10] Peng J-H, Zhang J-X and Chen F-M et al, 2006 *Journal of the Chinese Ceramic Society* vol 32(6), pp 723-727

IOP Conf. Series: Materials Science and Engineering **484** (2019) 012049 doi:10.1088/1757-899X/484/1/012049

Impact of Riverine Pollutants on the Water Quality of Lake Chaohu, China

Wang Zhang[1,2], Kun Lei[2], Libiao Yang[2,*] and Xubo Lv[2]

[1]Resources and Environmental Science College, Agricultural University of Hebei, Baoding 071000, China
[2]River and Estuary Coastal Zone Environmental Innovation Base, Chinese Research Academy of Environmental Sciences, Beijing 100012, China
Email: yanglb@craes.org.cn

Abstract. The water environmental capacities for COD_{Mn}, TP, and NH_3-N in Lake Chao were estimated with a zero-dimensional model using Class III water quality as the target. The effects of river pollutants transported by the eight major tributaries on the lake's water quality were then analyzed. The results showed that the water environmental capacities for COD_{Mn}, TP, and NH_3-N in Lake Chao for 2002–2008 were 6.21–6.74, 0.05–0.06, and $1.01–1.27\times10^4$ ton/yr, respectively. The safety margins for COD_{Mn}, TP, and NH_3-N, based on the lake's background reserves, were 5.17–5.86, 0.01–0.03, and $0.89–1.12\times10^4$ ton/yr, respectively. Annually, the river pollutants transported by the tributaries to the lake amounted to $3.58–4.49\times10^4$ ton of COD_{Mn}, $0.19–0.27\times10^4$ ton of TP, and $4.09–5.47\times10^4$ ton of NH_3-N. The total amount of COD_{Mn} exported by the tributaries was less than the water environmental capacity (accounting for 56–68%) and below the safety margin (accounting for 70–82%) of the lake. However, the total amounts of TP and NH_3-N were 3.78–14.75 times higher than the water environmental capacity and the safety margin of the lake. To prevent further deterioration of the lake's water quality, it is critical to strengthen pollution control in the rivers around the lake and to reduce the sources of river pollution.

1. Introduction

In China, all the lakes add up to about 12 million hm^2 of water surface, which accounts for 0.8% of the country's total land area. The water quality of the lakes directly affects the country's overall water quality [1, 2]. Social and economic developments in recent decades have led to large amounts of pollutants being transported to the lakes, causing heavy pollution of their waters. Currently, about 90% of the lakes are in the state of medium to high eutrophication [3, 4]. There have been considerable research and treatment efforts to restore the lake ecosystems and improve their water qualities [5, 6]. However, the level of pollution remains serious. In the long term, the treatment of lakes will continue to be a difficult issue that the official authorities will have to tackle [7, 8].

The treatment of lake pollution has been emphasized and strongly supported since the country's sixth 5-Year Plan [9, 10, and 11]. Although some progress has been achieved, the water quality remains questionable since it is difficult to control the total amount of pollutants being transported to the lakes [12]. In addition, the water environmental capacity of the lakes under varying conditions has not been clearly identified. In the country's eleventh 5-Year Plan, measures for the management of water environments were revised from total target to total capacity controls [13, 14]. Hence, the main task in the national management of lakes is to further research on the water environmental capacities. Pollutants in surface runoffs and point-source discharges are direct sources of lake pollutants [15, 16]. These enter the water of rivers and tributaries and are eventually transported to the receiving water bodies, including the lakes. Although river pollutants have direct and significant impacts on the water environmental capacities of lakes, quantitative research on the relationship between the two is lacking.

Lake Chao is China's fifth largest freshwater lake. It previously played an important role in agricultural irrigation and supply of potable water for the people of Anhui province [17]. In recent years, the lake has undergone serious eutrophication, resulting in frequent outbreaks of blue-green algal blooms [18, 19]. Hence, it has been earmarked as one of the three key lakes within China to be treated [20]. The local government has repeatedly enacted and amended controls over the total amount of pollutants entering the lake, in the bid to improve its water quality. However, its water environmental capacity for pollutants has not been quantified, thus proper treatment is difficult to administer [21]. In this study, the lake's water environmental capacities for COD_{Mn}, TP, and NH_3-N for 2002–2008 were estimated, using the target of Class III water quality. The impacts of river pollutants transported by the tributaries to the lake on its water environmental capacity were also quantitatively analyzed. These provided scientific data to support future pollution control measures for the lake.

2. Materials and Methods

2.1. Sites Description

Lake Chaohu is one of the five largest fresh lakes in China. Now it is one of the three most eutrophic lakes of China attracting more and more attention in the world-wide-concern. There are thirty three tributaries entering Lake Chaohu. The Nanfei (NFR), Hangbu (HBR), Fengle (FLR), Shiwuli (SR), Pai (PR), Dianbu (DBR), Ershibu (ER), and Zhao (ZR) Rivers are the major tributaries that contribute most of the total water discharge into the lake (about 90%). In this study, six sampling sites were selected throughout the lake, and one sampling site was selected in each of the eight tributaries (Figure 1). The concentrations of COD_{Mn}, TP, and NH_3-N in the sampling stations were monitored monthly during 2002 through 2008.

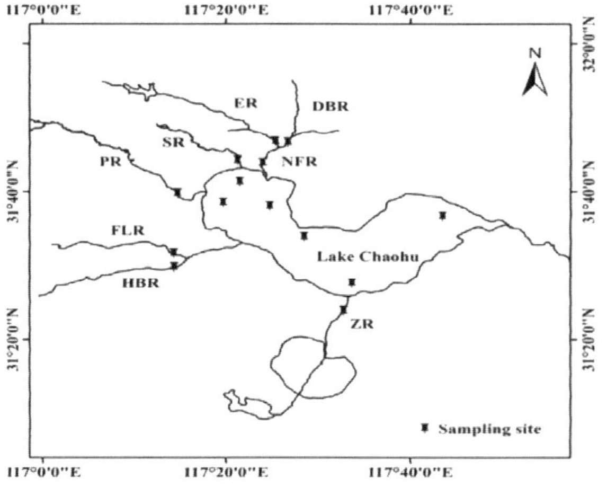

Figure 1. Location of sampling sites in Lake Chaohu and tributaries.

2.2. Calculation of Water Environmental Capacity of Pollutants

The water environmental capacity for COD_{Mn} in Lake Chaohu was calculated using the Vollenweider model [22, 23]:

$$\frac{Vdc}{dt} = Q_iC_i - Q_eC - kCV \tag{1}$$

Where, V is the water volume of the lake (m^3); Q_i is the total water input to the lake (m^3/yr), and Q_e is the total water output the lake (m^3/yr); C_i and C is the concentration of COD_{Mn} in water input to the lake and lake water(mg/L), respectively; k is the biochemical degradation coefficient (1/d).

The water environmental capacity for TP and NH_3-N in Lake Chaohu were calculated using the Dillon and Hetianjian models, respectively. The calculation equation of Dillon model [24, 25]:

$$W=Z*A*Cs\ (Qi/V)/\ (1-R) \qquad (2)$$

Where, Z is the average depth of lake (m); A is the surface water area of lake (km^2); R is the retention coefficient calculated as $R=1-\ Q_e/Q_i$. The meanings of other parameters were same as equation 1.

The calculation equation of Hetianjian model:

$$W=Cs*Z*A*(Qe/V+10/Z) \qquad (3)$$

The meanings of the parameters in equation 3 were the same as described above.

The water environmental capacities for COD_{Mn}, TP, and NH_3-N in Lake Chao were estimated based on Class III water quality, thus C_s was 6.00, 0.05,1.00 mg/L for COD_{Mn}, TP, and NH_3-N, respectively.

The total annual amounts of pollutants transported by the tributaries into the lake were calculated as equation 4:

$$F=Cp* \qquad (4)$$

Where, F (ton/yr) is the total annual amounts of pollutants transported by the tributaries into the lake, C_p (mg/L) is the annual mean concentration of pollutant, Q (m^3/yr) is the total water discharged to the lake by eight major tributaries.

2.3. *Calculation of Total Water Discharge of the Eight Studied Tributaries to Lake*

The total water discharge of the eight studied tributaries was estimated using rainfall data and the runoff coefficient within the watershed. The monthly rainfall ranged between 3.5 and 304.2 mm (Figure 2). The many-years' mean runoff coefficient was about 0.46 within the watershed. Then we calculated that the total water discharge to the lake would range from 41.2×10^8—56.2×10^8 m^3 during 2002--2008. This result is very close to the monitored value in other reports (40×10^8 m^3). Considering the water discharge of the eight tributaries to Lake Chaohu (about 90%), it is estimated that about 36.6×10^8--51.1×10^8 m^3 water was discharged to the lake during the study period.

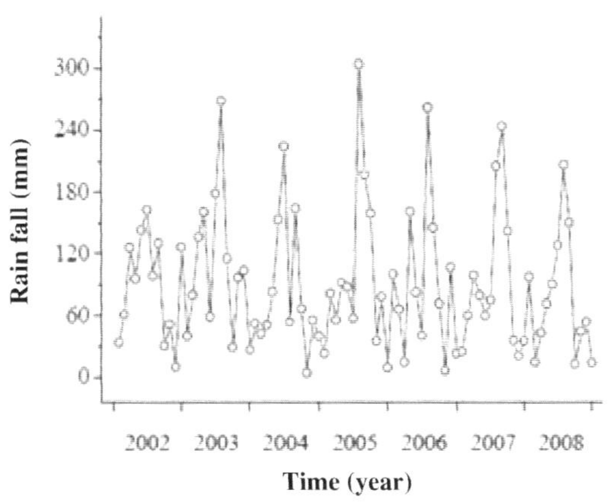

Figure 2. Monthly rainfall within the Lake Chaohu watershed.

3. Results and Discussion

3.1. *Water Environment Capacity of COD_{Mn}, NH_3-N, and TP*

The water environmental capacity for COD_{Mn} in the lake ranged 6.21-6.74×10^4 ton /yr from 2002 through 2008, with a mean value of 6.47 ± 0.18 (SD) $\times10^4$ ton /yr (Figure 3). The water environmental capacity for TP ranged 0.04-0.06×10^4 ton /yr calculated using the Dillon model and 0.06–0.07 $\times10^4$

ton /yr calculated using the Hetianjian model, respectively, with a mean value of 0.06 $\times10^4$ ton /yr of the two methodologies. The water environmental capacity for NH_3-N was similarly calculated with the two models; the results were 0.85–1.20 and 1.17–1.35 $\times10^4$ ton /yr, respectively, with a mean value of 1.14 $\times10^4$ ton /yr (Figure 4).

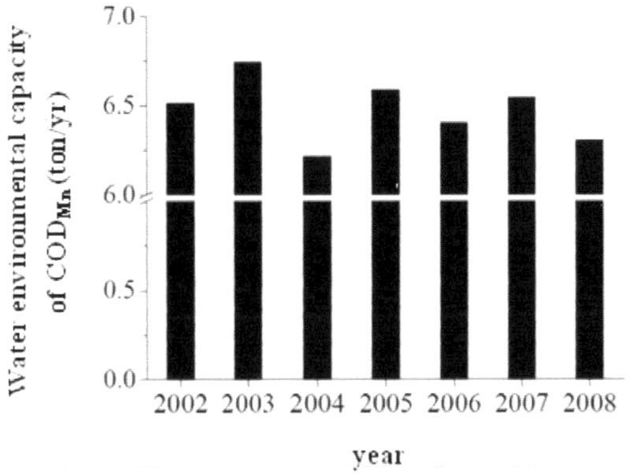

Figure 3. Water environment capacity of COD_{Mn} in Lake Chaohu

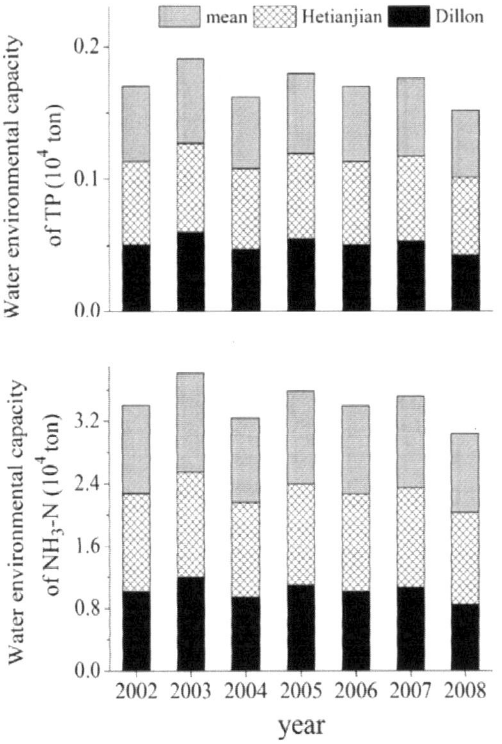

Figure 4. Water environment capacity of TP and NH_3-N in Lake Chaohu

In this study, the water environmental capacities for COD_{Mn}, TP, and NH_3-N in Lake Chao were different with other Chinese lakes under the same water quality target. For example, the study by Yang *et al.* showed that the water environmental capacity for COD_{Mn} was 1.65–2.80 ton /yr while was 58–97 ton /yr for TP in Dian Lake [10]. For Lake Bosten in the Xinjiang province, with the water level varying at 25-90%, the water environmental capacity for COD_{Mn} was 0.46–0.63$\times10^4$ ton /yr while was

0.08-0.12×10^4 ton /yr for TP [24]. This is because different types of lakes and pollution conditions may result in different water environmental capacities in various lakes. In addition, the interactions and relationships between pollutants, such as the improvement of the environmental capacity for one pollutant may lead to the reduction of other pollutants, may also result in the difference of water environmental capacity for nutrients in the lake [25].

3.2. Safety Margins and Background Reserves of COD_{Mn}, NH_3-N, and TP

The water environmental capacity of a lake refers to the theoretical assimilative capacity of its water under designated conditions. Here, the author is of the view that it is more scientific to use the lake's safety margin to establish the threshold for total emissions of pollutants. To estimate that safety margin, the background reserves of pollutants during the target year must be investigated. During 2002 and 2008, concentrations of COD_{Mn}, TP, and NH_3-N were monthly monitored in the six sampling sites cover the lake. Results showed that the concentrations of COD_{Mn}, TP, and NH_3-N ranged 4.14–13.14, 0.07–0.44 and 0.09–3.71 mg/L respectively, with mean values of 6.01±1.32, 0.19±0.07 and 0.75±0.62 (SD) mg/L respectively (Figure 5). All three pollutants showed higher concentration values in autumn and winter, while no significant difference was detected between the monitoring years. The water quality of Lake Chao maintained at Class IV–V during 2002 and 2008, with COD_{Mn} and TP being the major pollutants.

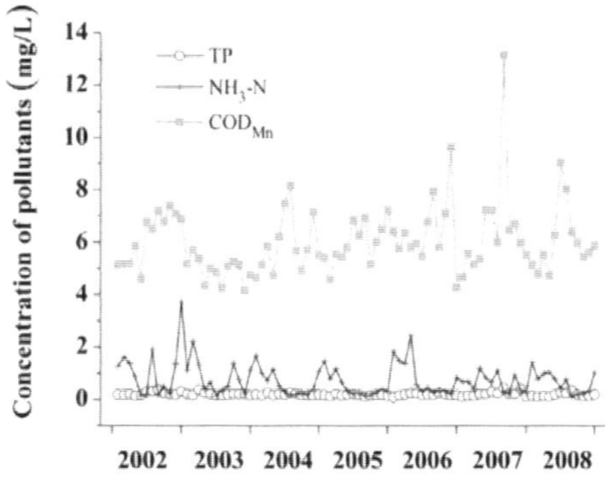

Figure 5. Monthly concentrations of COD_{Mn}, TP, and NH_3-N in Lake Chaohu

Based on the data of lake's storage capacity over 2002-2008, the annual reserves of COD_{Mn}, TP, and NH_3-N in the lake were estimated to be 0.89–1.21, 0.03–0.04, and 0.09–0.19 ×10^4 ton /yr, respectively. For each year, the annual reserves of COD_{Mn}, TP, and NH_3-N accounted for 13–18%, 56–73%, and 4–14% of the lake's total water environmental capacity, respectively (Figure 6). This means that the water environment has remaining assimilative capacity for all three pollutants. Then we can estimated that for the target of Class III water quality, the safety margins for COD_{Mn}, TP, and NH_3-N in the lake are 5.17-5.86, 0.01-0.03, and 0.89-1.12 ×10^4 ton /yr, respectively.

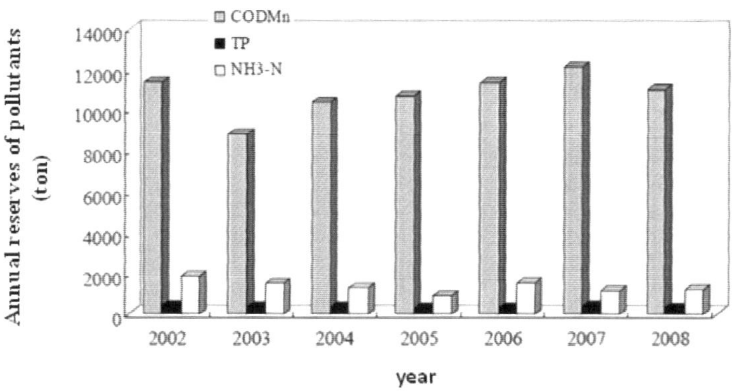

Figure 6. Annual reserves of COD_{Mn}, TP, NH_3-N in Lake Chaohu

3.3. Impact of Riverine Pollutants on the Water Quality of Lake Chaohu

The main and direct sources of pollutants for Lake Chao are the tributaries that transport pollutants to it. However, the river pollutants and their impact on water quality in the lake have not been quantified. In this study, the annual flux of pollutants transported by the tributaries to the lake was estimated, and the impact of river pollutants on the lake's water environmental capacity was quantitatively analyzed. The monthly changes in the concentration levels of COD_{Mn}, TP, and NH_3-N in the eight major tributaries for 2002–2008 are shown in figure 7. The monitoring data showed that the changes in concentrations of COD_{Mn}, TP, and NH_3-N were 2.2-75.2, 0.01-3.61 and 0.01-163.00 mg/L respectively, with mean values of 8.57±7.38, 0.59±0.68, and 8.08±16.35 (SD) mg/L, respectively.

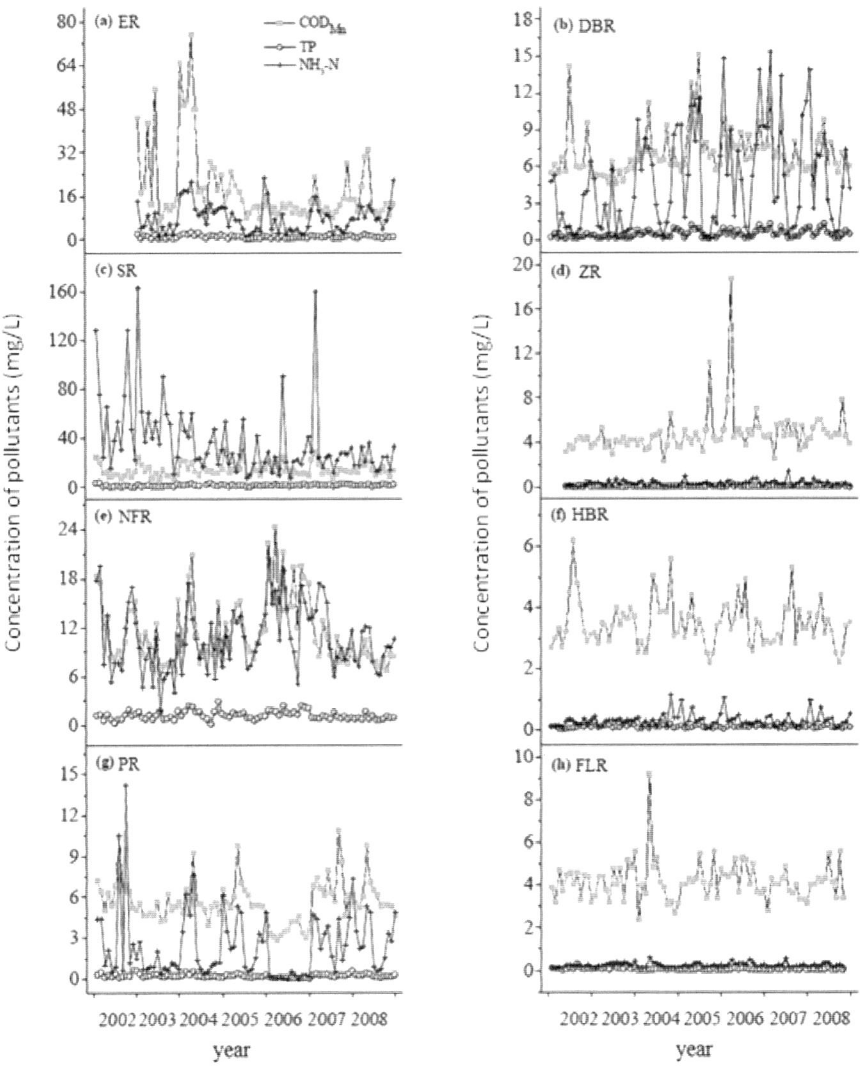

Figure 7. Monthly concentrations of COD_{Mn}, TP, and NH_3-N in eight tributaries

Among the tributaries, the pollution level of the Hangbu and Fengle Rivers was relatively low, with the water quality maintained at Class III perennially. However, the pollution level of the other six tributaries was more serious, with the water quality at Class IV–V (inferior). The annual mean concentrations of the three pollutants and the annual total water flowing from the tributaries into the

IOP Conf. Series: Materials Science and Engineering **484** (2019) 012049 doi:10.1088/1757-899X/484/1/012049

lake are shown in Table 1. Based on equation (4), it was estimated that the total annual amounts of COD_{Mn}, TP, and NH_3-N transported by the tributaries into the lake were 3.58–4.49, 0.19–0.27, and 4.09–5.47 $\times 10^4$ ton /yr during 2002 and 2008, respectively.

Table 1. Annual mean concentration and export flux of pollutants of eight tributaries

Time	annual mean concentration (mg/L) COD_{Mn}	TP	NH_3-N	total water entering the lake (10^8 m^3)	flux to lake (10^4 ton) COD_{Mn}	TP	NH_3-N
2002	8.96	0.52	10.40	42.2	3.78	0.22	4.39
2003	9.09	0.52	10.70	51.1	4.64	0.27	5.47
2004	9.28	0.49	10.46	39.1	3.63	0.19	4.09
2005	9.62	0.50	10.72	46.7	4.49	0.23	5.01
2006	9.94	0.51	10.76	42.6	4.23	0.21	4.59
2007	10.24	0.53	11.08	42.7	4.38	0.23	4.74
2008	9.78	0.55	11.33	36.6	3.58	0.21	4.14

A quantitative analysis comparing the total amounts of pollutants from the tributaries with the water environmental capacity and safety margin of Lake Chao was carried out. The results show that for 2002–2008, the total annual amount of COD_{Mn} transported was below the lake's water environmental capacity and safety margin, at the proportion of 56–68% and 70–82% respectively (Figure 8). However, the total annual amount of TP transported was 3.83–4.17 times the lake's water environmental capacity and 7.51–14.75 times its safety margin. For NH_3-N, the total annual amount transported was 3.78–4.29 times the lake's water environmental capacity and 4.30–4.87 times its safety margin. The total annual amounts of TP and NH_3-N transported by the tributaries far exceeded the lake's water environmental capacity. Thus, to control the pollution of Lake Chaohu, it is critical to reduce the yields of pollutants in the watershed.

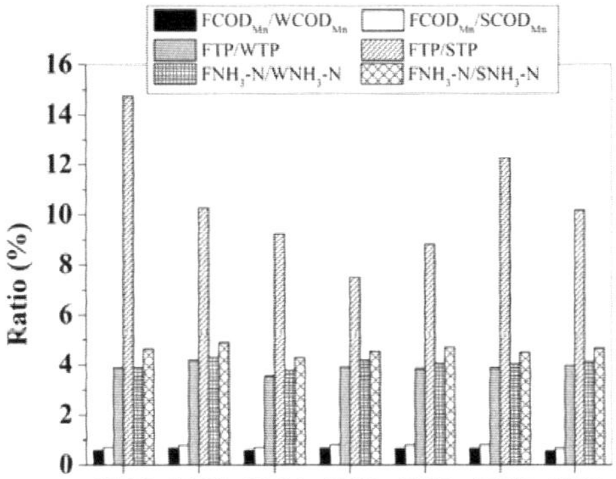

Figure 8. Impact of riverine pollutants fluxes on water environment capacity of Lake Chaohu

($FCOD_{Mn}$, FTP, and FNH_3-N indicate the fluxes of COD_{Mn},TP, and NH_3-N exported by the eight tributaries to the lake; $WCOD_{Mn}$, WTP, and WNH_3-N indicate the water environmental capacity of COD_{Mn},TP, and NH_3-N in the lake; $SCOD_{Mn}$,STP, and SNH_3-N indicate the safety margins of COD_{Mn},TP, and NH_3-N in the lake)

4. Conclusions

In order to analyze the impact of river pollutants transports on the water quality of Lake Chaohu, it is important to estimate the water environmental capacity and the riverine fluxes of pollutants. In this sense, the concentrations of COD_{Mn}, TP, and NH_3-N in Lake Chaohu and the eight major tributaries were monthly monitored during 2002 and 2008, and the water environmental capacities of the three pollutants in the lake were estimated with a zero-dimensional model using Class III water quality as the target. This is the first study that reports this issue for the region. Our results showed that the water quality of Lake Chao maintained at Class IV–V during 2002 and 2008, with CODMn and TP being the major pollutants. The total annual amount of TP transported was 3.83–4.17 times the lake's water environmental capacity and 7.51–14.75 times its safety margin. For NH_3-N, the total annual amount transported was 3.78–4.29 times the lake's water environmental capacity and 4.30–4.87 times its safety margin. This indicates that it is critical to strengthen pollution control in the rivers around the lake for the restoration of the lake.

5. Acknowledgements

This work was supported by the National Science-technology Support Project of China (No. 2010BAC69B02) and the National Natural Science Foundation of China (No. 41203080).

6. References

[1] Dong XD, Zhou Q and Zhou XD, 2004, Technology and development of pollution prevention and treatment on rivers and lakes in China, *Journal of Precious Metallic Geology*, **13(1)** pp 26-29.
[2] Liu RT, Xu JZ, Feng SY, Wang SF and Yao CM, 2002, Analysis of impact of non-point pollution from agriculture to lake water quality, *China Water Resources*, **6** pp 1-5.
[3] Liu ZD, Yu XG and Wang ZX, 2003, The current water pollution of Taihu drainage basin and the new management proposals, *Journal of Natural Resources*, **18 (4)** pp 465-473.
[4] Shang GP and Shang JC, 2007, Spatial and temporal variations of eutrophication in western Chaohu Lake, China, *Environmental Monitoring and Assessment*, **130(1-3)** pp 99-109.
[5] Qin BQ and Zhu GW, 2006, The nutrients forms, cycling and exchange flux in the sediment and overlying water system in lakes from the middle and lower reaches of Yangtze River, *Science in China*, **49 (Supp.1)** pp 1-13.
[6] Yang LB, Yan WJ, Ma P and Wang JN, 2011, Seasonal and diurnal variations in N_2O concentrations and fluxes from three eutrophic rivers in Southeast China, *Journal of Geographical Science*, **21(5)** pp 820-832.
[7] Fan CX, Zhang L and Qin BQ, 2004, Migration mechanism of biogenic elements and their quantification on the sediment-water interface of Lake Taihu: I .Special vari ammonium release rates and its source and sink fluxes, *Journal of Lake Science*, **16 (1)** pp 11-20.
[8] Yang LB, Lei K, Fu G and Yan WJ, 2013, Temporal and spatial changes in nutrients and chlorophyll-a in a shallow lake, Lake Chaohu, China: an 11-year investigation, *Journal of Environmental Research*, **25(6)** pp 1117-1123.
[9] Dong F, Liu XB, Peng WQ and Wu WQ, 2014, Calculation methods of water environmental capacity of surface waters: Review and prospect, *Advances in Water Science*, **25(3)** pp 451-463.
[10] Wang JY, Zhai QW, Guo Q and Tao YZ, 2017, Study on water environmental carrying capacity evaluation in Taihu lake Basin, *China Environmental Science*, **37(5)** pp 1979-1987.
[11] Yu JL, Yin L, Bao GQ and Li B, 2017, Research on water environmental carrying capacity of Tenggeli Lake based on BP neural networks, *China Rural Water & Hydropower*, **11** pp 83-86.
[12] Yin FC and Zhang ZY, 2003, Survey of Chaohu Lake eutrophication research, *Journal of Lake Sciences*, **15(4)** pp 377-384.
[13] Zhang F, Xu JX and Xu CG, 2010, Calculation and analysis of the water environment capacity for the typical watershed in Liaoning province, *Research of Soil and Water Conservation*, **17(5)** pp 231-234,237.
[14] Chu Y and Xia SX, 2009, Characteristics of pollutant export from Fengle River, *Journal of Anhui Agricultural University*, **36(3)** pp 476-482.
[15] Yan WJ, Zhang S, Sun P and Seitzinger S P, 2003, How do nitrogen inputs to the Changjiang basin impact the Changjiang River nitrate: A temporal analysis for 1968-1997, *Global Biogeochemical Cycles*, **17** p 1091.
[16] Wang R, 2009, Study on models for assessing and predieting water quality of the Chao Lake, Hefei Industrial University.

IOP Conf. Series: Materials Science and Engineering **484** (2019) 012049 doi:10.1088/1757-899X/484/1/012049

[17] Xu FL, Tao S, Dawson RW and Xu ZR, 2003, The distributions and effects of nutrients in the sediments of a shallow eutrophic Chinese lake, *Hydrobiologia*, **429 (1-3)** pp 85-93.

[18] Zhang LP, 2007, Diversity analysis of algae in Chaohu Lake, *Journal of Biology*, **24(6)** pp 53-54,72.

[19] Zhu GW, 2008, Eutrophication status and causing factors for a large, shallow and subtropical Lake Taihu, China, *Journal of Lake Science*, **20(1)** pp 21-26.

[20] Zhang ZY, Wang PH and Zhang CD, 1999, Study on the eutrophication in Chaohu Lake and its remediation, *Research of Environmental Sciences*, **12(5)** pp 45-48.

[21] Wang BZ, 1990, *Water Pollution Control* (Beijing: Academic Press).

[22] Li WP, Li CY, Wang L and Zhou LW, 2007, Analog calculation of nitrogen and phosphorus capacities in Wuliangsuhai water environment under different mathematic models, *Journal of Agro- Environment Science*, **26(Supp.)** pp 379-385.

[23] Wang YM, Zhou XD and Li JK, 2005, The research on water environment capacity in lake-taking Bosten Lake as an example, *Journal of Arid Land Resources and Environment*, **19(6)** pp 108-112.

[24] Fang ZY, 1988, *Water resources protection manual* (Nanjing: Hohai University press).

[25] Huang LF, Liu CH and Fan HY, 2010, Water quality analysis of different types of lakes based on SMS, *Safety and Environmental Engineering*, **17(5)** pp 37-40.

IOP Conf. Series: Materials Science and Engineering **484** (2019) 012050 doi:10.1088/1757-899X/484/1/012050

Research Progress in Water Quality Improvement

Changxin Cai, Canjun Liu and Baohui Men*

North China Electric Power University, Beijing, China,
E-mail: menbh@ncepu.edu.cn

Abstract. Rivers play an important role in flood control, sewage discharge, ecology, and so on. However, the intensification of water pollution leads to the river water quality to gradually decline, and pollution control cannot be delayed. This article analyzes the water environment from water quality improvement method. The water pollution treatment and improvement are analyzed and it points out the advantages and disadvantages from physics, chemistry and biology. The article also combs and summarize the issues and development trends in the river water quality improvement method in order to provide scientific reference for the improvement water quality assessment and comprehensive treatment of water pollution of rivers.

1. Introduction

Rivers play an important role in many aspects such as flood control, sewage discharge and ecology. With the rapid social and economic development of Beijing, the discharge of domestic sewage and industrial waste water increases, and the amount of sewage in rivers far exceeds its bearing range. The water quality is getting worse and worse. For example, Wenyu River, as the mother river of Being, which is a strategic backup water source and plays an important role in city development. However, its water quality is deteriorating and many river functions disappeared. In order to alleviate water stress, reduce water pollution in rivers, and realize the overall allocation of water resources, the governance and improvement of river water quality is of great significance.

Through understanding the present situation of water pollution and the achievements of scholars in water quality assessment and water pollution control, the methods of water quality improvement are discussed, and the existing problems and development trends are summarized. In order to improve the river water quality evaluation, comprehensive treatment of water pollution to provide a scientific reference.

2. Progress in Research on Water Quality Improvement

2.1. Water Pollution Current Status

In recent years, due to rapid population growth, water resources, land resources are becoming increasingly tight, river ecosystem is deteriorating day by day, the quality of water environment in China is not optimistic, the improvement of water quality is urgent. Water quality improvement refers to the treatment of polluted, domestic and produced waste water by physical, chemical and biological methods, so that its water quality index can reach the corresponding standard. As early as the 1950s and 1960s, there were water pollution incidents such as water mites caused by methyl mercury pollution and bone pain caused by cadmium pollution in Japan. In 1980, 80% of all diseases in the world were due to improper water supply or water is not clean. Before 1987, about 65.4% of the population had drinking water sources in China, and there were turbid, bitter, salty or fluorine-containing, arsenic poisons, or industrial pollution [1]. Wenyu river is also one of the polluted rivers, many water quality indicators of Wenyu river are excessive, and the environment around the

IOP Conf. Series: Materials Science and Engineering **484** (2019) 012050 doi:10.1088/1757-899X/484/1/012050

river became worse than befor. Ecological restoration of rivers and wetlands has become a worldwide phenomenon since 1990's. With the deep understanding of river function and the importance of urban water system to people's life, regional water quality improvement has been widely carried out at home and abroad. Among them, Chengdu Funan River Park in Sichuan Province was an early and successful case in China. It is an eco-environmental protection park with the theme of water quality improvement. It draws the polluted water from Funan River and carries on natural ecological purification through the constructed wetland system of the park, and then flows back to Funan River after the water quality reaches the standard.

The bulletin on environmental conditions in China pointed out in 2015 that the Yellow River, the Yangtze River, the Songhua River, the Pearl River, the Haihe River, the Huaihe River, the Liaohe River, and the northwest rivers, the southwest rivers, the Zhejiang and Fujian rivers, and other 700 state-controlled sections, the pollution of Haihe River Basin is the most serious, the proportion of inferior V water is close to 40%, and the pollution of Liaohe, Haihe and Huaihe River basins is also serious, the proportion of I-III water is about 50%. Secondly, with the rapid development of economy in China, the pollution problem of urban rivers is becoming more and more serious, and people need to improve the water quality of urban rivers more and more urgently. From the overall situation of water quality improvement in urban rivers in recent years, the problem of water pollution in urban rivers in China will still exist for some time in the future, and even the water quality of the local rivers will be further deteriorated.

2.2. Pollution Causes

According to the causes of water pollution, it can be divided into natural pollution sources and man-made sources of pollution, and physical sources of pollution (such as heat or radioactive substances) [2]. According to the types of harmful substances released by the sources it can be divided into chemical sources (such as inorganic or organic), biological sources (such as bacteria or toxins). According to the distribution characteristics of pollution sources it can be divided into point pollution sources such as municipal sewage, industrial and mining enterprises and sewerage vessels), surface pollution sources (e.g. surface runoff of rain water, soil erosion and drainage of large areas of farmland), diffuse pollution sources (air pollutants by sedimentation or sedimentation into water bodies, such as radioactive deposits, acid rain).

2.3 Brief Introduction of Water Quality Improvement Methods

With the development and improvement of water treatment technology, the purification and improvement of river water quality is becoming more and more mature. Through the experimental research and engineering practice of river pollution control, the effects of various water treatment techniques are stable and economic performance is good. These methods can be divided into three types according to their purification principles: physical method, chemical method and biological method. All kinds of methods have their own technical characteristics and application conditions, and the practical value of each technology is determined by the different conditions.

2.3.1. Physical Method.

a. Water Diversion Dilution. The water diversion dilution is mainly applied to the improvement of the eutrophic river water and lake water, and it is also an effective method for the water pollution control of the extremely slow rivers. In Lake Buffalo pound, Canada, the dominant species in the lake were transformed from green algae to macrophyte by adding water dilution with low nutrient content [3]. The principle can be summarized as three points: (1) The pollutants are transported to the lower reaches of the river in a short time through the diversion of water into the upstream channel, thus reducing the concentration of pollutants in the whole river and achieving the purpose of purifying the water quality; (2) In the process of water diversion and injection, the increase of river water velocity and the increase of water and air contact area in some areas make the content of dissolved oxygen increase, the water quality changes from hypoxia to aerobic, and the self-purification ability of the river is further improved; (3) The injection of foreign water enhances the fluidity of the river and exchanges the polluted rivers in different regions and reduces the concentration of pollutants in the seriously polluted areas.

It should be noted that the concentration of pollutants in the river is reduced by dilution, but the total amount of pollutants remains unchanged, which is only a physical process, and the pollutants in the lower reaches of the river will increase with the diversion of water. Therefore, before using this method to purify the water quality, it is necessary to carry out scientific calculation and analysis and implement it carefully on the premise of not exceeding the carrying capacity of the lower reaches of the river.

b. Artificial Aeration. Organic pollutants enter rivers and are oxidized and decomposed by aerobic microorganisms in water. The continuous development of this process makes the river reoxygenation rate less than the aerobic rate, the dissolved oxygen in the water is consumed, and finally the river water becomes hypoxia or anaerobic. Due to the serious shortage of dissolved oxygen, the ecosystem of rivers will be destroyed, aquatic organisms will die of hypoxia, and the purification capacity of rivers will be greatly reduced or even lost. Therefore, the artificial aeration of the river can effectively improve the dissolved oxygen concentration and improve the water quality of the river. In the treatment of the Rhine River in Germany, in addition to the river closure and pollution control, the tributary Mein River is also supplied with liquid oxygen to the section of about 60km of river[15]. The Suzhou River in Shanghai supplied oxygen to the tributaries in 1998, increasing the dissolved oxygen content from 0.2mg/L to 2mg/L [4-6].

c. Sediment Dredging. The sediment of the river bed is the main source of internal pollution of river water. Under certain conditions (certain temperature and pH), a large number of pollutants are released in the sediment, which seriously affects the water quality of the river. By dredging the river regularly, the polluted matter in the river sediment can be reduced effectively, and the water quality of the river can be improved. In the course of dredging, a reasonable and effective dredging method should be chosen according to the pollution condition of the river and the structural conditions of the river, so that both the river course can be dredged and the degree of pollutant diffusion can be reduced to the minimum. The ecological conditions of rivers should be considered to minimize the damage to river ecosystems. The experimental study on the sediment dredging by Zhong Chengsheng et al. showed that sediment dredging has a good control effect on the release of endogenous phosphorus and nitrogen pollutants in the river, at the same time; it has important influence on denitrification, microbial activity and community diversity of sediment [7].

2.3.2. Chemical Method.

a. Chemical algae removal. For eutrophic rivers and lakes, algae growth is controlled by adding chemical algae remover agent. Chemical algae removal can eliminate algae pollution in a short period of time, but at the same time, the input of chemical agent will also have a negative impact on the growth of other organisms in the water and cannot fundamentally eliminate algae pollution. Therefore, the chemical algae removal method can only be used for emergency use in water purification improvement. In the course of Dianchi's governance, chemical algae removal has shown a good effect on water quality improvement [8].

b. Flocculation Sedimentation. The coagulant is added to the polluted water body to precipitate the colloidal substance and suspended organic matter in the water body to achieve the purpose of purifying water. This method can effectively reduce the phosphorus-containing pollutants in water. The commonly used coagulants are inorganic flocculants (aluminum salts and iron salts) and high molecular flocculants (polyacrylamides).

c. Heavy Metal Fixation. The sediment in the river bed contains not only organic pollutants but also heavy metal pollutants. Under certain conditions, heavy metals will enter the water body in some specific forms and cause pollution to the water body. The principle of heavy metal fixation is to combine heavy metals in sediment, reduce the amount of diffusion to water, and reduce the pollution of heavy metals in water. The heavy metals in fixed water are mainly fixed by adjusting pH value, so that all kinds of heavy metals exist in the form of precipitation at the bottom of the river, and they are usually fixed by alkaline substances such as slag and steel slag.

2.3.3. Biological Method.

a. Inoculation Method. Microorganisms play an important role in the process of self-purification of rivers. If the effective microbes in the river are reduced or disappeared, the river's self-purification capacity will also decrease. Inoculation method is to put effective microorganisms into the river water to enhance the self-purification ability of the river, thereby promoting the degradation of organic pollutants in the river water body and achieving the purpose of improving the river water quality. Of course, the bacteria to be inoculated need not contain pathogens, do not harm other microorganisms, and can adapt to the environment of the river, which requires time to cultivate and screen efficient strains, which is also a disadvantage of the method. Zengyu et al. used photosynthetic bacteria to purify the water quality of Funan River. The results showed that the removal rate of CODcr reached more than 90% with the addition of bacteria. This shows that the use of microorganisms to control water pollution is promising [9].

b. Phytoremediation Technology. Phytoremediation technology is a method to use plant roots to absorb pollutants in water and finally harvest plants to achieve the purpose of water purification. Sun Lianpeng et al. used biological floating bed to remove nitrogen pollutants in Pearl River Basin [9]. This method has the advantages of good water purification effect, low cost, no need for energy consumption and operation cost, and cannot cause secondary pollution to water body. It can also improve the landscape and beautify the environment. It is a promising water purification technology.

c. Biological Manipulation Technology. The predation, competition, and interdependence between aquatic and plant populations and colonies play an important role in water quality regulation. Therefore, regulating and controlling the species and quantity of aquatic organisms can improve the purification ability of water and then improve the water quality [10]. The process of biological manipulation is to reduce the fish feeding on plankton in order to increase the number of zooplankton and to increase its intake of phytoplankton so as to reduce the number of phytoplankton and improve the water quality.

d. Constructed Wetland Technology. Constructed wetland is a wetland-like water purification system composed of fillers, plants and microorganisms. Through the synergistic action of physics, chemistry and biology in natural ecosystem, pollutants in water are filtered and absorbed to realize the purpose of water purification. For example, the microorganisms in the wetland system can effectively degrade the organic matter in the sewage. The plant roots can not only bring oxygen into the water, but also absorb the nitrogen and phosphorus pollutants from the sewage, and the soil layer can filter the suspended matter in the sewage [12].

3. Conclusion

The water quality detection of the river reach is a process of large amount of data, multi-level target and strong synthesis. It is necessary to correctly and accurately process and analyze the measured data and experimental results, and then establish the index system and determine the improvement method. In the background of increasing water demand and increasing water pollution, through consulting literature, this paper enumerates a variety of water quality improvement methods, and combs and summarizes the existing problems and development trends in the method. With the perfect implementation of monitoring, data accumulation and correct analysis, and finally make a reasonable evaluation and formulate specific treatment programs, it is believed that the river water quality will be greatly improved and the Beijing's mother river Wenyu will become better and better.

4. Acknowledgments

This work was supported by the funds for undergraduate innovative experiment plan of North China Electric Power University, and the Famous Teachers Cultivation planning for Teaching of North China Electric Power University (the Fourth Period).

5. References

[1] Yao Yunxian. Introduction of lake eutrophication water quality improvement project abroad [J].*The Administration and Technique of Environmental Monitoring*, 2002, **14**(2): 45-46.
[2] Zhou Lin, Li Yong. Study on water pollution status and water environment management strategy in

IOP Conf. Series: Materials Science and Engineering **484** (2019) 012050 doi:10.1088/1757-899X/484/1/012050

China [J]. Environment and Development, 2018, 4:51-52.

[3] Sun Junhou. Water Aeration Technology is An Effective Measure to Improve Water Quality of Urban Rivers and Lakes [J]. *Beijing Water Resources*, 2002, **4:35**-36.

[4] Zhong Jicheng, Liu Guofeng, Fan Xincheng. Environmental effect sediment dredging in lake (I): the role of sediment dredging in reducing internal phosphorous release [J]. *Journal of Lake Sciences*, 2009, **21** (1): 8 4 - 9 3.

[5] Zhong Jicheng, Liu Guofeng, Fan Xincheng. Environmental effect sediment dredging in lake (II): the role of sediment dredging in reducing internal nitrogen release [J]. *Journal of Lake Sciences*, 2009, **21**(3): 335-344.

[6] Zhong Jicheng, Liu Guofeng, Fan Xincheng. Environmental effect sediment dredging in lake (III): Influence of dredging on denitrification in sediments [J]. *Journal of Lake Sciences*, 2009, **21**(4): 465-473.

[7] Zhong Jicheng, Liu Guofeng, Fan Xincheng et al. Environmental effect sediment dredging in lake (IV): Influences of dredging on microbial activity and functional diversity of microbial community in sediments and its significance [J]. *Journal of Lake Sciences*, 2010, **22**(1):21-28.

[8] Wang Xiaoxiong, Jiang Chengchun, Zhu Jia. Research on the application of microbial techniques to algae removal [J]. *Industrial Water Treatment*, 2011, **31**(2):1-4.

[9] Zeng Yu, Qing Song. Application of PSB Method in Sewage Purification [J]. *Urban Environment & Urban Ecology*, 2000, **13**(6): 29-31.

[10] Sun Lianpeng, Feng Chen, Nitrogen Removal from Pearl River by Enhanced Ecological Floating Bed System [J]. *Acta Scientiarum Naturalium Universitatis Sunyatseni*, 2009,**48**(1):93-97.

[11] Yuan Xingcheng. Study on technology of water quality improvement and water environmental governance program for river in plain river network region [J].Zhejiang: Zhejiang University.

[12] Pengchao Ying, Zhu Gguohong, Yin Guo et al. Study on sewage treatment by artificial wetland [J]. Chongqing Environmental Science, 2000, 6:43-45.

IOP Conf. Series: Materials Science and Engineering **484** (2019) 012051 doi:10.1088/1757-899X/484/1/012051

A Study on the Application of Green Design Theory in Environmental Art Design

Kun Bai

Shandong Women's University, Jinan, Shandong, China
Email: bk3363918@163.com

Abstract. At present, human beings face three major problems: environment, resources and population. Modern design must take into account the basis of human survival - natural environment. The design community has put forward the slogan of green design: namely take conservation resources and environment protection as the subject, protect the ecology, and be friendly with the environment. The concept of green design has attracted more and more attention in the environmental art design. Combined with the background of green design in this paper, the connotation and characteristics of green design is analyzed to summarize the basic principles of green design and implementation countermeasure in environmental design. Finally, some examples of feasible green design are put forward for analysis.

1. Background of Green Design

With the rapid development of science and technology, on the one hand, it brings beautification and comfort to human life. On the other hand, it causes great waste and destruction of human resources and environment, and the contradiction between man and nature is increasingly prominent. In the face of many new problems, the establishment of "sustainable development" production model has been adopted by more and more countries as the development strategy, and the "green design theory" that uses advanced technology and takes into full consideration of product greenness and economy has emerged.

In China, the average living standard was low, and the people have relatively fewer material goods. With the development of human society, new technologies and new methods have been widely used in production practice. The ability of people to transform nature has been greatly enhanced, the economic situation has improved, the material resources have been enriched, and the purchasing power of the people has also increased. The government is adopting the proactive fiscal policy to expand domestic demand, and we are entering a period of consumption promotion. At the same time, China is a country with a lower average resource occupancy percentage, and the ecological and environmental problems are more serious.

In order to maximize profits, the company has made every endeavor to promote sales, thereby leading to a lot of waste. On the one hand, enterprises simply pursue economic benefits regardless of consumable materials such as "over-packaging" abundant in the field of packaging design; on the other hand, enterprises unrealistically expand production and accelerate the upgrading of products in order to stimulate consumption, and there are many phenomena of repeated production and repeated construction, such as VCD heat and "touch net heat", which have resulted in a great waste of resources. As a result of these phenomena, the degree of human damage to natural resources has increased, the ecological environment has become increasingly polluted, and the exhaust gas, waste water, waste residue and waste heat emitted from various industrial processes have polluted the atmosphere, water and soil. At the same time, noise and vibration may also be generated to endanger surrounding environment. In addition, the environmental damage caused by waste and domestic waste after industrial products have been used is also quite serious. Therefore, the protection of green environment of the earth has not only become the key topic of the world, but also the worldwide social problem. The field of design is also responding positively. More and more products are designed with the theme of environmental protection. While increasing productivity, the attention to environmental protection is changing the designer's thought and concept, thus a new trend of green design has emerged.

In the world, green design has become one of the hot spots in the field of design in recent years, especially in the United States, Japan, Europe and other countries, research is very active. The recycling laws and regulations formulated by the United States, Japan, and some European countries have attracted great attention from academia and industry. Many scholars have put forward their own theories on the recyclability of products. In addition, the University of California of the United States not only established a research institute on environmental awareness design, but also established a special web page for green design that can be systematically queried on the Internet. In Europe, there have been many papers on green design and green manufacturing in CIPR in recent years. In recent years, the International Organization for Standardization has proposed the ISO14000 industrial environmental management system standards, and thus as a basis for judging whether products can enter the international market. The emergence of a series of standards has created a strong impetus for manufacturers to promote the development of green products. At the same time, all developed countries and organizations have specialized institutions for green science and technology research, such as CenCITT in the United States, BMFT in Germany, ENEA in Italy, ETH in Switzerland, three national committees in the United Kingdom, and the Science and Technology Department in Japan.

2. Connotation and Characteristics of Green Design

"Green Design" (GD) was originally derived from a copy of environmental pollution regulation in the United States in the 1970s, which has the same meaning as current DFE (Design for the Environment)[1]. However, as a kind of new design concept, it comes from the rethinking of modern design art values by the Western design art community in the early 1980s. The so-called green design is the design to protect the green environment of the earth. It is the first step in building a green industry. When developing new products, designers must use the overall thinking (designers shall not only consider the environmental impact on production and use of new product, but also think of the way out after the failure of products in the whole process of product life cycle). The research scope of "Green Design" extends from product design and manufacture to product packaging, product promotion and product marketing that are closely related to products, to green service consciousness and green culture consciousness of the whole society. The 3RE principle - Reduce, Reuse, and Recycling is an important part of it.

Green design has two characteristics: One is that the entire life cycle of design object shall be considered in the design. The other is that the designer should give priority to the environmental properties of design object, prevent the environmental pollution fundamentally, and save the resources and energy while ensuring the normal function of the product.[2]

3. Basic Principles of Green Design

Compared with the past, the most important difference between modern design and the past is the transformation of design concept, namely adhering to the concept of "sustainable development". This transformation is mainly derived from the cognitive return of human beings to their relationship with nature. From the perspective of environmental protection, the green design includes three basic principles: Safety principle, economy principle and social principle.

3.1. Safety Principle - Protection of the Safety of Nature and Humans.

Natural ecological safety and human survival security are unified. In the environmental design, raw

1 Steven Ashley, Design for the Environment, Mechanical Engineering, 1993.3, P.52-55

2 Liu Beiguang: Green Concepts in Environmental Art Design, Decoration, 11/2005, Phase 151

materials that can be recycled or easy to regenerate are used as much as possible, which shall be easy to disassemble, recycle and reuse after use, hereby saving energy and not polluting the environment.
In the 27th Olympic Games held in Sydney in 2000, the concept of "Green Olympics" was used. In the construction of the Olympic venues in Sydney, the piece of wood and straw were crushed into recycled boards for partition board, and the solar battery was used for night illumination of Olympic street lights. The water was collected to water the garden, and even turning several rubbish hills into parks where flowers are in blossom. The Sydney people's understanding of the "Green Olympics" gives us a lot of inspiration.

3.2. Economy Principle - In the Environmental Art Design, the Simplified Design should be Implemented as Much as Possible to Avoid the Excessive Consumption and Occupation of Resources Due to the Complexity of Design and Increase the Use Efficiency and Number Of Resources.
Minimization of material use, maximization of product durability and minimization of energy consumption are our goals. However, simplified design is not equal to simple design, and does not mean to give up the pursuit of art aesthetics. Therefore, simplifying design, saving resources, and meeting the requirements of function and aesthetics have become important criteria for testing the quality of environmental art design. The building design the environmental load can be reduced and the lifetime resource consumption after the design is completed is minimized on the basis of creating comfortable communication space by coordinating the modeling with the surrounding environment and adjusting the setting distance appropriately.

3.3. Social Principle - Avoid Excessive Design and Waste of Resources, Even Environmentally Protection Materials should be used as much as Possible.
With the enforcement of awareness of environmental protection, environmental art designers have upgraded design practice and theory to the heights of lifestyle design, cultural model design and system design, that is, using system theory as the guiding ideology to create a reasonable and healthy survival mode and living environment, and oppose to pursue the so-called "pure art" of environmental art design at the expense of social resources.

4. Implementation Countermeasures of Green Design
The idea of vigorously promoting green technology and green design in all fields of social production activities plays an important role in future economic development. In order to better implement the thought of green design in the whole society, produce more green products and design more green buildings, and the following countermeasures are proposed:

4.1. Insist on Establishment of New Ecological Ethics.
Firstly, through environmental legislation, the development of green design ideas is promoted, the functions of internationally advanced green design technologies and products are carefully analyzed and studied, and technologies suitable for China's national conditions are actively introduced. The state should encourage and support the development of green products in terms of policy and economy, and acceleration of industrialization process and expansion of city scale have resulted in regional environmental pollution, ecological damage and resource depletion, bringing huge impacts on the limited resources and fragile ecological environment on the earth. Green design and environmental protection are complementary to each other.

4.2. Adhere to the Concept of Resource Conservation and Resource Recycling Use.
In order to reflect the concept of green resource conservation in building design, the recycling of resources shall be persisted. Resource conservation is not only a control of engineering cost, but also a practical way to save construction resources in the later stage of the project. It is very important for the application of green design to realize the rational utilization of resources and display the concept of green design while designing.

4.3. Adhere to Energy Conservation and Environmental Protection as the Basis.
In the green design concept, the emphasis is placed on the improvement of design concept. The former high consumption is transformed into low consumption. With the help of energy-saving technology, we can improve the effective utilization of resources, make innovations of building resources, avoid excessive reliance on existing resources and realize self-sufficiency in some resources. The current environmental pollution is closely related to building industry, which not only includes air pollution, but also light pollution and electromagnetic pollution. Therefore, the introduction of green design concept of the building needs to examine the entire construction cycle, assess the environmental pollution caused by the entire cycle of the construction process, improve the concept of resource conservation, strengthen the innovation of building materials, and enhance the protection effort of architectural environment.

5. Implementation Examples of Green Design

Now, the concept of green design has been widely recognized and promoted by countries all over the world. In this paper, the residence, restaurant and kindergarten design are used as an example for discussion.

5.1. Green Design in Residential Environment.
The construction of green residence environment will strive to create a green residential environment guided by "life, vitality and healthy development" with "the improvement of living environment in the residential environment" as the goal. The green residential environment can be understood not only as the urban residential environment with life and vitality, as well as optimized, high-quality living conditions and living space where people can develop in a sustainable and healthy manner, but also the residential space with high living quality, high performance, high living grade and cultural charm, which is characterized by low consumption of natural resources, less energy consumption, pollution-free, non- pollution.[3]

The design principle of green residential environment is the core of green design. All design factors are "people-oriented and nature-rooted", taking from nature and using in nature. Modern residential environment is not only a space for urban human living, but also condensed and divergent point of residential culture. Therefore, the emphasis on cultural communication in the planning and design of green settlement environment is to cultivate the healthy and noble sentiment of the residents in the residential area, and create a cultural atmosphere of the green environment of residential area at the same time.

5.2. Low Carbon Green Experience Restaurant Design.
The fluidity of air flow. The restaurant is the environment for dining, and the primary task is to reflect its own superiority to attract customers to experience low-carbon environment and green restaurants, while this restaurant atmosphere is extremely important as the first impression. Therefore, the fluidity of the air flow is of the utmost importance in the restaurant environment. The main reason of mature design on ventilation performance of the building is the particularity of restaurant and the mobility of restaurant population. If the air does not flow or the flowing performance is poor, it is easy to cause bacteria to grow and spread in the air. [4] Therefore, reasonable natural ventilation as well as supplementation and rational planning of artificial ventilation system greatly reflect the concept of green design.

About the issue of lighting. On the premise that it is possible to use natural light as far as possible, natural light is taken as the main body of lighting design, because natural light itself is softer relatively to artificial lighting, which is suitable for use in the environment like a restaurant. Secondly, the reasonable use of lighting can fully optimize the space of restaurant. Therefore, a large number of French window can be design to increase the lighting quantity, and the intelligent lighting systems

3 Zhou Lan. Urban Space Aesthetics [M]. Nanjing: Southeast University Press, 2001
4 Kong Xiaodan. Discussion on Design Principles of "Green" Interior Design [J]. Journal of Wenzhou Vocational & Technical College, 2014(02):38

such as the more popular optical control lamp are chose at the same time. The use of energy saving products can also reflect the design concept of low-carbon green experience restaurant.[5]
The selection of green and low carbon furniture. For the design of low carbon green experience restaurant, environmental protection and low carbon green furniture must be the first choice. In the premise of not considering the cost, green furniture can reflect the design concept and create a low-carbon atmosphere. As the demand for furniture in the restaurant is increasing, the furniture made of bamboo is the most practical and relatively low cost combined with cost consideration, which can play a very good role in promoting green and low carbon environmental protection propaganda at the same time.

5.3. Indoor Space Environment Green Design in Kindergarten.

The design of indoor space in kindergarten should take children philosophy to make children's activities harmonious with indoor space environment. The design of indoor space environment in kindergarten should be based on the needs of the children, take into full account the physiological, psychological and behavioral characteristics of the children, follow the principles of promoting the healthy development of children, and optimize the efficiency of indoor space environment for the purpose of safety, comfort, convenience, intelligence, green, environmental protection, health, hygiene and energy saving. In the design of indoor space in kindergarten, specificity of indoor space design, security of the space, color and wall decoration of the space should be considered emphatically. In the design of the indoor environment in kindergarten, the indoor air quality should be ensured by scientific selection of green decorative materials, green furniture, moderate decoration, scientific creation of physical environment, and reasonable selection of purification measures for indoor pollutants so that children can grow healthily.

6. Conclusions

Green design is a win-win game between environmental art design and natural environment. It is also a win-win game between designers and the public. Therefore, the advocacy of pursuing and promoting environmental art design based on three dynamic harmonies between people-to-people, people-to-nature and human themselves are in line with the trend of historical development and represents the progress of society. As a consequence, its wide application is inevitable. Nowadays, sustainable development is a major issue facing the whole world, and green design has become our only choice in today's deteriorating ecological environment. Although green design must rely on the support of high and new technology, there is still a long way to go to realize ecological architecture, green design has a very good development prospects, and we should be full of confidence in the future and development of green design.

7. References

[1] Xu Yuqing: Green Design, Beijing: Beijing Institute of Technology Press, second edition of 12/2013
[2] Zhou Xi and Li Zhandong: New Theory of Ecological Design: Rethinking and Re-understanding on Ecological Design, Nanjing: Southeast University Press, edition of 2002
[3] Wang Huiying: A Brief Analysis on Green Design Philosophy in Environmental Art Design, Journal of Shandong Agriculture and Engineering University, Issue No. 4 of 2012
[4] Qi Xinzheng, Zhao Songzheng, and Sheng Zhilin: Green Design——the Development Trend of Modern Design, Journal of Ningxia University (Humanities and Social Sciences), Issue No. 3 of 2010
[5] Sun Feifei and Zhang Hong: A Brief Analysis on Green Design and Environmental Protection, China High Technology Enterprises, Issue No. 4 of 2009
[6] Basson L, Petrie JG: An integrated app roach for the consideration of uncertainty in decision making supported by Life Cycle Assessment Environmental Modelling & Software, 2007

5 Gu Chaolin. Establishing Low-carbon City and Realizing Sustainable Development——Low-carbon City Planning and Development Model [J]. Urban and Rural Construction, 2016(11):70-72

IOP Conf. Series: Materials Science and Engineering **484** (2019) 012052 doi:10.1088/1757-899X/484/1/012052

An AHP-Bayesian Model for Operational Risk Evaluation of Integrated Energy Systems

Yongqi Yang[1], Wanlei Xue[1], Chaoyuan Li[2], Xin Zhao[1] and Nan Xu[1]

[1]Economic and Technology Research Institute, State Grid Shandong Electric Power Company, Jinan, China
[2]School of Finance, Shanghai University of Finance and Economics, Shanghai, China
Email: yangyongqincepu@163.com

Abstract. The rapid development of Integrated Energy Systems (IES) is of great significance in solving the increasingly serious environmental problems by promoting China's energy evolution. Under the huge policy support, China's IES technology has taken a great-leap-forward development. Meanwhile, a large number of IES projects will be carried out extensively. Therefore, operational risk evaluation for IES projects is of great urgency. In the first place, we will summarize the basic structure of IES and build operational risk evaluation index system for IES projects based on characteristics of China's IES systems Secondly, the improved Bayesian operational risk evaluation model will be established. In this improved model, AHP method will be used to calculate the weight of every index. Last but not least, a case study is implemented to prove the practicality of the model and provide references for the IES projects decision and optimization.

1. Introduction

As a key support of economic and social development, China's energy supply-utilization system, which relies on fossil fuel, has led to several serious environmental problems. China is now experiencing an important restructuring period, an energy revolution is of great urgency to change the present traditional extensive energy development mode to a more intensified mode[1]. IES, which is able to integrate multiple energy (including oil, coal, natural gas and power), is considered to be the key method to build the new energy system relies on clean energy[2,3]. The core idea of IES is multiple energy complementation and optimal coordination between "source-grid-load-energy storage", IES is able to enhance the energy efficiency and promote clean energy consumption.

IES was developed based "Internet plus initiative" and "Energy Internet". Since publication of *the Guidance for the Promotion and Development of Smart Energy based on Internet Plus Initiative* from National Development and Reform Commission, the related technologies of IES has developed rapidly[4]. [5] investigates multi-period optimal energy flow and energy pricing in carbon-emission embedded integrated energy systems, including electricity, natural gas, and district heating networks. A bilevel optimal coordinated control strategy for park-level integrated energy system is proposed in[6]. [7] has built a day-ahead scheduling optimization model, and the rigid coupling between heating network and power grid is improved by adjusting CHP output. [8] has proposed a typical integrated energy system which includes typical energy conversion technologies, after building energy demands and energy conversion processes in TIES are abstracted and simplified by a typical energy flow map, a dimensionless equation for calculating TIES energy and exergy efficiency is developed.

Now many industrial parks have established IES to fulfill users' energy demand, however, many studies focus on multiple energy conversion and system operation optimization. There are few researches in operational risk evaluation. Therefore, this paper will propose the basic structure of IES and establish operational risk evaluation index system at first. Second, AHP method will be introduced and build improved Bayesian Model. Last but not least, a case study will be carried out based on an IES project in North China, the operational risk of three different schemes of this project will be calculated to provide reference to the optimization and decision of project schemes.

2. Basic Structure of IES

Nowadays, several industrial parks are seeking to fulfill users' energy demand through IES. Basic structure of these IES is shown in Figure.1.

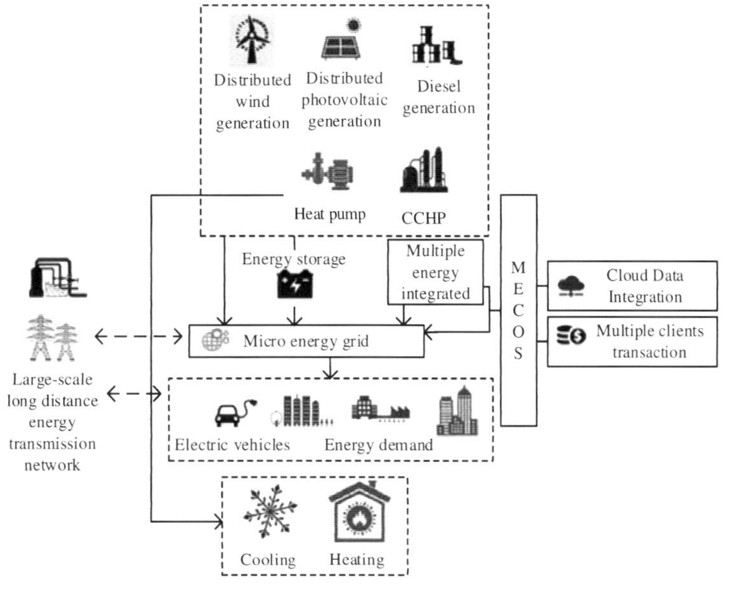

Figure 1. Basic structure of IES

The basic structure of IES can be divided into three parts:

For supply side, power supply mainly relies on distributed generation including distributed wind generation, distributed photovoltaic generation, etc. Cooling, heating and natural gas supply depend on CCHP and heat pump[9]. Along with the large-scale long distance energy transmission network, flexible coal-fired generation can be connected to IES to provide ancillary services for distributed generation. In addition, oil and gas can be transmitted through this long distance network.

For transmission side, micro energy grid is the main grid of IES. The structure of micro energy grid is established based on micro grid. Micro energy grid connects distributed energy generation and large-scale long distance energy transmission network, and it's able to set up multiple energy integrated devices on micro energy grid. Hence micro energy grid is able to realize the coordination between hydro energy, electric power, heat and gas. As for long distance energy transmission network, micro energy grid can be treated as a relatively independent energy supply-demand module. Multiple Energy Coordination Optimization System (MECOS), cloud data integration and multiple clients transaction platform should be built on transmission side to realize the optimized allocation of resources.

For demand side, MECOS is able to integrate demand side resources, in addition, MECOS can upgrade demand side management and demand side response to comprehensive energy utilization management and energy utilization response[10]. Therefore, demand side of IES is able to form controllable energy demand curve along with the organized charge-discharge energy storage operation mode, which can trace the energy supply curve to realize the supply-demand side coordination.

3. Evaluation Index System

According to the characteristics of IES projects in China, this paper has built the index system in three categories (supply, transmission and demand sides) as shown in Table 1.

Table 1. Investment risk evaluation index system for IES projects

Categories	Code of Indexes	Indexes
Supply R_1	R_{11}	Wind curtailment
	R_{12}	Solar energy curtailment
	R_{13}	SOx pollution rate
	R_{14}	NOx pollution rate
	R_{15}	COx pollution rate
	R_{16}	Dust pollution rate
	R_{17}	Primary energy utilization
Transmission R_2	R_{21}	Energy conversion efficiency
	R_{22}	Rate of energy loss during transmission
	R_{23}	Payback period
	R_{24}	Internal rate of return
Demand R_3	R_{31}	Energy demand forecast accuracy
	R_{32}	Energy consumption per unit of GDP
	R_{33}	Regional GDP growth rate

For one thing, energy conversion efficiency is included to demonstrate the conversion efficiency between multiple kinds of energy, renewable energy utilization rate, etc. For another, IES project should be able to achieve economic benefits as well as social benefits including pollution reduction, GDP growth rate, etc. Therefore, the index system is built in Table 1 according to the requirements above.

4. Improved Bayesian Model

4.1. AHP Method

AHP weighting approach is introduced and performed according to the following steps:

1) R represents the comprehensive benefit, which includes economic benefit R_1, energy efficiency R_2 and social benefit R_3. The set of comprehensive benefit is $R=\{R_1, R_2, R_3\}$, and their weight is $W=\{W_1, W_2, W_3\}$. As for the second level indexes, R_1 can be divided into wind curtailment R_{11}, solar energy curtailment R_{12}, etc. Their weight is $w_1=\{w_{11}, w_{12},...,w_{17}\}$, i represents decision indexes and its number is n, j represents different schemes and its number is m. The rest indexes can be done in the same manner[11].

2) The importance can be determined through pairwise comparison, scale meaning is used to demonstrate the importance of each index[12]. Scale meaning of AHP is shown in Table 2.

Table 2. Scale meaning of AHP

Values	Definition
1	Equally important
2	Experience and judgment slightly favor one index over another
3	Experience and judgment strongly favor one index over another
4	An index is favored very strongly over another
5	The evidence favoring one index over another is of the highest possible order of affirmation.

3) The judgement matrix can be built through pairwise comparison. Here we take the first level index as an example, $R=\{R_1, R_2, R_3\}$, and its judgement matrix is:

$$Q = \begin{bmatrix} q_{11} & q_{12} & q_{13} \\ q_{21} & q_{22} & q_{23} \\ q_{31} & q_{32} & q_{33} \end{bmatrix} \tag{1}$$

Where q_{12} is the comparison result between supply and transmission, if supply is more important than transmission, the value of q_{12} is higher than 1, otherwise it's less than 1. Obviously $q_{11}=q_{22}=q_{33}=1$.
4) Solve the following formula to get the weight of each index:

$$QW=\lambda_{max}W \tag{2}$$

Where λ_{max} is largest eigenvalue of matrix Q, W is the feature vector of λ_{max}, hence W_1, W_2, W_3 is the weight of R_1, R_2, R_3.
5) Consistency check should be performed, the consistency index is calculated as:

$$CI = \frac{\lambda_{\max} - n}{n-1} \tag{3}$$

$$CR = \frac{CI}{RI} \tag{4}$$

Where the value of RI varies with the order of judgement matrix Q, as it's shown in Table 3:

Table 3. The value of RI

n	1	2	3	4	5	6	7	8	9	10
RI	0	0	0.58	0.9	1.12	1.24	1.32	1.41	1.46	1.49

When CR=0, the consistency of judgement matrix is fully stochastic.
When CR>0.1, the consistency of judgement matrix do not meet the consistency requirements, the judgement matrix should be further modified.
When CR<0.1, the consistency requirements are fulfilled. Hence $W=\{W_1, W_2, W_3\}$ (after normalization) is the weight of R_1, R_2, R_3.

4.2. Bayesian Risk Investment Evaluation Model

The calculation process of Bayesian operational risk evaluation model is as follows:
In decision-making space $S=\{s_1,s_2,\ldots,s_n\}$, s_n represents decision scheme, S is the state space. Among those, δ_1, δ_2 and δ_3 separately reflect that the situation of the market is high risk, average risk and low risk. Therefore, $P(\delta_1)$, $P(\delta_2)$ and $P(\delta_3)$ demonstrates the probability of δ_1, δ_2 and δ_3. According to the authoritative analysis report, this paper sets $P(\delta_1)$=0.2, $P(\delta_2)$ =0.4, $P(\delta_3)$=0.4. Obviously, there is a risk in this prediction, so the risk factors should be added into the prediction of market situation[13]. In this model, μ_1, μ_2 and μ_3 means the prediction accuracy is low, average, high.
According to 3.1, the weight vector is $w_1=(w_{11},w_{12},\ldots,w_{17})$, $w_2=(w_{21},w_{22},\ldots,w_{24})$, $w_3=(w_{31}, w_{32}, w_{33})$. Therefore, the benefit function can be obtained, as shown in Equation (5).

$$B=w_{11}R_{11}+ w_{12}R_{12}+\ldots+ w_{17}R_{17}+w_{21}R_{21}+ w_{22}R_{22}+\ldots+ w_{24}R_{24}\ldots+w_{31}R_{31}+w_{32}R_{32}+w_{33}R_{33} \tag{5}$$

The risk decision table is shown in Table 4.

Table 4. Risk decision table

Market / Scheme	δ_1 $P(\delta_1)$	δ_2 $P(\delta_2)$	δ_3 $P(\delta_3)$
Scheme 1	α_{11}	α_{12}	α_{13}
Scheme 2	α_{21}	α_{22}	α_{23}
Scheme 3	α_{31}	α_{32}	α_{33}

Risk of every scheme is shown in Equation (6).

$$E(B_i)=P(\delta_1)\alpha_{i1}+P(\delta_2)\alpha_{i2}+P(\delta_3)\alpha_{i3} \tag{6}$$

The purpose of decision-making process is to calculate and get the operational risk of every scheme under every market situation. Therefore the basis of selection is demonstrated in Equation (7).

$$E(B_i)_{max}=max\ E(B_i) \quad (i \in [1,n]) \tag{7}$$

Besides, the risk factor of prediction will be introduced into the modeling, the table of relevance between the prediction of market situation and the prediction accuracy are shown in Table 5[14].

Table.5 Relevance between the prediction of market situation and the prediction accuracy

Prediction \ Accuracy	δ_1	δ_2	δ_3
μ_1	β_{11}	β_{12}	β_{13}
μ_2	β_{21}	β_{22}	β_{23}
μ_3	β_{31}	β_{32}	β_{33}

The Bayesian formula is shown in Equation (8). And use (8) we can calculate operational risk of each scheme.

$$P(\delta_i|\mu_j)=\frac{P(\delta_i\mu_j)}{P(\mu_j)}=\frac{P(\delta_i)P(\mu_j|\delta_i)}{\sum_{k=1}^{S}P(\delta_k)P(\mu_j|\delta_k)} \tag{8}$$

5. Case Study

Case study of this paper was performed by an IES project in North China. After the calculation process of AHP, the weight of indexes is shown in Table 6.

Table 6. Weight of indexes

Main index	First level indexes	Weight	Second level indexes	Weight	*CR*
Operational risk	Supply	0.468	Wind curtailment	0.220	0.032<0.1
			Solar energy curtailment	0.220	
			SOx pollution rate	0.106	
			NOx pollution rate	0.106	
			COx pollution rate	0.106	
			Dust pollution rate	0.106	
			Primary energy utilization	0.136	
	Transmission	0.329	Energy conversion efficiency	0.351	0.012<0.1
			Rate of energy loss during transmission	0.126	
			Payback period	0.224	
			Internal rate of return	0.299	
	Demand	0.203	Energy demand forecast accuracy	0.322	0.051<0.1
			Energy consumption per unit of GDP	0.426	
			Regional GDP growth rate	0.252	

Weight vector of indexes is able to be calculated through the AHP method. The result is W= (0.103, 0.103, 0.050, 0.050, 0.050, 0.050, 0.064, 0.115, 0.041, 0.074, 0.098, 0.065, 0.086, 0.051).

Based on the model established in Section 4, the benefit of all schemes is shown in Table 7.

Table 7. The results of benefit calculations

Market / Scheme	δ_1 $P(\delta_1)$=0.2	δ_2 $P(\delta_2)$ =0.4	δ_3 $P(\delta_3)$ =0.4	Expectations
Scheme A	0.313	0.327	0.313	0.319
Scheme B	0.451	0.429	0.438	0.437
Scheme C	0.412	0.427	0.420	0.421

Through the statistical analysis of the data and the collection of comprehensive information from all aspects, the relationship between prediction accuracy and market situation is shown in Table 8.

Table 8. Relationship between prediction accuracy and market situation

Market / Schemes	δ_1	δ_2	δ_3
μ_1	0.64	0.20	0.17
μ_2	0.21	0.68	0.13
μ_3	0.15	0.12	0.70

The prediction accuracy will be introduced into the Bayesian model, so the operational risk values can be calculated, the results are shown in Table 9.

Table 9. The results of operational risk calculations

Market / Scheme	δ_1 $P(\delta_1)$=0.2	δ_2 $P(\delta_2)$ =0.4	δ_3 $P(\delta_3)$ =0.4	Expectations
Scheme A	0.561	0.183	0.149	0.245
Scheme B	0.265	0.818	0.160	0.444
Scheme C	0.173	0.144	0.824	0.422

According to the results, the expectations of scheme B is 0.444, which means scheme B has the highest comprehensive benefit and lower risk, hence this project should be performed by scheme B. This result is in accordance with the practical implementation result of the projects.

6. Conclusion

IES, which is able to realize multiple energy complementary optimization and coordination between "source-grid-load-energy storage", will be a key method to carry out China's energy revolution. Nowadays many industrial parks are planning to provide energy supply through IES. However, related researches still focus on IES planning and dispatching technologies, there are few researches about operational risk evaluation of IES projects.

According to present research status, we have proposed the basic structure of IES including energy supply, transmission and demand. Evaluation index system of IES operational risk has been established based on IES characteristics. The improved Bayesian model with AHP as its weighting method has been designed and proved by the case study. As shown in the case study, the improved Bayesian model is able to evaluate the operational risk of IES projects accurately. In conclusion, the model and theory in this paper have great importance in realizing the optimization of IES system operation as well as resource allocation. The result of Bayesian analysis can provide valuable data and reference to the decision making process of IES operation scheme selection.

7. References

[1] Zeng M., Yang Y., Wang L. and Sun J., The power industry reform in China 2015: Policies, evaluations and solutions. *Renew. Sust. Energ. Rev.* 2016; 57: 94-110
[2] Yang Y., Zeng M., Xue S., Wang J. and Li Y., Coordination between clean energy generation and thermal power generation under the policy of "direct power-purchase for large users" in China. *Resour. Conserv. Recy.* 2018; 129: 402-419
[3] Xie W., Sheng P. and Guo X., Coal, oil, or clean energy: Which contributes most to the low energy efficiency in China? *Util. Policy.* 2015; 35: 67-71
[4] Gallagher K. P., Coal, oil, or clean energy: China's global energy finance: Poised to lead *Energy Research & Social Science.* 2018; 40: 89-90
[5] Jiang T., Deng H. and Bai L. Optimal Energy Flow and Nodal Energy Pricing in Carbon Emission-embedded Integrated Energy Systems. *CSEE Journal of Power and Energy Systems.* 2018; 4: 179-187
[6] Lihua Y., Zhao Y., Peng K., Xu B., Wang L. and Hu H., Bilevel optimal coordinated control strategy for park-level integrated energy system *Energy. Procedia.* 2018; 145: 44-53
[7] Xu J., Liu H., Xu K., Li J., Wang S., Yang Z. and Wang B., Synergistic Scheduling in Integrated Community Energy System Considering Wind Power Accommodation. *Energy Procedia.* 2018; 145: 211-216
[8] Huang Z., Yu H., Chu X., Peng X., Energetic and exergetic analysis of integrated energy system based on parametric method. *Energ. Convers. Manage.* 2017; 150: 588-598
[9] Wang L., Jing Z., Zheng J., Wu Q., Wei F., Decentralized optimization of coordinated electrical and thermal generations in hierarchical integrated energy systems considering competitive individuals. *Energy.* 2018; 158: 607-622
[10] Moslehi S., Reddy T., A. Sustainability of integrated energy systems: A performance-based resilience assessment methodology. *Appl Energ.* 2018; 228: 487-498
[11] Gnanavelbabu A., Arunagiri P., Ranking of MUDA using AHP and Fuzzy AHP algorithm *Mater Today. 2018*; 5(5): 13406-13412
[12] Sirisawat P., Kiatcharoenpol T., Fuzzy AHP-TOPSIS approaches to prioritizing solutions for reverse logistics barriers *Comput Ind Eng.* 2018; 117: 303-318
[13] Warne D. J., Baker R. E., Simpson M. J. Multilevel rejection sampling for approximate Bayesian computation. *Comput Stat Data An.* 2018; 124: 71-86
[14] Tang N., Wu Y., Chen D. Semiparametric Bayesian analysis of transformation linear mixed models. *J Multivariate Anal.* 2018; 166: 225-240.

IOP Conf. Series: Materials Science and Engineering **484** (2019) 012053 doi:10.1088/1757-899X/484/1/012053

Short-time Transport Stress Affected Biochemical Parameters in *Takifugu Rubripes*

Qingkui Wang[1,2], Lixiao Yang[1], Kezhi Xing[1], Xueliang Sun[1] and Jinsheng Sun[2]

[1]Tianjin Key Laboratory of Aqua-ecology and Aquaculture, College of Fisheries, Tianjin Agricultural University, 22 Jinjing Road, 300384 Tianjin, China.
[2]Tianjin Key Laboratory of Animal and Plant Resistance, Tianjin Normal University, 393 Binshuixi Road, 300387 Tianjin, China.
Email: kzxing6668@126.com; jinshsun@163.com

Abstract. Transport manipulation is a routine practice during the culture of fishes. Stress responses will occur during transport. In order to assess the impact of short-time transport on the health of juvenile *Takifugu rubripes*, a 30-minute simulated transportation was carried out. Result showed that Lactate and LDH in serum, muscle and liver of test fish respond rapidly after transport. Serum antioxidative and immune parameters, liver function parameters also changed significantly after transport. Most parameters recovered at 168 hour after transport. In conclusion, at least 168 hours recovery time is recommended for *Takifugu rubripes* suffered 30 minutes transport stress.

1. Introduction

Transport is an often occurred manipulation during fish culture. During transport, fish are exposed to many stressors including high density induced crowding, variation of dissolved oxygen, water temperature, and ammonia nitrogen. These stressors act on fish may evoke primary responses, such as increases in corticosteroid and catecholamine hormones, secondary responses such as metabolic changes, immune function changes, and tertiary responses such as modified behavioral patterns [1]. The changes of these physiological and biochemical parameters will damage the health status of fish. An extreme example is that transport stress will easily induce mass mortality of *Coilia nasus*, an economic migration fish between Yangtze River and estuary [2]. Furthermore, fish body surface, which act as physical barrier against infection, could easily be scratched during transport. The changes in body parameters and damage of body surface make fish more susceptible to adventitious infection. In addition, fish under transport may secrete more mucus as a stress reaction. This will worsen water quality and impair the overall health of fish.

Pufferfish *Takifugu rubripes*, which contains high content of essential amino acids, poly-unsaturated fatty acids and docosahexaenoic acid [3], is widely cultured and consumed in China, Korea and Japan. Before transport, pufferfish is usually packed into plastic bags with cold water and pure oxygen. This will relieve violent collision and bite elicited by transport, and will ensure high transport survival. However, people pay no attention to the health status of pufferfish subjected to transport. At present, evaluation of transport stress on health status of pufferfish is not found. In this paper, we simulated a short-time overland transport of pufferfish, assayed biochemical parameters during recovery to evaluate transport stress on its health status.

2. Materials and Methods

2.1. Fish and Acclimation

Juvenile pufferfish (100-150 g) were obtained in a cement tank (6.0 m × 6.0 m × 1.5 m) from Tianjin Haisheng Aquaculture Co. Ltd. (Binhai New Area, Tianjin, China). Pufferfish were acclimated in two cubic net cages (1 m^3) for seven days, with 84 individuals in each net cage. The two net cages were fixed in a cement tank, from which pufferfish were obtained. Pufferfish were fed twice a day. Seawater temperature was 21 °C, salinity was 20, dissolved oxygen was above 8 mg/L, ammonia nitrogen was below 0.01 mg/L, and nitrite was below 0.02 mg/L.

2.2. Simulated Transport

Two plastic tanks with 20 L seawater each were put on a cart with a portable blower and air diffusers. Seawater temperature in plastic tank was cooled to 17 °C with ice bag in advance. The plastic tanks were covered with sunshade net to avoid sunshine and disturbance. After fasted for 12 h, 84 pufferfish from one net cage were equally distributed into two plastic tanks (20 L). The cart was pulled to simulate overland transport for 30 min. Then, pufferfish were put back into the net cage immediately. During the simulated transport, air was pumped into the seawater continually and water temperature was monitored every 10 min. The 84 pufferfish in another net cage, fasted for 12 h likewise but not subjected to transport, served as control.

2.3. Sampling and Biochemical Parameters Assay

Pufferfish were sampled at 0 h, 8 h, 12 h, 24 h, 48 h, 96 h, and 168 h after 30 min simulated overland transport. Twelve transport fish and 12 control fish were promptly caught by a hand-held net from the two net cages at each sampling. Fish were euthanized by overdose of tricaine methanesulfonate (MS-222; Hangzhou Animal Medicine, Inc., China), then weighed. Blood was took from caudal vein. A 100 uL of blood was mixed with 20 uL 8% heparin to assay oxidative radical production. The rest blood was distributed in centrifuge tubes and stored at 4 °C overnight for serum separation. Dorsal muscle and liver were excised after blood sampling. All samples were stored at -20 °C for further use. Serum of four fish were pooled and make three iterations, then stored at -20 °C until further use. Dorsal muscle and liver of three fish were excised after blood sampling, and stored at -20 °C for further use. The remaining eight fish were returned to the company.

Blood oxidative radical production was assayed spectrophotometrically by means of nitroblue tetrazolium (NBT) reduction, according to Siwicki et al. [4] with some modifications. A stock solution of NBT was freshly prepared by dissolving 0.2 g NBT into 100 mL sterile saline water. The NBT working solution was prepared by equal volume mixing NBT stock solution with 1 mg/mL glucose (1 mg glucose dissolved into sterile 0.15 M phosphate buffer solution, pH 7.2). Fifty microliters of blood was added to 50 μL of NBT working solution in a centrifuge tube and incubated at 28 °C for 30 min. The centrifuge tube was shake every ten minute. Then, a 75 μL sample was taken from this mixture, added to 1.5 mL of N, N-dimethyl formamide (DMF;Amresco,America) to dissolve formed formazan crystals, and centrifuged at 1000 rpm for 15 min. The supernatant was determined in a spectrophotometer (UVmini-1240, Shimadzu, Japan) at 620 nm, with DMF as blank.

Catalase (CAT), Total superoxide dismutase (T-SOD), total antioxidant capacity (T-AOC), malondialdehyde (MDA), lysozyme (LZM), glucose (GLU), total glutathione (T-GSH), reduced glutathione (GSH), aspartate amino transferase (AST), and alanine aminotransferase (ALT) in serum were assayed. Lactate, lactate dehydrogenase (LDH) in serum, muscle and liver were determined. These parameters were assayed with commercial kits manufactured by Nanjing Jiancheng Bioengineering Institute, Nanjing, China.

2.4. Water Quality Monitoring

Water temperature, dissolved oxygen (DO), pH, salinity were monitored every day using YSI 556 m (Yellow Spring Instrument Co., Yellow Spring, Ohio). Ammonia nitrogen and nitrite were measured using commercial kits (Sunpu Biochemistry Co., Beijing, China).

IOP Conf. Series: Materials Science and Engineering **484** (2019) 012053 doi:10.1088/1757-899X/484/1/012053

2.5. *Statistical Analysis*
Data were subjected to independent-sample *T* test with SPSS 16.0 software (SPSS, Chicago, IL, USA).

3. Results

3.1. *Water Quality*
Before and after the simulated transport, water quality in the plastic tanks was monitored. Water temperature was 17-16 °C, DO was 8.5-8.2 mg/mL, pH was 8.5-8.4, salinity was 20, ammonia nitrogen was below 0.01 mg/mL, nitrite was below 0.01 mg/mL. After the simulated transport, pufferfish were transferred into net cage. Water quality in the net cages were as following: temperature 21 °C, DO 8.5 mg/mL, pH 8.5, salinity 20, ammonia nitrogen < 0.01 mg/mL, nitrite < 0.01 mg/mL.

3.2. *Lactate and LDH in Serum, Muscle and Liver after Transport*
Serum lactate level increased immediately after transport, maintained high level in 96 hours, and declined to control level at 168 h (Figure 1a). Serum LDH activity was significantly lower than control at 0 h, 8 h after transport, then increase at 12 h and 48 h, and get back to control level at 168 h (Figure 1b). Muscle lactate maintained significant higher level than control in 168 hours (Figure 1c), while muscle LDH fluctuated, increased at 0 h, 8 h, decreased at 12 h, 48 h, and then increased at 168 h (Figure 1d). Liver lactate content in transport group was near control group in 24 hours, then increased significantly at 48 h, and decreased to control level at 168 h (Figure 1e). Liver LDH activity in transport group elevated significantly after transport, reaching peak value at 12 h, then declined to control level after 96 h (Figure 1f).

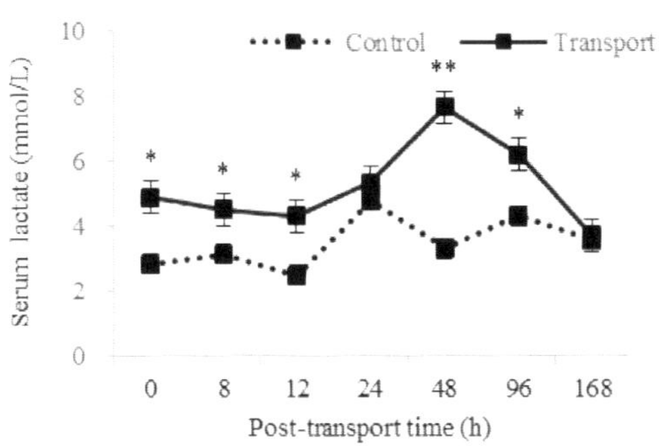

a

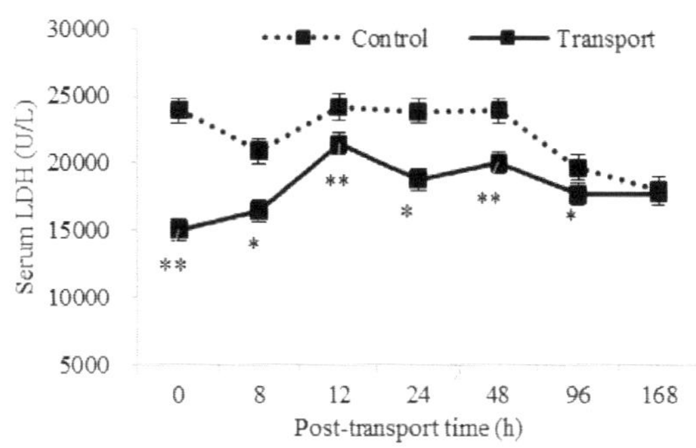

b

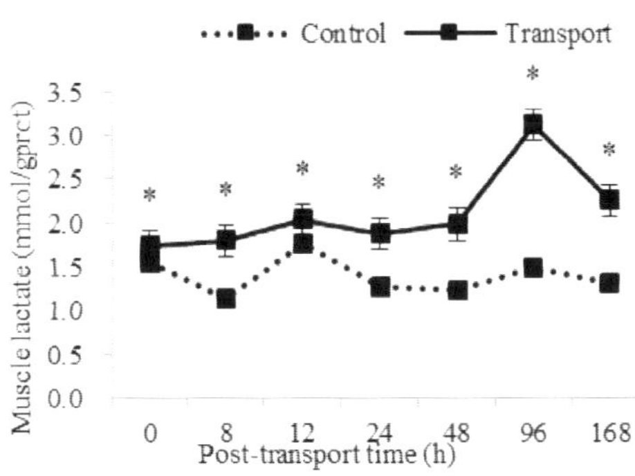

c

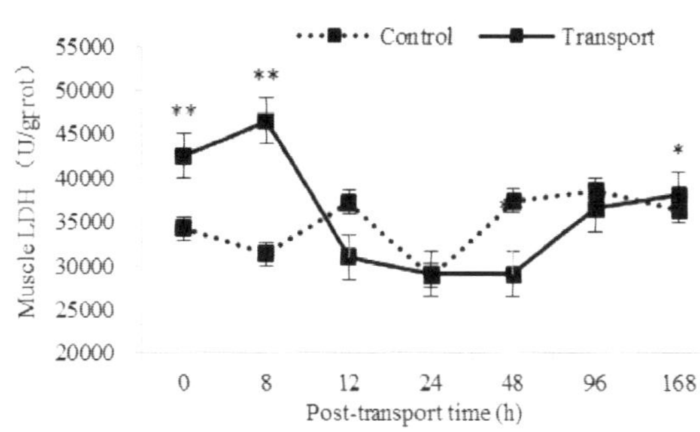

d

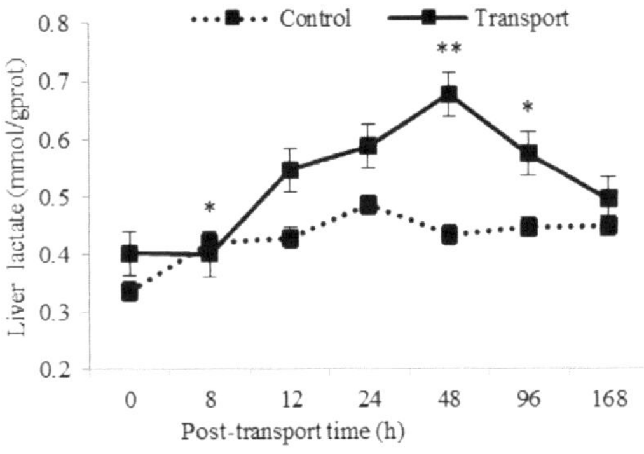

e

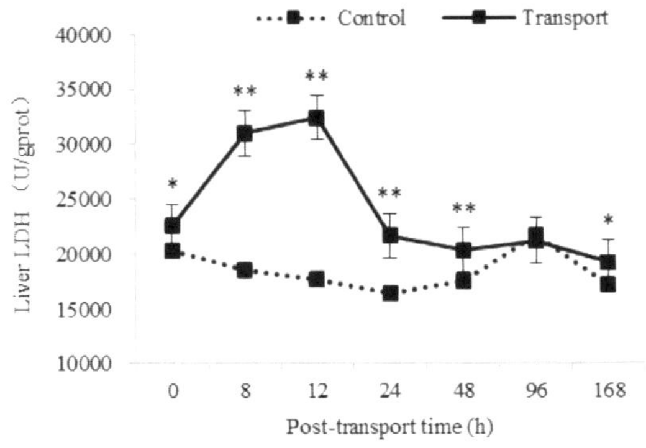

f

Figure 1. Variation of serum lactate (a), serum LDH (b), muscle lactate (c), muscle LDH (d), liver lactate (e), and liver LDH (f) in *Takifugu rubripes* after 30 minutes transport. Data between transport and control group were subjected to independent-sample *T* test. * indicates significant difference ($P<0.05$), ** indicates extremely significant difference ($P<0.05$). Data are means ± SD, n=3.

3.3. Antioxidative and Immune Parameters after Transport

Serum T-AOC in transport group was statistically higher after transport, reaching peak value, 4-fold of control, at 12 h (Figure 2a). Afterward, serum T-AOC decreased to control level.

Serum T-SOD in transport group was significantly lower than control at 8 h and 168 h, while there was no statistically difference at other sampling time (Figure 2b).

Serum CAT in transport group was statistically higher than control group at 8 h, 12 h, and 96 h after transport, while no statistically difference was observed at other sampling time (Figure 2c).

Figure 2d showed that serum MDA in transport group was significantly higher, 2.3-fold of control, at 0 h post-transport, then descended hastily at 8 h. Although serum MDA in transport group fluctuated and significant differences were observed at 24 h, 48 h, 96 h, it increased to control level at 168 h.

Figure 2e and Figure 2f showed than serum T-GSH and GSH varied greatly after transport. Serum T-GSH significantly increased at 0 h, decreased at 8 h, then increase to peak value at 24 h, and decreased rapidly to control level at 48 h. Serum GSH in transport group rose to top value at 8 h, then decreased gradually, returned to control level at 96 h.

Figure 3a showed that serum LZM in transport group varied statistically significant at 0 h and 8 h, notably higher at 12 h and 24 h, then decreased to the lowest value at 96 h, and returned to control level at 168 h.

Blood oxidative radical production dropped significantly after transport, reached the lowest value at 12 h, and rose to control level at 168 h (Figure 3b).

Serum glucose elevated significantly from 8 h post-transport, reaching peak value at 48 h, and declined to control level at 168 h (Figure 3c).

3.4. Liver Function Parameters after Transport

Serum aspartate aminotransferase (AST) was notably elevated at 0 h, 12 h, and decreased to control level at 24 h (Figure 4a) . However, serum alanine aminotransferase (ALT) was notably higher than control group at 12 h, 48 h, and 168 h (Figure 4b) .

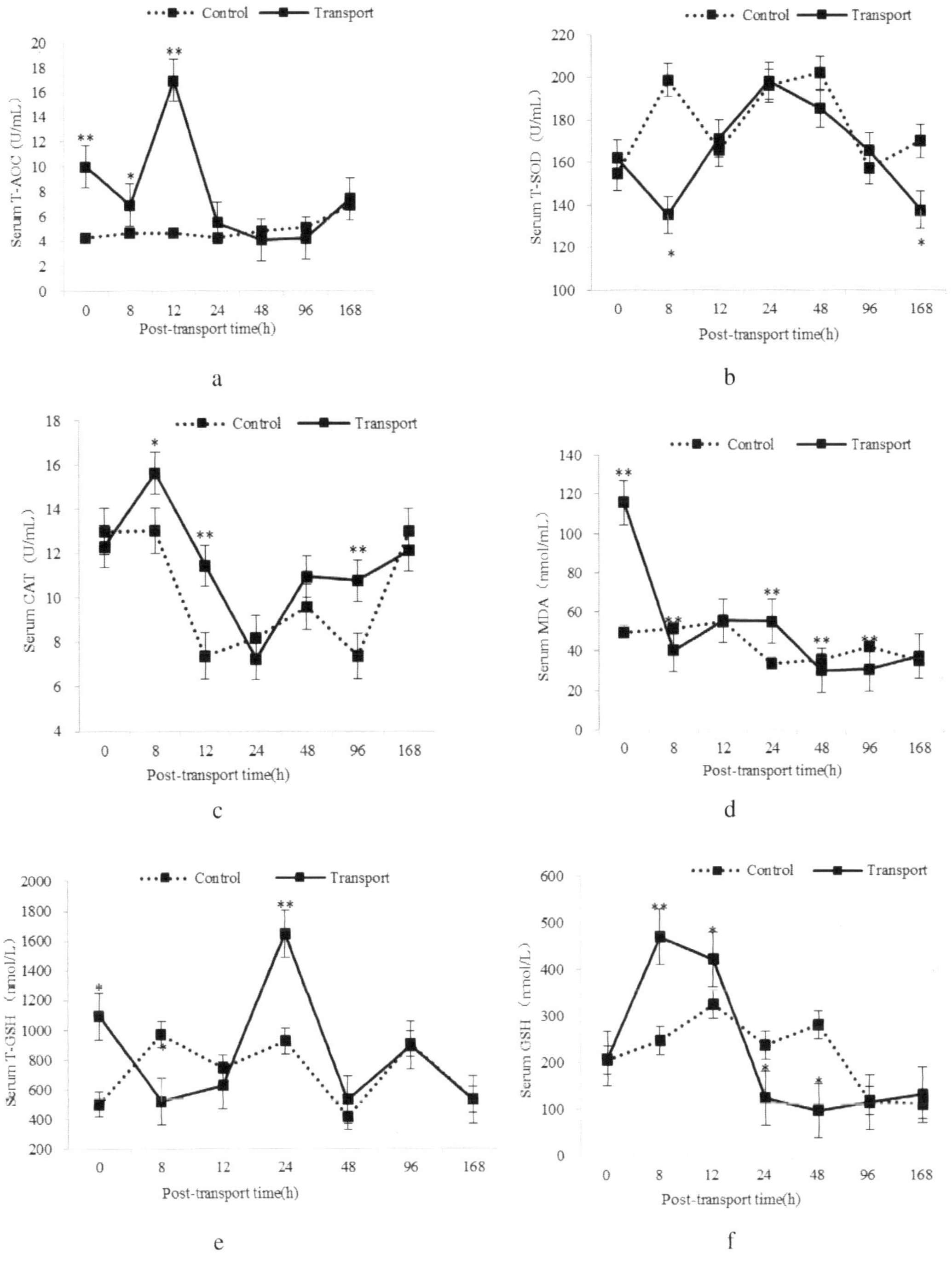

Figure 2. Variation of serum T-AOC (a), T-SOD (b), CAT (c), MDA (d), T-GSH (e), and GSH (f) in *Takifugu rubripes* after 30 minutes transport. Data between transport and control group were subjected to independent-sample *T* test. * indicates significant difference ($P<0.05$), ** indicates extremely significant difference ($P<0.05$). Data are means ± SD, n=3.

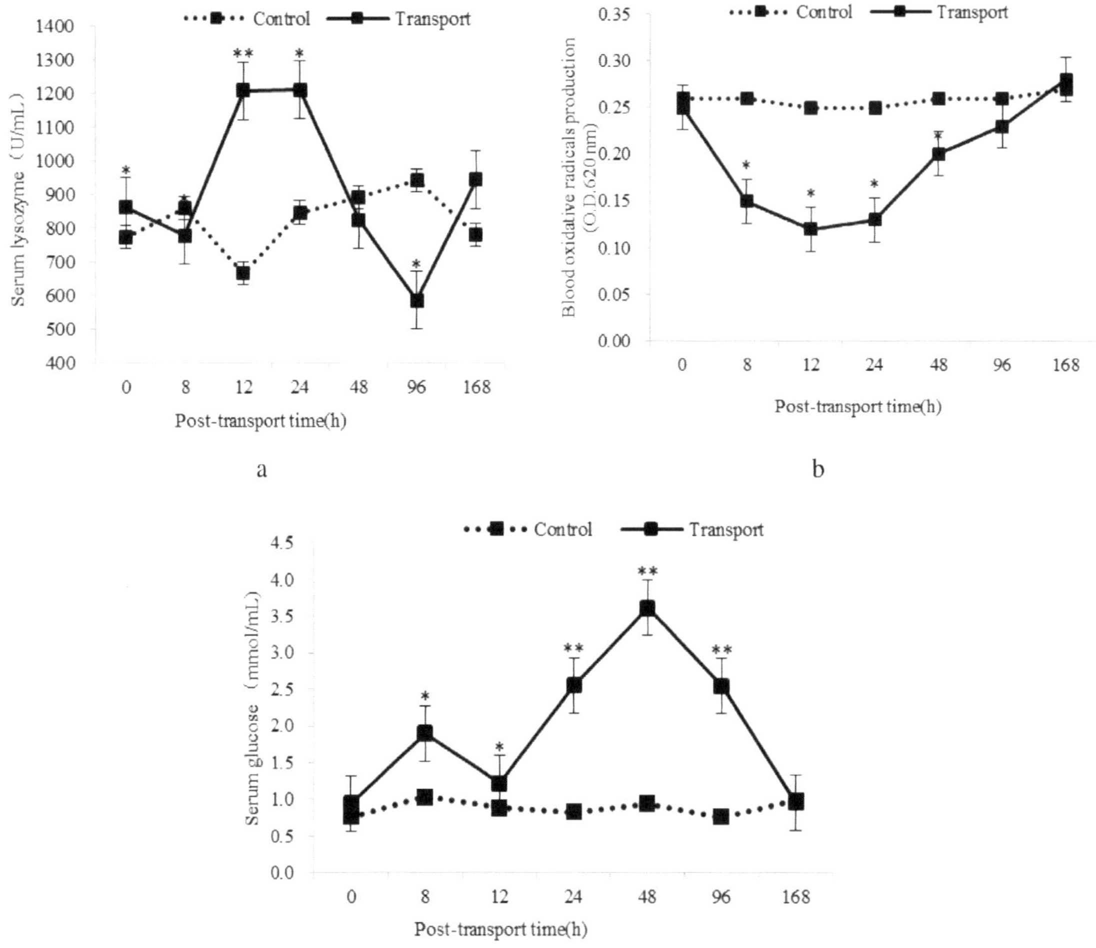

c

Figure 3. Variation of serum lysozyme (a), blood oxidative radicals production (b), serum glucose (c) in *Takifugu rubripes* after 30 minutes transport. Data between transport and control group were subjected to independent-sample *T* test. * indicates significant difference ($P<0.05$), ** indicates extremely significant difference ($P<0.05$). Data are means ± SD, n=3.

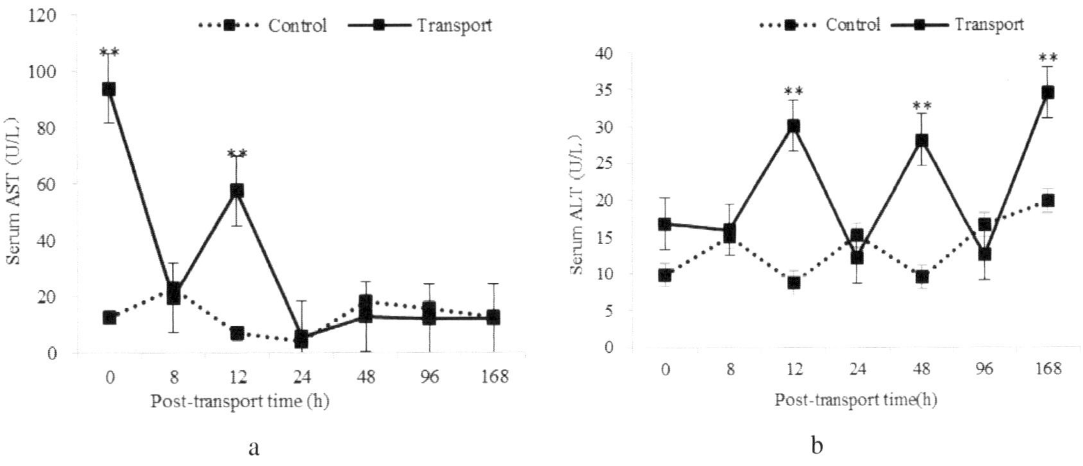

Figure 4. Variation of serum AST (a) and ALT (b) in *Takifugu rubripes* after 30 minutes transport. Data between transport and control group were subjected to independent-sample *T* test. * indicates significant difference (P<0.05), ** indicates extremely significant difference (P<0.05). Data are means ± SD, n=3.

4. Discussion

Prior works have documented that stressors will evoke physiological responses and impair the health of fish [1, 5]. These findings are very important reference for fish farmers to ensure the health of fishes during their culture process. In order to reduce transport induced mortality and ensure fish health, it is very important to reduce stress and its detrimental effects, provide favorable environment and enough recovery time for transported fishes. However, no related report about transport stress on pufferfish was found to date. The main purpose of this paper was to test the survival and find out suitable recovery time for juvenile pufferfish suffered 30 min transport stress. One hundred percent of survival rate was achieved during and after transport. As most parameters were recovered in 168 hours, it is the recommended recovery time for juvenile pufferfish suffered 30 min transport.

However, transport stress induced agonistic behaviors, physiological responses, survival, and recovery time varied greatly among different fish species. *Coilia nasus* is very sensitive to collision and transport. Almost a hundred mortality if it suffered these stress [2]. *Eleutheronema tetradactylum* suffered from 6 hours transport stress result in 27.5% cumulative mortality [6]. By testing the antioxidative and immune parameters, *Cynoglossus semilaevis* fry suffered from 11 hours practical transport, need at least 96 hours for recovery without death [7]. Hybrid grouper (*Epinephelus fuscoguttatus* ♀ × *E. lanceolatus* ♂) subadult suffered from 30 min transport stress, need 168 hours for recovery without death [8]. The survival and recovery time of this paper are in agreement with hybrid grouper mentioned above.

In this paper, serum lactate increased rapidly while serum LDH first decreased then increased significantly after transport. Lactate and LDH in muscle and liver increased simultaneously after transport. These results indicate that lactate and LDH are sensitive transport stress indicators for T. rubripes. With contrary to pufferfish, serum LDH of hybrid grouper elevated after 30-minute transport, and decreased to control level at 96 h [8]. In view of high level lactate and its long duration in serum, muscle and liver of transport pufferfish in this paper, it's preferable to take some measures to relieve the detrimental effect of lactate. Augmenting dissolve oxygen of aquaculture water may contribute to fish blood oxygen concentration, consequently, contribute to the oxidation of lactate into pyruvic acid by LDH. Pyruvic acid is catalyzed irreversibly into acetyl-ConA by pyruvate dehydrogenase complex. As pantothenic acid is component of coenzyme A (CoA), thiamine pyrophosphate (TPP), lipoic acid, CoA, flavin adenine dinucleotide (FAD), and nicotinamide adenine dinucleotide (NAD^+) are five cofactors of pyruvate dehydrogenase complex. So oral administration or bathing with pantothenic acid, thiamine, riboflavin and nicotinamide could contribute to the catalyzing of pyruvic acid. Further experiment should be carried out to test this hypothesis.

As a product of lipid peroxide, serum MDA level increased significantly at 0 h after transport. This indicated that 30 minutes transport induced oxidative stress in pufferfish. GSH can reduce lipid peroxide products to unsaturated fatty acid. High level of GSH and low level of MDA in transport group at 8h, 12h indicated that GSH exerted its reduction property. As ascorbic acid can reduce oxidized glutathione (GSSG) into GSH, hence contribute body to cope with oxidative stress.

We used NBT reduction to detect oxidative radical production of blood cells. Result in this paper showed that fish subjected to transport produced less oxidative radicals. This indicated that transport could induce high level of antioxidant in blood cells. Further experiment need to be carried out to assay antioxidant level in blood cells to verify this hypothesis.

The high level of serum AST and ALT in transport group indicated that transport damaged liver function of pufferfish. High level of serum glucose in transport group implied that more energy was aroused to cope with transport stress.

In conclusion, 30-minute transport stress induces significant changes of lactate, LDH, antioxidative and immune parameters, and liver function of pufferfish. At least 168-hour recovery time was recommended.

5. Acknowledgement

This paper was financially supported by open fund from Tianjin Key Laboratory of Animal and Plant Resistance, from Tianjin Key Laboratory of Aqua-ecology and Aquaculture, and research fund from Tianjin municipal government (fund number: 15JCZDJC33600, 15JCZDJC34000, ITTFRS2017004, TD13-5089). We would like to thank Jinliang Wang to help maintain the daily management.

6. References

[1] Adams S M. Biological indicators of aquatic ecosystems stress. Bethesda, Maryland: American Fisheries Society; 2002. 644 p.

[2] Gao J, Du F, Gu R, Xu G. Effect of transport stress on physiological-biochemical indices and expression of HPI genes and roles of glycrrhizin on transport response in *Coilia nasus*. Journal of Shanghai Ocean University. 2015;24(6):817-25.

[3] Gao L, Huang Y, Xia L, Lu J, Liu S. Comparison of flesh quality of farmed fugu, *Takifugu rubripes* from different culture models. J Fish China/Shuichan Xuebao. 2011;35(11):1668-76.

[4] Siwicki A K, Anderson D P, Rumsey G L. Dietary intake of immunostimulants by rainbow trout affects non-specific immunity and protection against furunculosis. Vet Immunol Immunopathol. 1994;41(1–2):125-39.

[5] Portz D E, Woodley C M, Cech Jr J J. Stress-associated impacts of short-term holding on fishes. Rev Fish Biol Fish. 2006;16(2):125-70.

[6] Liu Q, Wen J, Qu Y, Li J, Zhou H. Effects of transport stress on liver, gill and spleen tissue

structure of juvenile *Eleutheronema tetradactylum*. Journal of Southern Agriculture. 2017;48(9):1708-14.

[7] Wang Q, Chen C, Sun X, Xing K. The effects of transport stress on some antioxidative and immune parameters in *Cynoglossus semilaevis* Günther fry. Zhu K, Zhang H, editors. Sydney, Australia: Aussino Academic Publishing House; 2010.

[8] You H, Shi H, Jia L, Sun X, Li Y. Effect of short-distance transportation stress on serum enzyme activity and glucose content of pearl gentian grouper (*Epinephelus fuscoguttatus* ♀ × *E. lanceolatus* ♂). Journal of Economic Animal. 2017(10):1-9.

IOP Conf. Series: Materials Science and Engineering **484** (2019) 012054 doi:10.1088/1757-899X/484/1/012054

Microbiostatic and Synergistic Actions of Extract of Green Walnut Peel

Li Yang, Yao Gu, Rusi Wen and Lizhu Zhou*

Guangxi Key Laboratory of Special Non-wood Forest Cultivation and Utilization, Guangxi Zhuang Autonomous Region Forestry Research Institute, Nanning 530002, China

Email: 125103877@qq.com

Abstract. The content of juglone in green walnut peel reach the highest in the period of July to August Juglone, extracts of green walnut peel and tee tree oil were tested in antimicrobial experiments against *Escherichia coli*, *Pseudomonas aeruginosa*, *Staphylococcus aureus* and *Candida albicans*. The results demonstrated that extract from green walnut peel, and juglone were microbiostatic for all the tested strains. Distinct synergic effects were observed with mixtures of tee tree oil and extract of green walnut peel. Microbiostatic effects on the four microorganisms was highest with a 1 to 10 ratio of tee tree oil and green walnut peel extracts. This mixture was essentially nontoxic and showed good biocompatibility. There were no irritating reactions observed in the repeated dermal irritation test in rabbits, no allergic reactions in guinea pigs and a very low irritating effect in the vaginal mucosa of rabbits. The mixture is promising for use in medical antibacterial products.

Highlights

- Content of juglone in green walnut peel reach the highest in the period of July to August
- Green walnut peel extract was more microbiostatic than juglone.
- Tee tree oil and green walnut peel extract had synergic microbiostatic effects.
- The mixture of tee tree oil and green walnut peel extract was not toxic to animals.
- This mixture is promising for antibacterial products.

1. Introduction

Green walnut peel is the external layer, a thick green peel, of the immature walnut (*Juglans sigillata* Dode). As the walnut gradually matures, the green peel becomes black and, finally, sheds naturally [1][2] .The green peel contains many types of secondary metabolites with antibacterial activities[3]. Juglone is the primary toxic substance in green walnut peel [4] and walnut peel pigment is a stable dyeing agent [5]. Produce walnut 1 kg will remain walnut peel residue 1 kg, Most walnut peel residue is discarded as waste, walnut peel pigment can be dissolved in water and take a long time to be decomposed naturally. Juglone and walnut peel pigment can seriously pollute the environment by penetrating into ground water system [6]. Studies showed that juglone had significant antimicrobial [7][8], antitumor, antihypertensive[9] [10] and enzyme inhibitory activities [11][12]. Tee tree oil were extracted from *Melaleuca ahemifolia*, has showed significant antimicrobial ability. The variation of the content of juglone in green walnut peel was studied. It was showed that the content of juglone reach the highest in the period of July to August. Microbiostatic and synergistic actions of extract of green walnut peel, juglone and mixture of tee tree oil and green walnut peel extract were detected. Distinct synergic effects were observed with mixtures of tee tree oil and extract of green walnut peel. The study

IOP Conf. Series: Materials Science and Engineering **484** (2019) 012054 doi:10.1088/1757-899X/484/1/012054

provided a theoretical basis for applications of green walnut peel extracts in antibacterial products.

2. Materials and Methods

2.1. Materials and Reagents

Green walnut peel of *Juglans sigillata* Dode was provided by Purapharm Co., Ltd. Tee tree oil(≥98%) was provided by Nanning Wan Yao Tang Co., Ltd. PBS containing 0.5% lecithin, 0.5% histidine and 1.0% Tween-80 was used as neutralizer and obtained from Shanghai ML Biotechnology Co., Ltd. Seventh to ninth generation *Escherichia coli*, *Staphylococcus aureus*, *Candida albicans* and *Pseudomonas aeruginosa* were from CGMCC Mice and white rabbits were obtained from Guangxi Medical University Laboratory Animal Center and white guinea pigs from Kunming Medical University Laboratory Animal Center. Juglone standards and other materials and reagents were purchased from Nanning Lantian Experiment Equipment Co., Ltd.

2.2. Equipment

HH-B11-600-S-II Electrothermal thermostatic incubator was from Shanghai Longyue Equipment Co., Ltd., LRH-250A biochemical incubator was from Shaoguan Taihong Medical Equipment Co., Ltd, JY10001 electronic scales were from Shanghai Shuangxu Electronics Co., Ltd. and HNW-S-6 thermostatic water bath from Shanghai Hannuo Instrument Co., Ltd.

2.3. Experimental Methods

2.3.1. Green Walnut Peel Extract Preparation. Walnuts of *Juglans sigillata* Dode were harvested every 15 days and the peel were separated by manual stripping in the period of July 1st to September 1st. Green walnut peel powder (100 g) was macerated with 250 mL 95% ethanol and stirred at 200 rpm at room temperature for 2 h. The extract was filtered and the residue re-extracted twice. The pooled extracts were stored in 4°Cfor content determination, and then concentrated under vacuum to 40 mL at 45 °C, then fully dried under vacuum to obtain 18 mg of an orange acicular crystal. The primary components of the crystals were juglone (5-hydroxy-1, 4-naphthalenedione), 4-naphthoquinone and 7-methoxy-3, 4-dihydro-l (2-H)–naphthalenone at relative contents of 96.420, 0.704 and 2.875 %, respectively.

2.3.2. Determination of Juglone Content by HPLC. Chromatography was performed on a C18 column (4.6 mm × 150 mm, 5 μm particle sizes) eluted with methanol-water (1:1) mobile phase, with the water phase adjusted to pH 4 with phosphoric acid. The flow rate was 0.8 mL/min, with UV detection at 250 nm and the column temperature was 30 °C [13] [14].

2.3.3. Preparation of Test Solutions. The juglone and green walnut peel extract were each diluted separately to 300 mg/L in 95% ethanol.

2.3.4. Preparation of Bacterial or Fungal Suspensions. The *E. coli*, *S. aureus*, and *P. aeruginosa* were activated on the slant side of bacterial medium at 30–35 °C for 24 h and *C. albicans* was activated on the slant side of Sabouraud's agar medium at 20–25 °C for 48 h. Organisms were then diluted in normal saline to 105–110 cfu/mL (determined with a hemocytometer).

2.3.5. Determination of Microbial Growth Inhibition. Samples for testing were diluted with sterile water. Micro-organism suspensions (50 μL) (bacterial concentration at 1×10^6 cfu/mL and fungal concentrations at 1×10^5 cfu/mL) were spread onto each plate. A piece of filter paper was placed in the center of each plate and 30 μL sample solution was added. These plates were then cultured under the appropriate conditions (for bacteria, at 30–35 °C and for *C. albicans*, at 20–25 °C) and the diameters of the inhibition zones then measured. [15][16]

2.3.6. Determination of Microbicidal Actions. To each tube containing prepared sample, 2.5 mL in medium, was added 0.5 mL microbial suspension (1×10^7 cfu/mL) and tubes were incubated for 24 h with shaking at 150 rpm at the appropriate culture temperatures. [17][18]The three bacteria were cultured at 30–35 °C and *C. albicans* at 20–25 °C. Next, 50-μL aliquots of incubation mixtures were spread onto individual plates. Plates were observed when microorganisms had fully grown in the control plates [19].

2.3.7. Determination of Biocompatibility. Acute oral toxicity, skin irritation, vaginal mucosa irritation and dermal allergic reaction tests were performed as described in ISO 10993-10 (Biological evaluation of medical devices. Part10: Tests for irritation and skin sensitization) and Technical Standard for Disinfection (2002) [20].

2.3.8. Effects of *Mixtures of Green Walnut Peel with Tee Tree Oil.* Tee tree oil was combined with extract from green walnut peel at ratios of 1:10, 1:20, 1:30, 1:40 and 1:50. Microbicidal effects were determined as described above for individual extracts.

2.3.9. Evaluation Criteria for Synergic Actions. CTC was calculated as described by Sun Yun-pei. The equation used was CTC = (ATI /TTI) *100. Synergic action is defined as when CTC>120, an additive effect when CTC is within the range of 80–120 and an antagonistic effect when CTC<80. [21]

3. Results and Discussion

3.1. Variation of Juglone from July 1st to September 1st

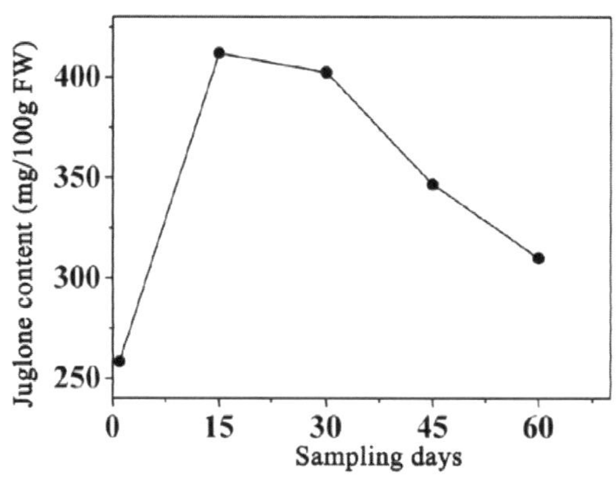

Figure 1. Juglone content of Walnut varieties in sampling date

3.2. Microbiostatic and Microbicidal Effects of Green Walnut Peel Extract and Juglone

Table 1. Comparison of antimicrobial effects

Bacteria to be examined	Diameter of inhibition zone /mm		
	Control	extract of green walnut peel	juglone
E.coli	0.0	25.3±0.3	20.5±0.2
P. aeruginosa	0.0	20.3±0.2	18.4±0.4
S. aureus	0.0	25.1±0.4	20.7±0.7
C. albicans	0.0	19.3±0.5	16.4±0.3

Microbiostatic and microbicidal effects of green walnut peel extract and juglone are shown in Table 1. The microbiostatic effects of the extract on the four microorganisms was superior to those of juglone. The effects of the two samples on *E. coli* and *S. aureus* were quite comparable.

Table 2. Lethal concentrations of two samples against various microorganisms

Volume of sample /%	extract of green walnut peel				juglone			
	E.coli	*P.aeruginosa*	*S.aureus*	*C.albicans*	*E.coli*	*P.aeruginosa*	*S.aureus*	*C.albicans*
50	-	-	-	-	-	-	-	-
25	-	-	-	-	-	-	-	-
10	-	-	-	-	-	-	-	+
5	-	-	-	-	-	+	-	+
1	+	+	+	+	+	+	+	+

Note: In plating results, + means growing, - means non-growing.

Both the extract and juglone completely inhibited the growth of the four microorganisms at concentrations at or above 25% (Table 2). Moreover, the minimal bactericidal concentrations of the two samples against *E. coli* and *S. aureus* were 5%, indicating comparable bactericidal effects. The microbicidal effect of juglone against *P. aeruginosa* was greater than that against *C. albicans*. However, the minimal toxic concentration of juglone against these two organisms was higher than that of the extract.

3.3. Microbiostatic and Synergic Effects of Extract Mixtures

The microbiostatic effects produced by mixing extract of green walnut peel (A), tee tree oil (B) at various ratios are shown in Table 3. These data showed that synergic effects of the mixture of tee tree oil and green walnut peel extract were superior to those of tee tree oil and green walnut peel extract. Synergic effects were greatest for Coptidis rhizoma and green walnut peel extracts mixed at a ratio of 1 to 10. This mixture was used for subsequent biocompatibility testing.

Table 3. Inhibitory effects of various extract mixtures on four microorganisms

Ratio of drugs	CTC (*E. coli*)	CTC (*S. aureus*)	CTC (P. aeruginosa)	CTC (*C. albicans*)
B:A=1: 10	160.2546	181.4568	142.5423	175.4586
B:A=1: 20	156.2456	177.1928	134.2165	173.5422
B:A=1: 30	154.3784	172.4326	125.4982	168.5362
B:A=1: 40	149.8546	168.843	122.2143	165.688
B:A=1: 50	145.5684	162.4152	116.8406	160.268

3.4. Biocompatibility of the Mixture

3.4.1. Acute Oral Toxicity Test .There were no obvious toxic symptoms in test animals and no animals died after the mixture was administered. The acute oral ID_{50} of the mixture in mice was over 5000 mg/kg body weight. According to specifications for the acute evaluation of disinfectants, this essentially belongs to the "nontoxic" class. [20]

3.4.2. Skin *Irritation Test* .No irritating effects were observed with repeated skin irritation testing of the mixture in rabbits.

3.4.3. Vaginal Mucosa Irritation test. In both treated and control groups, there was slight hyperemia observed in the vaginal mucosa of two rabbits. Mild edema was observed in the vaginal mucosa of two rabbits in the treated group. No edema was observed in the control group. The stimulation index of the vaginal mucosa was 0.33.

3.4.4. Dermal Allergic Reaction Test. As summarized in results of guinea pig skin allergic reaction test, no irritating reactions such as erythema or edema were observed in guinea pigs in the treated group. These were sensitized, then challenged by contact with the mixture for 24 or 48 h. Dermal sensitization in these animals was 0 %. No abnormalities were observed on the skin of the guinea pigs in the negative control.

4. Conclusions

Microbiostatic effects on the four microorganisms reach the highest with a 1 to 10 ratio of tee tree oil and green walnut peel extracts. This mixture was essentially nontoxic and showed good biocompatibility. There were no irritating reactions observed in the repeated dermal irritation test in rabbits, no allergic reactions in guinea pigs and a very low irritating effect in the vaginal mucosa of rabbits. The mixture is promising for use in food preservation and medical antibacterial products.

5. Acknowledgment

This work was financially supported by Science and Technology Major Project of Guangxi(Nos. AA17204058-21).

6. References

[1] Cosmulescu S, Trandafir I. (2011b). Variation of phenols content in walnut (Juglans regia L.). *South West J Hortic Biol Environ,* **2**: 25–33;

[2] Coder KD. (1983). Seasonal changes of juglone potential in leaves of black walnut (Juglans nigra L.). *J Chem Ecol,* **9:1203**–12;

[3] Colaric, M., Veberic, R., Solar, A., Hudina, M., Stampar, F.; Phenolic acids, syringaldehyde, and juglone in fruits of different cultivars of Juglans regia L.; *Journal of Agricultural and Food Chemistry*, (2005); **53**: 6390–6396;

[4] Ajay Thakur (2011).Juglone: A therapeutic phytochemical from Juglans regia L[J]. *Journal of Medicinal Plants Research Vol.* **5**(22), 5324-5330;

[5] Maryon P. Strugstad, 2012. A Summary of Extraction, Synthesis, Properties, and Potential Uses of Juglone: A Literature Review [J].*Journal of Ecosystems and Management*, 13(3): 1–16;

[6] LI Xiu- feng,2007.Progress of research on the chemical components and pharmaceutical action of walnut green husk [J],*Food Science and Technology*,(4)241~242;

[7] Kiran B.Aithal, Sunil Kumar M.R, B.Nageshwar Rao, Nayanabhirama Udupa, and Sadashiva Rao,2012.Tumor Growth Inhibitory Effect of Juglone and Its Radiation Sensitizing Potential:InVivo and In Vitro Studies[J]. *Integrative Cancer Therapies*,**11**(1):68-80;

[8] Yun-hua KONG, Liang ZHANG, Zheng-yi YANG, Cong HAN, Li-hong HU, Hua-liang JIANG, Xu SHEN,2008. Natural product juglone targets three key enzymes from Helicobacter pyloriinhibition assay with crystal structure characterization [J]. *Acta Pharmacol Sin,* **29** (7): 870–876;

[9] Carvalho, M., Ferreira, P.J., Mendes, V.S., Silva, R., Pereira, J.A., Jero´ nimo, C., et al.; Human cancer cell antiproliferative and antioxi- dant activities of Juglans regia L.; *Food and Chemical Toxicology*, (2010); **48**: 441–447;

[10] Saling SC, Comar JF, Mito MS, et al. (2011). Actions of juglone on energy metabolism in the rat liver. *Toxicol Appl Pharm*, **257**:319–27;

[11] Thilo Christopher Fischer, Christian Gosch, Beate Mirbeth, Markus Gselmann, Veronika Thallmair, and Karl Stich,2012.Potent and Specific Bactericidal Effect of Juglone (5-Hydroxy-1,4-naphthoquinone) on the Fire Blight Pathogen Erwinia amylovora [J]. J. Agric. *Food Chem.* **60**, 12074−12081;

[12] Michelle T. Paulsena, Mats Ljungman,2005. The natural toxin juglone causes degradation of p53 and induces rapid H2AX phosphorylation and cell death in human fibroblasts [J]. *Toxicology and Applied Pharmacology*, **209** , 1 – 9;

[13] LI Qing, Tan Hong, Yuan Xin, 2014. Determination of juglone content in green walnut peel by high performance liquid chromatography (HPLC) [J]. *Jiangsu Agricultural Science*,**42** (1): 259 – 261;

[14] Violeta Nour, Ion Trandafir and Sina Cosmulescu ,2012.HPLC Determination of Phenolic Acids,Flavonoids and Juglone in Walnut Leaves[J]. *Journal of Chromatographic Science*.November **6**:1–8;
[15] Alice Ferreira da Silva Hughes, Fabíola Guedes de Lima, Angélica Maria Lucchese, Aristóteles Góes Neto and Ana Paula Trovatti Uetanabaro(2013), Antimicrobial Activity of Syagrus coronata (Martius) Beccari[J]. *Brazilian Archives of Biology and Technology*, Vol.**56**, (2): 269-274;
[16] Thilo Christopher Fischer, Christian Gosch, Beate Mirbeth, Markus Gselmann, Veronika Thallmair and Karl Stich(2012). Potent and Specific Bactericidal Effect of Juglone (5-Hydroxy-1,4- naphthoquinone) on the Fire Blight Pathogen Erwinia amylovora[J].*Agric. Food Chem*, **60**, 12074−12081;
[17] Pereira JA, Oliveira I, Sousa A, et al. (2007). Walnut (Juglans regia L.) leaves: Phenolic compounds, antibacterial activity and anti- oxidant potential of different cultivars. *Food Chem Toxicol*, **45**: 2287–95;
[18] Daniel Teoh Chuan Tan, Hasnah Osman, Suriyati Mohamad and Azlina Harun Kamaruddin (2012).Synthesis and Antibacterial Activity of Juglone Derivatives. *J. Chem. Chem. Eng*. **6**: 84-89;
[19] Daniel Teoh Chuan Tan, Hasnah Osman, Suriyati Mohamad and Azlina Harun Kamaruddin, 2012. Synthesis and Antibacterial Activity of Juglone Derivatives [J]. *J. Chem. Chem. Eng*. **6**, 84-89;
[20] *Health Organization of the People's Republic of China*, 2012. Technical standard for disinfection [M]. Beijing: the ministry of health, health legal system and supervision department;
[21] Sun Yun-pei, Johnson E R. Analysis of joint action of insecticides against houseflies[J], *J Econ Entomol*,1960,**53**:887-892.

IOP Conf. Series: Materials Science and Engineering **484** (2019) 012055 doi:10.1088/1757-899X/484/1/012055

The Development and Application of the "DALY"-Based Environmental Risk Assessment Methods with a Case Study on the Impact of PM2.5 in Beijing

Shuo Yang[a,*], Xiuteng Wang[a], Huiting Guo[a], Jing Liu[a], and Jianfei Wang[b]

[a] Resource and Environmental Branch, China National Institute of Standardization, Beijing 100191, China

[b] Institute of Scientific and Technical Information, Shandong Academy of Agriculture Sciences, Jinan 250001, China

Email: yangshuo_1984@126.com

Abstract. The "Four-step" method is the well accepted procedure for risk assessment, which is also adopted in China. This method regards the human body as a benchmark and takes the effects on human health to characterize the risk. On the basis of the regular "Four-Step" method, the present study proposes a quantitative evaluation on the scenario that risk already occurred with the relevant impact represented by the indicator of "disability-adjusted life year" (hereinafter referred to as "DALY"). Based on the proposed methods, this study assessed the impact of "haze", namely, PM2.5, on the residents in Beijing in 2013. According to the assessment, the PM2.5-related death and Cardiovascular Disease are the most significant contributors in terms of health time losses of the residents. The total DALY caused by PM2.5 in 2013 is calculated to be 1557806 year, which is further converted to economic loss of RMB 15.58 Billion Yuan. This study also put forward the assessment procedure of environmental risk which can be used as the guideline for relevant research and works.

1. Introduction

Risk exists in all the human activities, while the activities in different sectors can cause the risks of variable natures. The relatively common risks include the disaster risk, investment risk, engineering risk, decision-making risk, health risk, environmental risk, and etc. Meanwhile, the risk assessment refers to the quantitative or qualitative assessment process that characterizes the risk resulted from the potential hazards related to a certain process or condition. Normally, risk assessment requires the analysis of the impacts upon human (such as residents, workers and tourists) health and ecological receptor (such as birds, fishes, and wild animals), e.g. chemical compounds leak and the hazardous substances are exposed to the environment and human beings. The risk managers need such information to facilitate their decision-making over how to protect human beings and environment from being harmed by the toxic substances or pollutants.

According to Hu Erbang, the environmental risk assessment is the evaluation of the hazard degree of the destruction, damage and other serious adverse events occurring to human society and environment due to the spontaneous natural causes or resulted from human activities and spread via the environmental media. However, Lu Yongsen [1] views the environmental risk assessment from both broad sense and narrow sense separately. The environmental risk assessment in the broad sense refers to the assessment of possible losses resulted from the risks regarding human health, social and economic development, ecological system and others caused by all sorts of human behaviors or the encountered hazards, as well as the management and decision-making process based on this. Whereas, the environmental risk assessment in the narrow sense refers to carrying out the probability estimation over the impact degree of toxic chemical substance's harm over human health, and also proposing the plans and countermeasures of reducing the environmental risks. As shown in these definitions, the environmental risk assessment in the broad sense includes two parts, namely, human health risk assessment and ecological risk evaluation.

Environmental risk control, also known as environmental risk management, requires effective control means and corresponding cost and efficiency analysis consistent with the environmental risk assessment results and the relevant laws and regulations. The policy and social analyses are also needed to determine the social willingness to pay for the reduction of risk. Nevertheless the ultimate goal of risk control is to protect the population health and safety of ecological system.

The study on environmental risk assessment began in 1970s. At the beginning, it was mainly conducted in developed countries, especially USA, where the relevant study and application were relatively prominent. In 1976, US Environmental Protection Agency specified the concept of risk assessment for the first time in *Carcinogen Risk Assessment Criteria.* Until now, the development of environmental risk assessment generally has gone through three phases. At phase 1, namely, from 1930s to 1960s, the risk assessment study was in its infancy. At that time, the connotation of risk assessment was still indefinite. The toxicological identification method was mainly applied to the human health risk analysis. The risk assessment model was not fully established yet, and the results were mainly in the form of qualitative description. Until 1960s, toxicologist proposed some quantitative methods to evaluate the health risk of being exposed to the low concentration toxic substances [2]. At phase 2 (from 1970s to 1980s), the risk assessment study was at the peak. At this phase, the risk assessment system was fully developed. US National Academy of Sciences issued the *Risk Assessment in the Federal Government: Managing the Process* in 1983, which proposed the "Four-Step Method" of risk assessment, namely, risk identification, dose-response assessment, exposure assessment and risk characterization [3]. This document has been viewed as a milestone in the development of risk assessment, and well accepted by France, Holland, Japan, China and many other countries and international organizations. The Phase 3 started from 1990s, In which the ecological risk assessment gradually draws more attentions. In 1990, US Environmental Protection Agency formally proposed the concept of ecological risk assessment. At first, it discussed the inclusion of the human health risk assessment into the ecological risk assessment. After going through the continuous exploration and studies, US Environmental Protection Agency officially enacted the *Guidelines for Ecological Risk Assessment* in 1998, proposing the framework of ecological risk assessment "Three-Step Method", including the development of problems, analysis and risk characterization. The framework has been extensively accepted and applied by researchers worldwide.

China's studies on environmental risk assessment began in 1980s, which mainly tried to follow up with the foreign progress at that time. With the fast economic and social growth, risk management and control became a pressing concern for China government. As a result, the environmental risk assessment system was gradually established. Generally, the development of China's study on environmental risk assessment went through three phases. At phase 1 (from mid-late 1980s to early 1990s), the relevant studies on environmental risk assessment were mainly about the establishment of the emergency countermeasure procedure in nuclear power plant, which took effect in the Qinshan Nuclear Power Plant leakage accident [3]. At phase 2 (from early 1990s to late 1990s), the environmental risk assessment was extensively carried out. In 1993, the former National Environmental Protection Agency enacted the *Technical Guideline for Environmental Impact Assessment*, which specified that the environmental impact assessment or environmental risk analysis must be carried out for the construction projects. Meanwhile, the risk assessment requirements are also put forward in the regulations and administrative system established by the relevant departments. At phase 3 starting from the 21st century, great progress was made for both the environmental risk assessment and relevant administration. In 2004, *Technical Guidelines for Environmental Risk Assessment on Projects* was issued, in which the purpose, principle, procedure, methods, compulsory contents and other requirements for the compilation of environmental risk assessment for the project were given. However, the following problems still exist in China's present environmental risk assessment system.

1) Obstacles still hindered the practice of risk assessment. Although China has recognized the importance of environmental risk assessment and carried out the compulsory environmental risk assessment via laws and regulations, there are still many shortcomings exist in the practice of risk assessment. For instance, the health risk and ecological risk are scarcely involved. Meanwhile, at present, the only available standards and criteria of environmental risk assessment are almost all about construction, whereas none have been developed for other subjects like industrial products.

2) Uncertainty problem is significant. One of the basic features of environmental risk assessment is uncertainty. The main reason is due to the complexity of objective world and limited knowledge about it. In China, because of the relatively short history of environmental risk study, there are very limited information and data accumulation in this field, resulting in a more non-negligible uncertainty issue in the risk assessment results in China.

3) The assessment cannot fully support the decision-making. On one hand, the environmental risk decision making need to consider not only the risk assessment results, but also other factors such as social and economic conditions, ethics ideology, public awareness, and even the ethnic group and other interest groups. On the other hand, the environmental risk assessment results are normally non-quantitative. Therefore, it is inadequate or difficult to direct the decision-making.

Furthermore, both the environmental risk assessment and control are end control approach, which is inconsistent with the current trend of process control, e.g. cleaning production[5]. The ultimate goal of environmental risk control is to protect the living environment so to guarantee the human health and ecological security. This study takes Beijing's PM2.5 as the subject, and evaluated its environmental risk by applying a standard environmental risk assessment procedure. The DALY indicator proposed by the UN was also adopted to manifest the risk impact. For relevant studies, the present research may provide some insights for environmental risk assessment method and interpretation.

2. Study Method

2.1. Human Health Impact Assessment Method

Human health impact assessment has been relatively well developed by relevant studies. To accurately describe the adverse health effects when the human beings are exposed to the environmental harmful factors, the standard procedure can be carried out as shown in Figure 1. Four basic steps, i.e. Four-step, are included as the sources analysis, hazard identification, dose-response assessment and exposure assessment, last but not least, risk characterization. Risk control strategy can only be made and conducted on the basis of the assessment results hereafter.

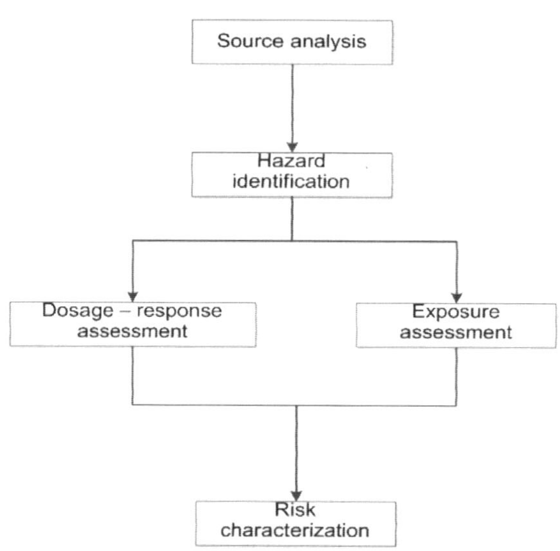

Figure 1. The "Four-step" procedure for human health impact assessment

IOP Conf. Series: Materials Science and Engineering **484** (2019) 012055 doi:10.1088/1757-899X/484/1/012055

Source analysis refers to the procedures that locate and determine the origin of the potential environmental risk. Though different countries and organizations may apply various methods for source analysis, the result must at least include the nature of risks, occurrence reason and occurrence frequency of risk.

For hazard identification, it associates the risk with human health and ecological safety. The conclusions obtained from DMPK test, short-term animal experimentation, long-term animal experimentation, and human epidemiology study provide useful information for this process. Based on the aforementioned researches and other surveillance means, the hazardous factors and their impacts can be unveiled. To facilitate the hazard identification, USA established the Integrated Risk Information System (hereinafter referred to as "IRIS"), which incorporated the pathological research data involving more than 500 common toxic elements and chemical compounds. Meanwhile, EU and other developed countries and regions also established the similar information database [6].

Dose-response assessment is the quantitative estimation process conducted over the relationship between the exposure level of harmful factors and occurrence rate of the adverse effects among the exposed population and ecological system. It provides a solid basis for the risk assessment afterwords. The present experimental data are mainly the dose-response of human health.

Exposure assessment involves survey, estimation, and sometimes, prediction of the amount, frequency, duration and nature (direct or indirect, airborne or waterborne, etc.) of the exposure. When it's about human exposure, the assessment also needs to describe the number, distribution, activity status, contact means and other factors of the exposed population (or biology). The assessment results generally include the information of the discharge amount, concentration, migration and transformation rules and other parameters of the hazardous substance on the basis of the given model.

Risk characterization is the last step of risk assessment. It combines the data and analysis results obtained from the abovementioned procedures, determines the occurrence probability of harmful effects, puts forward the acceptable risk level, and specifies the uncertainty of the assessment results. Meanwhile, risk characterization also works as the bridge connecting the risk assessment and risk management. In some models, the assessment need to classify the risk subjects into carcinogenic and non-carcinogenic (or lethal & non-lethal), and then get the risk results. However, two major problems exist for such model: firstly, although some pollutants do not have the carcinogenic or lethal effect, their impacts are also very servere, for example, the disability-caused substances; secondly, the assessment results of such method normally only have the statistic meaning, because the risk predicted from such assessment is based on the hypothesis that the human body is only exposed to a single risk factor, which is obviously not the typical case in reality.

2.2. Method of Risk Characterization

In order to better solve the risk characterization problem, Harvard University and WHO carried out the global burden of disease study, and put forward the "disability-adjusted life year" (hereinafter referred to as "DALY") indicator to quantify the health damage [7]. DALY stands for the health time loss due to sickness and early death (average loss compared to the life expectancy). It includes the "years of life lost" (hereinafter referred to as "YLL") resulted from early death and "years lived with disability" (hereinafter referred to as "YLD") resulted from the disability caused by the disease, as shown in eq.1.

$$DALY=YLL+YLD \quad (1)$$

DALY demonstrates the health problems of variable individuals in the form of years, which can be calculated and compared. For instance, the 1 year lived in the condition of total disability (including death) resulted from sickness equals to 10 years lived with 90% of full action competence.

As the definition of DALY shows, its calculation need to determine both the YLL and the YLD . The YLL can be calculated as per eq.2, while YLD is calculated as per eq.3.

$$YLL=N\times L \quad (2)$$

$$YLD=N_i\times W_i\times L \quad (3)$$

Where, N refers to the acute death number due to such pollutant; L refers to the current average life

IOP Conf. Series: Materials Science and Engineering **484** (2019) 012055 doi:10.1088/1757-899X/484/1/012055

expectancy (life expectancy-current mean age); In case that the pollutant may cause mutiple diseases, N_i refers to the occurrence number of disease i due to the specific pollutant; W_i refers to the health influence weight (or disability weight) of disease number i and L_i refers to the impact duration of disease number i. Therefore, the DALY is measured as below.

$$\mathrm{DALY} = \mathrm{N} \times \mathrm{L} + \textstyle\sum_i (\mathrm{N_i} \times \mathrm{W_i} \times \mathrm{L_i}) \tag{4}$$

3. Case Study

In this research, the case study is carried out to verify the proposed method. In this case study, the Beijing's air pollution of PM2.5 was selected as the assessment subject, due to its high concern and the availability of the relevant data.

With the rapid economic development, the urban population has been growing continuously in Beijing. Both the possession of automobiles as well as the total energy consumption have hit the record high in the recent years. As a result, the environmental quality and ecological conditions, specially the air condition in Beijing are deteriorating significantly. The suspended particles PM 2.5, which is commonly known as "haze", drawn widely concern. PM2.5 refers to the particles in the ambient air with the diameter less than or equal to 2.5 micron. They suspend in the air for a much longer time. The higher its concentration in the air is, the more serious the air pollution becomes. Although PM2.5 is only the minor element in the composition of the earth's atmosphere, it poses significant impact upon the air quality and visibility. Compared with the coarse atmospheric particulates, PM2.5 is small in grain size, but with large area coverage. More importantly, they easily bound with the toxic and harmful substances (for instance, heavy metal, microorganism, and etc), and have the long residence time in the air and long conveying distance. As a result, PM2.5 have much greater impact upon human health and atmospheric environment quality.

In 2013, the atmospheric environment quality in Beijing was overall poor. There were 25 haze days in January, while the average visibility was only 9.2km. The highest PM 2.5 concentration exceeded 800 ug/m^3, and the total haze days were 2.2 times of the same period of the other years (11.4 days). The statistic data of PM 2.5 in 2013 taken from weather record was shown in Table 1.

Table 1. PM2.5 Statistics in Beijing in 2013

Air Grade	Range of PM2.5 concentration ug/m^3	Computational parameter of PM2.5 ug/m^3	Days
Excellent	0~35	17.5	41
Good	35~75	55	135
Mild Contamination	75~115	95	84
Medium Contamination	115~150	132.5	47
Heavy Contamination	150~250	200	45
Severe Contamination	>250	250	13

PM2.5 and other small particles can enter into human body through inhalation. To measure the exposure, exposure concentration, exposure duration and inhalation rate must be considered. The exposed dose is calculated following the equation as in below.

$$D_{exp} = \int_{t_1}^{t_2} c(t) \times IR(t)dt \tag{5}$$

Where, D_{exp} refers to particle inhalation exposure amount (mg); t_1 & t_2 stand for the start and final time of the exposure duration, respectively; c(t) refers to the mass concentration of particle matter which varies with the exposure time (mg/m^3); IR(t) refers to the human inhalation rate which changes

IOP Conf. Series: Materials Science and Engineering **484** (2019) 012055 doi:10.1088/1757-899X/484/1/012055

with the exposure time (m^3/h); and t refers to exposure time (h). This exposure assessment model can be applied to the cases with much smaller scope, fewer population and shorter time. However, it is obviously not applicable to the whole-year resident health impact assessment in Beijing. Therefore, this study assumes the exposure process of PM 2.5 over all urban citizens in Beijing as a black-box process, and then conducted the dose-response assessment.

In previous study, Xie et al. has found that for every increase of the PM2.5 concentration by 10ug/m^3, the corresponding resident death number in China will increased by 0.38%[8, 9]. Meanwhile, PM 2.5 may also cause many diseases. Four diseases are considered herein, namely, chronic obstructively pulmonary disease (hereinafter referred to as "COPD"), cardiovascular disease, cerebrovascular disease and acute respiratory infection. If the concentration of PM 2.5 is increased by 10ug/m^3, the occurrence rates of COPD, cardiovascular disease, cerebrovascular disease and acute respiratory infection will be increased by 6%, 6%, 8% and 13.5% (γ_i), respectively.

It is assumed that when the air quality grade is good, the residents will not be affected by PM2.5, so the death rate and prevalence rate at that condition are regarded as control. The R_{total} of the annual death rate, and the incidence rate of COPD, cardiovascular disease, cerebrovascular disease and acute respiratory infection of Beijing's residents at present are 4.52%, 1.55%, 1.43%, 0.66% , 4.41% (*China Health Statistical Yearbook*), respectively. R_{total} includes both the cause of PM2.5 and other causes. If the regional variation and individual difference are ignored, the death and prevalence rates due to PM2.5 can be obtained by the following equations.

$$R_{total} = \left(\frac{c_{PM2.5}-c_0}{10}\cdot\gamma + 1\right)\cdot R_{other} \tag{6}$$

$$R_{PM2.5} = R_{total} - R_{other} \tag{7}$$

Where, $C_{PM2.5}$ is the PM2.5 concentration in 2013; C_0 is the concentration when the concentration condition of PM2.5 is considered good, namely, 17.5 ug/m^3. γ_i refers to the enhancement of Death rate or pathogenicity rate caused by the higher concentration of PM2.5. The death and disease due to higher PM2.5 (>55ug/m^3) of residents in Beijing were calculated and are shown in Table 2.

Table 2. The numbers of the death and disease cases caused by high PM2.5

	PM2.5-Related	Population Affected (Person. Year)	Health Weight	Duration Time
Death Rate	0.06%	12000	1	—
COPD	0.32%	64000	0.15	10a
Cardiovascular Disease	0.30%	60000	0.24	Lifetime
Cerebrovascular Disease	0.17%	34000	0.2	Lifetime
Acute Respiratory Infection	1.64%	328000	0.08	14d

The calculation results demonstrate that the PM2.5-related acute death number in Beijing in 2013 was 12000 persons, compared with Kan's study on the relationship between air pollution in 2000 and Shanghai's resident health problem which unveiled 8220 cases of resident death in that year were due to air pollution. Besides, there were also 16870 cases of newly infected chronic bronchitis, 5240 cases of respiratory hospitalization, 2,690 cases of cardiovascular disease hospitalization, 386,600 times of internal medicine department, 40,040 times of pediatric outpatient department, 540,300 cases of acute bronchitis and 9,990 cases of asthmatic attack [10]. For the Great Smog in 1952, more than12000 death cases occurred among the 5 million London urban citizens. As shown in the study of IEPOA and School of Public Health, Peking University, PM2.5 in 2013 caused 257000 excess number of deaths among 31 provincial capitals and municipalities nationwide. Among which, 79 excess number of deaths exist among every 100000 persons in Beijing (If the total population in Beijing is taken as 20 million, there are 15800 excess number of death due to PM2.5) [11]. Hence, the results obtained by the present study are within the reasonable scope.

Meanwhile, the health weight of all diseases related to PM2.5 is also given in Table 2, which represents the severity of the impact the specific disease pose upon the human action competence. In the global burden of disease study carried out by WHO, 22 sorts of indicative symptom and their disability weight are determined. Among which, 0 represents completely healthy, while 1 stands for death or complete loss of body function caused by serious disease (for instance paralysis, human vegetable and etc). For the recoverable disease, the duration time of the disease must be considered. For the disease that cannot be completely recovered, or sequel exists after the recovery or chronic disease, the duration time is regarded as lifetime, namely, the present life expectancy.

When the results in Table 2 were further combined with the life expectancy of Beijing residents, the PM2.5-related DALY may be calculated. As per the relevant data, it is known that the average life expectancy of Beijing residents is 81.5 years. According to the Sixth National Population Census, it is estimated that the present average age of Beijing residents is 37.5 years. Therefore, the present life expectancy of Beijing residents is 44 years. For PM 2.5, as per the DALY calculation results, it is found that cardiovascular disease is the primary threat for the overall health condition of Beijing residents, while death and cerebrovascular disease rank the second and third respectively. The corresponding economic loss were also measured with the association with the per capita GDP in Beijing (RMB 100000 (person. year) in 2013) and shown in Table 3.

Table 3. DALY and Corresponding Economic Loss of Residents in Beijing Resulted from PM 2.5 in 2013

	DALY (person. year)	Economic Loss ($\times 10^8$)
Death	528000	52.8
COPD	96000	9.6
Cardiovascular Disease	633600	63.4
Cerebrovascular Disease	299200	29.9
Acute Respiratory Infection	1006	0.1
In Total	1557806	155.8

4. Conclusion

The DALY-based pollutant environmental risk assessment process flow is as shown in Figure 2. As shown in the case study, it is found that based on this method, variable pollutants may cause different impacts upon human health, including pathopoiesia, disability, carcinogenesis and Death, which can be quantified and summed up via the bridge of "time". In this manner, the integrated assessment and analysis of environmental risk under the influence of multi-factors may be realized. Meanwhile, the risk assessment results are demonstrated in the manner of DALY, instead of possibility or statistical probability. Therefore, they may be further converted to economic loss and other easily understood manner, which also facilitates the application and comparison of environmental risk assessment results.

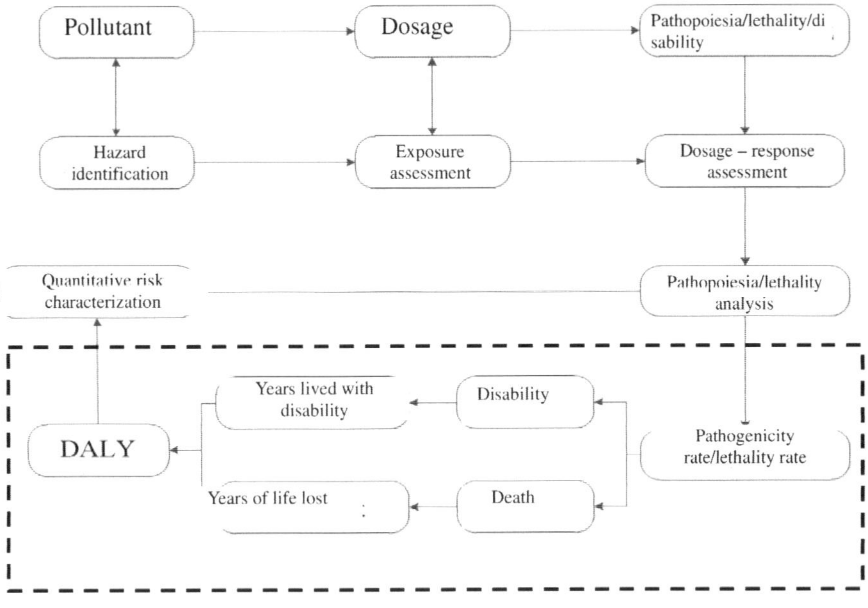

Figure 2. DALY-based Pollutant Environmental Risk Assessment Process Flow

As shown from such assessment process flow, the more complex the assessment object is, the larger the population base is (such as the case of health impact of Beijing haze provided in this paper), then the lower the credibility of assessment results becomes. On one hand, many required data cannot be directly obtained; they shall be calculated via the indirect data or take the average value. On the other hand, large gap may exist between the model adopted by the calculation and the reality. For instance, the aforesaid case does not consider the age impact and weight when calculating the death rate and life loss, which is obviously inappropriate. When the assessment is conducted over the object with a smaller range, the national average level may be directly taken for reference. For instance, when the study is conducted over certain occupational disease (such as the mining worker's silicosis resulted from dust), the relevant health data of patients may be compared with the national average level. Thus, the lethal and pathogenic data of a given pollution factor may be directly obtained. The DALY data obtained in this manner will be more precise, while the assessment results may also be more credible.

5. Acknowledgments

The authors disclosed receipt of the following financial support for the research, authorship, and/or publication of this article: This work was financially supported by Chinese National Institute of Standardization Funding Program (grant number 542017Z-5453).

6. References

[1] LU Yongsen, Environmental Assessment (Second Version) [C], Shanghai: Tongji University Press, 1999: 531-558.
[2] NRC, Science and Judgment in Risk Assessment [M], Washington, DC: National Academy Press, 1994.
[3] HU Erbang, Practical Technology, Methods and Cases of Environmental Risk Assessment [M], Beijing: China Environmental Science Press, 2009: 2-4.
[4] XU Li, The Basic Means and Developing Trends of Environmental Management in Our Country [J], Journal of Lanzhou Railway University (Social Sciences), 2002, 21(2): 39-42.
[5] WEI Zhirong, Problem of Environment Management in Our Country and Countermeasures [J], Northern Economy and Trade, 2006, 6: 210-211.

[6] YANG Wei, Lang Yinhai, Bai Jie, et al. Quantitative Evaluation of Carcinogenic and Non-carcinogenic Potential for PAHs in Coastal Wetland Soils of China. Ecological Engineering, 2015, 74: 117-124.
[7] Stein C., Global Burden of Disease (GBD) Approach and the Use of Disability Adjusted Life Years (DALY) at the World Health Organization World Health Organization, Geneva, Switzerland.
[8] XIE Peng, LIU Xiaoyun, LIU Zhaorong, et al., Impact of Exposure to Air Pollutants on Human Health Effects in Pearl River Delta [J], China Environmental Science, 2010, 30(7): 997-1003.
[9] XIE Peng, LIU Xiaoyun, LIU Zhaorong, et al., Exposure-Response Functions for Health Effects of Ambient Particulate Matter Pollution Applicable for China [J], China Environmental Science, 2009, 29(10): 1034-1040.
[10] KAN Haidong, Health-based Risk Assessment and Economic Valuation on Energy Options and Air Pollution in Shanghai [D], Fudan University Doctoral Thesis, 2003.
[11] PAN Xiaochun, et al., Dangerous Breath 2: The Impacts of Ambient PM 2.5 Upon China's Urban Public Health Effects.

IOP Conf. Series: Materials Science and Engineering **484** (2019) 012056 doi:10.1088/1757-899X/484/1/012056

Study on Environmental Risk Assessment and Management Methods for the Recycle of Industrial Solid Waste

Shuo Yang *, Huiting Guo and Xiuteng Wang

[a]Resource and Environmental Branch, China National Institute of Standardization, Beijing 100191, China

Email: yangshuo_1984@126.com

Abstract. Industrial solid waste demonstrates both the resources and pollution attributes when be used as raw materials for production. Hence, it is critical to prevent the industrial solid waste from endangering the human beings and environment when it is recycled, so to balance the resource efficiency and safety. The present study reviewed the development of relevant theories and methods in the relation with environmental risk assessment and management worldwidely. The characteristics of industrial solid waste are also taken into consideration. The pollutants in the industrial solid waste are classified as the human-health hazard type and product-quality damaging type depending on whether the pollutant can pose negative effect on human health, and only the human-health hazard type is regarded as the subject for risk assessment and management. Based on the classification and simplification, this study proposed the environmental risk assessment and management methods for the recycle process of the industrial solid. The aluminum extraction from coal fly ash was analyzed and the risk was assessed by applying the "Four-step Method". The corresponding risk surveillance scheme was developed based on the risk assessment. Both the method developed by the study as well as the case analysis may provide some insight for relevant research.

1. Introduction

The recycle and reutilization of solid waste can realize the purposes of resources saving, environmental protection and economic development at the same time. However, compared with the regular raw material from nature source, the environmental risk of solid waste in the cause of recycle and its potential affect during using can't be ignored. To secure the safety of environment and human beings, reliable risk assessment and management system must be established.

The Kalundborg industry park is a very typical case. At the beginning of 1970s, the enterprises (for instance, the power plant, refinery, pharmaceutical factory, and etc) in the park tried to collaborate to achieve better waste management and more efficient usage of fresh water. Therefore, they spontaneously formed an "Industrial Symbiosis" system, which consisted of 5 enterprises and the surrounding facilities at first, then more part of the cascade matter and energy flow was incorporated. Among which, the gypsum resulted from the desulfurization of Asnaesvaerket Coal-fired Power Plant was supplied to GyProc Gypsum Materials Company for the production of gypsum board. However, it was found very soon during the conventional analysis that the gypsum from the GyProc Gypsum Materials Company contained a large amount of vanadium, which may endanger the human health. After more thorough investigation, it was found that Asnaesvaerket Coal-fired Power Plant used the fuel containing vanadium. The vanadium migrated and transformed along the Kalundborg Industrial Symbiosis System, which caused the accumulation of heavy metal in the recycled products.

China's solid waste recycling industry has been developing rapidly in recent years. New solid waste recycling technology has been developed and applied to replace the nature resource for the production process. However, compared with the rapid technical development, China's risk management and surveillance work still lags behind. The toxic and harmful substances in the solid waste migrate along the industrial chain, which will cause the potential or actual hazards. For instance, the Zijin Mining Pollution Incident in 2010 and Guangxi Cadmium Pollution Incident in 2012 caused very severe environmental and social impacts. In 2014, more than 1000 environmental criminal cases occurred in China. The lack of risk assessment and management have obstructed future development of the recycle and reutilization of industrial solid waste in China. As a result. The establishment of relevant method and system become more important than ever especially in the context of the ever-worsen environmental and resource crisis in China.

The toxic and harmful substance contained in the solid waste may enter into the recycling process and even exist in the recycling products. The corresponding environmental risks will greatly affect the acceptance of the market and customers over the solid waste recycling products, restricting the development of solid waste recycling industry. Meanwhile, the analysis of environmental risk impact factors of industrial solid waste recycling industry may facilitate preventing from the secondary pollution occurred during the industrial solid waste recycling process, make the industrial solid waste recycling process conform to the standards of relevant environmental regulations, and also safeguard the quality and usage safety of solid waste recycling products.

The solid waste contains many kinds of toxic and harmful substances, while the environmental risks can be affected by the total amount, chemical form as well as the mitigation of these pollutants. For the environmental risk assessment & control of such a complex system, this study adopts the analytic hierarchy process to classify the pollutants depending on whether the they endanger the human health. The human hazard category refers to the substance kind that may cause the direct harm to the human health, for instance, the radionuclide that may impose the death, disability, disease and other impacts upon human body, organic and inorganic pollutants, strong acid and alkaline substances, heavy metal element, and etc. The quality damaging category, on the hand, mainly includes any substance that may not or indirectly affect human health but may lead to the quality problems of products, such as, the chlorine and sulfur contained in the concrete that will accelerating the metallic corrosion and shortening the lifespan of materials. For the quality damaging category substances, the quality standards of relative products may be directly adopted, which simplifies the complexity of risk assessment and control works.

For environment risks, this study selects its narrow sense definition, mainly due to the following reasons: firstly, for the subject of study, namely, bulk industrial solid waste recycling, the human hazard category pollutants shall be the major subject of environmental risk control during the industrial solid waste recycling process. For both the recycling process and products, human beings are not only the participant of recycling process, but also the user of recycling products. Therefore, the direct impact subjects of such pollutants are both human beings. Secondly, the human reaction against external environment is predictable, especially in medicine and toxicology, where a large number of cases and information are accumulated. Therefore, it is possible to establish the much more reliable risk assessment and control methods. Thirdly, great differences exist in the natural environment, while the pollution or risk bearing capacity of natural environment is jointly determined by many factors (such as geography, climate, biodiversity, and etc). Therefore, it is difficult to form the generally applicable risk assessment and control mechanism. As a result, this study finally simplifies the environmental risk assessment of industrial solid waste recycling as the human health impact evaluation works of the specific sort of pollutant contained in the evaluation subject, as per the analytic hierarchy process.

IOP Conf. Series: Materials Science and Engineering **484** (2019) 012056 doi:10.1088/1757-899X/484/1/012056

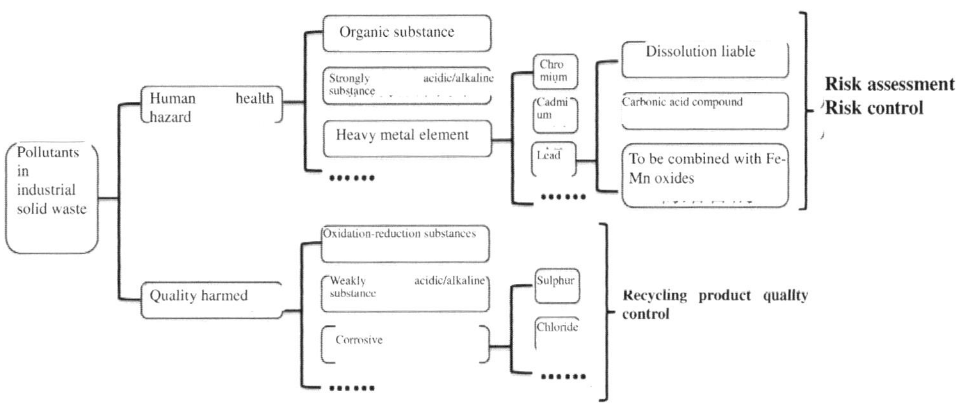

Figure 1. Analytic Hierarchy Process of Pollutant Impact during the Industrial Solid Waste Recycling Process

2. Study Method

The relevant study regarding the human health impact evaluation has already become very mature, almost developing the methodology that describes the adverse human health effects resulted from exposure to the environmental hazard factors. The basic evaluation process flow is as shown in Figure 2, mainly including the source term analysis, hazard identification, dose-response assessment, exposure assessment and risk characterization. The risk control will be carried out on the basis of evaluation results [1].

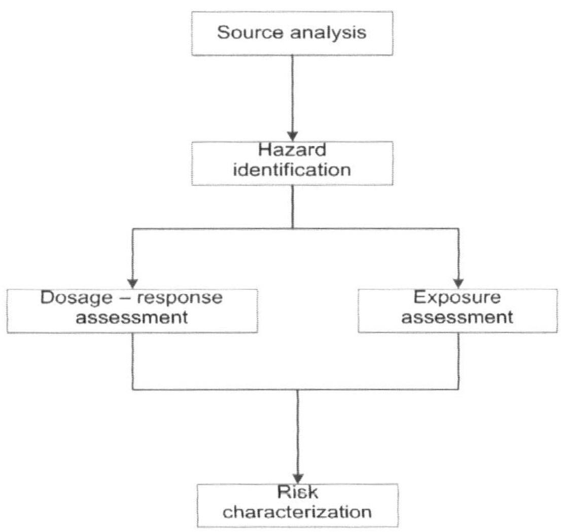

Figure 2. Process Flow of Human Health Impact Evaluation

The environmental risk assessment carried out in this study is mainly against the solid waste recycling process. Prior to the occurrence of risk incidents, it will prepare the surveillance and response plan of risk factors as per the actual conditions of the subject. Due to the variable features of industrial solid waste, there are different recycling technologies accordingly. Therefore, it is very difficult to develop a universal recycling process risk control process, while it is required to prepare the targeted environmental risk control plan in combination with the specific conditions. Based on the "Four-step Method" and combined with the features of solid waste recycling, this study provides the general idea of preparing the environmental risk control plan against the industrial solid waste recycling environment.

1) Risk Factor Identification

This study takes the substance that is harmful to human body as the subject of risk surveillance, while the identification of such substances may refer to *Catalogue of Hazardous Chemicals, Standard for Pollution Control on Hazardous Waste Storage, Standard for Pollution Control on the Security Landfill Site for Hazardous Wastes, Standard for Pollution Control on Co-processing of Solid Wastes in Cement Kiln*, and other laws, regulations and standards. Meanwhile, it is required to take the substances featured in the carcinogenicity, acute toxicity, skin corrosion/irritation, serious eye damage/eye irritation, respiratory or skin sensitization, germ cell mutagenicity, reproductive toxicity and aspiration hazard, as the subjects of risk surveillance and response. When determining the subject of risk surveillance, it is also required to consider the existence method and features of toxic element. Generally, more attention shall be paid to the unstable and easily released substances (as shown in Table 1).

Table 1. Existence Features of Toxic Elements and their Hazard/ Attention Degree

Existence Features of Toxic Elements	Hazard/ Attention Degree
Volatile (e.g. Hg)	★★★
Less volatile, but reactive (e.g. Cr)	★★
Less volatile & reactive (e.g. Cu)	★

2) Migration & Transformation Analysis and Risk Control Strategy of Toxic Elements

During the solid waste recycling process, both the quality and quantity of the toxic elements may change along with the process. Their migration & transformation may be analyzed depending on the inlet, outlet and accumulation along with the production process. The specific risk surveillance and control strategy can be adopted accordingly (as shown in Table 2).

Table 2. Migration & Transformation Analysis and Risk Control Strategy of Toxic Elements

Migration & Transformation Methods of Toxic Elements	Risk Control Strategy
With the main inlet, main outlet and accumulation point	Quality surveillance of key raw materials, operating condition surveillance of key points, and quality surveillance and risk assessment of key products or byproducts
With the main inlet & main outlet, but without the accumulation point	Quality surveillance of key raw materials, and quality surveillance and risk assessment of key products or byproducts
With the main inlet and accumulation point, but without main outlet	Quality surveillance & risk assessment of key raw materials, operating condition surveillance of key points, and quality surveillance and risk assessment of key products or byproducts
With the main inlet, but without the main outlet and accumulation point	Quality surveillance of key raw materials, and quality surveillance and risk assessment of key products or byproducts
Without the main inlet, but with the main outlet and accumulation point	Quality surveillance of raw materials, operating condition surveillance of key points, and quality surveillance and risk assessment of key products or byproducts
Without the main inlet and accumulation point, but with the main outlet	Quality surveillance of raw materials, and quality surveillance and risk assessment of key products or byproducts

3) Determination of Reference Dose

Based on the migration & transformation analysis of toxic elements, it is required to judge the likely exposure manner, mainly including the oral ingestion and inhalation ingestion, while the relevant dose limit may be determined with reference to the "Integrated Risk Information System" (hereinafter referred to as "IRIS") and other medical and toxicological study results.

e) Development of the Risk Surveillance Plan

The risk surveillance plan of the industrial solid waste recycling shall include but not be limited to five parts, namely, technical process flow introduction, risk element identification, risk element migration & transformation analysis, reference dose and surveillance & control measures, as shown in Table 3.

Table 3. Composition of Risk Surveillance Plan along with the Solid Waste Recycling Process

Technical Process Flow	➢ Basic information introduction of raw materials, auxiliary materials and products; ➢ Module partition of key links ……
Risk Element Identification	➢ Name, contents and existence methods (raw materials, auxiliary materials…) of risk elements; ➢ Human health impacts (lethal, carcinogenic, pathogenic…) ➢ Judgment basis (laws & regulations, standards and literature…) ……
Risk Element Migration & Transformation Analysis	➢ Element flow and substance flow analysis of recycling process ➢ Environmental risk source, destination, and enrichment trend analysis ➢ Determination of risk surveillance key points ……
Determination of Reference Dose	➢ Analysis of risk element exposure means (oral ingestion or inhalation ingestion) ➢ The actual exposure dose detection results (If any) ➢ Reference dose (RFD and RCF limits, and data source) ……
Risk Surveillance Response Technology	➢ Setting of key risk surveillance point ➢ Risk control & separation technology ➢ Risk response measures ……

3. Case Study: Aluminum extraction from Coal Fly Ash

3.1. Technical Process Flow

The recycling of aluminum extraction from coal fly ash is a typical solid waste recycling industrial chain, which can properly absorb the solid waste from the power plant, namely, coal fly ash to make the aluminum oxide product, by using the soda lime sintering-dominant new technology, while the process flow is as shown in figure 3. Among which, at the pre-desilication section, the adjusting coal fly ash coming from the thermal power plant will react with the pre-adjusting circulating mother liquor (with the major component of NaOH), carrying out the primary separation of aluminum & silicon. Thus, the desiliconized coal fly ash will be obtained after washing the pre-desiliconized coal fly ash. At the sintering section, the desiliconized coal fly ash will be mixed with other raw materials (such as limestone, anthracite, and etc) and then be added with the appropriate carbon recycling mother liquor for adjusting, which will produce the raw slurry. Then, the raw slurry will enter into the ignition kiln where the bitumite is added to carry out under high temperature to generate the clinker. At the dissolution section, ignition clinker will have the dissolution reaction with the adjusting dissolution liquid, achieving the secondary separation of aluminum & silicon. After multi-washing of the slag phase, the silicon-calcium slag will be produced. The silicon-calcium slag may be used to make the cement & building materials and etc. After the decomposition process (inclusive of carbonation and seed precipitation), the aluminum hydroxide and decomposition mother liquor will be produced from the dissolved pregnant liquor. At the active calcium preparation section, the refined desilication liquor resulted from the pre-desilication will react with the calcium hydroxide to prepare the active calcium silicate.

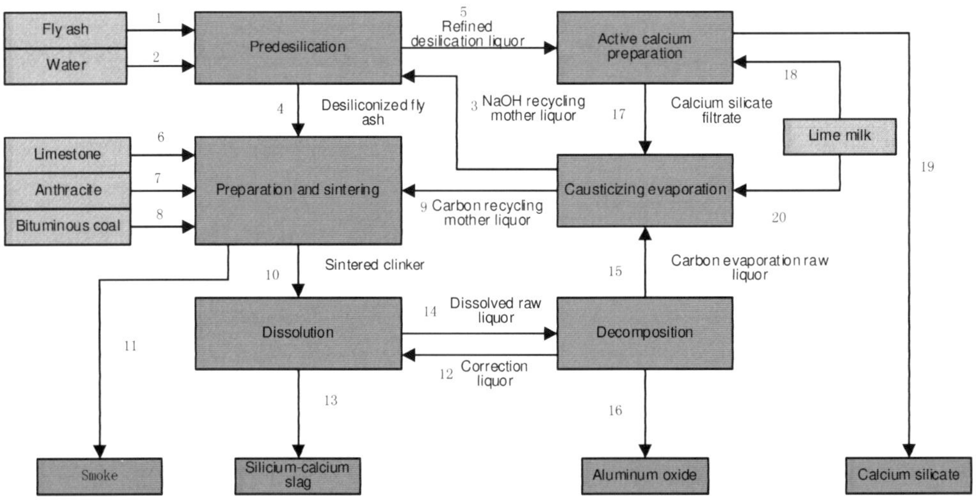

Figure 3. Process Flow of Aluminum Extraction from Coal Fly Ash

3.2. Risk Element Identification

As shown from the process flow of aluminum extraction from coal fly ash, it is found that the major factor that may cause the impacts upon human health in this industrial chain is heavy metal. As per the requirements of *Standard for Pollution Control on Hazardous Waste Storage* (GB18597-2001), *Standard for Pollution Control on the Security Landfill Site for Hazardous Wastes* (GB18598-2001) and *Standard for Pollution Control on Co-processing of Solid Wastes in Cement Kiln* (GB 30485-2013), and the heavy metal average value of raw materials and cement clinker, this study determines the environmental risk threshold of heavy metal in silicon-calcium slag, and screens the heavy metal element that needs special attention in the industrial chain of aluminum extraction from coal fly ash.

Table 4. Coal Fly Ash Environmental Risk Element Analysis

	Element	Average Content ppm	Threshold ppm	Relative Distance %
Coal Fly Cash	Cr	90.4	66	36.97
	Mn	210.37	95.1	121.21
	Pb	80.26	51	96.78
	Zn	89.13	44.5	100.29

Table 5. Silicon-Calcium Slag Environmental Risk Element Analysis

	Element	Average Content ppm	Threshold ppm	Relative Distance %
Silicon-Calcium Slag	Cr	143.92	66	118.06
	Mn	235.22	95.1	147.32
	Pb	70.55	51	119.25
	Zn	88.58	44.5	99.06

As per the average content of heavy metal elements of coal fly ash and silicon-calcium slag, it is found that the average contents of four heavy metal elements, namely, Cr, Mn, Pb & Zn exceed the relevant standards and regulated thresholds. Among which, Pb & Cr have the significant negative impact upon the human health. Meanwhile, with the ongoing production process, significant enrichment of Cr occurs to the silicon-calcium slag, while the content of Pb is greatly reduced, which means that it is likely to be released to the external environment. Therefore, this study takes Cr and Pb as the focus of enviornmental risk surveillance, and separately researches their dose-response and migration & transformation process.

3.3. Dose-Response of Pb & Cr

In order to support the implementation of environmental risk assessment, US Environmental Protection Agency establishes the "Integrated Risk Information System" (hereinafter referred to as "IRIS"), and also includes the pathology study data of 500 plus common toxic elements and chemical compound into the database. The hazard degrees of all sorts of toxic substances are demonstrated in the form of Reference Dose (hereinafter referred to as "RFD"), while it represents the Acceptable Daily Dose (hereinafter referred to as "ADI"), with the unti of mg/kg-day. When the human body's exposure dose of a certain toxic substance exceeds the RFD, it is thought that there is the risk of human health damage. The calculation equation of its quantification coefficient Harzard Quotient (hereinafter referred to as "HQ") is as shown below:

$$HQ = \frac{\text{Actual Daily Exposure(mg/kg-day)}}{\text{RFD}} \tag{1}$$

In IRIS system, the human health impacts of all sorts of element and chemcial compound fall into three categories, namely, oral type, inhalation type and carcinogenicity type. The dose-response information of lead and its compounds recorded in IRIS is as shown in Figure 4.

STATUS OF DATA FOR Lead and compounds (inorganic)

File First On-Line 03/01/1988

Category (section)	Assessment Available?	Last Revised
Oral RfD (I.A.)	qualitative discussion	07/08/2004
Inhalation RfC (I.B.)	not evaluated	
Carcinogenicity Assessment (II.)	yes	09/26/1988

Figure 4. Dose-Response Information of Lead and its Compounds

The lead and its chemical compounds have the impacts upon human bodies, such as neurotoxicity, developmental retardation, hypertension, hearing impairment, affecting the hemoglobin synthesis, causing the male reproductive damage and etc. However, different from other elements, lead does not have the significant toxicity symptoms over human impacts. Therefore, the quantitive RfD value is not given in the relevant studies. However, lead and its compounds may have the carcinogenicity effects. The animal test results show that the ingestion of some variable solubility lead salt via the food and subcutaneous manner may lead to the tumor. The lead carcinogenicity data are still not perfect. Although there are some toxicologic studies on the workers of lead smelter and battery plant, the very significant direct connection between lead and carcinogenicity is still not found.

The dose-response information of Cr recorded in IRIS is as shown in Figure 5.

STATUS OF DATA FOR Chromium (VI)

File First On-Line 03/31/1987

Category (section)	Assessment Available?	Last Revised
Oral RfD (I.A.)	yes	09/03/1998
Inhalation RfC (I.B.)	yes	09/03/1998
Carcinogenicity Assessment (II.)	yes	09/03/1998

Figure 5. Dose-Response Information of Cr

Chromium III has very little harmful impact upon human body, while the industrial poisoning report is not found. However, Chromium III mainly has the chronic toxicity impact upon human body. It may invade the human body through the digestive tract, respiratory tract, skin and mucosa, and is mainly accumulated in the liver, kidney and endocrine gland of human body. If entering into human body via the respiratory tract, Chromium III is prone to be accumulated in the lung. Chromium III has the strong oxidation function. Therefore, generally, the chronic intoxication will begin with the local damage and then gradually becomes incurable. When invading the human body via the respiratory tract, it may begin to invade the respiratory tract, causing the rhinitis, pharyngitis, laryngitis and bronchitis. The animal test dominates. For the oral ingestion, Hexavalent chromium RFD is 3E-3mg/kg-day. For the inhalation ingestion, the foggy hexavalent chromium RFD is 8E-6mg/m^3, while the granular hexavalent chromium RFC is 1E-4mg/m^3. For the inhalation ingestion, the lung cancer may be resulted. Mancuso has tracked the health conditions of 332 chrome exposure workers. Among the 283 death cases, there are 66 nos. of lung cancer patients (accounting for 23.3% of death number and 64.7% of cancer number). Therefore, it is believed that the chrome has the significant carcinogenicity effects.

3.4. Pb & Cr Migration & Transformation during the Aluminum Extraction from Coal Flying Ash

The heavy metal elements in the industrial chain of aluminum extraction from coal flying ash that exceeds the standard threshold and has the enrichment or discharge include the Pb and Cr. Among which, the exposure of Pb is through oral ingestion, while the exposure of Cr can be both oral ingestion and inhalation ingestion. Therefore, it is required to analyze the migration & transformation rules of Pb and Cr during the process of aluminum extraction from coal flying ash, while the corresponding risk control measures shall be taken.

1) Migration & Transformation of Pb

The Pb contents in the raw materials, output and emissions during the process flow of aluminum extraction from coal flying ash are as shown in Table 6.

Table 6. Mass Balance Calculation of Pb during the Process Flow of Aluminum Extraction from Coal Flying Ash

		Heavy Metal Flow Maximum (g/h)	Heavy Metal Flow Minimum (g/h)	Percentage Content of Heavy Metal %
Input	Coal Flying Ash	5839.79	5297.96	51.18%~61.28%
	Limestone	1759.77	1148.81	11.17%~18.33%
	Bitumite 21.3%	2499.87	1914.69	18.58%~26.12%
	Water	0.00	0.00	0.00%
	Calcium Hydroxide	314.74	272.03	2.51%~3.49%
	Anthracite 16.5%	478.67	353.62	3.28%~5.25%
Output	Silicon-Calcium Slag	6448.99	5175.64	53.81%~62.66%
	Calcium Silicate	1554.09	1254.06	11.84%~16.68%
	Smoke	2759.46	2486.23	23.41%~29.69%
	Aluminum Oxide	103.55	129.87	0.95%~1.43%

During the process of industrial chain of aluminum extraction from coal flying ash, the heavy metal Pb mainly comes from the coal flying ash, while the main outlet is silicon-calcium slag and two accumulation points exist in this process. The more heavy metal content is brought in by the coal flying ash, the higher the heavy metal concentration is contained in the silicon-calcium slag, and the larger the heavy metal environmental risk would become. Therefore, it is required to monitor the heavy metal Pb in the coal flying ash and silicon-calcium slag, to prevent from their impacts upon the environment. The migration law of heavy metal Pb is relatively general, while the major migration and transformation working procedure is pre-desilication, adjusting sintering and dissolution. Among which, the enrichment phenomenon occurs to the adjusting sintering section, while the heavy metal Pb flow in the ignition clinker is the largest among all the mass flow, and the heavy metal environmental risk is also much higher. Therefore, it is required to monitor the Pb at the sintering section, so as to reduce the environmental risk. Meanwhile, the enrichment phenomenon also occurs at the causticizing evaporation operating conditions of Pb. The Pb content in the recycling NaOH mother liquor is gradually accumulated over time, which may have a certain impact upon the production system. Therefore, it is required to monitor and control the Pb element in the recycling NaOH mother liquor and operating conditions of causticizing evaporation section. The Pb at the sintering section mainly comes from the desiliconized coal flying ash (48%~60%), and partially comes from the bitumite, anthracite and limestone. After the sintering process, 65%~75% Pb enters into the ignition clinker and 30% enters into the smoke. At the dissolution section, most Pb (≈90%) in the ignition clinker enters into the silicon-calcium slag. Other Pb in the raw materials enter into the calcium silicate and smoke, while partial Pb is recycling in the system together with the NaOH mother liquor.

2) Migration & Transformation of Cr

The Cr contents in the raw materials, output and emissions during the process flow of aluminum extraction from coal flying ash are as shown in Table 7.

Table 7. Mass Balance Calculation of Cr during the Process Flow of Aluminum Extraction from Coal Flying Ash

		Heavy Metal Flow Maximum (g/h)	Heavy Metal Flow Minimum (g/h)	Flow Percentage of Heavy Metal%
Input	Coal Flying Ash	5231.65	4911.14	45.03%~51.50%
	Limestone	2786.91	2273.86	21.22%~26.92%
	Water	0.00	0.00	0.00%
	Bitumite 21.3%	2135.67	1682.59	15.62%~20.75%
	Calcium Hydroxide	119.73	103.66	0.92%~1.22%
	Anthracite 16.5%	952.87	866.46	7.78%~9.60%
Output	Silicon-Calcium Slag	5992..23	5573.59	54.63%~60.96%
	Calcium Silicate	1279.65	1021.11	9.85%~13.23%
	Smoke	3237.15	2735.28	27.03%~32.66%
	Aluminum Hydroxide	111.52	81.26	0.77%~1.18%

During the whole process of industrial chain of aluminum extraction from coal flying ash, the heavy metal Cr mainly comes from the coal flying ash, while the main outlet is silicon-calcium slag and two accumulation points exist in this process. The heavy metal Cr is mainly taken out by the silicon-calcium slag, while the environmental risk is mainly concentrated on the silicon-calcium slag and smoke. The higher the heavy metal concentration is contained in the silicon-calcium slag, and the larger the heavy metal environmental risk becomes. Therefore, it is required to monitor the heavy metal Cr in the silicon-calcium slag, to prevent from the occurrence of risks. Meanwhile, the Cr in the smoke may cause the direct harm to the human body through the inhalation process. Therefore, it is also the key subject of risk control. The major migration and transformation working procedure of heavy metal Cr is pre-desilication, adjusting sintering, dissolution and causticizing evaporation. Among which, the enrichment phenomenon occurs to the causticizing evaporation section. The Cr in this section respectively comes from the calcium silicate filter liquor, calcium hydroxide and carbon evaporation raw liquor. Among which, the Cr in the calcium silicate filter liquor is about 35.17%~54.30% of total input amount. After passing through the causticizing evaporation process, almost all the Cr will enter into the NaOH recycling mother liquor. Together with the continuously recycling of NaOH mother liquor, Cr will also be accumulated. Therefore, it is required to take risk control measures over the Cr in NaOH.

4. Environmental Risk Control Measures

Based on the study of Pb & Cr migration & transformation during the production process, this study confirms the key links, subjects, heavy metal element category, potential exposure means and reference dose of environmental risk surveillance (as shown in Table 8).

Table 8. Environmental Risk Surveillance & Limits of Aluminum Extraction from Coal Flying Ash

Key Links	Subjects	Elements	Potential Exposure Means	Reference Dose
Pre-desilication	Coal Flying Ash (Main inlet of Pb/Cr)	Pb & Cr	Oral ingestion after dissolution Inhalation ingestion of dust	
Adjusting Sintering (Pb Enrichment)	Ignition clinker Smoke	Pb	Oral ingestion after dissolution Inhalation ingestion	Cr(VI): RFD=3E-3mg/kg-day RCF=1E-4mg/m^3 Pb: cannot be detected
Dissolution	Silicon-calcium slag (Main outlet of Pb/Cr)	Pb & Cr	Oral ingestion after dissolution	
Causticizing Evaporation (Cr Enrichment)	Recycling NaOH mother liquor	Cr	Oral ingestion	

For the aluminum extraction from coal flying ash, special attention shall be paid to prevent from the heavy metal entering into human body via the inhalation ingestion of dust, smoke and others during the production process. Meanwhile, the anti-dissolution measures shall be taken during the storage of silicon-calcium slag and coal flying ash. Due to the definite carcinogenicity effect of Cr, it is required to be taken as the key subject of risk surveillance. Especially, when the Cr in the recycling NaOH mother liquor is accumulated to a certain level, it shall be substituted, while it is also required to properly handle the old liquor.

5. Acknowledgments

The authors disclosed receipt of the following financial support for the research, authorship, and/or publication of this article: This work was financially supported by Chinese National Institute of Standardization Funding Program (grant number 542017Z-5453).

6. Reference

[1] USEPA. Risk Assessment Guidance for Superfund Volume I: Human Health Evaluation Manual (Part E, Supplemental Guidance for Dermal Risk Assessment). 2004.
[2] Yang W, Lang Y H, Bai J, et al. Quantitative Evaluation of Carcinogenic and Non-carcinogenic Potential for PAHs in Coastal Wetland Soils of China. Ecological Engineering, 2015, 74: 117-124.

IOP Conf. Series: Materials Science and Engineering **484** (2019) 012057 doi:10.1088/1757-899X/484/1/012057

Antialgal Activity of Glycoglycerolipids Derived from a Green Macroalgae *Ulva prolifera* on Six Species of Red Tide Microalgae

Yingying Sun [1,2,3,*], Shasha Dong [1], Ganlin Guo [2,3], Lei Guo [1,3] and Yinfang Pu [1]

[1] Jiangsu Key Laboratory of Marine Bioresources and Eco-environment, Huaihai Institute of Technology, Lianyungang 222005, China;
[2] Co-Innovation Center of Jiangsu Marine Bio-industry Technology, Lianyungang, 222005, China;
[3] Jiangsu key laboratory of marine biotechnology, Huaihai Institute of Technology, Lianyungang 222005, China
Email: syy-999@163.com

Abstract. 1-O-palmitoyl -3-O-β-D-galactopyranosyl glycerol, 1-O-octadecanoic acid-3-O-β-D- galactopyranosyl glycerol, and 1-O-palmitoyl-2-O-oleoyl-3-O-β-D-galactopyranosyl glycerol were isolated from *Ulva prolifera* for the first time in our previous research. There are growth inhibiton of these three glycoglycerolipids against six species of red tide microalgae (*Amphidinium carterae*, *Heterosigma akashiwo*, *Karenia mikimotoi*, *Phaeocystis globosa*, *Prorocentrum donghaiense*, and *Skeletonema costatum*) investigated. Results showed that they have selective antialgal activity against six species of red tide microalgae, and antialgal activities against test red tide microalgae obviously enhanced with the increase of concentration of glycoglycerolipids. Among them, 1-O-octadecanoic acid-3-O-β-D-galactopyranosyl glycerol and 1-O-palmitoyl-2-O-oleoyl-3-O-β-D-galactopyranosyl glycerol exhibited more extensive antialgal activities, and the growth of *Phaeocystis globosa*, *Prorocentrum donghaiense*, and *Skeletonema costatum* (or *Karenia mikimotoi*, and *Skeletonema costatum*) was inhibited by these two glycoglycerolipids. Further, $EC_{50\text{-}96h}$ values of these three glycoglycerolipids for six red tide microalgae were obtained for the first time. By analyzing and comparing $EC_{50\text{-}96h}$ values, it has been determined that 1-O-octadecanoic acid-3-O-β-D- galactopyranosyl glycerol and 1-O-palmitoyl-2-O-oleoyl-3-O-β-D-galactopyranosyl glycerol showed the superior application potential than potassium dichromate as a characteristic antialgal agent against *Phaeocystis globosa*; More importantly, 1-O-octadecanoic acid-3-O-β-D- galactopyranosyl glycerol exhibited the superior application potential than potassium dichromate and other reported compounds as a characteristic antialgal agent against *Prorocentrum donghaiense*.

1. Introduction

Green algae *Ulva prolifera* has a wide distribution in Chinese coastal waters, and its natural biomass is very abundant even excessive. Researchers have found a lot of solvent extracts from *Ulva prolifera* has anti-oxidant [1], anti-tumor [2], anti-inflammatory [3], anti-bacteria [4], anti-hepatic injury activity [5], antialgl [6], and other biological activities [7, 8].

Allelopathic effects of *Ulva prolifera* on growth of *Karenia mikimotoi* was studied in early time [9]. Also, allelopathic effects of *Ulva prolifera* on *Isochrysis galbana* [10], *Heterosigma akashiwo* [11], *Karenia mikimotoi* and *Alexandrium tamarense* [12] and *Skeletonema costatum* [12, 13] were found. Sun et al. (2010) reported growth inhibition of solvent extracts from *Ulva prolifera* for *Amphidinium carterae*, *Karenia mikimotoi*, *Alexandrium tamarense* and *Skeletonema costatum* [14]. The antialgal principles of *Ulva prolifera* against *Heterosigma akashiwo* were found to be fatty acids [15]. Sun et al. (2014) showed that antialgal effect of *Ulva prolifera* on the growth of *Karenia mikimotoi*, *Alexandrium tamarense* and *Skeletonema costatum* were probably resulted by lactone and coumarins compounds [16]. Recently, three glycoglycerolipids (1-*O*-palmitoyl -3-*O*-β-D-galactopyranosyl glycerol, 1-*O*-octadecanoic acid-3-*O*-β-D- galactopyranosyl glycerol, and 1-*O*-palmitoyl-2-*O*-oleoyl-3-*O*-β-D-galactopyranosyl glycerol) with antialgal activities against some red tide microalgae from *Ulva prolifera* were isolated for the first time [17].

Glycoglycerolipids are important components of cell membrances and play an important role in cellular biology. A number of glycoglycerolipids has been already isolated from higher plants [18, 19], freshwater and marine microalgae [20, 21] and marine macroalgae [22-33]. Glycoglycerolipids derived from microalgae and marine macroalgae display various bioactivities, such as anti-tumor [34], anti-HIV [35], anti-bacteria [25, 34], anti-infection [36], influenza virus-neutralizing [36] and other biological activities [26, 30]. There is currently considerable interest in glycoglycerolipids as a source of biologically active substances. However, these are very few studies that antialgal activity of glycoglycerolipids against red tide microalgae [37]. In our previous research, we only studied antialgal activities of these three glycoglycerolipids against *Alexandrium tamarense, Heterosigma akashiwo, Karenia mikimotoi, Prorocentrum donghaiense* and *Skeletonema costatum* at the concentration of 28.8 μg/mL [17], but failed to analyze quantitative relationship between the inhibition of algal growth and the concentration of glycoglycerolipids and determine half effective concentration (EC_{50}). Thus, this paper will be carried out to study antialgal activity of these three glycoglycerolipids on *Amphidinium carterae, Heterosigma akashiwo, Karenia mikimotoi, Phaeocystis globosa, Prorocentrum donghaiense* and *Skeletonema costatum*, to determine a quantitative relationship between the inhibition of algal growth and the concentration of each glycoglycerolipid and to obtain important parameters (EC_{50}) for future practical HAB control. Antialgal activities of these three glycoglycerolipids on six red tide microalgae (*Amphidinium carterae, Heterosigma akashiwo, Karenia mikimotoi, Phaeocystis globosa, Prorocentrum donghaiense*, and *Skeletonema costatum*) previously has not been reported.

2. Materials and Methods

2.1. HAB algae and Compound Samples

Red tide microalgae (*Amphidinium carterae, Heterosigma akashiwo, Karenia mikimotoi, Phaeocystis globosa, Prorocentrum donghaiense*, and *Skeletonema costatum*) were provided by the Microalga Research Laboratory of the Ocean University of China, Qingdao, China. All nine microalgae were cultured aseptically in Erlenmeyer flasks with f/2 medium. Cultures were incubated at 20°C and illuminated with fluorescent lamps in a 16/8 dark/light cycle, at an irradiance level of 60 μmol photons m^{-2} s^{-1}. All the flasks containing microalgae were shaken twice at a set time every day to prevent wall growth. Microalgae were cultured to exponential phase before subsequent inoculation.

Three glycoglycerolipids samples were prepared as the method reported by Sun et al. [17]. Each glycoglycerolipid sample was dissolved in methanol at a concentration of 5 mg/mL and filtered through a 0.22 micron syringe filter as a stock solution.

2.2. Seawater for Experiments

Seawater was obtained from the coast of Lianyungang. Aged natural seawater was filtered with cotton sheets, boiled, cooled, and filtered through glass fiber papers (Whatman GF/C, 0.22 μm poresize) to eliminate organic particles and debris of organisms. The pH and salinity of the seawater were adjusted to 8.0 and 30 ppt, respectively. This seawater was used for culture seawater of microalgae in all experiments of this study.

2.3. Antialgal Activity Assays

To determine antialgal ability of three glycoglycerolipids, 10 μL of each glycoglycerolipid sample was added to teat glass containing 0.5 mL of algal inoculant and 4.5 mL of culture medium (initial concentration of glycoglycerolipid sample in suspensions of microalgae: 0.4, 2, 10, and 50 μg/mL), growth inhibition of algal species in six red tide microalgae (*Amphidinium carterae*, *Heterosigma akashiwo*, *Karenia mikimotoi*, *Phaeocystis globosa*, *Prorocentrum donghaiense*, and *Skeletonema costatum*) was measured. Controls received the same volume of methanol. Potassium dichromate was used as a positive control. There were four replicates for every treatment used in this experiment, and the culture conditions were the same as mentioned above. This experiment lasted for 4 days. $EC_{50\text{-}96h}$ was calculated as the method reported by Marklund and Marklund [38]. Cells of microalgae were counted by hemocytometer, while morphology of red tide microalgae was observed under an Olympus optical microscope.

2.4. Data Process And Statistical Analysis
All the data of the growth assays in this study were analyzed by ANOVA and Tukey's test.

3. Results

3.1. Antialgal Activity of Three Glycoglycerolipids Against Six Species of Red Tide Microalgae
A significant increase of antialgal activity was observed at the concentration range (0.4~50 μg/mL) of glycoglycerolipid (Fig. 1).

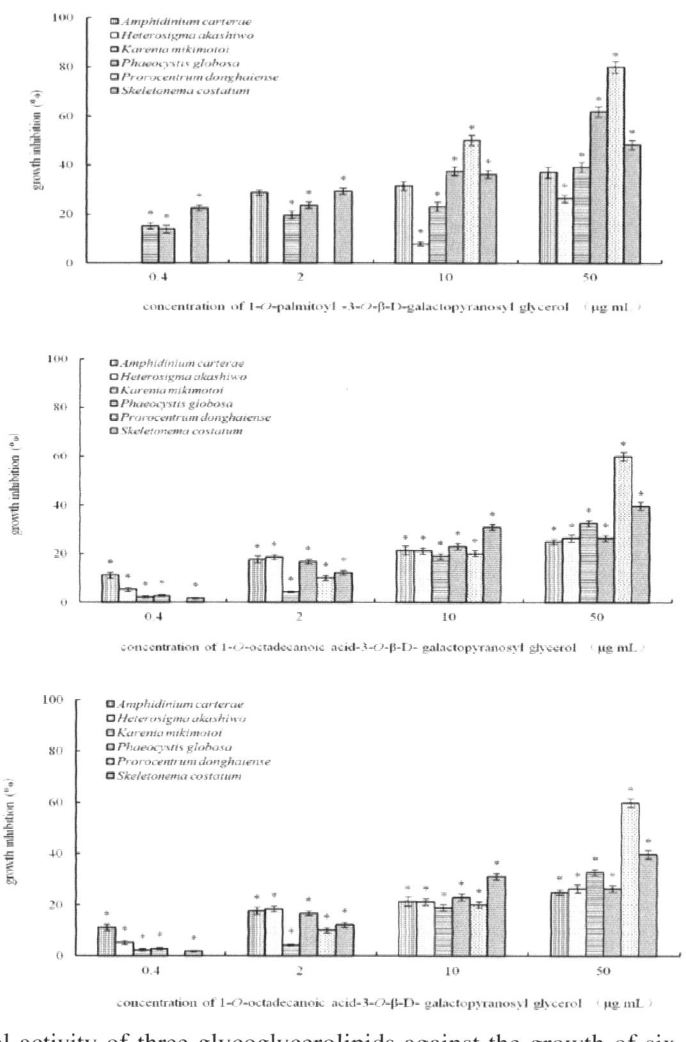

Figure 1. Antialgal activity of three glycoglycerolipids against the growth of six species of red tide microalgae. Data represent average values (*n*=4)

At 50 μg/mL, 1-*O*-palmitoyl -3-*O*-β-D-galactopyranosyl glycerol only significantly ($p<0.05$) inhibited the growth of *Prorocentrum donghaiense*. 1-*O*-octadecanoic acid-3-*O*-β-D- galactopyranosyl glycerol stronger ($p<0.05$) inhibited the growth of *Phaeocystis globosa* and *Prorocentrum donghaiense*. 1-*O*-palmitoyl-2-*O*-oleoyl-3-*O*-β-D-galactopyranosyl glycerol showed stronger ($p<0.05$) antialgal activity against others five red tide microalgae besides *Amphidinium carterae*. Among three glycoglycerolipids, 1-*O*-palmitoyl-2-*O*-oleoyl-3-*O*-β-D-galactopyranosyl glycerol exhibited more extensive antialgal activity.

And Growth inhibition of 1-*O*-octadecanoic acid-3-*O*-β-D- galactopyranosyl glycerol for *Prorocentrum donghaiense* at the concentration of 2 μg/mL was more than 50%; growth inhibition of 1-*O*-palmitoyl-2-*O*-oleoyl-3-*O*-β-D-galactopyranosyl glycerol for *Prorocentrum donghaiense* at the concentration of 10 μg/mL was 50%. Three glycoglycerolipids don't show significantly growth inhibition for *Amphidinium carterae*. Results showed that different algal species had variable sensitivity to same compounds. Among six red tide microalgae, *Prorocentrum donghaiense* was more sensitivity to three glycoglycerolipids; and *Amphidinium carterae* was most insensitive to three glycoglycerolipids. These considerable differences in antialgal activity of glycoglycerolipids against red tide microalgae may be due to species-specific differences.

3.2. $EC_{50\text{-}96h}$ of Three Glycoglycerolipids for Six Red Tide Microalgae

$EC_{50\text{-}96h}$ of potassium dichromate and three glycoglycerolipids for six red tide microalgae is shown in Table 1. As indicated in the literature [39], the grading standards for the toxicity of the algae growth inhibition experiment: $EC_{50\text{-}96h}$ < 1 μg/mL is extremely high toxic; 1~10 μg/mL is high toxic; 10~100 μg/mL is medium toxicity; > 100 μg/mL is low toxicity. Thus, 1-*O*-palmitoyl -3-*O*-β-D-galactopyranosyl glycerol was medium toxicity for *Heterosigma akashiwo*, *Phaeocystis globosa* and *Prorocentrum donghaiense*. It showed low toxicity for other three red tide microalgae; 1-*O*-octadecanoic acid-3-*O*-β-D- galactopyranosyl glycerol was high toxic for *Prorocentrum donghaiense*, and its $EC_{50\text{-}96h}$ for *Phaeocystis globosa* and *Prorocentrum donghaiense* was significantly ($p<0.05$) less than that of potassium dichromate. It exhibited medium toxicity for *Phaeocystis globosa* and *Skeletonema costatum*, and low toxicity for *Karenia mikimotoi*; 1-*O*-palmitoyl-2-*O*-oleoyl-3-*O*-β-D-galactopyranosyl glycerol showed medium toxicity for *Karenia mikimotoi*, *Phaeocystis globosa*, *Prorocentrum donghaiense* and *Skeletonema costatum*, and low toxicity for *Amphidinium carterae* and *Heterosigma akashiwo*. And its $EC_{50\text{-}96h}$ for *Phaeocystis globosa* was less than that of potassium dichromate. Results showed 1-*O*-octadecanoic acid-3-*O*-β-D- galactopyranosyl glycerol (or 1-*O*-palmitoyl-2-*O*-oleoyl-3-*O*-β-D-galactopyranosyl glycerol) possessed good application potential as a characteristic antialgal agent against *Phaeocystis globosa* and *Prorocentrum donghaiense* (or *Prorocentrum donghaiense*).

Table 1. EC_{50} (μg/mL) of potassium dichromate and three glycoglycerolipids against several red tide microalgae.

	Potassium dichromate	1	2	3
A. carterae	3.90	–	–	753
H.akashiwo	36.1	80.0	–	186
K.mikimotoi	16.2	156	133	41.0
P. globosa	38.3	55.4	14.0	24.5
P. donghaiense	5.00	37.0	2.28	10.0
S. costatum	2.70	989	66.1	10.3

Note: "–" no calculated.

4. Discussions

Three glycoglycerolipids isolated from *Ulva prolifera* were found to be 1-*O*-palmitoyl -3-*O*-β-D-galactopyranosyl glycerol, 1-*O*-octadecanoic acid-3-*O*-β-D- galactopyranosyl glycerol and 1-*O*-palmitoyl-2-*O*-oleoyl-3-*O*-β-D-galactopyranosyl glycerol, two of which, 1-*O*-palmitoyl -3-*O*-β-D-galactopyranosyl glycerol and 1-*O*-palmitoyl-2-*O*-oleoyl-3-*O*-β-D-galactopyranosyl glycerol, were the same compounds that had been isolated from *Gracilaria lemaneiformis* [37]. Other researchers have also reported that brown algae such as *Laminaria japonica* [40, 41], *Ecklonia kurome* [40], *Ectocarpus fasciculatus* [42], *Sargassum hemiphyllum* var. Chinense [23], *Sargassum carpophyllum* [24], *Ishige okamurai* [25], *Fucus vesiculosus* [43], *Sargassum horneri* [26], *Sargassum wightii* in India's coast [25], *Sargassum thunbergii* [27, 33], *Zonaria diesingiana* [44], *Sargassum fusiforme* [30] in Dalian, *Sargassum fulvellum* [31], *Sargassum muticum* [32] and *Sargassum pallidum* [41]; red algae such as *Chondiria dasyphylla* [45], *Gracilaria verrucoca* [40], *Eucheuma muricatum* [40], *Hypnea charoides* [46], *Laurencia majuscula* [46], *Chondria armata* [47], *Laurencia tristicha*[29], *Chondria crasscaulis* [29, 48], *Ahnfeltia tobuchiensis* [41] and *Porphyra haitanensis* [48]; green algae such as *Codium iyengarii* [22], *Caulerpa racemosa* [50], *Tydemania expeditionis* [51] and *Ulva fenestrata* [41] contain glycoglycerolipids. Glycoglycerolipids derived from marine macroalgae have been reported to have a few biological effects. Monogalactosyl diacylglycerol (MGDG) from brown algae *Sargassum carpophyllum* is shown activity causing morphological anbormality of *Pyricularia oryzae* mycelia and weak anti- tumor activity [34]. Digalactosyl diacylglycerol (DGDG) derived from brown algae *Sargassum horneri* showed inhibitory effect on the proliferation of CacO-2 cells *in vitro* [26]. Sulfoquinovosyl diacylglyccerol (SQDG) isolated from red algae *Gigartina tenella* inhibits DNA-polimerase and HIV-reverse transcriptase [34]. Sulfoquinovosyl diacylglyccerol (SQDG) derived from *Sargassum wightii* was active against *Xanthomonas oryzae* pv. oryzae which causes bacterial blight of rice [25]. Two glucopyranosyldiacylglycerols obtained from brown algae *Sargassum fulvellum* showed fibrinolytic activities in the reaction system of pro-u-PA and plasminogen *in vitro* [31]. In this work, we proposed new biologically active glycoglycerolipids isolated from *Ulva prolifera* for antialgal agent (Fig. 1). 1-*O*-octadecanoic acid-3-*O*-β-D- galactopyranosyl glycerol (or 1-*O*-palmitoyl-2-*O*-oleoyl-3-*O*-β-D-galactopyranosyl glycerol) possessed good application potential as a characteristic antialgal agent against *Phaeocystis globosa* and *Prorocentrum donghaiense* (or *Prorocentrum donghaiense*) (Table 1). Among them, 1-*O*-palmitoyl-2-*O*-oleoyl-3-*O*-β-D-galactopyranosyl glycerol exhibited more effective antialgal activity against tested red tide microalgae than others two compounds; and antialgal effects of 1-*O*-palmitoyl -3-*O*-β-D-galactopyranosyl glycerol on tested red tide microalgae were more stronger than that of 1-*O*-octadecanoic acid-3-*O*-β-D- galactopyranosyl glycerol (Fig. 1 and Table 1). These three glycoglycerolipids derived from *Ulva prolifera* are monogalactosyl diacylglycerols. Their structures showed that there are different in the composition of fatty acid segments [17]. Thus, we think that the inhibitory effects of three glycoglycerolipids correlated with their fatty acid segments. The unsaturated fatty acid had higher inhibitory activity than the saturated fatty acid did. The antialgal actions of the saturated fatty acids were related to the carbon chain length. The longer the carbon chain length was, the less the inhibitive effect was. These results are supported by these experiments by [52], Yin [29], Luo and Zeng [53]. Zhang et al. showed that the inhibitory effect of fatty acids correlated with their chemical structures, the more unsaturated linkages in fatty acid, the stronger the algal growth inhibited; the shorter the carbon chain of fatty acid, the better the algal growth inhibited [54].

Sugar chain of glycolipids can interaction with protein receptors, but also signaling molecules. Glycolipids are the main group of the outer cell membrane, cells provide structural stability and strength. Therefore, when the outer edge of the glycolipid compounds adhesion of algal cells, glycolipids can effectively and cytoplasmic membrane integration, the destruction of the cell membrane of algal cells, thereby affecting algal cell growth regulation, apoptosis, and ultimately affect the growth and proliferation of cells. Further research on the precise mechanism and mode of action of the antialgal glycoglycerolipids.

5. Acknowledgments

This work was supported by National Natural Science Foundation of China (41606120), Special Foundation for Excellent Young Teachers and Principals Program of Jiangsu Province, China, A Project Funded by the Priority Academic Program Development of Jiangsu Higher Education Institutions, and the project of Jiangsu key laboratory of marine biotechnology of Huaihai Institute of Technology of China (HS16004).

6. References

[1] Okai Y, Higashi-Okai K, Nakamura S, Yano Y and Otani S 1994 *Cancer Lett.* **87**(1): 25-32
[2] Lee J M, You S G and Kim S M 2005 *J. Korean Soc. Food Sci. Nutr.* **34**(8):1124-1129
[3] Okai Y and Higashi-Okai K 1997 *Int. J. lmmunopharmac.* **19**(6): 355-358
[4] Hellio C, De L B D, Dufosse L, Le Gal Y and Bourgougnon N 2001. *Mar. Environ. Researh.* 52(3): 231-247
[5] Hou X 2013. *The study on chemical components and anti- hepatic injury activity of n-butanol extracts from Enteromorpha prolifera* (Press: Ocean University of China) (in Chinese)
[6] Hou J N 2016 *The preliminary study of the allelopathic substances of Enteromorpha on three HABs* (Press: Ocean University of China) (in Chinese)
[7] Hudson J B, Kim J H, Lee M K, Dewreede R E and Hong Y K 1999 *J. Appl. Phycol.* **10**: 427- 434
[8] Wang C, Yu R C and Zhou M J 2011 *Journal of Fishery Sciences of China.* **18**(1): 202-207 (in Chinese)
[9] Huo YZ, Tian Q T, Xu S N, Wang Y Y, Feng Z H, Fang Y and He P M 2010. *Mar. Environ. Sci.* **29**(4): 496-9 (in Chinese)
[10] Zhang J H, Huo Y Z, Wang Y Y, Jia R, Zhu P, Hu X, Yang J Q, Fang J M and He P M 2011 *Journal of Shanghai Ocean University.* **2**: 211-6 (in Chinese)
[11] Jia Rui, Wu Min, Cai Chuner, Hui Y Z and He P M 2012 *Journal of Fisheries of China.* **36**(4): 562-567.(in Chinese)
]12] Sun Yingying, Shenzhen Xu, Jing Zhang, Yan B L and Wang C H 2012 *Mar. Sci. Bull.* **31**(4): 433-440 (in Chinese)
[13] Cui F 2014 *Studies on allelopathy effects of Ulva prolifera on red tide microalgae and allelochemicals identification* (Press: Shanghai Ocean University)
[14] Sun Y Y, Liu X X and Wang C H 2010 *Environ. Sci.* **31**(6): 1662-1669 (in Chinese)
[15] Cui F, Tu W B, Wang Y B, Wu H L, Zhang J H, Huo Y Z, Jia R and He P M 2014 *Journal of Tropical Oceanography.* **33**(5): 28-34 (in Chinese)
[16] Sun Y Y, Dong X K, Li G C, Yan B L and Wang C H 2014 *Environ. Sci.* **38**(6): 53-59 (in Chinese)
[17] Sun Y Y, Wang H, Guo G L, Pu Y F, Yan B L and Wang C H 2016 *Environ. Sci. Pollut. R.* **23**(2): 1449-1459
[18] Jung J H, Lee H and Kang S S 1996 *Phytoehemistry.* **42**(2): 447-452
[19] Gao Z, Ali Z and Khan I A 2008 *Phytoehemistry.* **69**(16): 2856-2861
[20] Rho M C, Matsunaga K, Yasuda K and Ohizumi Y 1996 *J. Nat. Prod.* **59**(3): 308-309
[21] Parrish C C, Bodennec G and Gentien P 1998 *Phytochemistry.* **47**(5):783-787
[22] Ali M S, Saleem M, Ahmad V U and Shameel S 2001 *Zeitschrift füer Naturforschung, B: Chemical Sciences.* **56**(8): 837-841
[23] Cui Z, Li Y H, Liu H B, Yuan D and Lu B R 2001 *J. Asian Nat. Prod. Res.* **3**(2): 117-122
[24] Tang H F, Yi Y H, Yao X S, Wu J H, Zhang S Y and Xu Q Z 2002 *China Journal of Chinese Materia Medica.* **27**(4): 269-273 (in Chinese)
[25] Arunkumar K, Selvapalam N and Rengasamy R 2005 *Bot. Mar.* **48**: 441
[26] Hossain Z, Kurihara H, Hosokawa M and Takahashi K 2005 *In vitro cell Dev-An.* **41**(5): 154- 159
[27] Kim Y H, Kim E H, Lee C, Kin M H and Rho J R 2007 *Lipids.* **42**(4): 395-399
[28] Sun J, Shi D Y, Li S, Han LJ, Fan X, Yang Y C and Shi J G 2007 *J. Asian Nat. Prod. Res.* **9**(8): 725-734
[29] Yin S W, Shi Y P, Li X M and Wang B G 2007 *Natural Product Research and Development.* **19**(6): 44-46 (in Chinese)
[30] Wang W, Li H Y, Wang Y Y, Xia Y, Xia X, Okada Y and Okuyama T 2008 *Chinese Traditional and Herbal Drugs.* **39**(5): 657-661 (in Chinese)
[31] Wu W H, Hasumi K J, Hui P and Bin B 2009 *Mar. Drugs.* **7**(2): 85-94
[32] Plouguerné E, Ioannou E, Georgantea P, Vagias C, Roussis V, Hellio C, Kraffe E and Stiger P V 2010 *Mar. Biotechnol.* **12**(1): 52-61
[33] Jin J, Shao C L, Cui Y D, Guan H S, Wei Y X and Wang C Y 2011 *Periodical of Ocean University of China.* **41**(5): 369-373(in Chinese)

[34] Tang H F, Yi Y H, Yao X S, Zhang S Y, Zhou Z R and Li L 2002 *Chinese Journal of Marine Drugs*. **5**: 5-9 (in Chinese)

[35] Ohta K, Mizushina Y, Hirata N, Takemura M, Sugawara F, Matsukage A, Yoshida S and Sakaguchi K 1998 *Chem. Pharm. Bull.* **46**(4): 684-686

[36] Nakata K, Guo C T, Matsufuji M, Yoshimoto A, Inagaki M, Higuchi R and Suzuki Y 2000 *J. Biochem(Tokyo)*. **127**(2): 191-198

[37] Bruno A, Rossi C, Marcolongo G, Lena A D, Venzo A, Berrie C P and Corda D 2005 *Eur. J. pharmacol.* **524**: 159-168

[38] Lu H M 2011 *Chemical constituents of the seaweed Gracilaria lemaneiformis and their allelopathic effects on Skeletonema costatum* (Press: Jinan Univesity)

[39] Marklund S, Marklun G 1974 *Eur. J. Biochem.* **47**: 169-174

[40] OECD 1984. *Alga growth inhibition test. Test Guideline No. 201. OECD Guidelines for Testing of Chemicals* (Paris: Organization for Economic Cooperation and Development)

[41] Xu W X, Zheng D B, She G Z and Lin H 1992 *Journal of Shantou University (Natural Science)*. **7**(2): 80-84 (in Chinese)

[42] Sanina N, Kostetsky E Y, Shnyrov V L and Bogdanov M 2012 *Biochimie*. **94**: 1048-1056

[43] Makewicz A, Gribi C and Eichenberger W 1997 *Plant Cell Physiol.* **38**: 952-60

[44] Deal M S, Hay M E, Wilson D and Fenical W 2003 *Oecologia*. **136**: 107-114

[45] Meng L Q, Ma B, Zhou B H and Zeng L M 2008 *Acta Scientiarum Naturalium Universitatis Sunyatseni*. **47**(3): 137-9 (in Chinese)

[46] Dembitsky V M, Pechenkina-Schubina E E and Rozentsvet O A 1991 *Phytochemistry*. **30**: 2279-2283

[47] Yang J P, Xu S H and Guo S H 2005 *Chinese Journal of Spectroscopy Laboratory*. **22**(2): 440-442 (in Chinese)

[48] Al-Fadhli A, Wahidulla S and D'Souza L 2006 *Glycobiology*. **16**: 902-915

[49] Song Y Y, Li M, Wu Z Z and Yan X J 2015 *Journal of Chinese Mass Spectrometry Society*. **5**: 425-433 (in Chinese)

[50] Rui W 2011 *Study on the partial chemical constituents and the bioactivities of marine algae Grateloupia turuturu and Caulerpa racemosa* (Press: Jinan University)

[51] Jiang Renwang, Hay M E, Fairchild C R and Kubanek J 2008 *Phytochemistry*. **69**: 2495-2500 [52] Kamaya Y, Kurogi Y and Suzuki K 2003 *Environ. Toxicol.* **1**: 289-294

[53] Luo W F and Zeng R Q 2012 *Journal of Southwest China Normal University (Natural Science Edition)*. **37**(8): 92-95 (in Chinese)

[54] Zhang T T, Zhen C Y, He M, Wu A P and Nie L W 2009 *China Environmental Science*. **29**(3): 274-279 (in Chinese)

Analysis of Iron and Manganese Release Effect in Drinking Water Systems from a New Perspective

Huiyan Tong* and Zhongyue Li

School of Food and Environment, Dalian University of Technology, NO.2 Dagong Road, New District of Liaodong Bay, Panjin City, Liaoning Province, 124221, China.
Email: tonghuiyan@dlut.edu.cn

Abstract. In recent years, with the continuous attention to the quality of drinking water and the intensive research on the corrosion of water distribution systems, the problem of metal ions release has attracted more and more attention by researchers. This paper studies metal ions release in drinking water systems from the perspective of occluded water. It has been found that the concentration of iron and manganese ions in the bulk water increases with the concentration of chloride and sulfate ions when only the effect of occluded water on metal ions release is considered. This result is consistent with previous studies. The concentration of iron and manganese ions in the bulk water does not decrease as the pH increases. At the same time, the presence of pipe scales is also a potential threat to the quality of drinking water. It has been found that the concentration of iron and manganese ions in the bulk water also increases as the amount of pipe scales increases. Fe_3O_4 has an adsorption effect on iron, and a single component pipe scale cannot hinder the migration of manganese in the occluded water.

1. Introduction

All water quality parameters of the finished water meet the health and hygiene requirements for drinking water; however the tap water which the customers exactly used might be contaminated, including an increase in turbidity, chromaticity, iron concentration, bacteria and other water quality problems [1]. The drinking water experiences a long hydraulic retention time (HRT) of storage and transportation in water distribution systems (WDS) [2], therefore, the pipes in WDS as well as the interaction between pipe surfaces and bulk water are crucial to the drinking water quality.

Unlined cast iron pipe is still widely existed in current WDS all over the world, particularly in some less developed and large cities. At present, iron pipes still hold a large proportion in WDS worldwide. According to the surveys in Italy and Beijing, China, in WDS, cast iron, ductile iron and steel pipes account for three-quarters of the total, mainly for water pipes, while service pipes are mainly made of galvanized iron and plastic [3, 4]. This situation is estimated to last for long as the pipe networks are expensive to be upgraded [5]. Internal corrosion of metal pipes is a common problem in WDS.

In order to alleviate the shortage of water resources and over exploitation of groundwater, the implementation of a variety of water sources for comprehensive utilization and even long-distance water transfer across river basins is an important strategy for solving water resources problems in many cities at home and abroad. However, sudden changes in the chemical composition of water, such as pH, SO_4^{2-}, Cl^-, alkalinity, and hardness of the pipe network caused by different water sources, can easily break the balance between the scale-water in WDS, resulting in a large amount of scale components, especially iron corrosion products, to the bulk water [6-9]. As a result, turbidity of water and color increase, causing serious "Yellow Water" problems.

IOP Conf. Series: Materials Science and Engineering **484** (2019) 012058 doi:10.1088/1757-899X/484/1/012058

Corrosion of the pipe network occurs on the inner surface of the pipe, and the solution medium in contact with it is the solution inside the pipe scale. This solution is occluded water. There is a great difference between occluded water and bulk water. The occluded water is weakly acidic and contains a lot of ions such as iron, manganese, chlorine and sulfate [10]. The influence of different factors on the release of metal ions is analyzed for bulk water, but their effect on the occluded water is not considered. A large amount of iron and manganese ions, which are abundant in occluded water, are likely to migrate among occluded water, corrosion scale and bulk water under certain conditions. Iron and manganese ions are thus released into the bulk water and lead to secondary pollution of drinking water. Therefore, it is very necessary to introduce the occluded water into the study of secondary pollution of drinking water. This work sought to identify and evaluate the change of metal ions in bulk water and occluded water under different conditions of water quality (sulfate, chloride, pH), hydraulic conditions (flowing and stationary), and scale component.

2. Experiment Set-up and Operation

A detailed schematic of the simulated occluded cell is shown in figure 1. The volume of the occluded water is 8 mL and the volume of the bulk water area is 110 mL. In the experiment, the anode electrode material is ductile iron. In addition to the scale composition experiment, the simulation corrosion scale of all experiments is a mixture of Fe_2O_3, Fe_3O_4, $Fe(OH)_3$ in the proportion of 6.5:3.5:1.

Experiment 1: The cathode working electrode is a glassy carbon electrode, and the anode working electrode material is ductile iron. Before the experiment, the surface of the anode working electrode was sanded from No. 360 to No. 2000, and then rinsed with ultrapure water, then the oil was removed with absolute ethanol. The experiment was operated under the condition of the flow velocity of 400 mL/min and the current of 0.1 mA. The water quality parameters of bulk water and occluded water are shown in Table 1. The occluded water was deoxygenated by nitrogen for 2 hours. When adjusting one of the water quality parameters, the other parameters remain unchanged. The experiment ran for 24 hours. The experimental operation of the flow rate on the release of iron and manganese in the occluded water is the same as the above experimental operation. The bulk water flow rate is designed to be 400 mL/min and 0 mL/min respectively.

Experiment 2: According to the purpose of this experiment, quantitative analytical pure chemicals such as Fe_2O_3, Fe_3O_4, $Fe(OH)_3$, $CaCO_3$, SiO_2 and other simulated scales were filled in the container. The electrodes used in the experiment and the operation are the same as above. The peristaltic pump was adjusted to a speed of 25 rpm to allow the solution to flow at a constant rate in the bulk water. Adjust the voltage of the external battery so that the current is maintained at around 0.10 mA. During the experiment, samples were taken at 0 h and 48 h after the operation of the device. The water quality of experiment 2 is shown in Table 2.

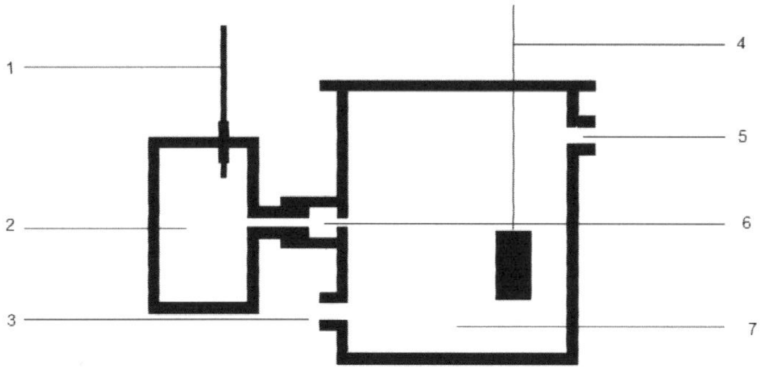

Figure 1. Schematic of a simulated occluded cell: 1) Occluded anode electrode; 2) Occluded cavity; 3) Bulk water inlet; 4) Cathode electrode; 5) Bulk water outlet; 6) Simulated scale area; 7) Bulk water area.

Table 1. Bulk water and occluded water quality parameters in experiment 1

Water quality parameters	Bulk water	Occluded water
pH	7.5	4.1
Fe(mg/L)	0.83	1001.63
Mn(mg/L)	0.01	15.345
Cl^-(mg/L)	89.15	475.20
SO_4^{2-}(mg/L)	185.39	1348.31

Table 2. Bulk water and occluded water quality parameters in experiment 2

Water quality parameters	Bulk water	Occluded water
pH	7.0	4.0
Mg(mg/L)	20.12	15.53
Ca(mg/L)	0.61	0.38
Fe(mg/L)	0.012	1011.71
Mn(mg/L)	0.005	16.35
Cu(mg/L)	0	1.03
Cr(mg/L)	0.002	0.074
Zn(mg/L)	0	0.016
Al(mg/L)	0	0.27
NO_3^-(mg/L)	9.09	0
Cl^-(mg/L)	71.54	552.47
SO_4^{2-}(mg/L)	137.15	1454.66
DO(mg/L)	7.95	≤0.5

3. Experiment Results and Discussion

Experiment 1 designed the effect of three gradients and three experimental factors on the release of iron and manganese in pipelines. The three gradient experimental factors are as follows: rank 1 pH, chloride, sulfate are 6.5, 89.15 mg/L, 185.39 mg/L respectively, rank 2 pH, chloride, sulfate are 7.5, 169.12 mg/L , 327.95 mg/L respectively, rank 3 pH, chloride, sulfate are 8.5, 228.76 mg/L, 457.31 mg/L respectively. The experiment used the control variable method to study the effect of single factor on the release of iron and manganese from drinking water systems. When one of the factors is changed, the other factors maintain pH, chloride ion and sulfate ion respectively of 7.5, 89.15 mg/L and 185.39 mg/L. The release of iron and manganese ions is substantially proportional to the chloride and sulfate ions. In this experiment, the effect of pH on the release of iron and manganese in the pipeline is different from that of previous studies. The iron and manganese released at a pH of 7.5 are the most, and the release amount of iron and manganese when the bulk water at a pH of 8.5 is higher than the release amount of iron and manganese when the bulk water at a pH of 6.5 (Figure 2). This indicates that iron and manganese are more easily released from the occluded water and the scale to the bulk water when the initial pH of the bulk water is 8.5. The influence of water quality parameters in the bulk water on the concentration of iron ions in the occluded area is similar to that of the bulk water. The iron in the occluded area increases with the concentration of chloride and sulfate. First, the concentration of manganese ions in the occluded water is the highest at pH 7.5, and the concentration of manganese ions in the other two gradients is substantially equal (Figure 2). The manganese ion in the occluded area was the smallest at a chloride ion concentration of 130 mg/L and the largest at 65 mg/L (Figure 2). It can be seen from figure 3 that the concentration of iron and manganese in the bulk water in the flowing state is higher than the concentration of iron and manganese in the stagnant state. This may be due to the violent water disturbance in the flow state causing the release of iron and manganese from the scale and occluded water into the bulk water. The concentration of iron and manganese in the occluded water has little to do with the state of the water flow. The effect of sulfate ion on the manganese ion in the occluded area is opposite to that of the bulk water. As the concentration of sulfate ion increases, the manganese ion gradually decreases. The influence of chloride and sulfate on the manganese ion in the occluded area is inconsistent with previous studies.

IOP Conf. Series: Materials Science and Engineering **484** (2019) 012058 doi:10.1088/1757-899X/484/1/012058

Experiment 2 designed the effect of different scales on iron and manganese in the occluded water. Iron and manganese in the occluded area increase as the amount of pipe scale increases (Figure 4.). This indicates that part of the iron and manganese in the occluded area may come from the dissolution of the pipe scale. The scale can not only serve as a source of iron and manganese in the bulk water and occluded water, but also the adsorption of the pipe scale can also cause changes in the composition of metal ions in the bulk water and the occluded water. We designed the following sets of experiments to study the effects of different scale components on bulk water and occluded water. It can be seen from figure 5(a) that in the bulk water iron ion concentration is the highest in the Fe_2O_3 group, followed by SiO_2, $CaCO_3$, Fe_3O_4 and $Fe(OH)_3$. For the metal pipe scale, the iron in the Fe_2O_3 is more easily released into the bulk water, and the iron ions in the Fe_3O_4 and the $Fe(OH)_3$ are not easily dissolved and released into the bulk water under the experimental conditions. It can also be considered that in the theory of occluded cells, the presence of pipe scale plays a role in the selective permeability, and in the scale, $Fe(OH)_3$ and Fe_3O_4 play a major hindrance to avoid iron release in the occluded water. Manganese is released directly into the bulk water. However, SiO_2 and $CaCO_3$ release have no obstruction effect on the metal ion, and the iron and manganese in the occluded water are directly released into the bulk water, so the concentration of iron and manganese in the bulk water is higher. It can be seen from figure 5(b) that the concentration of iron ions in the occluded area of the five experimental experiments is increased. It can be judged that the iron ions in the occluded area are mainly from the corrosion of the ductile iron, rather than the dissolution of iron in the pipe scale and released to the occluded area. From the figure 5(c), we can see that under different conditions of scale composition, the concentrations of manganese ions in the bulk water after 48h are all increased, which means that the manganese in the main water is not mainly from the scale. It can be seen from figure 5(d) that under different pipe scale conditions, the manganese ion concentrations in the occluded area of the five groups of experiments are reduced, because that a large amount of manganese in the occluded area is released into the bulk water. This also shows that the source of manganese in the bulk water is mainly occluded water.

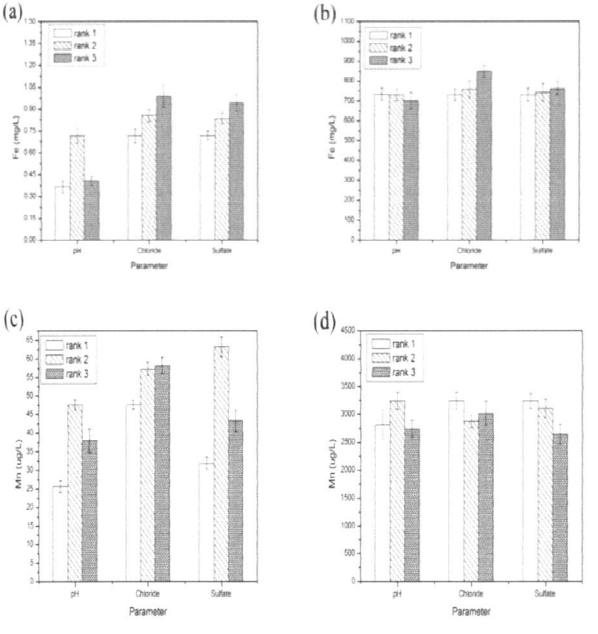

Figure 2. Metal ion concentration under different water quality parameters and ranks (a) Iron ion concentration in bulk water, (b) Iron ion concentration in occluded water, (c) Manganese ion concentration in bulk water, (d) Manganese ion concentration in occluded water.

IOP Conf. Series: Materials Science and Engineering **484** (2019) 012058 doi:10.1088/1757-899X/484/1/012058

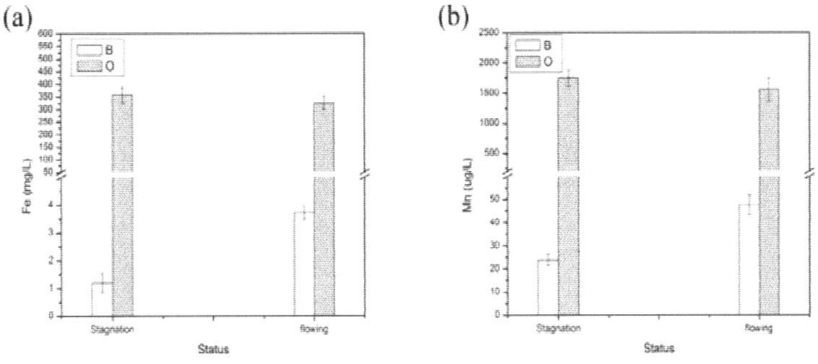

Figure 3. (a) Concentration of iron ions in bulk water and occluded water under different flow conditions, (b) Concentration of manganese ions in bulk water and occluded water under different flow conditions. B-bulk water, O-occluded water.

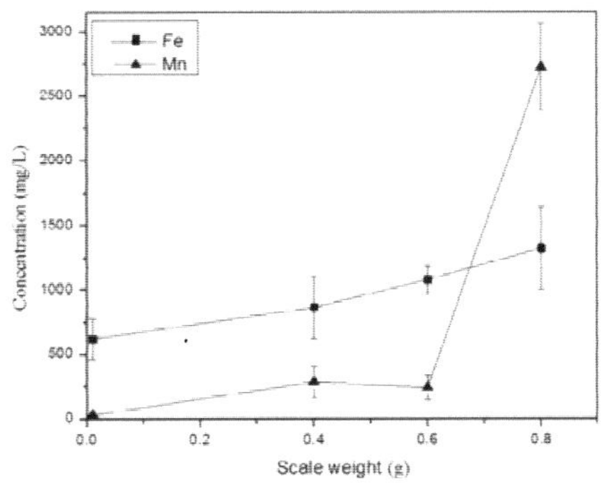

Figure 4. Effect of different scale content on iron and manganese concentration in occluded water.

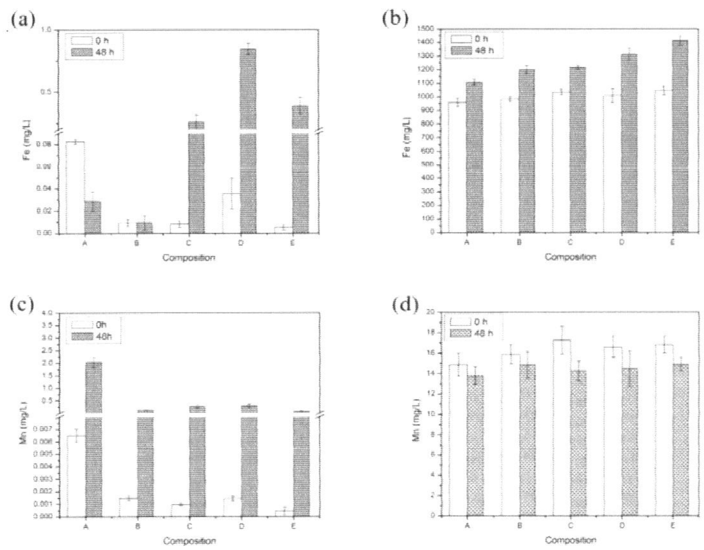

Figure 5. Effect of different scale components on iron (a, b) and manganese (c, d) ion concentration in bulk water (a, c) and occluded water (b, d). A-Fe_3O_4, B-$Fe(OH)_3$, C-$CaCO_3$, D-Fe_2O_3, E-SiO_2.

4. Conculsions

This paper cites a special area of occluded water to study the effects of various parameters on iron and manganese release from the perspective of occluded water. In this paper, the effects of water quality parameters, hydraulic parameters and the composition and content of pipe scale on the concentration of iron and manganese ions in occluded water and bulk water were studied. It was found that in the three states of pH 6.5, 7.5, and 8.5, when the pH of the bulk water was 7.5, the corrosion rate of the cast iron electrode and the release rate of iron and manganese reached the maximum at the same time. The flow rate has a certain influence on the release of manganese, and the flow state is more likely to cause the release of iron and manganese than the stagnant state. Within a certain concentration range, SO_4^{2-} and Cl^- in the bulk water promote the release of iron and manganese. The concentration of iron and manganese ions in the occluded water increases with the increase of the pipe scale content. When the amount of pipe scale increases to 0.8g, the concentration of iron and manganese ions in the occluded water is the largest, which also proves that some of the iron and manganese in the occluded water come from dissolving of the scale. In the bulk water iron content of the Fe_3O_4 group is lowered due to the adsorption of the scale, and other scale components do not act as adsorption. The concentration of manganese ions in the five scales of $Fe(OH)_3$, Fe_3O_4, Fe_2O_3, $CaCO_3$ and SiO_2 increased, indicating that a certain kind of pipe scales cannot hinder the migration of manganese in occluded water. In order to control the content of iron and manganese in drinking water distribution systems, this paper proposes control measures from the perspective of occluded water. Reducing the hydraulic disturbance, pH, sulfate and chloride ion concentration is beneficial to reduce the metal concentration in the bulk water.

5. Acknowledgments

This work was supported by the National Natural Science Foundation of China (No. 51608093) and "the Fundamental Research Funds for the Central Universities" (DUT18RC (4)062).

6. References

[1] X. Cheng, E.D. Peterkin, G.A. Burlingame, Study of the effect of DMSO on VOS odour production in a wastewater plant, Water Science & Technology A Journal of the International Association on Water Pollution Research, 55 2007 327-333.
[2] M. Li, Z. Liu, Y. Chen, Y. Hai, Characteristics of iron corrosion scales and water quality variations in drinking water distribution systems of different pipe materials, Water Research, 106 2016 593-603.
[3] L.I. Zhen-Hai, R. Zhao, Q.M. Zhu, Primary Environmental Impact and Countermeasures of the Beijing Section of the Middle Route of the South-to-North Water Transfer Project, South-to-North Water Transfers and Water Science & Technology, 8 2010 19-23.
[4] E. Veschetti, L. Achene, E. Ferretti, L. Lucentini, G. Citti, M. Ottaviani, Migration of trace metals in Italian drinking waters from distribution networks, Toxicological & Environmental Chemistry Reviews, 92 2010 521-535.
[5] C. Davies, L.F. Dan, P.C. Hertzler, R.T. Jones, USEPA's infrastructure needs survey, Journal, 89 1997 30-38.
[6] P. Sarin, J.A. Clement, V.L. Snoeyink, W.M. Kriven, Iron Release from corroded, unlined cast-iron pipe, Journal, 95 2003 85-96.
[7] B. Shi, J.S. Taylor, Iron and copper release in drinking-water distribution systems, Journal of Environmental Health, 70 2007 29-36, 44, 46.
[8] Y. Xie, D.E. Giammar, Effects of flow and water chemistry on lead release rates from pipe scales, Water Research, 45 2011 6525-6534.
[9] H. Sun, B. Shi, F. Yang, D. Wang, Effects of sulfate on heavy metal release from iron corrosion scales in drinking water distribution system, Water Research, 114 2017 69-77.
[10] H. Tong, P. Zhao, H. Zhang, Y. Tian, X. Chen, W. Zhao, M. Li, Identification and characterization of steady and occluded water in drinking water distribution systems, Chemosphere, 119 2015 1141-1147.

IOP Conf. Series: Materials Science and Engineering **484** (2019) 012059 doi:10.1088/1757-899X/484/1/012059

Study on the Existence Form and Removal of Boron Acid

Rongrong He, Tao Yu, Jiajia Du, Chengtun Qu*

Shaanxi Oil and Gas Pollution Control and Reservoir Protection Key Laboratory, Xi'an Shiyou University, 710065, Xi'an
E-mail: 1805762285@qq.com

Abstract. Boron, as a microelement, in the form of boric acid, borate, borosilicate in aqueous solution, has advantages and disadvantages for human production and life. When the concentration of boron is too high, it will affect the surrounding environment and biological survival. The chemical precipitation, adsorption (mainly resin method) and membrane separation method are the main methods for removing boron from sewage. The paper introduces the existence form of boron acid and the principle of these methods, but with the development of science and technology, we need to upgrade and optimize the original method to achieve efficient, energy-saving, environmentally friendly environmental governance.

1. Introduction

There are three kinds of forms, boric acid, borate and borosilicate, of boron existed in nature. Meanwhile, there is no elemental boron. Boron plays an important role in human's production and life. It is widely used in steel and alloy industry, chemical industry, agricultural and other fields [1-2]. However, with the rapid development of boron-related industries, the boron content of environment increased dramatically, which caused serious environmental pollution and waste of resources. In addition, the increasing of boron in environment affected human health directly or indirectly [3]. For example, the boron content of fracturing fluid must be controlled when the fracturing fluid is reused in the oil field, otherwise the reaction of crosslinking of boron with guar will occur, the excessive crosslinking ratio will lead to the block of the liquid distribution pipeline [4]. Furthermore, long-term exposure in atmosphere of boron-containing substances can damage human's nervous system, reproductive system, and even cause death [5]. However, the regulations on boron content vary from different regions of different countries. "Standards for Drinking Water Hygiene" of China strictly stipulates that the boron content in drinking water is less than 0.5 mg/L, the maximum value of boron concentration is 1 mg/L in South Korea, Japan, Britain and other countries. And thus, how to effectively remove boron from water medium becomes an inevitable problem. In this paper, the existence form of boric acid in solution is mainly introduced. In addition, various boron acid removal methods such as chemical precipitation, adsorption (especially resin) and membrane separation have been compiled.

2. The Existence Form of Boric Acid in Solution

The form of boric acid in solution can be illustrated by the following chemical equation:

$$H_3BO_3 + H_2O \rightleftharpoons [B(OH)_4]^- + H^+ \tag{1}$$

It can be seen that $[B(OH)_4]^-$ is stable in an alkaline environment, and H_3BO_3 is dominant under acidic conditions. Theoretically, when the pH is 12, the molar ratio of $[B(OH)_4]^-$ and H_3BO_3 in the solution is $10^{2.76}$:1, and pH decreasing, the equilibrium proceeds to the left. When the pH is 6, the molar ratio of $[B(OH)_4]^-$ and H_3BO_3 in the solution becomes 1:$10^{3.24}$, at which time the boron in the solution exists almost in the form of H_3BO_3.[4,6]

In the solution, when the concentration of boric acid is below 1500 mg/L, it mainly exists in the form of H_3BO_3 and $[B(OH)_4]^-$. From equation (1), it is known that the lower the pH, in the form of boron is mainly H_3BO_3, $[B(OH)_4]^-$ is very low. With the pH value increases, the $[B(OH)_4]^-$ begins to increase and the H_3BO_3 begins to decrease, but the pH=9.4, the $[B(OH)_4]^-$ accounts for 80%; when the boric acid concentration is above 1500 mg/L, a variety of polyborate ions can be formed, such as $[B_3O_3(OH)_4]^-$, $[B_4O_5(OH)_4]^{2-}$, $[B_3O_3(OH)_5]^{2-}$ and $[B_5O_6(OH)_4]^-$, etc. While the presence of boric acid in aqueous solution, the tetrahedral configuration of $[B(OH)4]^-$ allows it to form a stable complex with the ortho-cis hydroxy group, which provides the way to remove boron acid from solution[7].

3. The Method of Boron Acid Removal

3.1. Chemical Precipitation

In the 1970s and 1980s, some researchers studied the removal of boron acid by a series of inorganic and organic precipitants [8]. The principle of boron removal by chemical precipitation is to convert boron in solution into borate precipitate or boric acid with low solubility. The precipitant is divided into organic precipitant (polyvinyl alcohol, hydroxyl carboxylic acid, etc.) and inorganic precipitant (oxides and hydroxides containing elements of calcium, magnesium, iron and aluminum). Because of the difference of acid properties of precipitant, the precipitation method can be divided into acid precipitation and alkali precipitation [9].

The acid precipitation is based on the principle that the solubility of H_3BO_3 in the inorganic acid is small. So hydrochloric acid or sulfuric acid was added into the solution to convert boron into boric acid with lower solubility. Xiao et al [10] used hydrochloric acid as precipitant to achieve preliminary boron removal from brine. The alkali precipitation means that the metal oxide reacts with boron acid in solution to form an insoluble borate precipitate under weak alkaline conditions. For example, ferrous hydroxide can react with boron acid to form boron-ferrate that is poorly soluble in water [11]. In addition, Lu et al.[12] removed boron acid from water medium by the mixture of calcined product obtained from magnesite with ammonium chloride under heating condition. The removal rate of boron was more than 90%. Remy et al.[13] used calcium hydroxide powder to remove boron from wastewater. When the content of calcium hydroxide powder was 50 g/L, the temperature was 90 °C, and the treatment time was 2 h, the boron removal was better, and the boron acid content was reduced from 700 mg/L to 50 mg/L. Complexes can be formed between hydroxyl-carboxylic acids and boric acid. The stoichiometry of the complex is dependent on the number of groups (hydrous or carboxylic) available for binding to boron, the pH value of the solution, the relative concentrations of ligand and borate, and the mode of preparation [14]. The chemical precipitation is mostly used for the removal of high concentration boron. The raw materials of precipitant are easy to be obtained and its cost is low. However, the large consumption of the precipitating agent produces a large amount of sludge, which increases the difficulty of subsequent processing.

3.2. Adsorption

The adsorption utilizes different kinds of porous adsorbents to remove boron from the solution. Guan et al.[15] summarized several adsorbents, including chelating resins, activated carbon, fly ash and industrial waste, oxides and hydroxides, and layered double hydroxide et al, for removing boron. Li et al.[16] studied the adsorption properties of layered double hydroxide Mg/Al-LDO towards boron and explored the effects of solution pH, adsorbent dosage, adsorption time and adsorption temperature on adsorption properties of boron. The results showed that the removal of boron was 94.47% and the adsorption amount was 47.24 mg/g. The amount of adsorbent of 0.20 g, optimum pH of 9.5, initial boron concentration of 100 mg/L, reaction time of 100 minutes and solution temperature of 303 K were kept constant in the experiment.

The remove principle of resin towards boron concludes complexation and adsorption. The complexation reaction is the complexation of polyhydroxy functional group with $[B(OH)_4]^-$ and acidic polyhydroxy functional group with H_3BO_3[5,17]. So the resin can also be used as a special adsorption method. Yu et al [18] introduced research status of ion exchange resin of boron from meglumine, polyol, saccharide, polystyrene, polyacrylate, inorganic-organic composite resin and natural polymer and also summarized the application of boron ion exchange resin in water treatment, environmental protection, extraction and separation. Joanna [19] modified the Purolite A170 resin into GLY-resin by using glycerol. The GLY-resin was applied to the removal of oxoborate in solution and post-crystallization lye. When pH of solution was 9.5, the maximum adsorption capacity was 1.6 mg/g and the maximum removal rate was 98.5%. The adsorption is also mostly used for the removal of high concentration of boron. Although the effect of adsorption is relatively high, the disadvantages, large consumption of adsorbent and difficulty in regeneration, may cause the expensive cost of boron removal. The resin is mostly suitable for the removal of low concentration boron. Due to its high selectivity and high adsorption, most studies concentrate this field of boron removal.

3.3. Membrane Separation

The membrane separation is that the raw material components are separated selectively through a separation medium of selective semipermeable membrane and the driving force (such as concentration difference, pressure difference, potential difference) on both sides of semipermeable membrane. The membrane separation includes reverse osmosis, forward osmosis, nanofiltration and electrodialysis.

3.3.1. Reverse osmosis. Reverse osmosis is to apply an external force on the high concentration side, so that the high concentration zone penetrates into the low concentration zone through the selective semipermeable membrane under the action of external force. Wu et al[20] introduced the principle and application status of removing boron about reverse osmosis process at domestic and abroad, and the removal characteristics of boron and scope of application also were summarized, concluding SWRO (conventional seawater reverse osmosis), SWRO+BWRO (brackish water reverse osmosis), SWRO+BWRO-BSR (boron selection resin) and SWRO+EDR (electrodialysis). However, reverse osmosis is easily affected by the pH of the solution, the applied pressure, and the species of the membrane. D. Prats et al[21] studied the effect of pH and pressure on the boron removal when the conventional reverse osmosis membrane was used into the treatment of brackish water. The results showed that the removal of boron was about 50% when pH was below 9.5; but when pH was 10.5, the removal rate of boron is almost 100%, and the boron removal increases with the increase of pressure.

3.3.2. Forward Osmosis. Forward osmosis is that the solution from a low concentration zone automatically permeates to a high concentration zone through selective semipermeable membrane. It is the advantages that are no external force addition, low energy consumption, low environmental pollution and high water recovery rate compared with the reverse osmosis. So forward osmosis has been attracted much attention in recent years. Ding et al [22] analyzed the difference between forward and reverse osmosis technology and introduced the application of forward osmosis in the fields of energy recovery and wastewater treatment, seawater desalination. Meanwhile, Ding also pointed out the defects of forward permeation in membrane performance and extracting liquid. This method is also affected by the pH of the solution on removing boron. W. Fam et al. [23] utilized this method to remove boron acid. Under neutral conditions, the boron removal was only 45%; however, the rate of boron removal was 94% when the pH value of solution was 11, because boron in the solution contains larger-size borate species, and thus increases boron rejection rate up to 94% by electrostatic repulsion.

3.3.3. Nanofiltration. Nanofiltration is a pressure-driven membrane separation technique, whose filtration accuracy is between ultrafiltration and reverse osmosis. It is used for the reduction of hardness and the removal of small molecules matter in solution. Liu [24] studied seawater desalination and its performance of boron removal by the technology of nanofiltration, when the initial concentration of boron was determined as 5.59 mg/L. The results indicated that the final effluent boron concentration is about 1.0 mg/L under the conditions of two-stage nanofiltration and second-stage influent pH 10.3; the final effluent boron concentration is below 0.1 mg/L in the case of two-stage nanofiltration, complexation reaction and second-stage influent pH 10.3.

3.3.4. Electrodialysi.s Because of the selective permeability of the ion exchange membrane, the substance separation can be completed by electrodialysis under the action of an external DC electric field. The ion exchange membrane can be divided into cation exchange membrane that only allows cations to pass through and anion exchange membrane that allows anions to pass through and repel the cations. The boron removal of electrodialysis is mainly affected by voltage, membrane's types, solution pH and initial boron concentration. N. Kabay et al.[25] used TS-l-l0 electrodialyzer and a Neosepta cation and anion exchange membrane to carry out boron removal in boric acid solution. Meanwhile, the effects of pH and boron concentration on boron removal were carried out. When the pH of solution is increased from 9 to 10.5, the boron removal is increased from 20% to 80%. Furthermore, the rate of boron removal increased as the increasing of initial boron concentration when pH value of solution is 10.5. Electrodialysis can be applied difficultly due to the large consumption of electric energy. While the membrane separation is used to remove boron, the boron removal is higher, but it usually needs a higher pH (10-11), which causes an increase in alkalinity and facilitates precipitation of calcium, magnesium, aluminum, and iron and so on in the solution, thereby causing damage to the membrane and causing membrane pollution.

4. Conclusion

In summary: if the boron content is high in solution, the boron content can be reduced by pretreatment such as precipitation and adsorption so that reducing the economic cost in the subsequent treatment, and then the resin or the membrane separation is used to remove boron deeply to meet the requirements of drinking water and industrial water. Due to boron is present in the aqueous solution in the form of H_3BO_3 or $[B(OH)_4]^-$. It can be seen from the foregoing that the introduction of polyhydroxy functional groups on the materials for modification is a thought worthy of further investigation, thereby improving the boron removal and achieving a high-efficiency, energy-saving and environmentally-friendly concept.

5. References

[1] Ran Liu, Xiangxin Xue, Jiang Tao, et al. The Current Situation and Development of Boron and Boride [J]. Materials Review, 2006, 20(6): 1-4.
[2] Chunyan Yan, Xiaochuan Deng, Jianzhi Sun, et al.. The Development of Separation Methods of Boron [J]. Sea-lake Salt and Chemical Industry, 2005, 34(5): 27-30.
[3] Öztürk N, Kavak D. Boron removal from aqueous solutions by adsorption on waste sepiolite and activated waste sepiolite using full factorial design[J]. Adsorption, 2004, 10(3): 245-257.
[4] Renlan Zha. Study on residual boron crosslinker in fracturing backflow fluid [D]. Southwest Petroleum University, 2016.
[5] Ruozhen Hou, P Gu, Xiaozhu Wei, et al. Research progress on boron removal methods in water [D]. CNOOC Tianjin Chemical Research and Design Institute, 2012
[6] W Chen, T Bu, Y Wu, et al. Removal of residual boron in fracturing and returning liquid and reuse of return liquid [J]. Petrochemical Technology s, 2018, 47(1): 67-73.
[7] Kochkodan V, Darwish N B, Hilal N. The Chemistry of Boron in Water[M]// Boron Separation Processes. 2015:35-63.
[8] Xu Y, Jiang J Q. Technologies for boron removal[J]. Industrial & Engineering Chemistry Research, 2008, 47(1): 16-24.
[9] Xiaowei Liu, g TuqiaoZhan, W Shao, et al. Research progress on control of boron in desalinated seawater [C]//Drinking Water Safety Control Technology Conference and Proceedings of the 13th Annual Meeting of Water Supply Committee of China Civil Engineering Society Water Industry Branch 2013.
[10] Jingbo Xiao, Juling Chen. Study on Boron Removal Process of Brine[J]. Henan Chemical Industry, 2013, 30(16): 33-35.
[11] Songhe Wang. A method for removing boron from water bodies [P]. China: CN106554063A, 2017.
[12] Xuchen Lu, Zhimin Zhang. A method for removing boron from water by magnesite [P]. China: CN105293652A, 2016.
[13] Remy P, Muhr H, Plasari E, et al. Removal of boron from wastewater by precipitation of a

sparingly soluble salt[J]. Environmental Progress, 2005, 24(1): 105-110.
[14] Farmer, J. B. The Removal of Borante from Solution the Complexation of Borate by Hydroxycarboxylic Acids; Technical Report TR-74-7; Borax Technical, Ltd.: London, 1974.
[15] Guan Z, Lv J, Bai P, et al. Boron removal from aqueous solutions by adsorption—A review[J]. Desalination, 2016, 383: 29-37.
[16] Xiaoyan Li, Yibao Liu, Weimin Zhang, et al. Study on the performance of boron in layered bimetallic oxide adsorption solution[J]. Journal of Functional Materials, 2017, 48(2): 2020-2025.
[17] Yanzhen Chen, Junwei Ye, Guiling Ning. Progress in the removal, separation and detection of boron in water[J]. Inorganic Salt Industry, 2015, 47(8): 5-9.
[18] Shiping Yu, Hongjie Yao, X Han, et al. Research progress in boron ion exchange resin[J]. Ion Exchange and Adsorption, 2012, 28(4): 375-384.
[19] Kluczka J, Korolewicz T, Zołotajkin M, et al. Boron removal from water and wastewater using new polystyrene-based resin grafted with glycidol [J]. Water Resources and Industry, 2015, 11: 46-57.
[20] Shuibo Wu, Xianhui Pan, Xizhang Chu. Status and progress of reverse osmosis desalination and boron removal[J]. Journal of Environment and Health, 2008, 25(10): 928-930.
[21] Liberman B, Liberman I. Method of boron removal in presence of magnesium ions: U.S. Patent 7,097,769 [P]. 2006-8-29.
[22] L Ding, Liangcheng Shao. Research progress on key technologies of positive osmosis water treatment [J]. Guangzhou Chemical Industry, 2018, 46(1): 1-3.
[23] Fam W, Phuntsho S, Lee J H, et al. Boron transport through polyamide-based thin film composite forward osmosis membranes [J]. Desalination, 2014, 340: 11-17.
[24] J Liu. Study on desalination and deboronation performance of nanofiltration seawater [D]. Tianjin: Tianjin University, 2010.
[25] Kabay N, Arar O, Acar F, et al. Removal of boron from water by electrodialysis: effect of feed characteristics and interfering ions [J]. Desalination, 2008, 223(1-3): 63-72.

IOP Conf. Series: Materials Science and Engineering **484** (2019) 012060 doi:10.1088/1757-899X/484/1/012060

A Review of the Mechanism of Microbial Degradation of Petroleum Pollution

Xiaokang Li, Hong Li and Chengtun Qu*

Shaanxi Oil and Gas Pollution Control and Reservoir Protection Key Laboratory, Xi'an Shiyou University, 710065, Xi'an

E-mail: xianquct@163.com

Abstract. A large amount of petroleum hydrocarbon pollutants will be produced during the oilfield exploitation process. If traditional treatment methods of physical or chemical methods are used, problems such as incomplete treatment and secondary pollution will occur. In order to achieve better treatment effects, the microbial degradation method is currently used to treat petroleum hydrocarbon pollutants. Studying the mechanism of microbial degradation of petroleum hydrocarbon pollutants is conducive to exploring the path of petroleum pollution remediation technology and improving the efficiency of pollution control. In this review, the degradation mechanism of alkanes, alkenes and aromatics is discussed in detail, and the research directions of microbial pollution control are analyzed and discussed.

1. Introduction

Petroleum exploitation, transportation and refining process may pollute the environment, and the leakage of crude oil during mining, transportation and refining is the main cause of environmental pollution. The leakage of crude oil is scattered in soil and water environment. If the scattered crude oil was not treated in time, it will seriously affect the normal life of people in the surrounding area, and also will have a toxic effect on the soil and water environment [1]. The serious harm caused by oil leakage has been widely reported at home and abroad [2-3]. For the treatment of petroleum hydrocarbon pollutants, physical methods, chemical methods and the combination of the two methods have been used widely. However, these methods often cannot completely remove the emulsified oil and dissolved oil in water and soil, and the treatment cost is higher. At the same time, it will cause repeated pollution to the surrounding environment, so physical or chemical methods can only be used in emergency handling [4]. Biological methods were used to degrade petroleum hydrocarbon pollutants by some oil-loving microorganisms in the environment. This method does not cause repeated pollution and has low processing cost. There are many kinds of microorganism that can degrade petroleum hydrocarbon pollutants in nature. By using suitable petroleum degrading microorganisms, some toxic and harmful petroleum hydrocarbon pollutants can be degraded into harmless substances, and even some hydrocarbon substances can be thoroughly degraded [5]. However, the existing micro-biotreatment technology is not efficient in the treatment and remediation of crude oil leakage. In order to improve the degradation efficiency of petroleum degrading bacteria to petroleum hydrocarbon pollutants, we must have an in-depth understanding of the mechanism of petroleum degradation bacteria to degrade petroleum hydrocarbon pollutants, and lay a good foundation for subsequent research and application.

2. Mechanism of Microbial Degradation of Petroleum Pollutants

The degradation of petroleum hydrocarbons by microorganisms is mainly caused by the catalysis of

IOP Conf. Series: Materials Science and Engineering **484** (2019) 012060 doi:10.1088/1757-899X/484/1/012060

intracellular enzymes. The process of microbial degradation of petroleum hydrocarbons has four main steps: First, petroleum pollutants are emulsified by surfactants secreted by microorganisms; Second, the emulsified petroleum hydrocarbon is adsorbed by the surface of the microorganism; Then, the petroleum hydrocarbon adsorbed on the surface of the cell membrane enters the cell membrane through active transport or passive transport, endocytosis. Finally, the petroleum hydrocarbon entering the cell undergoes an enzymatic reaction with the corresponding enzyme to achieve the purpose of degrading the pollutant.

The degradation ability of petroleum hydrocarbons in microorganisms is a critical factor in pollution remediation. The degradation pathway of alkanes is divided into single-end oxidation, double-end oxidation and secondary-end oxidation. These three oxidation modes are implemented in different positions on the carbon chain to form alcohol, aldehyde, acid, acid and CoA. And then the β oxidation is completed to provide energy for microbial life activities [6-7].

The degradation mode of cycloalkane is slightly different with paraffin. First, the cycloalkane is oxidized by the hydroxylase to form a cycloalkanol according to fig 1. The cycloalkanol takes off the hydrogen to produce naphthenic ketone. Under the action of the ketooxygenase, the naphthenic ketone is converted into caprolactone. And then a H_2O molecule was inserted into the ring of caprolactone to break the ring. Whose progress produces the 6-hydroxycaproic acid that is further oxidized to 6-oxohexanoic acid by 6-hydroxycaproate dehydrogenase. The dicarboxylic acid is formed through the oxidization by aldehyde dehydrogenase readily. Finally, the dicarboxylic acid is further oxidized through the tricarboxylic acid cycle to maintain the life activity of the microorganism [8-9]. However, it is also said that the degradation of naphthenes also has the degradation caused by co-metabolism and symbiosis, but this is generally only for the degradation of terpenes, which is more difficult to degrade than alicyclic hydrocarbons.

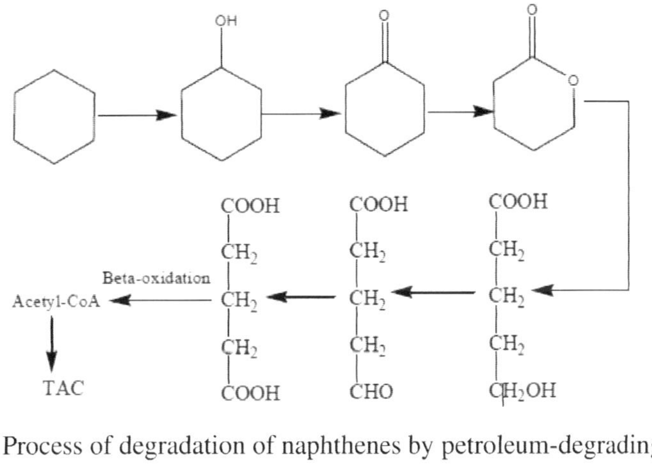

Figure 1. Process of degradation of naphthenes by petroleum-degrading bacteria

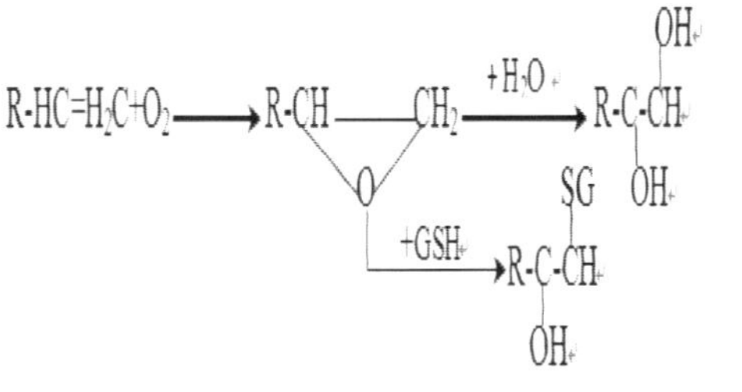

Figure 2. Degradation of olefins by petroleum-degrading bacteria

There is more than one degradation method for olefin degradation. The double bond is reduced to a secondary alcohol, and the secondary alcohol is oxidized to a ketone. The ketone is then esterified by the bacteria to form an ester. The ester bond is then cleaved to form two parts, namely primary alcohol and fatty acid [10]. According to the Figure 2, microorganisms directly oxidize epoxides formed by the addition of olefins to aldehydes or ketones, and then gradually oxidize them to be used by microorganisms without generating intermediate alcohols [11]. There are also some linear paraffins directly dehydrogenated in the microorganism to form olefins, and then further oxidized into alcohols, aldehydes, and acids by the catalytic action of the enzyme to generate energy by β oxidative decomposition.

The degradation of aromatic hydrocarbons is more complicated than other hydrocarbons. The same aromatic hydrocarbons will have different degradation sites due to their different sites of degradation. For example, Arthrobacter and Mycobacterium have more than one degradation pathway for phenanthrene. They are opened at different positions from the phenanthrene ring, and may form products such as 2-hydroxy-1-naphthoic acid or 2, 2-biphenyldioic acid. [12].The degradation of different polycyclic aromatic hydrocarbons is basically the same. Polycyclic aromatic hydrocarbons are degraded. First, a ring of polycyclic aromatic hydrocarbons is oxidized to form a trans-dihydro-dihydroxy compound or catalyzed by a microenzyme in a microorganism. a benzene ring initial oxide of a cis-dihydro-dihydroxy compound, which is dehydrogenated to form a diol epoxide. The epoxidative cleavage of the diol forms an alcohol with less than one ring and produces salicylic acid, catechol and The product such as gentisic acid then continues to oxidize, and the falling chain hydrocarbon undergoes an alkane oxidation pathway to form a carboxylic acid, which reacts with CoA to form acetyl-CoA, which provides energy for self-metabolism through the TAC cycle [13-14].For polycyclic aromatic hydrocarbons with more than four rings, as the number of benzene rings increases, it is difficult to be degraded by microorganisms, and can only be oxidatively degraded by co-metabolism. Complex polycyclic aromatic hydrocarbons are also opened and produced in various ways. Intermediate products such as trans-diols, phenols, naphthalenes, and epoxides are all toxic to microorganisms, so existing biotechnology can only be applied to the degradation of low-concentration polycyclic aromatic hydrocarbons.

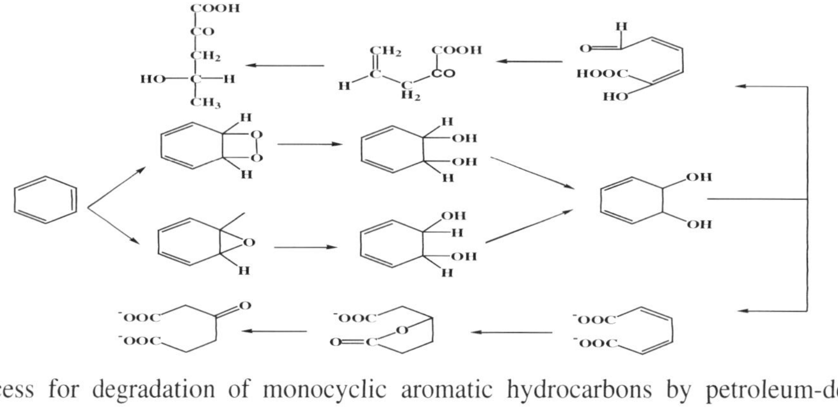

Figure 3. Process for degradation of monocyclic aromatic hydrocarbons by petroleum-degrading bacteria

3. Summary

In recent years, the state has strongly advocated the use of clean energy, but the use of diesel, gasoline and petroleum-based products has not decreased, so the amount of oil pollutants is also high, which has a negative impact on people's production, life and environmental protection. .As the country pays more and more attention to pollution control, the use of high-efficiency, no secondary pollution, economic microbial combined treatment technology has become the main means of petroleum degradation, and it is constantly developing, but due to the limitations of microbial characteristics, biotechnology There are also many drawbacks .For example, the characteristics of refractory degradation of high-concentration petroleum pollutants cannot be repaired by microorganisms, and the microorganisms themselves are susceptible to environmental influences, and the degradation effect will be greatly reduced with slight changes.

(1) Construct genetically engineered bacteria with high efficiency and environmental adaptability.

Bioengineering methods, such as genetic recombination, protein modification, and molecular integration techniques, are used to completely renovate petroleum-degrading bacteria to construct ideal strains for the specialization of microbial pollution control.

(2) Bio-enhancement and biostimulation of complex petroleum-degrading bacteria. Further degradation of petroleum contaminants is achieved by adding appropriate nutrients and changing F/M values or by adding exotic microorganisms to stimulate the oil-degrading flora.

(3) Optimization and combination of processes. Design a more complete and sophisticated treatment process and combine with the original treatment process to reduce the degradation cycle of petroleum degrading bacteria and increase the degradation efficiency. Strengthening the optimization and innovation of the petroleum refining process, such as bio-electrochemical treatment combined with biological and electrochemical methods to treat petroleum pollution, can not only improve the degradation efficiency, but also convert petroleum pollutants into electricity and hydrogen for people to use.

(4) Improve the environmental risk level and evaluation system of bioremediation technology, reduce the unnecessary risks arising in the process of governance, and formulate the treatment process in the appropriate environment.

4. Reference

[1] Zhang Lei, Zhao Qi, Wu Weinan, et al. Current status and prospects of bioremediation technology for petroleum contaminated soils [J]. Modernization, 2018, 38 (01): 18-22.

[2] China Pipeline Business Network official website. Colombian oil company 600 barrels of crude oil leakage - resulting in 70 households suffering. [R/OL]. (2014-4-11) [2018-03]. https://www.china5e.com/m/news/news-1025182-1.html.

[3] Wu Jiayi. Analysis of Compensation System and Administrative Punishment Related to Marine Pollution-Taking Kangfei Case as an Example[J]. Chinese civil business, 2018 (1).

[4]O'Brien P L, Desutter T M, Casey F X M, et al. Evaluation of Soil Function Following Remediation of Petroleum Hydrocarbons—a Review of Current Remediation Techniques[J]. Current Pollution Reports, 2017, 3(3):192-205.

[5] Sakthipriya N, Doble M, Sangwai J S, et al. Bioremediation of Coastal and Marine Pollution due to Crude Oil Using a Microorganism Bacillus subtilis[J]. Procedia Engineering, 2015: 213-220.

[6] Varjani S J. Microbial degradation of petroleum hydrocarbons[J]. Bioresource Technology, 2017: 277-286.

[7] Wentzel A, Ellingsen T E, Kotlar H, et al. Bacterial metabolism of long-chain n-alkanes[J]. Applied Microbiology and Biotechnology, 2007, 76(6): 1209-1221.

[8] Song Zhiwen, Xia Wenxiang, Cao Jun et al. Microbial degradation and bioremediation of marine petroleum pollutants[J]., 2004, 23(03): 99-102.

[9] Comandini A, Dubois T, Abid S, et al. Comparative Study on Cyclohexane and Decalin Oxidation[J]. Energy & Fuels, 2014, 28(1): 714-724.

[10] Yang Liqin, Jiang Jihui. Degradation mechanism of petroleum hydrocarbons by microorganisms[J]. Environmental Protection of Oil & Gas Fields, 2011, 21(02): 24-26+61.

[11] Oelschlagel M, Groning J A, Tischler D, et al. Styrene Oxide Isomerase of Rhodococcus opacus 1CP, a Highly Stable and Considerably Active Enzyme[J]. Applied and Environmental Microbiology, 2012, 78(12): 4330-4337.

[12] Seo J, Keum Y, Hu Y, et al. Phenanthrene degradation in Arthrobacter sp. P1-1: initial 1,2-, 3,4- and 9,10-dioxygenation, and meta- and ortho-cleavages of naphthalene-1,2-diol after its formation from naphthalene-1,2-dicarboxylic acid and hydroxyl naphthoic acids.[J]. Chemosphere, 2006, 65(11): 2388-2394.

[13] Maigari A U, Maigari M U. Microbial metabolism of polycyclic aromatic hydrocarbons (PAHs): a review[J]. Int J Sci Eng, 2015, 6: 1449-1459.

[14] He Jingjing. Analysis of Microbial Diversity and Polycyclic Aromatic Hydrocarbon Degrading Bacteria in Petroleum Hydrocarbon Contaminated Soil [D]. Urumqi: Xinjiang Agricultural University, 2008.

IOP Conf. Series: Materials Science and Engineering **484** (2019) 012061 doi:10.1088/1757-899X/484/1/012061

Brief Analysis of Seasonal Correlations in the Power Generation Data of Photovoltaic Power Stations

Weihao Yu

Department of Electrical Engineering, North China Electric Power University , School of Electrical and Electronic Engineering, room 8A111, No.2, Beinong Road, Changping District, Beijing, P.R.China..

Email: ywh0131@163.com

Abstract. Correlations in the power generation of photovoltaic (PV) power stations could improve the PV power utilization in electric power systems and reduce the occurrence of PV power dumping and throttling. In this study, the seasonal correlations in the PV power generation data of the Hanergy Xinhe PV power station in Xinjiang are investigated from a temporal perspective. Firstly, the PV output curves of the power station in typical weather conditions, i.e., the sunny day and non-sunny days, are analyzed. And then, after the presentation for the basic description of correlation theory, the correction of PV generation in the viewpoint of season is presented. The resulted with the field data shows that, for the sunny days, the correlation between autumn and spring data is much stronger than the correlation between winter and summer data, and for the non-sunny days, the correlation may be small.

1. Introduction

In the face of fossil fuel depletion, environmental pollution, and climate change, renewable energies have become an important part of global sustainable energy strategies. Photovoltaic (PV) power generation draws from an abundant source of energy, i.e., the sun, and is also relatively cheap and efficient. With the maturation of PV technologies, PV power generation is now expanding at a breakneck pace.

In China, PV power generation has reached 106.9 billion kWh, with a year-on-year growth of 72%. The annual PV power generation of China has already exceeded 100 billion kWh as of January 2, 2018. However, the randomness and intermittency of PV power generation induces large-scale fluctuations in power. These fluctuations significantly alter the distribution of power in electric power systems and thus have a significant impact on load regulation and frequency modulation. These effects will threaten the safety of the electric power system and subsequently lead to PV power dumping and throttling.

The PV power generation characteristics of different PV power stations are similar as they are spatiotemporally related to each other. These correlations could help electric power systems take pre-emptive measures to minimize PV power dumping and throttling. There are numerous studies about the correlations of PV power generation.

Studies on the correlations of PV power generation generally focus on correlations with sunshine intensity or sunshine duration. Although significant progress has been achieved by these studies, the requirements of electric power systems have yet to be satisfied. At present, studies about the correlations of PV power generation have yet to account for the seasonal characteristics of PV power generation, which have an immense impact on the characteristics of PV power generation. Furthermore, reports about the seasonal yield of PV power generation are almost non-existent. It is

IOP Conf. Series: Materials Science and Engineering **484** (2019) 012061 doi:10.1088/1757-899X/484/1/012061

therefore necessary to perform an in-depth study about the characteristics of large-scale PV power stations.

Based on our recognition of the above problems, the correlations of PV power generation were investigated from a temporal perspective in this work, using the on-site PV power generation data of the Hanergy Xinhe PV power station.

The rest of this paper is organized as follows. The characteristic curves of the PV power generation data are provided in the section 2. And then, the briefly introduction of concept of correlation and application in the PV generation is showed in section 3. Finally, the conclusion and discussions are given in section 4.

2. Analysis of the Typical Daily Characteristics

The Hanergy Xinhe PV Power Station lies in Xinhe County (Aksu Prefacture, Xinjiang), in the southwestern region of the Xinjiang Uyghur Autonomous Region. The power generation data indicate that the effective power output is sometimes a negative value. In the data for January and June (and the data of other months), it was observed that the effective power output is a constant value during unlit hours. The negative values were therefore balanced by simply adding an offset.

2.1. Sunny Days

To ensure that representative sunny days were selected for this analysis, we selected a period that contains several sunny days in succession, and thereafter selected a few days from this period. This ensures that the selected data are representative of sunny days in this region. Using this approach, we performed our preliminary analysis using the data of 7/13, 7/14, 7/15, 7/16, 7/21, 7/22, and 7/23 in July.

The weather, sunrise/sunset times, and PV output curves of these days are shown below.

Table 1. Weather of the Xinhe County in July (partial).

Date	Highest temperature	Lowest t temperature	Weather	Sunrise	Solar noon	Sunset	PV output curve
2015-07-13	34	17	Sunny	07:06:51	14:35:17	22:03:43	C1
2015-07-14	37	18	Sunny	07:07:38	14:35:24	22:03:11	C2
2015-07-15	36	18	Sunny	07:08:25	14:35:31	22:02:36	B1
2015-07-16	37	17	Sunny	07:09:14	14:35:36	22:01:59	A
2015-07-21	37	19	Sunny	07:13:30	14:35:58	21:58:26	B2
2015-07-22	38	21	Sunny	07:14:24	14:36:01	21:57:38	C3
2015-07-23	38	21	Sunny	07:15:18	14:36:03	21:56:48	B3

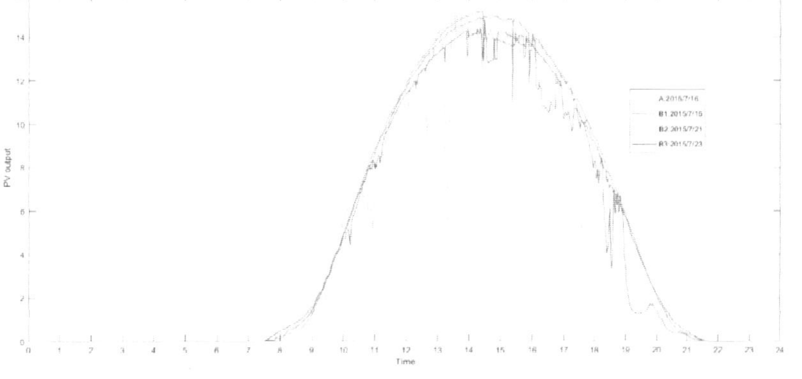

Figure 1. PV output curves of Group A and Group B.

IOP Conf. Series: Materials Science and Engineering **484** (2019) 012061 doi:10.1088/1757-899X/484/1/012061

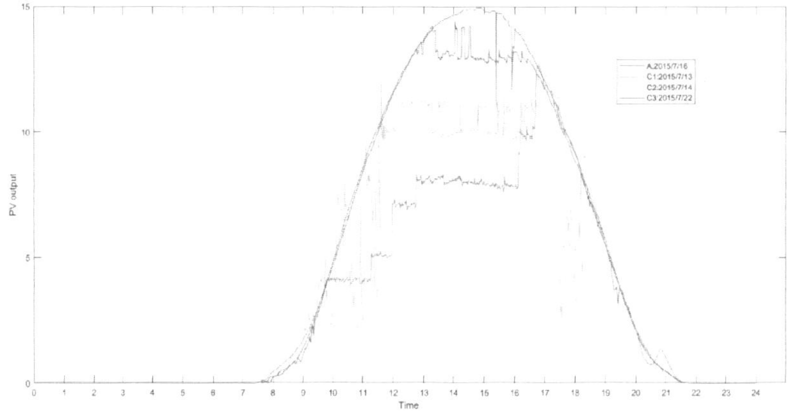

Figure 2. PV output curves of Group A and Group C.

The following conclusions were drawn from a preliminary analysis of figures 1 and 2 and tables 1.

(1) The PV output curves are generally parabolic in shape, and the peak occurs around 14:00–15:00. This is caused by the effects of sunshine on PV power generation, and the peak is related to the solar noon (the point when the solar elevation angle is at its highest). This result is consistent with our expectations.

(2) In the C1 (2015/7/13) data, it may be observed that the second half of the curve still shows signs of a parabolic shape. However, there are several periods of time in the first half of this data, where the numerical value stays at a constant value for a while before jumping to another value. Data of this type do not only appear in a single day. Owing to the abnormal stability of these numerical values, we suspect that this is the result of PV power dumping and throttling. Results of this type could produce a significant uncontrollable effect in the data of this work.

2.2. Non-sunny Days

As non-sunny days are atypical around the area of study, they were only briefly analysed in this work.

The data corresponding to 2015/7/28 (cloudy) and 2015/1/28 (light snow followed by cloudy weather) were used to plot the following figure.

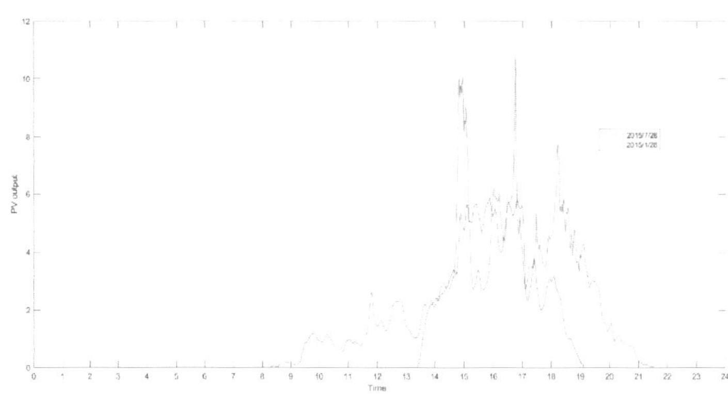

Figure 3. PV output curve on 2015/7/28.

The figure indicates that the PV output curves on non-sunny days do not exhibit any significant features. Therefore, the correlation analyses in the later sections of this paper simply focused on sunny days.

Thus, we analysed single-day data from summer and winter and obtained a simple understanding of PV power generation in these seasons.

3. Analysis of Seasonal Correlations

3.1. Correlation analysis on Summer and Winter Data

Please note that this analysis was performed using a sampling-based approach as there was an excessive amount of data. The data corresponding to the following dates (which had good weather) were included in this case analysis: 1/4, 1/11, 1/18, 1/25, 2/8, 2/15, 2/22, 6/7, 6/21, 6/28, 7/12, 7/19, 7/26, 8/9, 8/16, 8/23, 8/30, 12/13, 12/20, and 12/27. Here, December, January, and February are defined as winter months, whereas June, July, and August are defined as summer months.

The mathematical expression of the correlation coefficient is

$$r = \frac{\sum(x_i-\bar{x})(y_i-\bar{y})}{\sqrt{\sum(x_i-\bar{x})^2}\sqrt{\sum(y_j-\bar{y})^2}}. \tag{1}$$

The value of the correlation coefficient, r, varies between -1 and 1. If $r > 0$, the two variables are positively correlated; if $r < 0$, the two variables are negatively correlated. The value of $|r|$ reflects the closeness of the linear relationship between these variables. The linear correlation between the two variables becomes closer as $|r|$ approaches 1.Conversely; the linear correlation between the two variables becomes weaker as $|r|$ approaches 0.

Correlation analysis was performed after the selected data were preprocessed. Based on statistical tools, the correlation coefficient of these data is

$$r_1 = 0.5440.$$

This indicates that the selected data are correlated to an extent, but the correlation is not very strong. However, Section 4 indicates that there is significant similarity in the PV output trends of each day, despite the presence of large differences between summer and winter in terms of effective sunshine duration and sunshine intensity. Therefore, we believe that higher correlation coefficients can be obtained by processing the data.

The selected data were sorted into a continuous arrangement, as shown in figure 4..

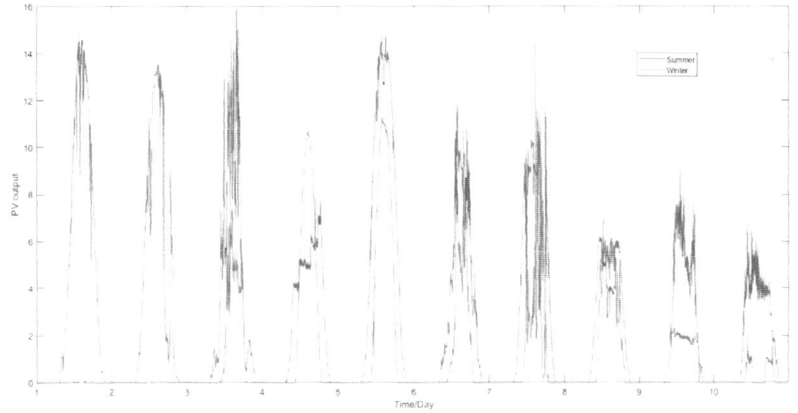

Figure 4. Continuous plot of (partial) summer and winter PV power generation data.

In this figure, it is shown that the days selected in December have very low PV output levels and are excessively different from the data for June, which exhibit very high PV output levels. Figures 5 and 6 are plotted to highlight the trends and numerical values of the selected December data.

IOP Conf. Series: Materials Science and Engineering **484** (2019) 012061 doi:10.1088/1757-899X/484/1/012061

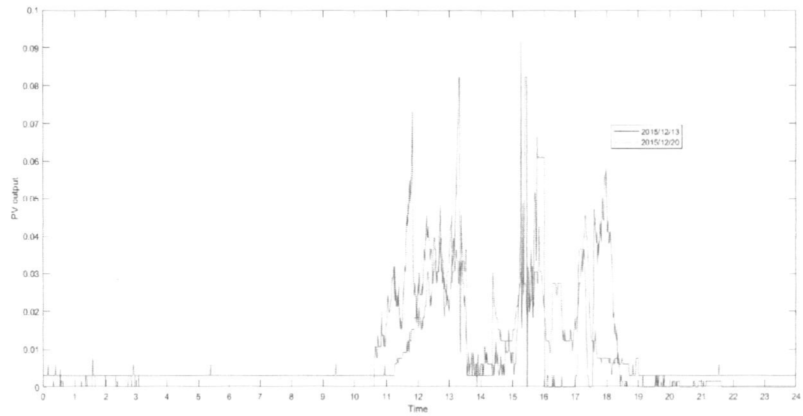

Figure 5. PV output curves on 12/13 and 12/20.

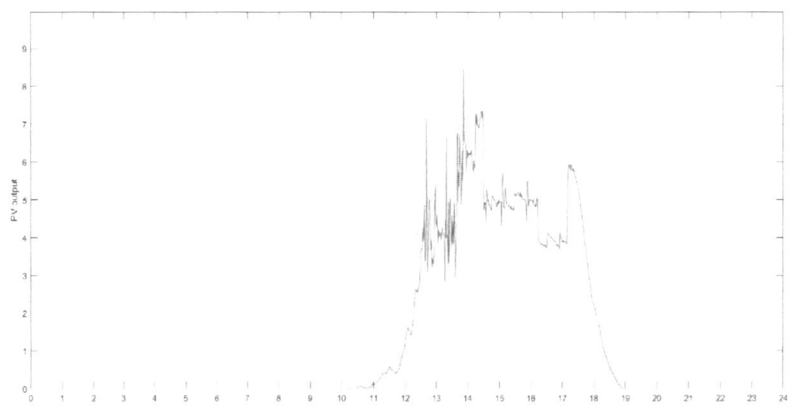

Figure 6. PV output curve on 12/27.

Please note that the vertical axes of figure 5 and 6 are very different from each other, and this is likely to have a detrimental and uncontrollable impact on our analysis. As this issue was previously discussed in the first subsection of Section 4, no further elaboration will be provided here.

The data for 12/13 and 12/20 were treated as bad data and thus excluded. The correlation coefficient thus becomes

$$r_2 = 0.7361.$$

The daily PV output trends obtained from the processed data are shown below:

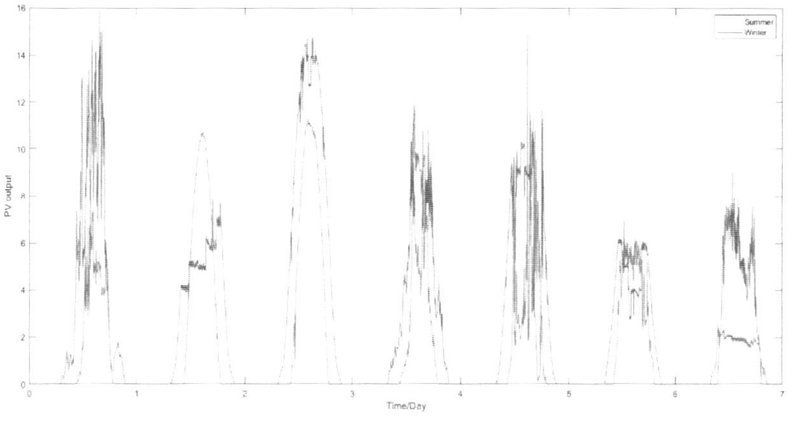

Figure 7. Continuous plot of processed summer and winter power generation data.

It may be observed that the data are now strongly correlated, and the shape of the plot in figure 7 confirms our hypothesis.

3.2. Analysis of Correlations between Spring and Autumn Data

As in the previous section, this analysis was performed using a sampling-based approach. The data corresponding to the following dates (which had good weather) were selected for this analysis: 3/1, 3/8, 3/15, 3/22, 3/29, 4/5, 4/12, 10/25, 11/1, 11/8, 11/15, 11/22, and 11/29. Here, March, April, and May are defined as spring months, whereas September, October, and November are defined as autumn months.

The following correlation coefficient was obtained after the data were processed (using the same procedures as in the previous section):

$$r_3 = 0.8674.$$

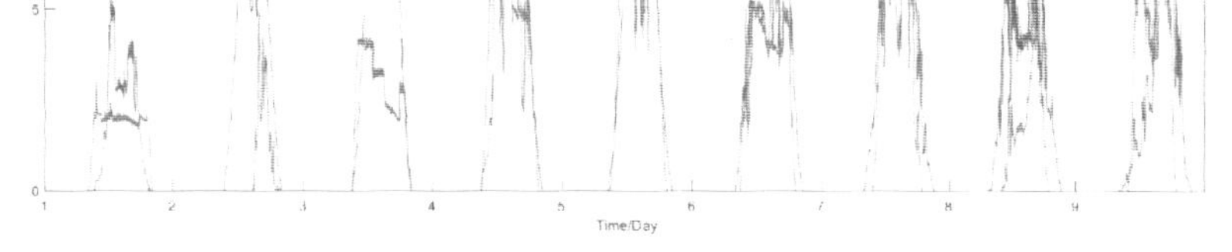

Figure 8. Continuous plot of processed spring and autumn power generation data.

The red line corresponds to the data for September, October, and November (autumn), whereas the blue line corresponds to the data for March, April, and May (spring).

4. Conclusion and Discussions

Summer and winter are extremely different in Xinjiang. Nonetheless, the PV outputs of the power station in summer and winter are still strongly correlated under normal circumstances, and this will be very useful for PV output predictions. Although a considerable number of data points do not show any correlation, this appears to have been caused by human factors (PV power dumping and throttling). In correlation analyses and PV output predictions, it is necessary to account for this fact and take the corresponding data-processing measures

The correlation between autumn and spring data is much stronger than the correlation between winter and summer data. The possible causes for this difference are given below:

The correlation between spring and autumn weather is much stronger than the corresponding correlation between summer and winter. In this case study, the weather was sunny over long periods of time in summer, which is conducive for PV power generation. In winter, the weather was not sunny for several days in a row, and abnormal weather also occurred occasionally. These differences led to a lower level of correlation between the summer and winter PV power generation data. As the weather in spring and autumn is relatively similar, a much stronger level of correlation was observed between their PV power generation data.

Human-controlled PV power dumping and throttling did not occur very frequently in spring and autumn, which is beneficial for the data analyses of this work.

In addition to sunshine duration and sunshine intensity (which have been discussed previously in this work), we suspect that data deviations may be caused by other factors, such as temperature. Related work will be carried out in future research.

5. Acknowledgments

This work was supported in part by the Fundamental Research Funds for the Central Universities (2018ZD01).

6. References

[1] Baike entry: Photovoltaic power generation https://baike.baidu.com/item/%E5%85%89%E4%BC%8F%E5%8F%91%E7%94%B5/269917?fr=aladdin

[2] People's Daily Online: *PV power generation in China exceeds 100 billion kWh for the first time*, http://finance.people.com.cn/n1/2018/0102/c1004-29741194.html

[3] Zhu HL and Liu ZH *PV System Output Analysis of Environmental Factors Affect*

[4] Zhang YP *An overview of research progress of short-term photovoltaic forecasts*

[5] Wang D and Sun YJ *Progress in prediction of PV power generation systems*

[6] Ni Q *Study on Some Photovoltaic Ultra Short-term Output Power Probability Forecasting*

[7] Yang J *Numerical Study on Weather Types in the Prediction of Power Generation by Photovoltaic Power Systems*

[8] Cui Y, Chen ZH and Sun PJ *Comparative Analysis of Short-Term Photovoltaic Power Generation Predictions in Regions with Different Latitudes in the Presence of Photovoltaic Power Dumping and Throttling*

[9] Dai X, Xiao WB, Hu FY and Wu SY *Research of Photovoltaic Prediction Model*

[10] Hu Y *the Output Forecast of Photovoltaic Power Based on Kernel Partial Least Squares*